Blockchain and Digital Twin Applications in Smart Agriculture

Blockchain and Digital Twin Applications in Smart Agriculture presents scenarios, solutions, applications to optimize crop management, improve resource allocation, and enhance sustainability in modern farming practices. It examines how blockchain and digital twin technology could enable farmers and stakeholders to create virtual, data-driven replicas of their agricultural operations while leveraging the security and transparency of blockchain technology. Highlights of the book include:

- An immutable blockchain ledger for storing and sharing agricultural data securely
- Decentralized data management to share data among such stakeholders as farmers, suppliers, distributors, and consumers
- Supply chain transparency to ensure product authenticity, reduce food fraud, and protect consumers
- Solutions to improve the allocation of such resources as water, fertilizer, and labor to reduce costs and improve productivity

The book helps researchers and academics develop solutions that can transition the agricultural industry into a more data-driven, sustainable, and transparent future. It explains how creating digital twins of agricultural operations and securing data through blockchain can enhance the ability to make informed decisions, reduce waste, and ensure that agriculture practices are more efficient and environmentally responsible. Featuring practical applications and case studies, the book provides a roadmap to farming that is smart, data-driven, and more.

Amit Kumar Tyagi is an assistant professor, Department of Fashion Technology, National Institute of Fashion Technology, New Delhi. He earned a Ph.D. from Pondicherry Central University, India. He has worked as assistant professor and senior researcher at the School of Computer Science and Engineering, Vellore Institute of Technology, Chennai, India.

Shrikant Tiwari received his Ph.D. from the Department of Computer Science & Engineering at the Indian Institute of Technology (Banaras Hindu University), Varanasi, India. Currently, he is an associate professor in the School of Computing Science and Engineering (SCSE), Galgotias University, Greater Noida, India. He has authored or co-authored more than 50 national and international journal publications, book chapters, and conference articles. He has five patents filed to his credit. His research interests include machine learning, deep learning, computer vision, medical image analysis, pattern recognition, and biometrics. Dr. Tiwari is a FIETE, a senior member of the IEEE, and member of ACM, IET, CSI, ISTE, IAENG, and SCIEI.

Blockchain and Digital Twin Applications in Smart Agriculture

Edited by
Amit Kumar Tyagi and Shrikant Tiwari

CRC Press
Taylor & Francis Group
Boca Raton London New York

CRC Press is an imprint of the
Taylor & Francis Group, an **informa** business

First edition published 2025
2385 NW Executive Center Drive, Suite 320, Boca Raton FL 33431

and by CRC Press
4 Park Square, Milton Park, Abingdon, Oxon, OX14 4RN

CRC Press is an imprint of Taylor & Francis Group, LLC

ISBN: 978-1-032-81746-0 (hbk)
ISBN: 978-1-032-83034-6 (pbk)
ISBN: 978-1-003-50739-0 (ebk)

DOI: 10.1201/9781003507390

Typeset in Times LT Std
by Apex CoVantage, LLC

Contents

Preface....xix
List of Contributors....xxi

Chapter 1 Introduction of Blockchain in Smart Agriculture....1

Roopa Devi E. M., Shanthakumari R, Dhanushya S, and Kiruthika G

1.1 Introduction....1
1.1.1 Challenges in Traditional Agriculture....2
1.1.2 The Emergence of Smart Agriculture....3
1.1.3 Benefits of Smart Agriculture Compared to Traditional Agriculture....3
1.1.4 Current Technologies Used in Agriculture....4
1.2 Introduction to Blockchain Technology, Cryptocurrencies, and Smart Contracts....5
1.2.1 Blockchain Technology....5
1.2.2 Cryptocurrencies....6
1.2.3 Smart Contracts....6
1.3 Why Blockchain in Agriculture?....6
1.4 Blockchain Smart Agriculture....7
1.5 The Various Ways in Which Blockchain Technology Could Transform the Agriculture Industry....7
1.6 Uses of Blockchain Technologies in Agriculture....8
1.7 Problems in Agricultural Supply Chains....9
1.8 Overcoming Challenges and Adoption Barriers....10
1.9 How Can Agriculture Production Be Revolutionized by Blockchain Technology?....10
1.10 Stage 1: IoT Devices Generating Data....11
1.11 Stage 2: Cleansing and Enhancement of Gathered Data....11
1.12 Incorporating Metadata....11
1.13 Stage 3: Applying Machine Learning Algorithms to Enhance the Data's Intelligence....11
1.14 Stage 4: The Blockchain Stores Data....12
1.15 Blockchain in Weather Monitoring....12
1.16 Role of Blockchain in Revolutionizing Agricultural Finance....13
1.17 Access to Capital....13
1.18 Transparent and Efficient Transactions....13
1.19 Risk Management and Insurance....13
1.20 Investment in Sustainable Agriculture....13
1.21 Future Perspectives and Opportunities....14

Chapter 2 Smart Agriculture and Digital Twins: A Vision of Sustainability....17

B. Santhoshini and M. Shanmugam

2.1 Introduction....17
2.2 Literature Review....17
2.3 Obstacles Confronting Conventional Farming Methods....19
2.3.1 Limited Access to Information and Technology....20

2.3.2 Weather and Climate Impacts on Agriculture 20
2.3.3 Soil Health and Fertility Concerns 20
2.3.4 Addressing Water Scarcity in Agriculture 20
2.3.5 Market Access Challenges and Price Fluctuations 20
2.4 Applications of Blockchain in Agriculture 21
2.5 Evolving Landscape of Agriculture 24
2.6 Digital Twins in the Agriculture Sector 26
2.7 Conclusion 26
References 27

Chapter 3 Digital Twins for the Agriculture Sector 29

Hrishikesh Kokate, Abhijit Vhatkar, and Alex Khang

3.1 Introduction 29
3.2 Advantages of Digital Twins for the Agriculture Sector 30
3.2.1 Farm Apparatus and Machinery 31
3.2.2 Farms and Croplands 33
3.2.3 Livestock 33
3.2.4 Controlled Environments and Greenhouses 35
3.2.5 Logistics and Supply Chains 37
3.2.6 Water and Soil Resources 39
3.2.7 Control of Diseases and Pests 40
3.2.8 Biotechnology in Agriculture 41
3.2.9 Packaging and Processing of Food 42
3.2.10 Rules and Policies about Agriculture 43
3.3 Limitations of Digital Twins in Agriculture 44
3.4 Conclusion 45
References 47

Chapter 4 AI-Enabled Blockchain for Livestock Tracking and Health Management in Smart Agriculture 52

Ruchi Jakhmola Mani, Naman Kapoor, Harsh Kashyap, Abhishake Shukla, and Deepshikha Pande Katare

4.1 Overview of Artificial Intelligence 52
4.2 Understanding Livestock Tracking 52
4.2.1 Advantages of AI in Livestock Tracking 54
4.3 Evolution of Livestock Tracking with AI 55
4.3.1 Benefits and Use Cases of AI-Enabled Livestock Tracking 56
4.4 Blockchain Technology in Livestock Tracking 57
4.5 Integration of AI and Blockchain for Livestock Tracking and Health Management 58
4.5.1 Development of Smart Contracts for Livestock Management 59
4.5.2 Data Privacy and Security Considerations in AI–Blockchain Integration 60
4.6 Real-World Applications and Practical Case Studies 61
4.7 Future Directions and Challenges 63
4.7.1 Emerging Trends in AI–Blockchain Integration for Smart Agriculture 63
4.8 Conclusion 66
References 67

Chapter 5 A Blockchain-Enabled Framework to Enhance Cybersecurity Challenges in Smart Agriculture 69

Joseph Bamidele Awotunde, Abidemi Emmanue Adeniyi, Peace Busola Falola, and Temitayo Caroline Adeniran

5.1 Introduction 69
5.1.1 General Overview and Concepts of Smart Agriculture 71
5.2 Key Concepts of Smart Agriculture 72
5.3 Security and Privacy Challenges in Smart Agriculture Systems 76
5.4 Attacks on Hardware in Smart Agriculture 76
5.5 Privacy Protection in Smart Agriculture Systems 77
5.6 Attacks on Networks and Equipment in Smart Agriculture 78
5.7 Attacks on Supply Chain in Smart Agriculture 79
5.8 Attacks on Sensor Devices in Smart Agriculture 80
5.8.1 General Framework Using Blockchain-Based Infrastructure in Smart Agriculture 82
5.9 Research Issues and Future Work 83
5.9.1 Research Issues 83
5.10 Latency 83
5.11 Energy Consumption 85
5.12 Data Privacy and Security 85
5.13 Interoperability 85
5.14 Cost 85
5.15 Regulatory and Legal Issues 85
5.16 Future Work 85
5.17 Conclusion 86
References 86

Chapter 6 Transformative Potential: IoT, Cloud, Big Data, and AI-Driven Blockchain Digital Twins in Agriculture 92

Garima Jain, Ankush Jain, Anuj Jaiswal, Shakeel Ahmad, Vinod M. Kapse, and Sovers Singh Bisht

6.1 Introduction 92
6.1.1 The Significance of Smart Farming 92
6.1.2 IoT, Cloud Computing, and Big Data in Agriculture 92
6.1.3 Objectives of the Chapter 93
6.2 IoT in Agriculture 94
6.2.1 Understanding IoT and Its Application in Farming 94
6.2.2 Elevating Agriculture through IoT: Exploring Its Application in Farming 94
6.2.3 Integration of IoT with Farming Technology 95
6.2.4 Challenges and Benefits of Implementing IoT in Agriculture 95
6.3 Cloud Computing for Agriculture Data Management 95
6.3.1 Introduction: Harnessing Cloud Computing for Enhanced Agricultural Practices 95
6.3.2 Challenges and Considerations in Cloud Computing for Agriculture 96
6.3.3 Addressing Security and Privacy Concerns in Cloud-Based Agricultural Systems 97
6.4 Big Data Analytics for Precision Farming 98
6.4.1 Introduction 98

6.4.2 Smart Agricultural Importance of Big Data in the Industry..............98
6.4.3 Big Data Analytics for Crop Management in Precision Farming......99
6.5 Smart Sustainable Farming Framework.. 100
6.5.1 Introduction to Smart Sustainable Farming.................................... 100
6.5.2 Exploring Technologies Used in Smart Farming............................. 100
6.5.3 Embracing Sustainability in Agricultural Practices 101
6.5.4 Developing a Sustainable Farming Framework............................... 102
6.6 Smart Farming: IoT, Cloud, Big Data, AI, and Blockchain Integration........ 103
6.6.1 Cultivating Connectivity: Unleashing the Power of IoT in Agricultural Practices 103
6.6.2 Benefits of IoT in Agriculture 103
6.6.3 Harnessing the Power of Case Studies and Success Stories 104
6.6.4 Challenges and Opportunities... 104
6.6.5 Cloud Cultivation: Optimizing Farming Operations through Cloud Computing 105
6.6.6 Benefits of Cloud Computing in Agriculture 106
6.6.7 Exemplary Case Studies and Success Narratives 106
6.6.8 Challenges and Considerations 106
6.6.9 Future Outlook 106
6.6.10 Data Harvest: Maximizing Yield with Advanced Big Data Analytics 106
6.6.11 Benefits of Big Data Analytics in Agriculture 106
6.6.12 Exemplary Case Studies and Success Narratives 107
6.6.13 Challenges and Considerations 107
6.6.14 Future Outlook 107
6.6.15 Cultivating Intelligence: Revolutionizing Agricultural Systems with AI Integration.. 107
6.6.16 Benefits of AI Integration in Agriculture... 107
6.6.17 Exemplary Case Studies and Success Narratives 108
6.6.18 Blockchain Blossoms: Ensuring Transparency and Security in Agriculture through Blockchain Technology.................................. 108
6.6.19 Understanding Blockchain in Agriculture 108
6.6.20 Advantages of Blockchain in Agriculture....................................... 108
6.6.21 Applications of Blockchain in Smart Farming 109
6.6.22 Ensuring Data Privacy and Security 109
6.7 Challenges and Future Directions of Smart Farming 109
6.8 Conclusion 110
References 110

Chapter 7 Blockchain Implementation in IoT-Digital Twin-Based Crop Production..............112

Ushaa Eswaran, Vivek Eswaran, Keerthna Murali, and Vishal Eswaran

7.1 Introduction 112
7.1.1 Overview of the Importance of Agriculture and Crop Production 112
7.1.2 Challenges Faced in Traditional Crop Production Practices 112
7.1.3 The Need for Digital Transformation in Agriculture....................... 113
7.1.4 Objectives of the Chapter... 113
7.1.5 Organization of the Chapter... 113
7.2 Internet of Agriculture Things 114
7.2.1 Definition and Principles of IoT 114
7.2.2 IoT Applications in Agriculture 114

7.2.3 Benefits of IoT in Crop Production 115
7.3 Digital Twins in Agriculture 116
7.3.1 Definition and Concept of Digital Twins 116
7.3.2 Components of a Digital Twin System 116
7.3.3 Applications of Digital Twins in Crop Production 117
7.4 Blockchain Methods 118
7.4.1 Introduction to Blockchain Technology 118
7.4.2 Key Features of Blockchain 118
7.4.3 Blockchain Use Cases in Agriculture 118
7.5 Integrating Blockchain with IoT and Digital Twins for Crop Production 119
7.5.1 Challenges in Traditional Crop Production Systems 119
7.5.2 The Role of Blockchain in Addressing These Challenges 119
7.5.3 Architecture and Components of a Blockchain-Enabled IoT-Digital Twin System for Crop Production 120
7.6 Data Management and Security 121
7.6.1 Types of Data Collected in IoT-Digital Twin Crop Production Systems 121
7.6.2 Data Privacy and Security Concerns 122
7.6.3 Blockchain-Based Solutions for Secure Data Storage and Sharing 122
7.7 Smart Contracts and Automation 123
7.7.1 Introduction to Smart Contracts 123
7.7.2 Applications of Smart Contracts in Crop Production 123
7.7.3 Integration of Smart Contracts with IoT and Digital Twins 124
7.8 Case Studies and Real-World Examples 124
7.8.1 Successful Implementations of Blockchain-Enabled IoT-Digital Twin Systems 124
7.8.2 Lessons Learned and Best Practices 126
7.9 Challenges and Future Directions 127
7.9.1 Technical Challenges 127
7.9.2 Regulatory and Legal Challenges 128
7.9.3 Future Research Directions and Potential Advancements 129
7.10 Conclusion 129
7.10.1 Final Remarks and Recommendations for Adoption and Implementation 130
References 130

Chapter 8 Digital Twin Applications in the Agriculture and Farming Domain 132

Sivaretinamohan R

8.1 Introduction: Background 132
8.2 Prerequisites for Agricultural Digital Twins 133
8.3 How Do Digital Twins Work in Smart Agriculture? 133
8.4 Potential Digital Twins in Smart Agriculture 134
8.4.1 Digital Twins in Within-Field Management Zoning 134
8.4.2 Digital Twins in Happy Cow 136
8.4.3 Digital Twins in Chain-Integrated Greenhouse Production 138
8.4.4 Digital Twins in Added-Value Weeding Data 139
8.4.5 Digital Twins in Pig Farm Management 140
8.4.6 Digital Twins in Soil and Irrigation 142
8.4.7 Digital Twins in Crop Production and Post-Harvesting 142

8.5 Emerging Digital Twin Applications in Agriculture and Digital Farming ... 143
8.6 Conclusion ... 143
References ... 144

Chapter 9 Blockchain-Enabled Security Framework for Smart Agriculture: Harvesting Smart Agriculture Supply Chains Transformed by Blockchain Efficiency ... 146

Maheswari R, Pattabiraman V, and Basim Alhadidi

9.1 Introduction ... 146
9.1.1 Agricultural Supply Chain ... 146
9.1.2 Smart Contract ... 147
9.1.3 Supply Chain in Agriculture ... 148
9.2 Literature Review ... 149
9.3 Objective ... 150
9.4 Scope of the Work ... 150
9.5 Case Study ... 150
9.6 Implementation ... 153
9.7 Working Model ... 155
9.8 Novelty ... 156
9.9 Results and Discussion ... 157
9.10 Conclusion ... 158
References ... 158

Chapter 10 A Systematic Review of Plant Leaf Disease Identification Using Image Search Engines ... 160

D Kavitha, Sanjay M, Pillaram Manoj, Benjamin Roy, and Parth Vadera

10.1 Introduction ... 160
10.2 Background Study ... 160
10.2.1 Digitizing Agricultural Practices in Africa ... 160
10.2.2 Advancements in Apple Leaf Disease Identification ... 161
10.2.3 Advancements in Grape Leaf Disease Detection ... 162
10.2.4 Literature Review: Leaf Disease Classification in Crops ... 163
10.3 Proposed System ... 164
10.4 Module Decomposition ... 165
10.4.1 VGG19 Model Module ... 165
10.4.2 Dataset Loading and Pre-Processing Module ... 166
10.4.3 Feature Extraction Module ... 166
10.4.4 Similarity Comparison and Search Module ... 166
10.5 Results and Discussion ... 166
10.5.1 Rice (Farmers) ... 166
10.5.2 Tomatoes (Home Gardening) ... 170
10.6 Conclusion ... 175
10.7 Future Work ... 176
References ... 176

Chapter 11 Precision Agriculture: Navigating the Future of Farming with Artificial Intelligence ... 178

Peace Busola Falola, Joseph Bamidele Awotunde, Abidemi Emmanuel Adeniyi, and Folashade Titilope Ogunajo

11.1 Introduction ... 178
11.1.1 Key Contributions of This Chapter ... 179
11.1.2 Chapter Organization ... 180
11.2 The Evolution of Precision Agriculture ... 180
11.2.1 Role of AI in Agriculture ... 181
11.3 Fundamentals of AI in Agriculture ... 182
11.3.1 AI Technologies ... 182
11.3.2 Data Sources ... 183
11.3.3 AI Models and Algorithms in Precision Agriculture ... 183
11.4 Applications of AI in Precision Agriculture ... 185
11.5 Challenges and Solutions ... 186
11.6 The Future of AI in Precision Agriculture ... 187
11.7 Conclusion ... 188
References ... 188

Chapter 12 Smarter Yields, Sustainable Practices: Predictive Modelling for Informed Decision Making in Smart Farming ... 192

Ruchi Jakhmola Mani, Saba Masood, Piyush Bhalerao, and Deepshikha Pande Katare

12.1 Introduction to Predictive Modelling in Agriculture ... 192
12.1.1 Evolution of Predictive Modelling in Farming ... 192
12.1.2 Importance of and Benefits in Smart Farming ... 194
12.1.3 Overview of Neo4j in Predictive Modelling ... 194
12.2 Role of Neo4j in Predictive Modelling ... 194
12.2.1 Features Supporting Predictive Modelling ... 195
12.2.2 Role of Neo4j for Predictive Modeling in Agriculture ... 196
12.3 Integration of Agricultural Data into Neo4j ... 196
12.3.1 Sources of Agricultural Data ... 197
12.3.2 Data Quality and Pre-Processing Techniques ... 197
12.3.3 Data Integration in Smart Farming Using Neo4j ... 198
12.4 Building Predictive Models Using Graph Algorithms with Neo4j ... 198
12.4.1 Path Finding and Graph Search Algorithms ... 199
12.4.2 Centrality Algorithms ... 200
12.4.3 Community Detection Algorithms ... 200
12.5 Predictive Applications in Smart Farming ... 201
12.5.1 Crop Yield Prediction ... 201
12.5.2 Disease Outbreak Forecasting ... 202
12.5.3 Livestock Monitoring ... 202
12.6 Real-Time Decision Support Systems ... 204
12.6.1 Monitoring Agricultural Systems ... 204

12.6.2 Detecting Anomalies and Adapting Strategies 204
12.6.3 Utilizing IoT Devices and Sensor Networks 205
12.7 Scalability and Performance Considerations 205
12.7.1 Handling Large-Scale Agricultural Data 205
12.7.2 Optimizing Performance of Predictive Models Using Neo4j 207
12.7.3 Scalable Deployment Architectures 207
12.8 Challenges and Limitations 208
12.8.1 Data Privacy and Security Concerns 208
12.8.2 Overcoming Technological and Resource Constraints 209
12.9 Future Directions and Emerging Trends 209
12.9.1 Digital Twins in Smart Farming 209
12.9.2 Robotics in Smart Farming 210
12.9.3 Edge Computing in Smart Farming 210
12.10 Conclusion 211
References 211

Chapter 13 Tech-Driven Harvest: Blockchain's Role in Sustainable Food Systems and Waste-Free Future 215

Ruchi Jakhmola Mani, Suyashi Singh, Subham Sonu, and Deepshikha Pande Katare

13.1 Introduction 215
13.1.1 The Global Food Waste Crisis: Scope and Impact (Environmental, Economic, Social) 215
13.1.2 The Need for Innovation: Smart Technology as a Solution 216
13.2 Understanding Food Waste: A Breakdown by Stage 217
13.2.1 Production and Storage 217
13.2.2 Processing and Distribution 217
13.2.3 Retail and Consumption 218
13.3 Leveraging Smart Technology for Food Waste Reduction 218
13.3.1 Smart Farming and Precision Agriculture (Optimizing Yield and Quality, Minimizing Spoilage in Pre-Harvest Stages) 218
13.3.2 Intelligent Inventory Management in Supply Chains (Real-Time Tracking and Monitoring, Demand Forecasting, and Dynamic Logistics) 219
13.3.3 Smart Packaging and Food Storage Solutions (Extending Shelf Life, Monitoring Freshness and Spoilage) 219
13.4 Applications of Smart Technology in Food Waste Reduction 220
13.4.1 Smart Refrigeration Systems 220
13.4.2 Automated Inventory Management Systems 221
13.4.3 Dynamic Pricing and Discounting Strategies 221
13.4.4 Food Waste Tracking and Analytics Platforms 222
13.5 Building a Sustainable Food System: Collaboration and Education 222
13.5.1 Integrating Smart Technologies across the Chain 222
13.5.2 Consumer Awareness and Behaviour Change 223
13.5.3 Policy and Regulatory Incentives 223
13.6 Case Studies and Success Stories 224
13.6.1 Implementations of Smart Technology in Food Waste Reduction 224

13.6.2 Real-World Applications in Retail, Hospitality, and Food Service Industries 225
13.6.3 Impact Assessment and Lessons Learned 226
13.7 Challenges and Limitations of Smart Technology Adoption 227
13.7.1 Technological Barriers and Infrastructure Requirements 227
13.7.2 Cost Considerations and Return on Investment 227
13.7.3 Cultural and Behavioural Shifts 227
13.8 Future Directions and Opportunities 228
13.8.1 Emerging Trends in Smart Technology for Food Waste Reduction 228
13.8.2 Innovations in Food Packaging and Preservation 229
13.8.3 Collaborative Initiatives and Partnerships 230
13.9 Ethical and Environmental Considerations 231
13.9.1 Social Equity and Food Access 231
13.9.2 Environmental Impacts of Food Waste 231
13.9.3 Ethical Dilemmas in Redistribution and Donation Programs 231
13.10 Conclusion and Recommendations 232
13.10.1 Summary of Key Findings and Insights 232
13.10.2 Implications for Future Research and Practice 232
13.10.3 Recommendations for Stakeholders and Policymakers 233
References 234

Chapter 14 Digital Twins for the Agriculture Sector 237

Savita Sheoran and Ritika

14.1 Introduction 237
14.1.1 Digital Twins 237
14.1.2 Background 239
14.2 Architecture of Digital Twins 239
14.3 Components of Digital Twins 240
14.4 Types of Digital Twins 241
14.5 Applications of Digital Twins 242
14.6 DT-Developed Technologies for Agriculture 245
14.7 Challenges and Opportunity 247
14.8 Conclusion and Future Scope 248
References 248

Chapter 15 Smart Farming with "Blockchain": Revolutionizing Medicinal Plant Cultivation 251

Rati Kailash Prasad Tripathi

15.1 Introduction 251
15.2 Smart Farming in Medicinal Plant Cultivation 252
15.2.1 Components of Smart Farming 252
15.2.2 Applications of Smart Farming in Medicinal Plant Cultivation 254
15.2.3 Benefits of Smart Farming in Medicinal Plant Cultivation 255
15.3 Blockchain Technology in Medicinal Plant Cultivation 256
15.3.1 Fundamentals and Features 256
15.3.2 Applications of Blockchain in Medicinal Plant Cultivation 256

15.3.3 Benefits of Blockchain Integration in Medicinal Plant Cultivation ... 257
15.4 Case Studies and Real-World Applications of Smart Farming Integrated with Blockchain in Medicinal Plant Cultivation ... 258
15.5 Challenges of Smart Farming and Blockchain in Medicinal Plant Cultivation ... 261
15.5.1 Limitations of Smart Farming ... 261
15.5.2 Limitations of Blockchain ... 263
15.6 Emerging Trends in Smart Farming and Blockchain in Medicinal Plant Cultivation ... 265
15.6.1 Precision Agriculture and IoT Integration ... 265
15.6.2 Blockchain for Traceability and Quality Assurance ... 266
15.6.3 Advanced Robotics and Automation ... 266
15.6.4 AI and Big Data Analytics ... 266
15.6.5 Sustainable Farming Practices ... 266
15.6.6 Vertical Farming and Controlled Environment Agriculture ... 267
15.6.7 Collaborative Platforms and Digital Ecosystems ... 267
15.7 Future Prospects ... 267
15.7.1 Integration of Biotechnological Advances with Smart Farming of Medicinal Plants ... 267
15.7.2 Integration of Renewable Energy with Smart Farming of Medicinal Plants ... 269
15.8 Conclusion ... 271
References ... 272

Chapter 16 Insights into Coral Reefs of the Western Indian Ocean: A Comprehensive Analysis of Habitat Classification, Forecasting Taxa Changes, and Explainable AI-Driven Health Assessment ... 275

Tanvi Bahedia, Sparsh Wabhale, S A Sajidha, and Aakif Mairaj

16.1 Introduction ... 275
16.1.1 Coral Reefs ... 275
16.1.2 Environmental Effects on Coral Reefs ... 275
16.1.3 Efforts to Conserve Coral Reefs Using Technology ... 276
16.1.4 Recent Trends and Statistics ... 276
16.2 Related Work ... 277
16.3 Methodology ... 279
16.4 Dynamic Coral Taxa Prediction ... 279
16.5 AI-Powered Decision Support System ... 279
16.6 Explainable AI for Conservation ... 280
16.7 Predictive Mapping ... 280
16.8 Data Set ... 281
16.9 Overall Architecture ... 281
16.10 Implementation and Results ... 282
16.10.1 Dynamic Coral Taxa Prediction ... 283
16.10.2 Time Series Forecasting ... 284
16.10.3 AI-Powered Decision Support System ... 285
16.10.4 Confusion Matrix ... 286

16.10.5 Explainable AI for Conservation 288
16.10.6 Predictive Risk Mapping 290
16.11 Conclusion 290
References 291

Chapter 17 Farm Easy-Smart Farming Practices for Sustainable Agriculture 292

N. C. Brintha, J. T. Winowlin Jappes, Arivendan Ajithram, and Yegnanarayanan Venkataraman

17.1 Introduction to Smart Agriculture 292
17.1.1 Challenges in Traditional Farming Practices 293
17.1.2 History of Smart Agriculture 294
17.2 Factors Driving Innovation in Smart Agriculture 296
17.2.1 Technological Advancements 296
17.2.2 Economic Drivers 297
17.2.3 Environmental Concerns 297
17.2.4 Social and Demographic Factors 297
17.2.5 Policy and Regulatory Drivers 298
17.2.6 Global Trends 298
17.2.7 Research and Development 298
17.3 Benefits of Sustainable Agriculture 298
17.4 Automation in Agricultural Practices 299
17.4.1 Key Areas of Automation in Smart Agriculture 299
17.4.2 Benefits of Automation in Smart Agriculture 301
17.4.3 Challenges and Considerations 301
17.5 Technologies in Smart Farming for Sustainable Development 301
17.5.1 Barriers to Implementing Smart Farming Technologies 302
17.5.2 Addressing the Barriers 304
17.6 Current Challenges and Future of Smart Farming Practices 304
17.6.1 Current Challenges in Smart Farming 304
17.6.2 Future Expectations in Smart Farming 305
17.7 Conclusion and Future Scope 306
References 306

Chapter 18 Blockchain and Digital Twin Applications in Precision Agriculture: A Comprehensive Approach 308

Kanthavel R, S. Krithikaa Venket, Anju A, Adline Freeda R, Dhaya R, Frank Vijay, and Joseph Fisher

18.1 Introduction to Precision Agriculture 308
18.2 Blockchain in Precision Agriculture 309
18.3 Case Studies and Real-World Applications 310
18.3.1 Digital Twin Technology in Precision Agriculture 311
18.3.2 Real-Time Monitoring and Simulation 311
18.3.3 Predictive Analytics and Decision Support 311
18.3.4 Resource Optimization 311
18.3.5 Enhancing Supply Chain Transparency 311
18.3.6 Improved Risk Management 312

18.4 Integration of Blockchain and Digital Twins in Precision Agriculture......... 312
18.5 Challenges 312
18.5.1 Scalability........ 312
18.5.2 Data Security and Privacy........ 313
18.5.3 Cost and Accessibility........ 313
18.5.4 Complexity of Integration 313
18.6 Future Directions........ 313
18.6.1 Integration of AI and Advanced Analytics 313
18.6.2 Edge Computing........ 313
18.6.3 Blockchain Interoperability........ 313
18.6.4 Internet of Things Integration 313
18.6.5 Decentralized Autonomous Organizations 314
18.7 Conclusion 314
References 314

Chapter 19 Effective Deep Learning-Based Agricultural Models for Classifications of Plant Diseases 316

Pramit Brata Chanda, Subir Kumar Sarkar, and Surajit Naskar

19.1 Introduction 316
19.2 Literature Review 316
19.3 Proposed Work 317
19.3.1 Dataset Collection and Description........ 318
19.3.2 Importing Libraries 318
19.3.3 Loading the Data........ 318
19.3.4 Data Preprocessing........ 318
19.3.5 Splitting Data into Training and Test Data 318
19.3.6 Data Augmentation........ 318
19.3.7 Model Selection........ 318
19.4 Results and Discussion: Plant Disease Classification System........ 322
19.4.1 Results of Different Algorithms........ 322
19.4.2 Accuracy Comparison........ 326
19.4.3 Making Predictions 327
19.5 Conclusion and Future Work........ 330
References 331

Chapter 20 Enabling Blockchain in Digital Twins for Decentralized Supply Chain Traceability and Crop Growth Optimization........ 332

Naveen Raj Y, Sudha G, and J. Jayanthi

20.1 Introduction 332
20.1.1 The Growing Pains of the Modern Food System........ 332
20.1.2 Blockchain: The Unbreakable Chain of Trust........ 333
20.1.3 Digital Twins: The Virtual Greenhouse for Crop Optimization........ 333
20.1.4 Converging Technologies for a Sustainable Future........ 334
20.2 Literature Survey........ 334
20.2.1 Improving Innovation, Reducing Crime, and Empowering Consumers 334
20.2.2 Building Trust and Transparency in Food Systems 334

20.2.3 Smart Farming and Precision Agriculture 335
20.2.4 Addressing Agricultural Challenges with Data Models and Interoperable Platforms 335
20.2.5 Integration of Blockchain and Digital Twins 335
20.2.6 Blockchain Enhancements for Agricultural Supply Chains 336
20.2.7 Digital Twin Advancements for Crop Optimization 336
20.2.8 Synergistic Integration of Blockchain and Digital Twins 337
20.2.9 Blockchain's Expanding Role in Agricultural Supply Chains 338
20.2.10 Digital Twin Advancements for Crop Optimization 338
20.2.11 Synergistic Integration of Blockchain and Digital Twins 339
20.2.12 Challenges and Future Directions 339
20.2.13 Blockchain in Agriculture: Promises and Perils 340
20.2.14 Digital Twins in Agriculture: Opportunities and Challenges 340
20.2.15 Ethical Frameworks and Governance 341
20.2.16 Social Impact and Equity 341
20.2.17 Research Gaps and Opportunities 341
20.3 Limitations 341
20.3.1 Technical Challenges 341
20.3.2 Adoption Barriers 342
20.3.3 Digital Twin Model Limitations 342
20.3.4 Nascent Research Landscape 343
20.3.5 Exacerbation of Power Imbalances 343
20.3.6 Potential for Unintended Consequences 343
20.3.7 Cultural Sensitivity and Contextual Relevance 343
20.3.8 Difficulty in Measuring Social Impact 343
20.4 Proposed Solutions and Models 344
20.4.1 Addressing Technical Challenges 344
20.4.2 Overcoming Adoption Barriers 345
20.4.3 Addressing Digital Twin Model Limitations 346
20.5 Conclusion 346
References 347

Index 349

Preface

Advancements in blockchain technology, digital twins, and Internet of Things (IoT) are likely to increase the market's interest in books such as this one as these technologies become more central to smart agriculture. As the agriculture industry increasingly adopts technology-driven solutions for efficiency and sustainability, there is a growing market for books addressing the integration of blockchain and digital twins in agriculture. This book starts with an innovative approach that merges digital twin technology with blockchain to revolutionize the agricultural industry. This concept enables farmers and stakeholders to create virtual, data-driven replicas of their agricultural operations while leveraging the security and transparency of blockchain technology. It promises to optimize crop management, improve resource allocation, and enhance sustainability in modern farming practices. A few key components of this book will be digital twin technology in agriculture, blockchain security (ensures data security and integrity by providing an immutable ledger for storing and sharing agricultural data), decentralized data management (agricultural data can be shared among various stakeholders, including farmers, suppliers, distributors, and consumers), precision agriculture, sustainable farming practices, supply chain transparency (ensures product authenticity, reduces food fraud, and enhances traceability for consumers), improved resource allocation (including water, fertilizers, and labor; reducing costs; and improving overall productivity), and many more. Hence, this book offers the agricultural industry an opportunity to transition into a more data-driven, sustainable, and transparent ecosystem. Hence, by creating digital twins of their agricultural operations and securing data through blockchain, farmers can make informed decisions, reduce waste, and ensure the future of agriculture is efficient and environmentally responsible.

This book gives a detailed description of the integration of blockchain technology and digital twin solutions in modern farming. This book will provide concise information for the agriculture sector on how these technologies empower farmers with real-time data, precision insights, and secure supply chain management. This book covers the core principles, practical applications, and some real-world case studies, offering a practical roadmap for implementing smart agriculture practices. This book will be an essential resource to navigate the future of sustainable, data-driven farming, whether you are a farmer or researcher in the agriculture sector.

The rapid advancements in blockchain technology, digital twins, and the Internet of Things are set to transform smart agriculture, making this book a timely and valuable resource. As the agriculture industry increasingly adopts technology-driven solutions to enhance efficiency and sustainability, there is a growing demand for comprehensive guides on integrating these cutting-edge technologies into farming practices.

This book presents an innovative approach by merging digital twin technology with blockchain to revolutionize the agricultural sector. This synergy allows farmers and stakeholders to create virtual, data-driven replicas of their agricultural operations, leveraging the security and transparency inherent in blockchain technology. The result is optimized crop management, improved resource allocation, and enhanced sustainability in modern farming practices.

Key components covered in this book include digital twin technology in agriculture, blockchain security, decentralized data management, precision agriculture, sustainable farming practices, supply chain transparency, and improved resource allocation. By integrating digital twin and blockchain technologies, this book provides the agricultural industry with a pathway to a more data-driven, sustainable, and transparent ecosystem. Farmers can leverage these technologies to make informed decisions, reduce waste, and ensure efficient and environmentally responsible agricultural practices.

This comprehensive guide details the integration of blockchain technology and digital twin solutions in modern farming. It provides concise information on how these technologies empower farmers with real-time data, precision insights, and secure supply chain management. Covering core principles, practical applications, and real-world case studies, the book offers a practical roadmap for implementing smart agriculture practices. Whether you are a farmer or a researcher in the agriculture sector, this book is an essential resource for navigating the future of sustainable, data-driven farming.

Contributors

Temitayo Caroline Adeniran
Department of Telecommunication Science
University of Ilorin
Kwara State, Nigeria

Abidemi Emmanuel Adeniyi
Department of Computer Science
College of Computing and Communication Studies, Bowen University
Iwo, Nigeria

Adline Freeda R
Department of Information Technology
KCG College of Technology
Chennai, India

Shakeel Ahmad
Department of Data Science
Noida Institute of Engineering and Technology
Greater Noida, India

Arivendan Ajithram
Department of Mechanical Engineering
Shine-Ming School of Intelligent Engineering, South China University of Technology (SCUT)
China

Basim Alhadidi
Al-Balqa Applied University
Jordan

Anju A
Department of Information Technology
KCG College of Technology
Chennai, India

Joseph Bamidele Awotunde
Department of Computer Science
Faculty of Communication and Information Sciences
University of Ilorin
Ilorin, Nigeria

Tanvi Bahedia
Vellore Institute of Technology
Chennai, India

Piyush Bhalerao
Sinhgad Institutes's, Sinhgad College of Engineering
Pune, Maharashtra, India

Sovers Singh Bisht
Department of Data Science
Noida Institute of Engineering and Technology
Greater Noida, India

N. C. Brintha
Department of Computer Science and Engineering
Kalasalingam Academy of Research and Education
Anand Nagar, Krishnankoil, India

Pramit Brata Chanda
Electronics And Tele Communication Engineering
Jadavpur University
Jadavpur, West Bengal, India

Dhanushya S
Department of Information Technology
Kongu Engineering College (Autonomous)
Perundurai, Tamil Nadu, India

Dhaya R
Department of Information Technology KCG College of Technology
Chennai, India

Ushaa Eswaran
Principal and Professor
Department of ECE, Mahalakshmi Tech Campus
Anna university
Chennai, Tamilnadu, India

Vishal Eswaran
Senior Data Engineer
CVS Health Centre
Dallas, Texas, United States

Vivek Eswaran
Staff Software Engineer
Tech Lead at Medallia
Austin, Texas, United States

Peace Busola Falola
Department of Computer Science
Faculty of Pure and Applied Sciences
Precious Cornerstone University
Ibadan, Nigeria

Joseph Fisher
Department of Electrical and Communication Engineering
University of Technology
Lae, Papua New Guinea

Ankush Jain
Department of Computer Science and Engineering
Netaji Subhas University of Technology
Dwarka, New Delhi, India

Garima Jain
Department of Computer Science and Business System
Noida Institute of Engineering and Technology
Greater Noida, India

Anuj Jaiswal
Department of Architecture and Planning MANIT
Bhopal, India

J. T. Winowlin Jappes
Department of Mechanical Engineering
Kalasalingam Academy of Research and Education
Anand Nagar, Krishnankoil, India

J. Jayanthi
Department of Computer Science and Engineering
Sona College of Technology
Salem, Tamil Nadu, India

Kanthavel R
Department of Electrical and Communication Engineering
University of Technology
Lae, Papua New Guinea

Naman Kapoor
Proteomics and Translational Research Lab, Centre for Medical Biotechnology
Amity Institute of Biotechnology, Amity University
Noida, India

Vinod M. Kapse
Department of Electronics and Communication Engineering
Noida Institute of Engineering and Technology
Greater Noida, India

Harsh Kashyap
Proteomics and Translational Research Lab, Centre for Medical Biotechnology
Amity Institute of Biotechnology, Amity University
Noida, India

Deepshikha Pande Katare
Proteomics and Translational Research Lab, Centre for Medical Biotechnology
Amity Institute of Biotechnology, Amity University
Noida, India

D Kavitha
School of Computer Science and Engineering
Vellore Institute of Technology
Chennai, India

Alex Khang
Department of AI and Data Science
Global Research Institute of Technology and Engineering
Fort Raleigh, North Carolina, USA

Kiruthika G
Department of Information Technology
Kongu Engineering College (Autonomous)
Perundurai, Tamil Nadu, India

Hrishikesh Kokate
Global Business School & Research Centre
Dr. D. Y. Patil Vidyapeeth
Pune, India

Maheswari R
Information Technology
Sri Eshwar College of Engineering
Coimbatore, India

Aakif Mairaj
Department of Electrical Engineering and Computer Science
Indiana University
Bloomington, United States

Ruchi Jakhmola Mani
Proteomics and Translational Research Lab, Centre for Medical Biotechnology
Amity Institute of Biotechnology, Amity University
Noida, India

Pillaram Manoj
School of Computer Science and Engineering
Vellore Institute of Technology
Chennai, India

Saba Masood
Proteomics and Translational Research Lab, Centre for Medical Biotechnology
Amity Institute of Biotechnology, Amity University
Noida, India

Keerthna Murali
Dell
Austin, Texas, United States

Surajit Naskar
Computer Science and Engineering
Kalyani Government College
Kalyani, West Bengal, India

Naveen Raj Y
Department of Computer Science and Engineering
Sona College of Technology

Folashade Titilope Ogunajo
Department of Computer Science
Atiba University
Oyo, Nigeria

Pattabiraman V
School of Computer Science and Engineering
Vellore Institute of Technology
Chennai, India

Ritika
Indira Gandhi University
Meerpur, Rewari, India

Roopa Devi E. M.
Department of Information Technology
Kongu Engineering College (Autonomous)
Perundurai, Tamil Nadu, India

Benjamin Roy
School of Computer Science and Engineering Vellore Institute of Technology
Chennai, India

S A Sajidha
Vellore Institute of Technology
Chennai, India

Sanjay M
School of Computer Science and Engineering
Vellore Institute of Technology
Chennai, India

B. Santhoshini
Department of Computer Science & Engineering
Vijay Rural Engineering College
Nizamabad, Telangana, India

Subir Kumar Sarkar
Electronics and Telecommunication Engineering
IIEST Shibpur
Howrah, West Bengal, India

M. Shanmugam
Department of Computer Science
School of Engineering & Technology
Pondicherry University, Pondicherry, India

Shanthakumari R
Department of Information Technology
Kongu Engineering College (Autonomous)
Perundurai, Tamil Nadu, India

Savita Sheoran
Indira Gandhi University
Meerpur, Rewari, India

Abhishake Shukla
Amity Institute of Organic Agriculture (AIOA), AUUP
Noida, India

Suyashi Singh
Proteomics and Translational Research Lab, Centre for Medical Biotechnology
Amity Institute of Biotechnology, Amity University
Noida, India

Sivaretinamohan R
Symbiosis Centre for Management Studies, Bengaluru Campus
Symbiosis International (Deemed University)
Karnataka, India

Subham Sonu
Patna Campus
Amity University
Patna, Bihar, India

Sudha G
Department of Computer Science and Engineering
Sona College of Technology
Salem, Tamilnadu, India

Rati Kailash Prasad Tripathi
Department of Pharmaceutical Sciences
Sushruta School of Medical and Paramedical Sciences, Assam University (A Central University)
Silchar, Assam, India

Parth Vadera
School of Computer Science and Engineering
Vellore Institute of Technology
Chennai, India

Yegnanarayanan Venkataraman
Department of Mathematics
Kalasalingam Academy of Research and Education
Anand Nagar, Krishnankoil, India

S. Krithikaa Venket
Department of Information Technology
KCG College of Technology
Chennai, India

Abhijit Vhatkar
Global Business School & Research Centre
Dr. D. Y. Patil Vidyapeeth
Pune, India

Frank Vijay
Department of Information Technology
KCG College of Technology
Chennai, India

Sparsh Wabhale
Vellore Institute of Technology
Chennai, India

1 Introduction of Blockchain in Smart Agriculture

Roopa Devi E. M., Shanthakumari R, Dhanushya S, and Kiruthika G

1.1 INTRODUCTION

The agricultural sector has experienced a notable shift in the last several years, fueled by advancements in technology. From precision farming techniques to Internet of Things (IoT) applications, farmers are increasingly embracing innovative solutions to enhance efficiency, productivity, and sustainability. Among these emerging technologies, blockchain stands out as a disruptive power that may completely transform smart agriculture.

Prior to exploring blockchain's potential in agriculture, it's important to understand the problems that the sector is currently facing. Agriculture still faces challenges like ineffective supply chains, a lack of transparency, food fraud, and unsustainable practices even with technological advancements. Access to markets is a challenge for small-scale farmers, and consumers want more clarity about the source and caliber of the goods they buy. Furthermore, there is a risk to public health and substantial financial losses due to fraud and contamination in the global food supply chain. Smart agriculture, also known as precision agriculture, leverages technology to optimize the efficiency and sustainability of farming practices. It encompasses a range of technologies, including IoT, AI, drones, and data analytics, to gather and evaluate data for informed decision-making. By employing sensors and data analytics, farmers can monitor various parameters such as crop health, weather, and soil moisture content in real time. Precise resource management is made possible by this data-driven strategy, which minimizes environmental effect while lowering waste and raising yields.

Although smart agriculture has improved farm management significantly, the agriculture ecosystem gains more security, trust, and transparency with the addition of blockchain technology. Initially designed as the foundational technology for cryptocurrencies such as Bitcoin, blockchain has developed into a flexible instrument with uses in other industries. Peer-to-peer transactions are made possible by distributed ledger technology, or blockchain, which eliminates the need for intermediaries. Each transaction or data entry is recorded as a "block" and arranged chronologically in a chain, creating a tamper-proof and transparent ledger. Consensus techniques improve security and trust by ensuring that all network users concur on the legitimacy of transactions. It is perfect for addressing long-standing challenges in agriculture, such as supply chain traceability, food safety, and data management, because of its decentralized, immutable, and transparent nature. Traditional agricultural practices are facing increasing pressure to adapt to evolving global challenges, including population growth, climate change, and resource constraints. By leveraging blockchain's immutable ledger and decentralized consensus mechanism, stakeholders across the agricultural value chain can securely record and verify transactions, ensuring transparency and traceability from farm to fork. This transparency fosters trust among stakeholders, mitigating issues such as food fraud, counterfeit products, and supply chain inefficiencies. Smart agriculture, enabled by technologies like artificial intelligence, big data analytics, and IoT, offers promising solutions to optimize resource utilization, minimize environmental impact, and ensure food security. However, existing systems often lack transparency, interoperability, and trust, hindering the industry's ability to fully capitalize on these technological advancements.

DOI: 10.1201/9781003507390-1

Blockchain can be applied in smart agriculture in different ways. Farm products may now be tracked from start to finish due to blockchain technology, from farm to fork. By recording each stage of the supply chain, such as manufacturing, processing, shipping, and retail, stakeholders can verify the authenticity and quality of products. With blockchain, farmers can track and record vital information such as crop inputs, cultivation practices, and surroundings. This information can be utilized to ensure compliance with regulations, identify potential hazards, and implement targeted interventions to mitigate risks.: Blockchain-based platforms can facilitate transparent and secure transactions, enabling smallholder farmers to access financial services, insurance, and markets previously out of reach since there is no collateral or credit history.

There are certain challenges that should be considered when introducing blockchain in smart agriculture. Scalability continues to be a major concern as blockchain networks get bigger and more complicated. The technology must evolve to deal with the rising number of transactions and data without compromising speed or efficiency. Interoperability between various blockchain platforms and traditional systems must be achieved in order to facilitate data interchange and teamwork throughout supply chains. While blockchain offers enhanced security through encryption and consensus mechanisms, ensuring data privacy and protection against cyber threats remains paramount.

Despite the challenges, the advantages of blockchain technology for agriculture are too great to ignore. As technology advances and standards emerge, we anticipate broad adoption across the industry. By fostering transparency, trust, and efficiency, blockchain has the power to transform smart agriculture into a more resilient, sustainable, and inclusive ecosystem. Smart agriculture has undergone a paradigm leap with the incorporation of blockchain technology in how we produce, distribute, and consume food.

1.1.1 Challenges in Traditional Agriculture

Traditional agriculture faces numerous challenges that threaten its sustainability and efficiency. One of the primary hurdles lies in supply chain management, where inefficiencies abound due to long and convoluted supply chains involving multiple intermediaries. These intermediaries add layers of complexity and cost to the process, making it difficult for farmers to efficiently track their products from farm to fork. Consequently, delays, inefficiencies, and potential losses plague the system, hindering its overall effectiveness. Transparency within the supply chain is another pressing issue. Today's consumers are increasingly demanding detailed information about the origin and journey of their food, from production to distribution. However, the current system often fails to provide adequate transparency, leaving consumers uninformed about crucial aspects of their food's journey. This opaqueness damages the integrity of the entire agricultural system in addition to undermining customer confidence. Concerns about food safety exacerbate the problems that traditional agriculture faces. Foodborne illness outbreaks present serious dangers to public health and safety, resulting in a great deal of recall activity, financial losses, and harm to businesses' and farmers' reputations. Food safety is already a complicated subject, and it gets considerably more complicated when it comes to guaranteeing the traceability of products in the event of a recall. An additional crucial challenge facing traditional agriculture is environmental sustainability. Unsustainable farming methods exacerbate climate change and environmental degradation by causing soil erosion, water pollution, deforestation, and greenhouse gas emissions. The environmental effects of conventional agriculture are evident and call for immediate action in the direction of more sustainable alternatives. These effects range from monocropping and excessive pesticide usage to overgrazing and deforestation. Additionally, traditional agriculture contributes to economic inequality, especially for small-scale farmers. Due to their restricted access to markets, credit, and resources, smallholder farmers find it difficult to compete in an increasingly globalized industry that is controlled by huge agribusinesses. This discrepancy makes them more susceptible to marginalization and exploitation, which widens the socioeconomic gap in the agriculture industry. A comprehensive strategy that incorporates technical innovation, legislative changes, and community involvement is needed to address these issues. We can build a more robust and just agricultural system for the future by encouraging greater openness, boosting food safety standards, optimizing supply chain efficiency, encouraging sustainable farming methods, and empowering smallholder farmers.

1.1.2 The Emergence of Smart Agriculture

Smart farming, commonly referred as precision agriculture, represents the next evolution in farming practices by leveraging advanced technologies to optimize agricultural production. At its core, smart agriculture integrates IoT devices, sensors, drones, AI, and data analytics to monitor, manage, and automate farming operations. These technologies give farmers access to real-time insights into soil conditions, crop conditions, weather trends, and pest activity, giving them the ability to make data-driven choices that maximize yields while minimizing resource waste. One of the main benefits of smart agriculture is enhanced productivity through optimized resource management. By precisely controlling inputs like insecticides, fertilizers, and water, farmers can enhance agricultural yields and quality. Additionally, smart agriculture promotes sustainability by reducing environmental impact through efficient resource use and targeted interventions. For example, drones and sensors can identify nutrient deficits or early disease signs, allowing for timely interventions and reducing the need for broad-spectrum treatments.

Furthermore, smart agriculture fosters innovation and adaptability in farming practices. With AI and data analytics, farmers can adapt quickly to changing conditions, mitigate risks, and optimize farming operations for better outcomes. This data-driven strategy increases farm productivity while also enhancing market access and profitability by producing higher-quality crops that meet market demands. In essence, smart agriculture is revolutionizing the agricultural sector by offering farmers the tools and insights needed to prosper in a more difficult and cutthroat environment.

1.1.3 Benefits of Smart Agriculture Compared to Traditional Agriculture

1. Enhanced Effectiveness:
   - Utilizing data to inform decisions and automate processes enhances output while maximizing resource usage.
   - Planting, irrigation, and pest management decisions are influenced by real-time study of crop health, weather patterns, and soil conditions.
2. Resource Optimization:
   - Precision farming techniques reduce waste and environmental impact by applying inputs more precisely.
   - Tailoring inputs to specific crop and field needs minimizes resource use while maximizing output.
3. Improved Crop Management:
   - Real-time monitoring and analysis enable early detection of pests, diseases, and nutrient deficiencies.
   - Timely action based on IoT sensor data mitigates risks and protects crop yields, ensuring food security.
4. Enhanced Sustainability:
   - Reduction of reliance on chemical inputs and adoption of eco-friendly practices promote soil health and biodiversity.
   - Smart agriculture fosters long-term sustainability and resilience to climate change through conservation tillage and integrated pest management.
5. Increased Profitability:
   - Optimization of inputs and productivity reduces costs and increases yields.
   - Premium prices for high-quality, sustainably produced crops enhance financial viability and long-term success.

6. Greater Resilience to Climate Change:
   - Climate-smart agriculture techniques, such as drought-resistant crops and precision irrigation, help farmers adapt to changing environmental conditions.
   - Incorporating resilience into farming practices enables farmers to maintain productivity and livelihoods in the face of climate-related challenges.
7. Empowerment of Smallholder Farmers:
   - Access to information, markets, and resources through mobile technology and digital platforms improves decision-making and livelihoods.
   - Smallholder farmers in developing countries benefit from weather forecasts, market prices, and agricultural inputs, enabling informed decisions.
8. Enhanced Food Safety and Quality:
   - Transparent and traceable supply chains enabled by blockchain technology ensure authenticity and quality of food products.
   - Tamper-proof record of each step in the supply chain reduces the risk of food fraud, contamination, and recalls, ensuring food safety for consumers.

1.1.4 Current Technologies Used in Agriculture

Modern agriculture is undergoing a technological revolution, with advancements aimed at enhancing efficiency, productivity, and sustainability. A diverse range of technologies is being deployed across the agricultural value chain, from field operations to logistics of the supply chain. Some of the current technologies revolutionizing agriculture are the following.

1.1.4.1 Precision Agriculture

Precision agriculture leverages technology to optimize insecticides, fertilizers, and water inputs while optimizing crop yields and minimizing environmental impact. Key technologies include:

Global Positioning System (GPS): GPS technology enables farmers to accurately map field boundaries, track machinery movements, and create precision application maps for inputs.

Remote Sensing: Satellite imagery, drones, and sensors provide real-time information about pest infestations, soil moisture content, and crop health. Farmers can use this information to make well-informed decisions on pest control, fertilizer, and irrigation.

Variable Rate Technology (VRT): This systems adapt the input application rates according to spatial variability within fields, optimizing resource utilization and reducing costs.

1.1.4.2 Robotics and Automation

Robotic systems are increasingly being used to automate repetitive tasks in agriculture, ranging from planting and harvesting to weeding and spraying. Key technologies include:

Autonomous Tractors: With the aid of GPS, sensors, and computer vision systems, self-driving tractors are capable of precisely and effectively completing operations like tillage, sowing, and plowing.

Harvesting Robots: Precision-picking ripe fruits and vegetables and sorting them according to quality standards are all possible using robotic harvesters that are outfitted with cameras and artificial intelligence algorithms.

Weeding Robots: Autonomous robots equipped with cameras and mechanical or chemical weed control mechanisms can selectively remove weeds while minimizing herbicide usage.

1.1.4.3 Internet of Things

IoT involves connecting sensors, devices, and machinery to the internet to collect and exchange data in real time. Key applications in agriculture include:

Smart Sensors: IoT sensors monitor environmental conditions, providing farmers with actionable insights to optimize growing conditions.

Connected Machinery: Tractors, irrigation systems, and other farm equipment have IoT sensors installed and telematics devices to track usage, performance, and maintenance needs.

Livestock Monitoring: In order to improve feeding and breeding procedures and enable early diagnosis of health problems, IoT-enabled devices, such as smart ear tags and collars, monitor the location, activity, and health of animals.

1.1.4.4 Vertical Farming and Controlled Environment Agriculture

Vertical farming and controlled environment agriculture (CEA) technologies enable year-round production of crops in controlled indoor environments using techniques such as hydroponics, aeroponics, and aquaponics. Key technologies include:

LED Lighting: Energy-efficient LED lights provide the optimal spectrum and intensity for plant growth, allowing vertical farms to operate without natural sunlight.

Climate Control Systems: In order to give plants the best growing conditions possible, automated climate control systems adjust humidity, temperature, and CO_2 levels.

Hydroponic Systems: Soilless hydroponic systems supply nutrient-rich water straight to the roots of plants, maximizing water and nutrient efficiency while minimizing water usage.

1.2 INTRODUCTION TO BLOCKCHAIN TECHNOLOGY, CRYPTOCURRENCIES, AND SMART CONTRACTS

1.2.1 Blockchain Technology

Distributed ledger technology, or blockchain, allows transactions to be recorded on a network of computers in a transparent and safe manner. A list of transactions is present on every block in the series of blocks that comprise this system. By connecting these blocks with cryptographic hashes, an unchangeable record of the data is produced. Fundamentally, blockchain technology is characterized by several key features:

Decentralization: In contrast to conventional centralized systems, blockchain functions through a decentralized network of computers called nodes. To provide redundancy and fault tolerance, each node keeps a copy of the blockchain. By doing away with the requirement for a central authority, this decentralized system lowers the possibility of single points of failure and permits peer-to-peer transactions.

Immutability: Data cannot be changed or tampered with once it is recorded on the blockchain. A chain of blocks that is impervious to change is created by each block having a cryptographic hash of the one before it. The blockchain is perfect for applications where data integrity is crucial because of its immutability, which guarantees the reliability and integrity of the data stored on it.

Transparency: By allowing all transactions to be seen by network users, blockchain promotes transparency. A visible and verifiable record of every transaction is kept on the blockchain, enabling stakeholders to monitor the movement of assets and confirm the legitimacy of transactions. This openness lowers the possibility of fraud or manipulation while fostering participant trust.

Security: Blockchain technology secures transactions and shields data from unwanted access by using cryptographic algorithms. To maintain the integrity and security of the blockchain, network users validate and verify transactions using consensus techniques like proof of work (PoW) or proof of stake (PoS). Furthermore, because blockchain is decentralized, it is immune to hacking attempts and cyberattacks.

Numerous sectors can benefit from the extensive applications of blockchain technology. Blockchain is used in finance to provide safe and transparent transactions, while in supply chain management, it enables traceability and accountability. In healthcare, blockchain facilitates secure and interoperable sharing of patient data, while in government, it enhances transparency and efficiency in public services.

In agriculture, blockchain technology is transforming the way supply chains operate. It enables transparent and immutable record-keeping of transactions and data across the whole supply chain of agriculture. Every stage of the process, from planting and harvesting to processing and distribution, can be tracked on the blockchain, providing stakeholders with instantaneous insight into the source and path of agricultural goods. This transparency helps improve trust, quality assurance, and traceability in the agricultural sector.

1.2.2 Cryptocurrencies

A cryptocurrency is a type of virtual or digital money that runs on decentralized networks built on blockchain technology and employs encryption for security. It enables secure peer-to-peer transactions without the need for intermediaries like banks. Bitcoin, created in 2009, was the first decentralized cryptocurrency, followed by thousands of others, including Ethereum, Ripple, and Litecoin. Digital wallets are used to hold cryptocurrency, which may be traded for products and services or traditional currencies. They are known for their volatility, with prices fluctuating based on market demand and supply. Despite regulatory challenges, cryptocurrencies continue to gain popularity as a new form of digital asset and investment.

Cryptocurrencies are playing a role in agriculture by facilitating secure and efficient transactions. Farmers can use cryptocurrencies as a medium of exchange for buying and selling agricultural products, equipment, or services. Cryptocurrencies offer advantages such as lower transaction fees, quicker international payments, and greater financial inclusion, especially in areas where access to traditional banking services is restricted. Additionally, cryptocurrencies can enable direct peer-to-peer transactions between farmers and consumers, bypassing intermediaries and reducing costs.

1.2.3 Smart Contracts

A self-executing contract known as a "smart contract" has the conditions of the parties' agreement encoded directly into the code. When certain criteria are satisfied, these contracts—which are implemented on blockchain networks like Ethereum—automatically carry out and enforce the provisions of the contract. Smart contracts eliminate the necessity of intermediaries like lawyers or brokers, as the code itself enforces the rules and ensures trust between parties.

Smart contracts offer security and transparency because they are immutable, which means that once they are put on the blockchain, they cannot be changed. They are executed automatically without the necessity of human involvement, lowering the possibility of errors and fraud. Smart contracts can facilitate various types of transactions, including payments, asset transfers, and token exchanges, in a secure and efficient manner.

Smart contracts in agriculture automate and enforce agreements between different parties without the need for intermediaries. For example, smart contracts can be used to facilitate fair trade agreements between farmers and buyers, ensuring that farmers receive fair compensation for their products. They can also automate repetitive processes like processing payments and quality assurance. Smart contracts run on blockchain networks, providing security, transparency, and efficiency in agricultural transactions and contracts.

By enhancing the agricultural supply chain's efficiency, trustworthiness, and transparency, these technologies are revolutionizing agriculture. These innovations enable stakeholders to trace the source and caliber of agricultural products, streamline transactions, automate contract enforcement, ultimately leading to an agriculture sector that is more robust and sustainable.

1.3 WHY BLOCKCHAIN IN AGRICULTURE?

Blockchain technology offers a transformative potential for the agricultural industry, addressing a spectrum of challenges while unlocking new opportunities. Its core attributes of transparency, traceability, efficiency, market access, and data security promise to reshape how agricultural products are produced, distributed, and consumed. At its essence, blockchain's transparent and immutable ledger system revolutionizes stakeholders' ability to verify the authenticity and quality of agricultural goods, fostering accountability across the supply chain. By meticulously recording every transaction and event, blockchain ensures traceability, enhancing food safety and integrity. Moreover, blockchain's impact transcends mere record-keeping, streamlining supply chain operations through automated tasks such as payments and logistics. By reducing redundancies and friction, blockchain enhances operational efficiency, driving down costs and enhancing profitability for agricultural businesses. The democratizing potential of blockchain technology has the ability to greatly help small-scale farmers by giving them direct access to international markets and equitable recompense for their produce. Farmers are empowered by their increased market access, which fosters rural development and economic inclusion. Blockchain strengthens data security in tandem with market access, protecting private agricultural data from manipulation and cyberattacks. Data integrity and secrecy are guaranteed by its decentralized design and cryptographic protocols, which are essential for preventing unwanted access. In conclusion, blockchain technology has the potential to improve food safety and quality, streamline supply chains, and strengthen the agricultural ecosystem's resilience and sustainability. Using blockchain technology can enable stakeholders to successfully negotiate the challenges of contemporary agriculture, opening the door to a more open, effective, and just future.

1.4 BLOCKCHAIN SMART AGRICULTURE

The primary blockchain use case in the agriculture sector is smart farming, sometimes referred to as smart agriculture. To make precision farming more reliable and efficient, it integrates a number of recent technological developments. It combines sensors, machine learning technology, information and communications technology, and a variety of data collection and analytical tools. While farming has long had access to smart technology, implementing a strong and complete security system to manage and profit from the data gathered is a significant challenge.

The centralized nature of traditional smart technology management system processes leads to various distortions and errors in data collection. Additionally, there is a greater chance that the entire system will be the target of a cyberattack. But it is possible to store data securely with the state of blockchain technology today. Throughout the entire process, numerous stakeholders can produce the required environmental monitoring data at any stage, from the seed to the sale of diverse agricultural production. Professionals with experience in blockchain technology can implement it in ways that guarantee the immutability of all statistics and facilitate data transparency.

Decentralization is by far blockchain's greatest advantage in smart agriculture. This feature lowers data loss and distortion while simplifying the process of supplying data to the displays of a wide range of users. In a blockchain, every transaction is accurately documented for the purpose of transparency. To date, a great deal of smart farming models have been developed based on blockchain technology's value and promise in the agricultural sector. These models help to bridge the characteristics of Internet of Things and blockchain technology. One such solution was created especially for greenhouses, which need to use IoT sensors to operate as a locally managed, private blockchain that the farmer can simply maintain. Another strategy that makes use of blockchain and IoT technologies has been developed for widespread use. This structure's main goal is to facilitate the growth of trust among the numerous blockchain participants. Smartphones can be used by a variety of users to access data that is kept throughout the farming process, from seed to product sale.

In addition, a number of businesses are investing a substantial sum of time and resources in the rural growth of the smart agriculture supply chain. One company, "Filament," for instance, offers products that link many networks to tangible items using intelligent agricultural technology. The organization created a coin-sized technology to help users transact safely on a blockchain.

1.5 THE VARIOUS WAYS IN WHICH BLOCKCHAIN TECHNOLOGY COULD TRANSFORM THE AGRICULTURE INDUSTRY

1. Traceability and Transparency: Blockchain makes it possible to store data and transactions in an unchangeable and transparent manner throughout the whole agricultural supply

chain. Every stage, starting from cultivation to harvesting, processing, and distribution, can be meticulously documented on the blockchain, granting all involved parties immediate access to the origins and progression of agricultural goods. This enhanced transparency fosters greater trust, ensures quality control, and enables traceability, empowering consumers to validate the legitimacy and sustainability of their purchases.

2. Supply Chain Efficiency: Supply chain operations are revolutionized by blockchain technology because smart contracts automate jobs. These agreements codify specified conditions, enabling automatic payments, quality checks, and shipping. As a result, delivery is accelerated, waste is reduced, and supply chain efficiency in agriculture is increased. Furthermore, by providing a decentralized ledger of transactions, blockchain improves transparency. Real-time tracking of commodities by stakeholders promotes accountability and reduces fraud. Blockchain promises to revolutionize agriculture by promoting openness, efficiency, and sustainability throughout the supply chain while also making regulatory compliance even easier.
3. Quality Assurance: Quality assurance is facilitated through continuous monitoring of environmental factors such as temperature, humidity, and soil health throughout the production process. This data is securely recorded on the blockchain, instilling confidence in both farmers and consumers regarding the quality and safety of agricultural products. By ensuring compliance with stringent quality standards, blockchain minimizes the risk of contamination and foodborne illnesses, ultimately bolstering food safety and enhancing overall security within the agricultural sector.
4. Market Access and Fair Trade: Blockchain can empower small-scale farmers by giving them access to global markets and ensuring fair compensation for their products. By recording ownership rights and production data on the blockchain, farmers can establish a digital identity and build trust with buyers, leading to more equitable trading relationships. Additionally, blockchain-based platforms can facilitate direct peer-to-peer transactions between farmers and consumers, bypassing intermediaries and reducing costs.
5. Data Sharing and Collaboration: Secure data sharing and collaboration are made possible by blockchain technology among stakeholders in the agricultural ecosystem, including farmers, researchers, suppliers, and government agencies. This shared infrastructure fosters innovation and allows for the development of data-driven solutions to complex agricultural obstacles such as climate variability, limited water resources, and the assurance of food provision.

1.6 USES OF BLOCKCHAIN TECHNOLOGIES IN AGRICULTURE

Here are some different uses of blockchain technologies in agriculture with examples:

1. Crop Traceability: Blockchain technology offers the capability to trace the entire journey of crops from the farm to the consumer's table, furnishing detailed insights into their origin, farming techniques, and handling procedures. An illustration of this is IBM Food Trust, which employs blockchain to track the source of mangoes throughout the supply chain, empowering consumers to authenticate the fruit's quality and genuineness.
2. Smart Contracts for Fair Trade: Smart contracts can automate fair trade agreements between farmers and buyers, making certain that farmers obtain just recompense for their products. For instance, AgriDigital uses blockchain-based smart contracts to facilitate fair and transparent transactions between grain growers and buyers, reducing disputes and delays in payment processing.
3. Supply Chain Optimization: Blockchain technology has the potential to enhance supply chain operations by enhancing transparency, efficiency, and confidence among involved parties. An instance of this is the collaboration between the World Wildlife Fund (WWF)

and Bumble Bee Foods, where blockchain is employed to monitor the path of tuna from the sea to retail shelves. This ensures adherence to sustainable and ethical sourcing standards within the seafood supply chain.

4. Data Management and Sharing: Blockchain enables secure and decentralized data management and sharing among farmers, researchers, and other stakeholders in the agricultural ecosystem. For instance, AgriChain uses blockchain to securely store and share data related to crop productivity, atmospheric conditions, and market values, granting farmers the ability to make well-informed choices and improve productivity.
5. Fraud Prevention and Food Safety: Blockchain technology can help prevent fraud and ensure food safety by recording every transaction and event in the agricultural supply chain. As an illustration, Walmart integrated blockchain technology to track the source of leafy greens, enabling swift identification and to address foodborne illness outbreaks, thereby protecting consumer health and brand reputation.
6. Land Ownership and Property Rights: Blockchain can be used to record and verify land ownership and property rights, especially in regions where land tenure is unclear or disputed. For example, Bitland uses blockchain to create transparent and immutable land registries in Ghana, providing small-scale farmers with secure land tenure and access to financial services.
7. Decentralized Marketplaces: Blockchain-based marketplaces can connect farmers directly with consumers, eliminating intermediaries and reducing transaction costs. For instance, GrainChain utilizes blockchain to establish a decentralized marketplace for grain producers and buyers, enabling farmers to negotiate fair prices and access broader markets.
8. Crop Insurance and Risk Management: Blockchain can facilitate transparent and automated crop insurance programs, reducing the administrative burden and increasing trust among insurers and farmers. For example, Etherisc uses blockchain to offer parametric crop insurance products that initiate automatic payouts according to predetermined weather conditions, helping farmers mitigate financial risks associated with climate variability.
9. Certification and Compliance: Blockchain can streamline certification and compliance processes by securely storing and sharing records of organic certification, fair trade practices, and adherence to regulatory requirements. For instance, Provenance uses blockchain to verify authenticity and sustainability of agricultural products, providing consumers with assurance regarding ethical sourcing and production practices.
10. Carbon Credits and Sustainable Agriculture: Blockchain technology can support carbon offset projects and incentivize sustainable agricultural practices by enabling transparent and verifiable tracking of carbon credits. For example, Nori uses blockchain to create a marketplace for carbon removal credits generated through regenerative agriculture practices, rewarding farmers for sequestering carbon in soil while mitigating greenhouse gas emissions.

1.7 PROBLEMS IN AGRICULTURAL SUPPLY CHAINS

The agricultural supply chain, a complex network spanning farm to consumer, encounters numerous challenges that hinder its efficiency, sustainability, and profitability. Foremost among these challenges, the absence of transparency poses challenges in monitoring the source and standard of agricultural products. This opacity exposes the supply chain to risks such as fraud, counterfeiting, and food safety incidents, eroding consumer trust and posing significant economic and health hazards. Furthermore, the agricultural supply chain struggles with fragmented data management practices, where information is often siloed and poorly coordinated across stakeholders. This fragmentation exacerbates inefficiencies, complicates decision-making processes, and limits opportunities for innovation and optimization. Inefficient logistics and transportation also pose significant hurdles, leading to delays, spoilage, and loss of perishable goods during transit. These logistical challenges compromise product quality and incur substantial financial losses for producers and distributors, undermining the sector's overall profitability and competitiveness. Market access and price volatility further exacerbate the woes of the agricultural supply chain, leaving small-scale farmers particularly vulnerable to unfair pricing practices and income instability. Additionally, ensuring quality assurance and compliance with regulatory standards is a resource-intensive endeavor, particularly for smallholder farmers with limited resources. Failure to adhere to these standards not only jeopardizes consumer health but also exposes stakeholders to legal liabilities and reputational damage. Climate change and environmental risks also threaten agricultural productivity and sustainability, exacerbating food insecurity and undermining the sector's long-term viability. Financial constraints further compound these challenges, limiting investment in modern technologies, infrastructure, and sustainable practices. Labor shortages and workforce issues add another layer of complexity, disrupting production cycles and increasing operational inefficiencies. Addressing these challenges requires a comprehensive and collaborative approach, leveraging innovative technologies such as blockchain, IoT, AI, and data analytics. Furthermore, supportive policies, capacity-building initiatives, and investment in sustainable practices are essential for overcoming the hurdles facing the agricultural supply chain. By embracing innovation, collaboration, and sustainability, stakeholders can unlock the sector's full potential to feed the world's growing population while ensuring economic viability and environmental stewardship.

1.8 OVERCOMING CHALLENGES AND ADOPTION BARRIERS

Although blockchain technology holds promise for agriculture, its extensive implementation encounters numerous hurdles. Among these, scalability emerges as a significant barrier, with existing blockchain networks finding it challenging to manage the high transaction volumes demanded by large agricultural supply chains. Additionally, interoperability issues make it difficult to integrate blockchain with existing systems and technologies, hindering seamless communication and data exchange. Regulatory concerns also pose a significant barrier to adoption, as the legal framework surrounding blockchain and cryptocurrencies varies widely from country to country. Uncertainty about compliance, taxation, and data privacy regulations creates challenges for companies and organizations looking to implement blockchain solutions in agriculture. Technical limitations, such as high transaction costs, slow processing times, and energy consumption, further complicate the adoption of blockchain technology in agriculture. Improving scalability, reducing fees, and increasing transaction speeds are key areas of focus for blockchain developers and researchers working to overcome these challenges. Education and awareness play a pivotal role in fostering the uptake of blockchain technology in agriculture. Many farmers and producers are unfamiliar with blockchain and may exhibit reluctance to embrace novel technologies in the absence of a comprehensive grasp of their advantages and potential risks. Educating stakeholders about the advantages of blockchain, as well as providing training and support for implementation, can help overcome resistance and promote acceptance. Collaboration and partnerships are essential for overcoming challenges and fostering innovation in the agricultural sector. By bringing together stakeholders from across the industry, including farmers, producers, retailers, government agencies, and by collaborating with both industry stakeholders and technology providers, we can collectively develop and deploy blockchain solutions tailored to meet the distinct requirements and obstacles encountered within the agricultural value chain.

1.9 HOW CAN AGRICULTURE PRODUCTION BE REVOLUTIONIZED BY BLOCKCHAIN TECHNOLOGY?

Blockchain technology offers transparency, security, and decentralization in the food production process. One feature of blockchain is immutability. Transactions cannot be altered or concealed because every modification is tracked, recorded, and visible to the entire network of users who have access to the data. Additionally, it uses dispersed data storage. This suggests that multiple servers are used to store information. Instead, many copies of the information are kept on various networks so that everyone can view and access them. Because blockchain fosters such transparency, it does away with the need for both trust and the requirement that a central body mediate disputes between the parties.

Blockchain has the potential to revolutionize the process of crop or food production through four key stages.

1.10 STAGE 1: IOT DEVICES GENERATING DATA

With the global population expected to reach 9.6 billion by 2050, the agricultural industry is integrating IoT devices and sensors to meet the growing demand

Within smart farming enabled by IoT:

- A crop field monitoring system is designed to use sensors for temperature, pH, light, humidity, and soil moisture.
- Farmers can make well-informed decisions about the growth of their crops with the use of data produced by sensors and devices.
- Before being stored on a data storage device, the information obtained from these devices must be organized.

1.11 STAGE 2: CLEANSING AND ENHANCEMENT OF GATHERED DATA

Before being stored on the blockchain, information needs to be clear and well structured. Data enrichment aims to enhance the quality and value of captured information. Subsequently, data is refined through two additional steps before being stored on the distributed storage platform.

1.12 INCORPORATING METADATA

Information pertaining to the following should be added in order to efficiently structure data:

- type of timestamp demographic
- preparing data for legal requirements
- the blockchain's ability to store data facilitates more seamless compliance enforcement
- achieving compliance guarantees the protection and observance of security protocols for the data collected from IoT devices linked to personally identifiable information.

The enriched data is then placed in a format that is ready for machine learning.

1.13 STAGE 3: APPLYING MACHINE LEARNING ALGORITHMS TO ENHANCE THE DATA'S INTELLIGENCE

To extract meaningful insights from the sensor data, machine learning is applied. Several high-value use cases can be powered by predictive models, including:

- Crop Quality Guidelines
- Identification of Crops
- Crop Yield Forecast
- GrowScore: An Automated Measure of Crop Growth
- Crop Demand Prediction

Machine learning algorithms will provide information that farmers and other stakeholders can use to periodically enhance the irrigation system. To allow transparent access to the insightful data by producers, innovators, retailers, growers, and service providers in the agriculture market, it should be stored on the blockchain.

1.14 STAGE 4: THE BLOCKCHAIN STORES DATA

High-value data collected through machine learning is kept on the blockchain-encrypted distributed storage network interplanetary file system (IPFS), where addresses are hashed. A single point of failure could occur with the current method of storing crucial data on a centralized server. Blockchain, on the other hand, distributes the data among all of the network's nodes. As a result, it keeps a central authority from having systemic control. When the recorded data is integrated into the blockchain, smart contracts will automatically trigger the execution of predefined rules embedded within them. These smart contracts make it easier for the particular system stakeholders to exchange data that is stored on the blockchain. Every player in the agriculture market will be able to see the information, making it seamless to increase crop or food efficiency.

1.15 BLOCKCHAIN IN WEATHER MONITORING

Blockchain technology offers several advantages in weather monitoring systems integrated with various sensors like CO_2, pressure, temperature, smoke, illuminance, wind, air quality, and pollution. Coupled with IoT devices and smart agriculture practices, blockchain ensures data integrity, security, and transparency while facilitating efficient management and decision-making processes. Each sensor reading is recorded in a tamper-resistant blockchain ledger, ensuring the integrity of the data. This feature is crucial in critical applications like agriculture, where accurate weather information directly impacts crop yield and quality. By leveraging blockchain, stakeholders can trust the authenticity of the data, fostering collaboration and innovation in agricultural practices. Moreover, blockchain enhances data security by encrypting sensor data and distributing it across multiple nodes in the network. This decentralized architecture minimizes the risk of a single point of failure and prevents unauthorized access or manipulation of sensitive information. As a result, farmers and agricultural organizations can confidently rely on weather data for making informed decisions regarding irrigation, fertilization, and pest control.

Blockchain technology also facilitates transparent data sharing and incentivizes data contribution through smart contracts. Smart contracts are self-executing agreements coded on the blockchain, enabling automated transactions when predefined conditions are met. In the context of weather monitoring, smart contracts can reward farmers or individuals for sharing their sensor data with the network. This incentivization encourages widespread participation, leading to a more comprehensive and accurate dataset for weather forecasting and agricultural planning. Furthermore, blockchain enables efficient management of IoT devices and sensors through a decentralized management hub. This hub serves as a central platform for monitoring, configuring, and updating connected devices across different locations. By utilizing blockchain for device management, stakeholders can streamline operations, reduce maintenance costs, and ensure interoperability between heterogeneous sensor networks.

In addition to data integrity and security, blockchain enhances traceability and auditability in weather monitoring systems. Each sensor reading is timestamped and cryptographically linked to previous records, creating an immutable audit trail of data provenance. This feature is particularly valuable in supply chain management for tracking the origin and conditions of agricultural products from farm to fork. Consumers can verify the authenticity and quality of produce, fostering trust and transparency in the agricultural industry.

Furthermore, blockchain technology facilitates interoperability and data exchange between disparate systems through standardized protocols like constrained application protocol (COAP) for IoT communication. The COAP bench provides a lightweight and efficient communication framework for IoT devices, enabling seamless integration with blockchain-enabled weather monitoring platforms. This interoperability ensures compatibility with existing infrastructure and accelerates the adoption of innovative technologies in smart agriculture.

1.16 ROLE OF BLOCKCHAIN IN REVOLUTIONIZING AGRICULTURAL FINANCE

Blockchain technology is revolutionizing agricultural finance by providing innovative solutions to longstanding challenges in funding, lending, and investment within the agricultural sector. Through its decentralized and transparent nature, blockchain is empowering farmers, investors, and financial institutions to access, manage, and leverage financial resources more efficiently and securely.

1.17 ACCESS TO CAPITAL

Blockchain-based platforms enable farmers, particularly smallholders and rural producers, to access capital directly from a global pool of investors. Through tokenization of agricultural assets, such as land, crops, and livestock, farmers can create digital representations of their assets, which can be traded or used as collateral for loans on blockchain-based lending platforms. This democratization of access to capital reduces reliance on traditional banking systems and empowers farmers to invest in their operations, expand production, and improve livelihoods.

1.18 TRANSPARENT AND EFFICIENT TRANSACTIONS

Blockchain facilitates transparent and efficient financial transactions by recording all transactions on a decentralized ledger that is immutable and transparent to all parties involved. Smart contracts, self-executing contracts with predefined terms and conditions, automate and enforce agreements between borrowers and lenders, streamlining the lending process and reducing administrative costs. Farmers benefit from faster loan approval processes, lower transaction fees, and increased trust in financial transactions, while lenders gain greater transparency and reduced risks of fraud and default.

1.19 RISK MANAGEMENT AND INSURANCE

Blockchain technology enhances risk management and insurance solutions for agricultural finance by providing real-time access to data on weather patterns, crop yields, and market prices. By recording and verifying agricultural data on a blockchain platform, insurers can offer parametric insurance products that automatically trigger payouts based on predefined triggers, such as adverse weather conditions or yield fluctuations. This innovative approach to insurance reduces administrative costs, eliminates the need for manual claims processing, and provides farmers with timely and reliable financial protection against various risks.

1.20 INVESTMENT IN SUSTAINABLE AGRICULTURE

Blockchain enables transparent and traceable investment in sustainable agriculture by allowing investors to track the impact of their investments on environmental, social, and governance (ESG) criteria. Through blockchain-based platforms, investors can access detailed information about the agricultural projects they are funding, including practices related to soil health, water conservation, and carbon sequestration. This transparency promotes responsible investing in agriculture; incentivizes sustainable farming practices; and fosters collaboration between investors, farmers, and other stakeholders to achieve shared sustainability goals.

1.21 FUTURE PERSPECTIVES AND OPPORTUNITIES

As we gaze into the future, the potential of blockchain technology in agriculture seems boundless. Advancing technologies like machine learning, 5G, and edge computing stand ready to augment the capacities of blockchain-driven agricultural systems, facilitating instantaneous data analysis, independent decision-making, and tailored guidance. Machine learning algorithms have the capacity to scrutinize extensive datasets produced by IoT sensors, drones, and other agricultural innovations to discern patterns, anticipate shifts, and refine farming methodologies. By leveraging machine learning, farmers have the opportunity to acquire valuable insights into crop well-being, pest occurrences, and atmospheric conditions, empowering them to make decisions based on data analysis that maximize yield with minimum risk. The rollout of 5G networks promises to revolutionize connectivity in rural areas, enabling high-speed internet access and real-time data transmission on the farm. With 5G, farmers can deploy advanced technologies including drones, robotics, and applications powered by artificial intelligence to monitor and manage their operations with unprecedented speed and efficiency. Edge computing relocates processing power closer to the origin of data generation, facilitating instantaneous analysis and decision-making at the network's edge. Through the utilization of edge computing, farmers can deploy AI models and analytics tools directly on IoT devices, drones, and other edge devices, enabling autonomous decision-making and reducing latency in data transmission. In addition to technological advancements, international efforts like the United Nations Sustainable Development Goals (SDGs) are driving innovation and collaboration in the agricultural sector. By promoting sustainable practices, reducing food waste, and improving access to markets and resources, these initiatives aim to develop more resilient, inclusive, sustainable food structures for future generations. Entrepreneurial opportunities abound in the field of blockchain-powered agriculture, with startups and innovators developing new solutions to address the challenges facing

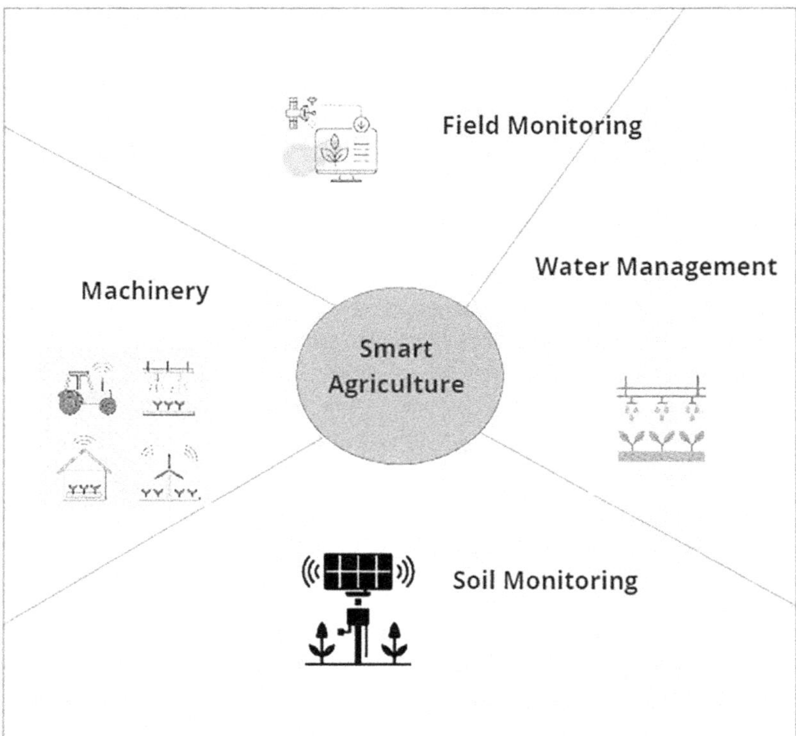

FIGURE 1.1 Smart agriculture.

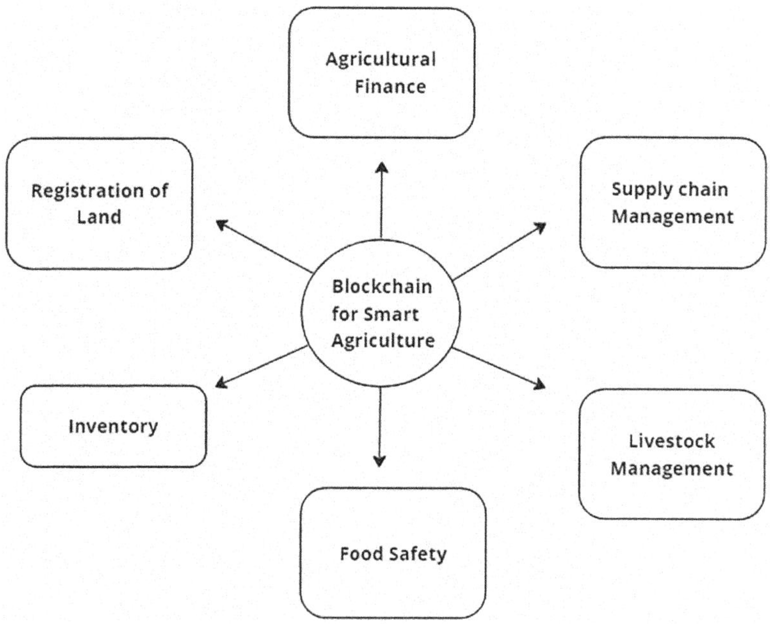

FIGURE 1.2 Blockchain in smart agriculture.

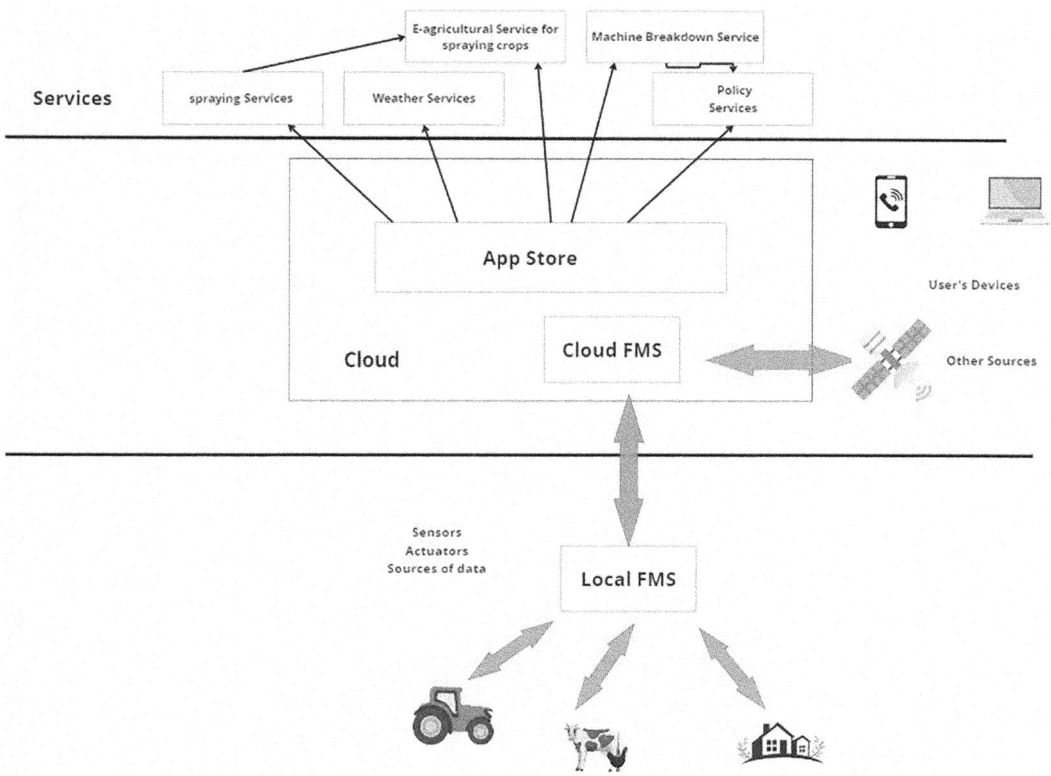

FIGURE 1.3 Working of blockchain in agriculture.

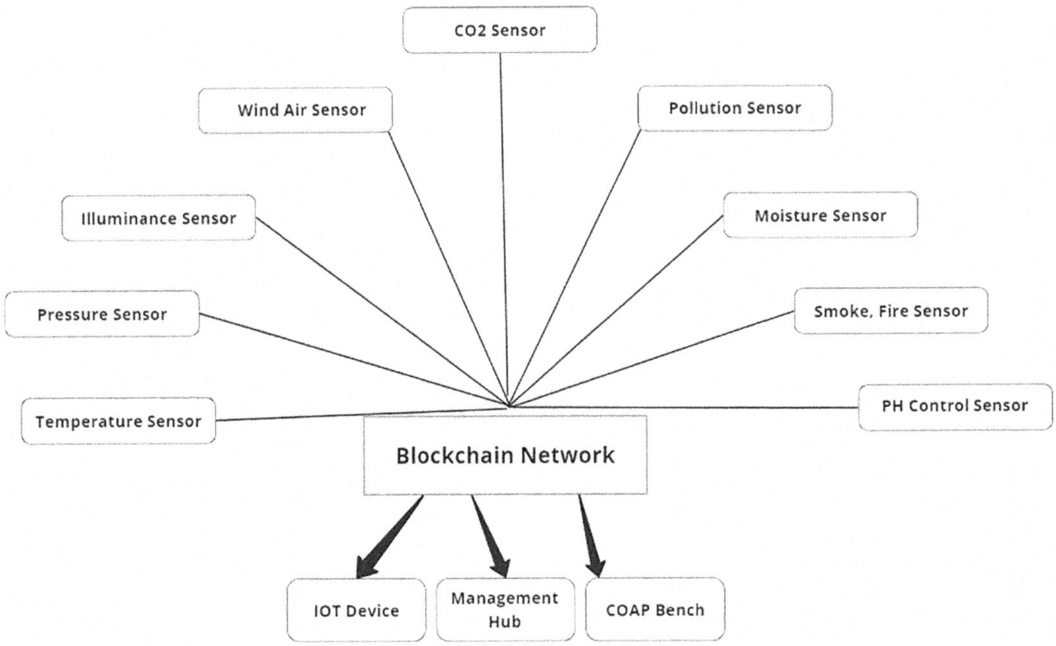

FIGURE 1.4 Blockchain in weather monitoring.

farmers and producers. From blockchain-based marketplaces and supply chain platforms to digital identity solutions and carbon offset programs, entrepreneurs are leveraging blockchain to create value and drive positive change in the agricultural industry. Blockchain technology is catalyzing a transformative shift in agricultural finance, empowering farmers, investors, and financial institutions to collaborate more effectively and efficiently in funding, lending, and investing in agriculture. By providing access to capital, enabling transparent and efficient transactions, enhancing risk management and insurance solutions, and promoting investment in sustainable agriculture, blockchain is unlocking new opportunities for growth, resilience, and sustainability in the agricultural sector.

To conclude, the outlook for blockchain in agriculture appears promising, as advancing technologies, worldwide endeavors, and entrepreneurial prospects contribute to a food system that is transparent, efficient, and sustainable. Through the adoption of innovation and cooperation, we can leverage blockchain's potential to revolutionize food production, distribution, and consumption, securing a better future for farmers, consumers, and the environment.

2 Smart Agriculture and Digital Twins

A Vision of Sustainability

B. Santhoshini and M. Shanmugam

2.1 INTRODUCTION

In response to the growing challenges facing the agricultural sector, such as population growth, climate change, and resource limitations, there has been a notable shift towards modernizing farming practices. Central to this transformation is the concept of smart agriculture, which harnesses digital technologies to optimize agricultural processes and improve sustainability. Key technological advancements driving this paradigm shift include data analytics, Internet of Things (IoT) devices, artificial intelligence (AI), and automation.

An emerging trend within smart agriculture is the integration of digital twins, which are virtual representations of physical assets or systems. By creating digital twins of farms, crops, and ecosystems, farmers gain valuable insights into the dynamics of agricultural processes. This allows for real-time monitoring, analysis, and optimization, empowering farmers to make informed decisions and enhance productivity while minimizing environmental impact.

Blockchain technology in agriculture has gained attention by maintaining cryptocurrencies in various sectors, including farming. It is an interconnection between various stakeholders. It allows information to be shared among different stakeholders, which improves safety in the food chain. It implements various technologies in farming, which enables traceability of data. Blockchain technology enhances and maintains the flow of information to store data among farmers, distributers, and retailers. Smart contracts on the blockchain make an agreement between farmers and buyers, which reduces the number of intermediate stakeholders.

Blockchain securely records and tracks every step of production and distribution of food processing. It maintains immutable records of every transaction in the food production and distribution process. Consumers can trace their food journey from farming to destination and maintain information, like how it is processed and when it will be delivered. It will also be easier to maintain the safety standards of agricultural products. At each stage of the supply chain, it checks information like temperature, humidity, and the handling process of the food chain.

This chapter explores the symbiotic relationship between smart agriculture and digital twins, presenting a vision of sustainability for the agricultural sector. Through the synergistic utilization of blockchain technologies and innovative approaches, farmers can address the complex challenges facing agriculture, ensuring food security and environmental stewardship for future generations.

2.2 LITERATURE REVIEW

Blockchain technology, initially popularized by cryptocurrencies, has garnered significant attention across various industries for its potential to revolutionize traditional processes. In agriculture, blockchain offers unique opportunities to address challenges related to transparency, traceability, and trust in food supply chains. This chapter explores the applications of blockchain technology in agriculture, highlighting its role in enhancing food safety, supply chain efficiency, and sustainability.

DOI: 10.1201/9781003507390-2

Wright et al. state that blockchain technology is characterized by its decentralized and immutable ledger system and enables transparent and tamper-proof recording of transactions. In the context of agriculture, blockchain facilitates the secure tracking and recording of various stages of the food supply chain, from farm to fork. By providing a verifiable record of each transaction or event, blockchain enhances transparency and accountability, thereby instilling greater trust among consumers, producers, and other stakeholders.

One of the key applications of blockchain in agriculture is the implementation of food traceability systems. By leveraging blockchain, stakeholders can track the journey of agricultural products from production to distribution, enabling swift identification and mitigation of food safety issues or contamination incidents. Furthermore, blockchain enables the authentication of organic or sustainable farming practices, empowering consumers to make informed choices about the products they purchase.

Ordonez et al. state the immutability of digital transaction is ensured extensively by disruptive technology. The digitalization and advancement of technology play a significant role in addressing overall sustainability challenges in agriculture. Blockchain is a trusted technology to trace products and supply chain management from farm to fork. The nodes in the network are called miners that verify transaction. They get a reward of some bitcoins. Proof of Work (PoW) is implemented, which is a consensus mechanism to solve mathematical puzzles. The miners who solves first can get authority to add a new block to the network.

Ferrag et al. state that there are security and privacy concerns in green Internet of Things (IoT) agriculture. They classified the threat model into five categories: privacy, authentication, confidentiality, availability, and integrity. Smart agriculture utilizes deployed Internet of Things (IoT) applications with wireless sensors, such as irrigation sensors, soil monitoring systems, and other smart farming technologies. We use various kinds of sensors for temperature, water, chemical, gas, and humidity. This enables the deployment of a large number of low-power, multifunctional devices for efficient data collection related to crops.

Ahmed and Yadav say that the adoption of remote sensing in agriculture faces several challenges like high cloud coverage, low quality of data, and spatial limits. At present, the industry collaborates closely with machines to get the relevant decision capability and sustainability. This article gives a comprehensive survey of remote sensing technologies and aspects related to that in various agriculture practices. Remote sensing technology with IoT and robotics plays a major role in precision agriculture. It also monitors diseases and pests in plants by identifying the health of a plant from distance. It helps to reduce time consumption.

Phang et al. review the usage of different satellites for collecting data. These satellites facilitate a data-driven approach for farming. Precision agriculture is a strategy that gathers processes and analyzes spatial and individual data for improving resource portability. Its main aim is to minimize the resources required per unit output. This paper discusses a direct solution for low-to-ground sensing technology, which operates at a lower cost. It also enables the collection of data on crops, microbial properties of a specific area, and weed control.

Chacko et al. explain the robust integration of IoT and blockchain technology, which will enable tracking of plant growth through the use of sensors. They also propose a mathematical formula to calculate the loss of infrastructure. This paper looks at various problems in integrating IoT in an agricultural context. Empirical analysis is conducted to analyze throughput, scalability, and latency. Various challenges, like proposing the architecture for an agroforestry supply chain, implementation of monitoring crop growth using IoT, and blockchain technology in an agroforestry scenario, are explained.

Nguyen et al. state that service provisioning, through a world of integrated services, provides ecosystem services for the next generation. This article gives full-fledged information regarding designing, construction of service-level agreements (SLAs) which is called pen chain. It is a mutual agreement between customers and a service provider. It investigates how dynamic service-level agreements make enforcement of penalty rules programmable. An agriculture data hub operates a trusted platform for data exchange.

Wang et al. explain the complexity of a supply chain, which leads to safety or quality issues in food. Blockchain technology has effectively solved existing problems. This paper proposes a consortium framework to track the flow of work of an agricultural food supply chain to improve traceability and shareability. In smart contracts, farmers will store information and record details of crop growth and climate-related data with hashing techniques to maintain security. Blockchain is a distributed ledger that contains blocks, each with a timestamp, in a point-to-point network. They explain the Merkel tree structure used to implement hashing at the leaf nodes and store the hash of data in the blocks and non-leaf nodes, which contain the encrypted data of the child nodes.

Oruma et al. state that an implementation framework for achieving food security in Nigeria's post–COVID-19 era is necessary. It integrates a precision agriculture in Nigeria's food production and enables digital technologies in the agrofood supply chain. They specified that some studies investigate household food security measurements in terms of location and specific food types.

Raza and Haq say that blockchain, smart contracts, and IoT are essential technologies for re-engineering and supply chains in the new era. The supply chain plays a crucial role in monitoring and providing secure transactions. A business process modeling (BPM) approach is incorporated with various layers among different technologies. Delay in transaction, poor storage practices in warehouses, and lack of communication among suppliers of food lead to trust issues among suppliers and customers. Blockchain is the solution for all these factors. An IoT, blockchain, and smart contract-based framework for automating intra-organizational and inter-organizational processes in the food supply chain.

Wu and Tsai explain current intelligent agriculture is actively promoting its development. They made a specific plan related to farming in an environment by considering various individual terms. They propose a study to apply dark web technologies for privacy in the server and blockchain. This study also monitors the frequency of packet transmission in agriculture to avoid Distributed Denial of Service (DDoS) attacks. The system includes key components such as a mechanism to identify authentication, secure information transfer, blockchain establishment, improvements in the blockchain authentication system, and measures to maintain resistance against DDoS attacks.

Singh et al. state that precision agriculture is a strategy to manage communication and information technology for farm management. There will be various challenges to develop agriculture like soil degradation, variation of climate, and cost increases. In order to face these challenges, we use various sensors for communicating and enabling technologies to crop yield. Machine learning and artificial intelligence have transformed precision agriculture at every level. This paper discusses various technologies in precision agriculture transformation and multidisciplinary approaches in integrating future precision agriculture. It also proposes multidisciplinary agriculture solutions for efficient and cost-effective agriculture.

Adeel Ahmed explains the optimized data fusion method, which involves clustering sensor segments. A cluster head is maintained to store data at the base station, allowing local processing and communication with farmers for immediate action on pest attacks. It also emphasizes the importance of controlling pest attacks to ensure successful crop production in the field. IoT with smart agriculture gives detailed information about crop health, soil moisture levels, and weather conditions. With this information, farmers can make decisions on their own. This paper also elaborates on the aggregation of data in real time, which makes use of advanced technologies like precision forming, IoT drones, and sensors. With this information, we can collect field-level data in which smart contracts give that information to all stakeholders in the network.

2.3 OBSTACLES CONFRONTING CONVENTIONAL FARMING METHODS

Generally, we see problems in agriculture like greater demand for food, as the population is increasing. Unfavorable weather conditions and climate fluctuations are expected to diminish crop yields, while approximately 25% of farmland has already experienced depletion due to a significant decline in underground water levels. Additionally, inefficiencies in market operations and inadequate supply chain logistics contribute to food wastage. Moreover, labor shortages and a lack of transparency pose challenges for consumers.

Currently, food supply chain companies encounter numerous challenges, including disconnects between suppliers and retailers, opacity in food production and distribution, constraints in financing resources, and an increased amount of third-party involvement. We categorize these challenges into five distinct areas for further examination.

2.3.1 Limited Access to Information and Technology

Traditional agricultural practices often lack access to modern information and technology, hindering productivity and efficiency.

2.3.2 Weather and Climate Impacts on Agriculture

Traditional farming is highly susceptible to the impacts of weather and climate variability, leading to crop failures and income loss. The global distribution of freshwater resources is not uniform, with approximately 20% of the world's surface freshwater concentrated in the North American Great Lakes.

Cao, S. et al highlighted that water scarcity impacts over 40% of the global population and is anticipated to escalate. In some of the most impoverished nations, one in ten individuals lack access to safe and readily available water sources. Human water consumption is on the rise alongside population growth. The authors provided insights into arid desert environments and methods of water collection from fog, drawing lessons from natural processes. Additionally, they discussed data concerning bioinspired surfaces designed for water collection.

2.3.3 Soil Health and Fertility Concerns

Continuous farming without implementing effective soil management practices can have detrimental effects on soil health. Soil degradation, erosion, and loss of fertility are common consequences of this unsustainable approach. These factors not only diminish crop productivity but also threaten the long-term viability of agricultural land. Therefore, it is essential for farmers to prioritize soil conservation methods such as crop rotation, cover cropping, and organic matter addition to maintain soil health and ensure sustainable agricultural practices.

2.3.4 Addressing Water Scarcity in Agriculture

Inadequate access to water resources and inefficient irrigation methods pose significant challenges to traditional agriculture, particularly in arid and semi-arid regions.

An analysis of the projected implications of water scarcity by 2050 examines potential country-level deficits in water availability for food self-sufficiency. Additionally, it highlights the significant reliance on virtual water trade for nearly half of the world's population. Furthermore, discussions center around the concept of a planetary freshwater boundary, proposed to ensure the sustainability of the global water circulation system on a planetary scale. Consideration is given to the remaining consumptive water use reserve, which must be allocated for global food production, fuelwood cultivation, and carbon sequestration purposes.

2.3.5 Market Access Challenges and Price Fluctuations

Farmers practicing traditional agriculture often face difficulties in accessing markets, resulting in limited bargaining power and vulnerability to price fluctuations.

Henceforth, the blockchain platform emerges as a promising solution, providing secure remedies to these obstacles via decentralized, automated, and reliable data and transportation management systems.

TABLE 2.1
Key Metrics and Their Impact on Sustainability

Metric	Description	Statistics/Values	Impact on Sustainability
Water Use Efficiency	Ratio of crop yield to water used.	20–30% increase with smart irrigation systems	Reduces water waste and conserves water resources.
Fertilizer Use Efficiency	Amount of crop produced per unit of fertilizer used.	15–25% improvement with precision agriculture	Minimizes runoff and environmental pollution.
Crop Yield	Amount of crop produced per hectare.	10–20% increase using digital twin modeling	Enhances food security and reduces the need for additional farmland.
Pesticide Reduction	Decrease in pesticide use due to early disease and pest detection.	20–40% reduction through sensor technology	Lowers chemical input, benefiting ecosystems and human health.
Greenhouse Gas Emissions	Emissions produced per unit of agricultural output.	10–15% reduction with optimized farming practices	Contributes to climate change mitigation efforts.
Labor Productivity	Crop output per labor hour.	30–50% increase with automation and robotics	Enhances economic sustainability and worker conditions.
Energy Consumption	Energy used per unit of agricultural output.	15–20% decrease with smart energy management systems	Lowers carbon footprint and operational costs.
Waste Reduction	Reduction in post-harvest and supply chain losses.	Up to 30% decrease using real-time monitoring and analytics	Reduces food waste and enhances food availability.
Adoption Rate	Percentage of farms using smart agriculture technologies.	25% of farms globally by 2025	Indicates the growing shift towards sustainable practices.
Return on Investment (ROI)	Financial return compared to the investment in smart agriculture technologies.	15–20% ROI within 2–3 years	Encourages adoption by demonstrating economic benefits.

2.4 APPLICATIONS OF BLOCKCHAIN IN AGRICULTURE

The application of blockchain technology in agriculture represents a significant advancement in modern farming practices, offering innovative solutions to longstanding challenges within the industry. Blockchain, originally introduced as the underlying technology behind cryptocurrencies, has transcended its initial purpose and found widespread application across various sectors, including agriculture. In recent years, there has been growing interest and exploration into the potential uses of blockchain technology to enhance transparency, traceability, and trust within agricultural supply chains. This introduction aims to provide an overview of the diverse applications of blockchain in agriculture, highlighting its transformative impact on key areas such as food safety, supply chain management, and sustainable farming practices. Table 2.1 is a numerical and stat-based table focused on smart agriculture and digital twins, highlighting key metrics and their impact on sustainability.

1. **Crop and food production:** IoT sensors or devices will monitor the crop or soil health, which generates information regarding temperature, soil PH, soil moisture, pests, and diseases. It helps farmers or other stakeholders to make decisions related to crop growth. That information should be stored in data storage. We apply machine learning techniques to data, which gives insights regarding crop quality, crop yield prediction, crop growth

factor, crop demand prediction, and irrigation requirements when this kind of information is stored in blockchain, a distributed storage platform where a hashing technique is implemented. It makes other stakeholders to access its transparency. The information is centralized and distributed across every node in the network, which prevents central authority.

2. **Food supply chain:** Once the crops are mature, food processing companies will bid on bidding platforms, and validation is done through smart contracts. With this process, companies will store all information while transferring information on the blockchain. It helps retailers and wholesalers identify the quality of work and helps in establishing trusted relations between customers and retailers. It also helps in getting accurate and reliable information.

   One of the key applications of blockchain in the food supply chain is in food traceability. By leveraging the blockchain, stakeholders can accurately track the journey of agricultural products from production to consumption. This enables swift identification and mitigation of food safety issues, contamination incidents, and product recalls, thereby safeguarding public health and enhancing consumer confidence.

   Furthermore, blockchain technology facilitates the authentication of product attributes such as origin, quality, and sustainability. By recording detailed information about each stage of the production and distribution process on the blockchain, consumers can access accurate and reliable data regarding the provenance and characteristics of the food they consume. This fosters greater transparency and accountability within the supply chain, enabling informed consumer choices and promoting ethical and sustainable practices.

3. **Agriculture insurance:** Insurance will vary depending on how losses are assessed, such as asymmetric information between farmers and insurers regarding production, payments, and other factors. Blockchain helps in overcoming these problems via smart agriculture and will help farmers understand crop behavior like soil temperature at different heights, rainfall, wind speed, and humidity. All these parameters are measured, recorded, and stored in the blockchain which make farmers and other authorized stakeholders to access its transparency. With this, farmers can make decisions and overcome problems by analyzing data. With blockchain's transparent and immutable behavior, it makes insurance companies to access the data captured by smart weather station easily with the help of smart contracts. With this, both the farmer and insurance person will get identical information.

4. **Improved quality control and food safety:** One of the main use of blockchain is accountability to the supply chain and elimination of inadequate processes. Agencies related to agriculture production would be certified and put on the blockchain, which leads to a better price for farmers' work and quality products obtained by customers. Agencies can regularly access accurate information regarding products, which prevents food fraud.

   Blockchain technology facilitates improved quality control and food safety in agriculture by providing a transparent and immutable ledger system for tracking the entire supply chain, ensuring traceability and accountability.

   By leveraging blockchain, farmers and consumers can access real-time information about the origin, production practices, and transportation of agricultural products, enhancing transparency and trust in the food system.

5. **Traceability and accountability:** Currently, involving third-party or high-level agents adds significant costs to the system and makes the process more time-consuming. With blockchain, all the information can be simplified by a single distribution ledger. Additional payments and third parties can be removed. A large share of sales can be received by farmers.

   In a study, the applications of blockchain technology in agricultural supply chains, focusing on its role in enhancing traceability and accountability. Through a comprehensive analysis, the authors highlight how blockchain facilitates transparent and tamper-proof recording of transactions, thereby improving supply chain visibility and accountability.

The potential of blockchain-based solutions in improving accountability in agricultural trade is significant. Through case studies and theoretical frameworks, the authors demonstrate how blockchain enables secure and transparent transactions, reducing the risk of fraud and enhancing accountability across the agricultural value chain.

Fiore and Mongiello provide an overview of blockchain technology's applications in agricultural product traceability. The authors discuss the technical aspects of blockchain implementation and highlight its benefits in ensuring the authenticity and integrity of food products throughout the supply chain.

Cao et al. examine the potential of blockchain solutions in enhancing traceability in organic farming. Through empirical research and case studies, the authors evaluate the effectiveness of blockchain in certifying organic produce, ensuring compliance with regulations, and enhancing consumer trust in organic farming practices.

Tao and Chao carried out a case study of blockchain-based solutions for sustainable agriculture, focusing on the Chinese agricultural sector. The authors analyze the implementation of blockchain technology in improving supply chain transparency, enhancing trust among stakeholders, and promoting sustainable agricultural practices.

6. **Logistics and payments:** Implementing blockchain in logistics can simplify deliveries where a long chain is a big challenge. With the help of smart contracts, the number of intermediaries can be decreased. Farmers can easily see information regarding products reaching a destination. Smart contracts help farmers get paid on time without delay.

   Particularly, Ashokkumar et al. explore how blockchain technology revolutionizes agricultural logistics by providing a transparent and secure platform for tracking and tracing agricultural products from farm to market. Through decentralized and immutable ledgers, blockchain ensures data integrity and authenticity, enabling stakeholders to verify the origin, quality, and condition of agricultural products during transportation. Additionally, smart contracts on the blockchain automate payment processes, reducing transaction costs and enhancing trust between buyers and sellers in agricultural supply chains. Their findings highlight the potential of blockchain to streamline logistics operations, improve supply chain efficiency, and reduce food wastage in the agricultural sector.

   Additionally, Lin et al. cover the potential of blockchain technology to enhance agricultural payments by providing a secure and transparent platform for financial transactions. Their study emphasizes the importance of adopting blockchain-based payment solutions to address existing challenges and unlock new opportunities for financial empowerment in agriculture.

   According to the findings from a study on auto manufacturers, blockchain technology is seen as having significant potential to lower operational expenses, ensure product integrity, and improve visibility and digitization within the automotive supply chain.

7. **Data security:** With cryptographic hash functions, data will be maintained securely. Since blockchain is fully decentralized with nodes, hacking is nearly impossible.

   Patel et al. introduced a blockchain-based system called AgriOnBlock to tackle the challenges prevalent in the agriculture sector. AgriOnBlock aims to connect different stakeholders using IoT devices and smart contracts on the Ethereum platform.

   Katsaliaki et al. conducted a comprehensive review and analysis of the predominant modeling methodologies pertaining to the subject, providing illustrative examples, as well as exploring the information technology (IT) tools that bolster resilience and mitigate the risks of disruptions.

Let's break down the UML class diagram (Figure 2.1) for smart agriculture and digital twins into its core components. We will focus on key classes and their relationships, incorporating smart agriculture, digital twins, and blockchain technology.

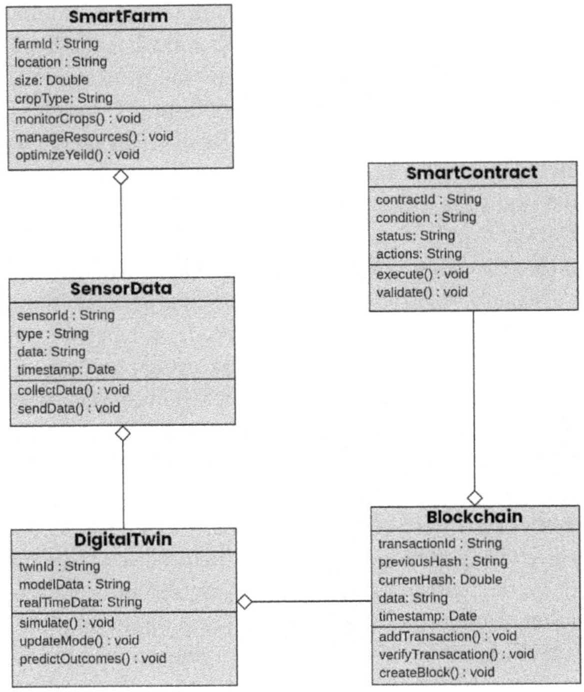

FIGURE 2.1 UML class diagram: smart agriculture and digital twins.

Here is a detailed description of the Unified Modeling Language class diagram components:

SmartFarm: Represents the overall farm management system. Monitors crops, manages resources, and optimizes yield. Aggregates data from SensorData, DigitalTwin, and interacts with blockchain.

1. SensorData: Collects and sends data from various sensors (e.g., soil moisture, weather conditions). Contains methods to collect and send sensor data.
2. DigitalTwin: A virtual replica of the farm. Simulates farm operations, updates models, and predicts outcomes based on real-time data. Uses data from SensorData.
3. Blockchain: Secures data transactions related to farm operations. Adds, verifies, and creates blocks of transactions. Ensures transparency and immutability of data.
4. SmartContract: Automates actions based on predefined conditions. Executes and validates contracts within the blockchain. Ensures compliance and automatic enforcement of agreements.
5. Relationships: SmartFarm aggregates SensorData, DigitalTwin, and blockchain components to manage farm operations efficiently DigitalTwin uses data from SensorData for simulations and predictions. Blockchain stores and verifies transactions, interacting with SmartContract for automated operations.

2.5 EVOLVING LANDSCAPE OF AGRICULTURE

From traditional farming to cutting-edge technology, the agriculture field has evolved to meet the demands of populations globally. Data analytics and sensors create optimization in various aspects of farming. Farmers can keep drones, GPS-guided tractors, satellite imaginary to assess soil health. Fertilizers and pesticides applicatory devices which enhances efficiency.

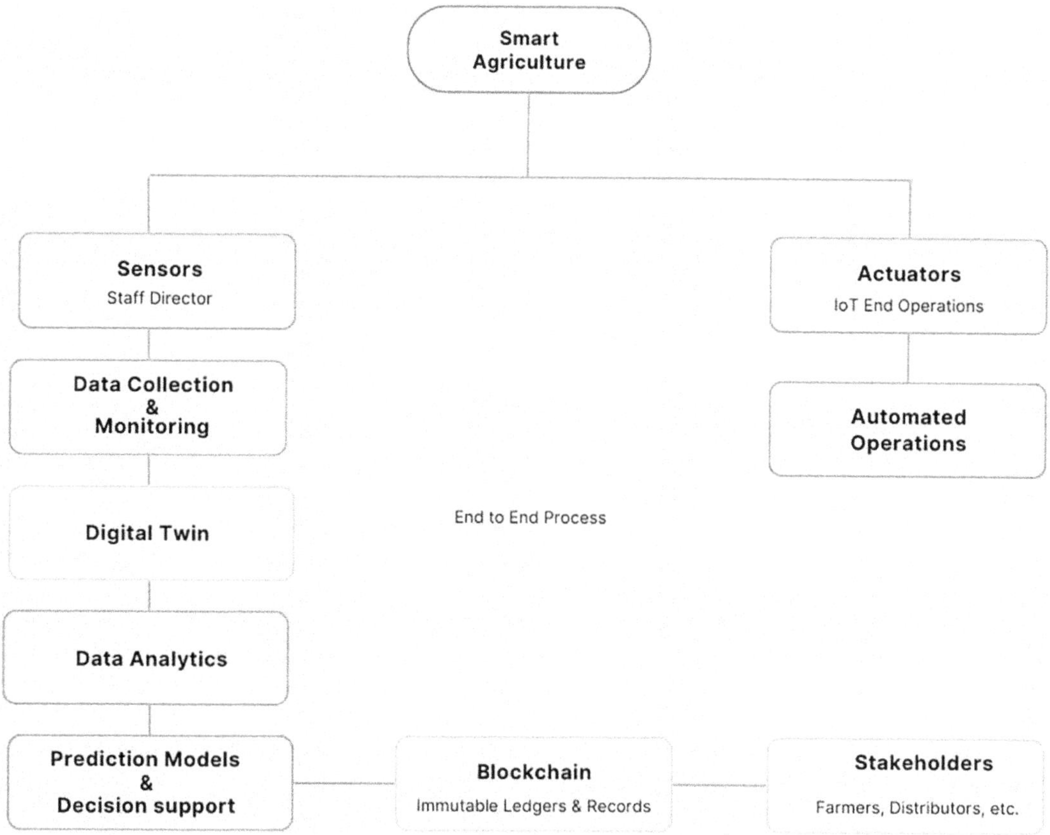

FIGURE 2.2 Integration of smart agriculture, digital twins, and blockchain for sustainable farming.

IoT in agriculture enables real-time monitoring and control of farming operations. It also helps in decision making for farmers to promptly respond to weather change conditions and optimize irrigation, which improves farming. The demand for organic farming is increasing as consumers prioritize environmentally friendly and chemical-free food. Robotics and automation increase harvesting, weeding, planting, and monitoring. Autonomous drones and robots can perform tasks such as soil sampling, spraying, and monitoring environmental impact.

Data analytics can help farmers in crop planning, field mapping, some inventory management, and financial analysis to maximize profits.

Figure 2.2 provides a generic structural view of the key technologies smart agriculture, digital twins, and blockchain for sustainable farming.

1. Smart Agriculture: This encompasses the overall system integrating technology into farming practices.
2. Sensors and Actuators: These components are crucial for data collection and automated responses.
   a. Sensors: Collect data on various environmental parameters such as soil moisture, temperature, and crop health.
   b. Actuators: Perform automated actions like adjusting irrigation systems based on sensor data.
3. Digital Twin: A virtual replica of the physical farm, utilizing data from sensors for real-time monitoring and simulation.

4. Data Analytics and Predictive Models: Analyze the collected data to create models that predict future conditions and outcomes.
5. Decision Support: Provides actionable insights and recommendations based on predictive models.
6. Blockchain: Ensures data integrity and security by recording transactions and data immutably.
7. Stakeholders: The end-users of this system, including farmers, distributors, and other involved parties.

Figure 2.2 clearly represents a high-level overview of how smart agriculture integrates with digital twins and blockchain technology to create a sustainable and secure farming ecosystem.

2.6 DIGITAL TWINS IN THE AGRICULTURE SECTOR

The use of digital technologies helps the agriculture sector by optimizing farming operations and improving decision-making to increase production. A digital twin is a virtual replica of a crop field or an entire farm. Digital twin can track crop growth based on real-time data collected from satellites or sensors. Farmers can see and make decisions on improving nutrient levels, temperature maintenance, and adding pesticides level to crops.

By comparing historical data with present real-time data, a digital twin can generate predictive analysis to forecast crop yields. It also helps farmers to use resources like water, fertilizers, and pesticides. With this, farmers can reduce waste and minimize environmental impact to improve resource efficiency. It can monitor performance and conditions of farming equipment such as tractors, harvesters, and irrigation systems. Table 2.1 provides a concise overview of how smart agriculture and digital twin technologies contribute to sustainability in farming.

The integration of real-time virtual representation positions the digital twin in a distinctive role to facilitate digitization within the agricultural sector. By amalgamating data, modeling, and what-if simulation, digital twins offer a viable approach to address existing constraints in decision-making support and automation across various agricultural operations.

Researchers have tried to formulate a comprehensive and resilient definition of the digital twin, one that remains applicable even amid the emergence of new roles, unforeseen applications, and technological advancements.

2.7 CONCLUSION

This chapter aims to merge insights from smart agriculture and digital twins with the innovative application of blockchain technology. This endeavor is particularly pertinent given the escalating frequency and magnitude of disruptions within the agricultural domain. By delving into four distinct yet interconnected streams, the challenges faced by traditional agriculture, the applications of blockchain in agriculture, and the evolving landscape of agriculture, with a specific focus on digital twins within the agricultural sector, our research seeks to offer a nuanced understanding of the subject matter.

First, we explore the enduring challenges confronting traditional agricultural practices, such as limited access to modern technology and information, dependence on unpredictable weather patterns, soil degradation, and market access constraints. These challenges underscore the urgent need for innovative solutions to enhance agricultural resilience and sustainability.

Second, we investigate the manifold applications of blockchain technology in agriculture, ranging from supply chain traceability and transparency to facilitating secure and transparent transactions between stakeholders. By leveraging blockchain's immutable ledger system, stakeholders can effectively track and authenticate the provenance of agricultural products, mitigate food safety risks, and ensure fair compensation for farmers.

TABLE 2.2
Overview of How Smart Agriculture and Digital Twin Technologies Contribute to Sustainability in Farming

Aspect	Description	Impact on Sustainability
Definition of Smart Agriculture	Integration of advanced technologies like IoT, AI, and robotics into farming practices.	Increases efficiency, reduces waste, and optimizes resource usage, leading to more sustainable farming.
Digital Twin Technology	A virtual replica of physical assets, processes, or systems that can be used for simulation and analysis.	Enhances decision-making by providing real-time data and predictive insights, improving resource management.
Resource Management	Efficient use of water, energy, and fertilizers through precise monitoring and control.	Minimizes environmental impact by reducing overuse and pollution.
Crop Monitoring	Use of sensors and drones to monitor crop health and growth conditions.	Early detection of diseases and pests, reducing the need for chemical treatments.
Predictive Analytics	Utilization of AI and machine learning to predict weather patterns, crop yields, and market trends.	Allows for proactive management, reducing losses and optimizing supply chains.
Supply Chain Optimization	Streamlined logistics and inventory management using real-time data.	Reduces food waste and ensures timely delivery, lowering the carbon footprint.
Labor Efficiency	Automation of repetitive tasks and augmentation of labor with smart machinery.	Reduces physical strain on workers and increases productivity, contributing to economic sustainability.
Sustainability Goals	Alignment with global sustainability goals such as reducing greenhouse gas emissions and conserving biodiversity.	Ensures agricultural practices contribute positively to environmental and social goals.
Challenges and Barriers	High initial investment, data privacy concerns, and the need for technical expertise.	Addressing these challenges is essential for widespread adoption and maximizing

Last, we examine the evolving landscape of agriculture, with a keen focus on the integration of digital twins. Digital twins, virtual replicas of physical assets or systems, offer immense potential in revolutionizing agricultural practices by enabling real-time monitoring, analysis, and optimization. By creating digital replicas of farms, crops, and ecosystems, farmers gain invaluable insights into agricultural processes, facilitating data-driven decision-making and enhancing productivity.

Through an extensive review and synthesis of existing literature, including well-cited reviews and surveys in the field, our chapter aims to contribute to a deeper understanding of the intersection between smart agriculture, digital twins, and blockchain technology. By elucidating key challenges, opportunities, and emerging trends, we seek to inform future research endeavors and practical applications in the agricultural sector.

REFERENCES

Ahmed, A., Parveen, I., Abdullah, S., Ahmad, I., Alturki, N., Jame, L. (2024). Optimized data fusion with scheduled rest periods for enhanced smart agriculture via blockchain integration. *IEEE Access*. https://doi.org/10.1109/ACCESS.2024.3357538

Ahmed, I., & Yadav, P. K. (2022, October 25). Plant disease detection using machine learning approaches. *Expert Systems*, *40*(5). https://doi.org/10.1111/exsy.13136

Ashokkumar, B., & Naik, A. (2021, June 23). Transforming Indian agriculture with digital technologies. *Asian Journal of Agricultural Extension, Economics & Sociology*, 76–90. https://doi.org/10.9734/ajaees/2021/v39i630596

Cao, S., Xu, H., & Bryceson, K. P. (2023, September 8). Blockchain traceability for sustainability communication in food supply chains: An architectural framework, design pathway and considerations. Sustainability, 15(18), 13486. https://doi.org/10.3390/su151813486

Chacko, N. M., Narendra, V. G., Balachandra, M., & Rathinam, S. (2023). Exploring IoT-blockchain integration in agriculture: An experimental study. *IEEE Access*, *11*, 130439–130450. https://doi.org/10.1109/access.2023.3334726

Ferrag, M. A., Shu, L., Yang, X., Derhab, A., & Maglaras, L. (2020). Security and privacy for Green IoT-based agriculture: Review, blockchain solutions, and challenges. *IEEE Access*, *8*, 32031–32053. https://doi.org/10.1109/access.2020.2973178

Fiore, M., & Mongiello, M. (2023). Blockchain technology to support agri-food supply chains: A comprehensive review. *IEEE Access*, 11, 75311–75324. https://doi.org/10.1109/access.2023.3296849

Katsaliaki, K., Galetsi, P., & Kumar, S. (2021, January 8). Supply chain disruptions and resilience: a major review and future research agenda. *Annals of Operation Research/Annals of Operations Research.* https://doi.org/10.1007/s10479-020-03912-1

Lin, W., Huang, X., Fang, H., Wang, V., Hua, Y., Wang, J., Yin, H., Yi, D., & Yau, L. (2020). Blockchain technology in current agricultural systems: From techniques to applications. *IEEE Access*, *8*, 143920–143937. https://doi.org/10.1109/access.2020.3014522.

Nguyen, T. V., Lê, L. S., Shah, S. A., & Draheim, D. (2023). PenChain: A blockchain-based platform for penalty-aware service provisioning. *SSRN Electronic Journal.* https://doi.org/10.2139/ssrn.4364516

Ordonez, J., Alexopoulos, A., Koutras, K., Kalogeras, A., Stefanidis, K., & Martos, V. (2023, January). Blockchain in agriculture: A PESTELS analysis. *AgriRxiv*, *2023.* https://doi.org/10.31220/agrirxiv.2023.00199

Oruma, S. O., Misra, S., & Fernandez-Sanz, L. (2021). Agriculture 4.0: An implementation framework for food security attainment in Nigeria's post-Covid-19 era. *IEEE Access*, *9*, 83592–83627. https://doi.org/10.1109/access.2021.3086453

Patel, A., Brahmbhatt, M. N., Bariya, A., Nayak, J. B., & Singh, V. K. (2023, June 1). Blockchain technology in food safety and traceability concern to livestock products. *Heliyon.* https://doi.org/10.1016/j.heliyon.2023.e16526

Patel, H., & Shrimali, B. (2023, April 1). AgriOnBlock: Secured data harvesting for agriculture sector using blockchain technology. *ICT Express.* https://doi.org/10.1016/j.icte.2021.07.003

Phang, S. K., Chiang, T. H. A., Happonen, A., & Chang, M. M. L. (2023). From satellite to UAV-based remote sensing: A review on precision agriculture. *IEEE Access*, *11*, 127057–127076. https://doi.org/10.1109/access.2023.3330886

Raza, Z., Haq, I. U., & Muneeb, M. (2023). Agri-4-all: A framework for blockchain based agricultural food supply chains in the era of fourth industrial revolution. *IEEE Access*, *11*, 29851–29867. https://doi.org/10.1109/access.2023.3259962

Singh, R. K., Berkvens, R., & Weyn, M. (2021). AgriFusion: An architecture for IoT and emerging technologies based on a precision agriculture survey. *IEEE Access*, *9*, 136253–136283. https://doi.org/10.1109/access.2021.3116814

Tao, Z., & Chao, J. (2024, February 1). The impact of a blockchain-based food traceability system on the online purchase intention of organic agricultural products. *Innovative Food Science and Emerging Technologies/Innovative Food Science & Emerging Technologies.* https://doi.org/10.1016/j.ifset.2024.103598.

Wang, L., Xu, L., Zheng, Z., Liu, S., Li, X., Cao, L., Li, J., & Sun, C. (2021). Smart contract-based agricultural food supply chain traceability. *IEEE Access*, *9*, 9296–9307. https://doi.org/10.1109/access.2021.3050112

Wright, C. S. (2008). Bitcoin: A peer-to-peer electronic cash system. *SSRN Electronic Journal.* https://doi.org/10.2139/ssrn.3440802

Wu, H. T., & Tsai, C. W. (2019, October). An intelligent agriculture network security system based on private blockchains. *Journal of Communications and Networks*, *21*(5), 503–508. https://doi.org/10.1109/jcn.2019.000043

3 Digital Twins for the Agriculture Sector

Hrishikesh Kokate, Abhijit Vhatkar, and Alex Khang

3.1 INTRODUCTION

The idea of the "digital twin" has become a game-changer for a variety of sectors within the rapidly changing field of technology. The designing, monitoring, and optimizing of complex systems in various industries has become possible because of digital twins. This chapter delves into the topic of agriculture and examines how the digital twin is changing and how we raise cattle, grow crops, and take care of our land in an environmentally responsible manner.

Fundamentally, a digital twin is an electronic copy or model of a real-world object, such as a machine, a process, or an ecosystem. This digital counterpart is a dynamic model that replicates real-time data and activity rather than just being a static depiction. Through the utilization of sophisticated sensor technology, data analytics, and simulation algorithms, digital twins provide an unparalleled understanding of the internal operations of intricate systems.

A wide range of industries are adopting digital twins, seeing their potential to boost production, efficiency, and creativity. Digital twins in manufacturing allow for predictive maintenance, which maximizes machine efficiency and reduces downtime. Digital twins that are unique to each patient make medical interventions and treatment approaches more individualized. Digital twins also expedite design iterations in the aerospace industry, enhancing aircraft performance and safety.

The agricultural industry is well positioned to capitalize on the revolutionary potential of digital twins given their widespread use. The world's population is expected to reach over 9 billion by the year 2050, and this will create a need to boost food production more than ever. To maximize yields, reduce risks, and encourage environmental stewardship, digital twins provide a comprehensive approach to agricultural management by fusing data-driven insights with conventional farming methods.

The entire agri-food value chain, from crop production and soil management to supply chain logistics and customer engagement, is using digital twins. Digital twins, which use real-time data on soil moisture, nutrient levels, and weather patterns to enhance pest management, fertilization, and irrigation, simplify precision agriculture at the farm level. Farmers can minimise resource loss and maximise yields by customizing inputs for specific crops.

Digital twins support larger programs that support environmental sustainability and resilience in agriculture, even outside of specific farms. Digital twins provide integrated land management plans that strike a balance between conservation and production goals by simulating ecosystems at the landscape scale. Moreover, digital twins assist with attempts to prevent climate change by informing policy decisions through the analysis of data on biodiversity, water usage, and greenhouse gas emissions.

Digital twins, however, have a plethora of fascinating potential applications in agriculture. Digital twins help farmers make informed decisions, manage resources efficiently, and adapt to changing environmental conditions by using collaborative platforms, simulation modelling, and data-driven insights. Additionally, by encouraging innovation and knowledge sharing, digital twins have the potential to bring about systemic change in the agriculture industry, ensuring food security, sustainability, and resilience for future generations.

DOI: 10.1201/9781003507390-3

3.2 ADVANTAGES OF DIGITAL TWINS FOR THE AGRICULTURE SECTOR

A new age of sustainability, efficiency, and accuracy in agriculture is being ushered in by the adoption of digital twin technologies. Digital twins provide a host of benefits that transform conventional farming methods by building virtual versions of fields, crops, and cattle. We examine some of the main advantages here:

- Improved Decision-Making: Digital twins give farmers access to data-driven, real-time insights that facilitate well-informed decision-making. Farmers may improve on planting dates, irrigation schedules, and pest management tactics by keeping an eye on soil moisture levels, weather projections, and crop health indicators. By minimizing risks and maximizing yields, this proactive approach eventually increases farm profitability.
- Precision Agriculture: Digital twins' capacity to support precision agriculture is one of its most important benefits. Farmers can customize inputs like water, fertilizer, and pesticides to the unique requirements of their crops by utilizing comprehensive geographical and temporal data collection. This focused strategy encourages sustainable farming methods, lessens the impact on the environment, and wastes less resources.
- Predictive Maintenance: Digital twins allow for the predictive maintenance of agricultural infrastructure and machinery in addition to crop management. Farmers may minimise downtime and maximise operational efficiency by proactively scheduling maintenance tasks regularly checking equipment performance and identifying early symptoms of wear and tear. By being proactive, this strategy lowers expenses and increases the life of important assets.
- Better Livestock Management: Digital twins have revolutionized livestock management techniques in addition to crop agriculture. Through the implementation of wearable sensors and RFID tags, farmers can track various health data, identify potential health issues, and enhance feeding practices. Digital twins also make it possible to forecast herd behaviour, which helps farmers foresee and stop problems like disease outbreaks and stress-related behaviours.
- Resource Optimization: Digital twins help farms allocate resources optimally by combining data from multiple sources, such as satellite photography, Internet of Things sensors, and historical records. Farmers may find inefficiencies, streamline procedures, and reduce waste in a variety of areas, from labour and inputs to water and energy. This resource optimization lowers operating expenses and its impact on the environment while simultaneously increasing production.
- Risk Mitigation: Farmers can recognize and reduce risks related to market changes, insect infestations, and unpredictable weather thanks to digital twins. To protect against unanticipated catastrophes, farmers should create backup plans and adaptable tactics by modelling various situations and evaluating their possible effects. In the face of uncertainty, this proactive approach to risk management guarantees company continuity and improves resilience.
- Data-Driven Insights: The capacity of digital twins to produce useful insights from massive volumes of agricultural data may be their most revolutionary feature. Farmers can spot patterns, improve procedures, and seize new opportunities by examining past trends, the environment, and market dynamics. Furthermore, digital twins can continuously learn and develop by utilizing machine learning algorithms, which promotes constant innovation and optimization on the farm.

(The values provided in Table 3.1 are estimates or projections based on various reports and studies from reputable sources. Actual results may vary depending on specific conditions and factors.)

Table 3.1 shows various indicators of the possibilities and difficulties in contemporary agriculture. It draws attention to the predicted rise in global food consumption by 2050 as well as

TABLE 3.1
Digital Twins in Agriculture—Potential Impact and Adoption Statistics

Metric	Value	Source
Projected increase in global food demand by 2050	60%	Food and Agriculture Organization (FAO)
Potential increase in crop yields with precision agriculture	10–25%	McKinsey & Company
Estimated annual cost of crop losses due to pests and diseases	$220 billion	United Nations
Potential reduction in pesticide use with targeted pest management	20–47%	European Commission
The projected growth of the digital twin market in agriculture by 2028	$5.7 billion	Markets and Markets
The current adoption rate of digital twin technology in agriculture	<10%	Deloitte
Potential reduction in greenhouse gas emissions with sustainable practices	20–30%	World Resources Institute
Potential increase in water use efficiency with precision irrigation	20–50%	Food and Agriculture Organization (FAO)
Estimated annual cost of equipment downtime in agriculture	$10–20 billion	Boston Consulting Group
Potential reduction in food waste with supply chain optimization	20–50%	World Resources Institute

the possible advantages of implementing precision agricultural technology, such as digital twins, targeted pest control, and precision irrigation, in terms of increased yield, lower input costs, and resource preservation. Furthermore, it offers approximations of the financial expenses linked to crop losses, equipment malfunctions, and food waste, underscoring the necessity of effective and sustainable methods in the agricultural industry.

Because technology makes it possible to create virtual models and simulations of actual farming systems, procedures, and assets, the digital twin idea is drastically changing the agriculture industry. The major usage of digital twin technology in the agriculture sector is as follows.

3.2.1 Farm Apparatus and Machinery

Tractors, harvesters, irrigation systems, and other agricultural machinery may all have digital twins made for them (Pylianidis et al., 2021). These virtual models may be used to optimize equipment utilization, forecast maintenance requirements, and monitor real-time performance.

In modern agriculture, farm machinery and equipment play a crucial role. This equipment ranges from tractors and harvesters to irrigation systems and speciality tools (Rahman et al., 2021). Digital twin creation for these assets can provide the following advantages:

- Predictive maintenance and condition monitoring: It relies on digital twins, which collect data from sensors and Internet of Things (IoT) devices attached to the machinery. This allows real-time monitoring of factors such as vibration levels, engine performance, fuel consumption, and component wear (Van Dinter et al., 2022). Through the analysis of this data, the digital twin may anticipate any malfunctions or maintenance requirements in advance, facilitating proactive maintenance and decreasing downtime.
- The predictive maintenance cycle is depicted in Figure 3.1. Data gathering comes first, then the examination of the data. Predictive maintenance activities are chosen based on

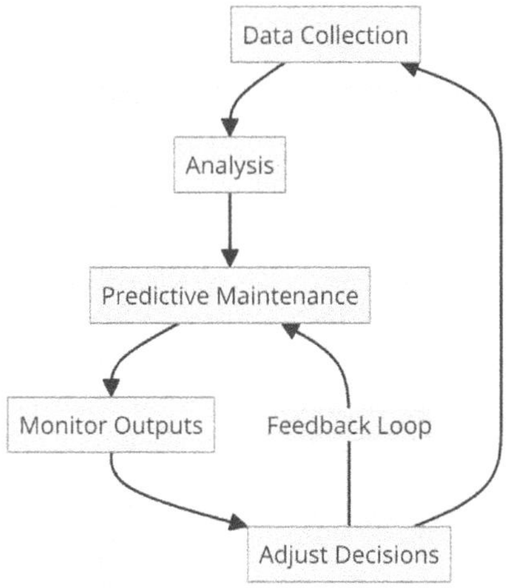

FIGURE 3.1 Predictive maintenance process.

the analysis. These activities' results are observed, and future maintenance choices are modified based on the observations' input. The predictive maintenance method may be continuously enhanced and optimized thanks to this feedback loop.

- Performance optimization: Digital twins can replicate how agricultural equipment would operate in various scenarios, including variations in weather, crop varieties, and topography (Cesco et al., 2023). To improve effectiveness, productivity, and fuel economy, this enables the optimization of equipment settings, routes, and operating methods.
- Guidance and training for operators: Before operating actual agricultural machinery, operators can practice and become acquainted with the apparatus in a secure and regulated virtual setting by using virtual versions of the machinery (Relf-Eckstein et al., 2019).
- Design and testing: Before building actual prototypes, digital twins can be used to simulate and assess various designs, configurations, and components during the development and testing stages of new agricultural machinery (Peladarinos et al., 2023). This can guarantee that the finished product satisfies the required performance standards and specifications while also quickening the development process and cutting expenses.
- Fleet management: Digital twins can help with centralized fleet management for large-scale agricultural enterprises with several pieces of equipment. Farm managers may improve scheduling, logistics, and resource allocation by keeping an eye on each machine's current state, position, and performance (Nasirahmadi & Hensel, 2022).
- Data analytics and integration: To give a complete picture of agricultural operations, digital twins may combine data from several sources, including historical records, crop yield maps, and weather forecasts (Kalyani et al., 2023). The data can be analyzed using advanced analytics tools to uncover trends, patterns, and areas in need of enhancement.

Several technologies are used to build and manage precise digital twins for farm machinery and equipment, including:

- Sensors and Internet of Things devices: These gather data in real time from the actual machines, including operating data, GPS positions, and engine characteristics (Kheder & Mohammed, 2024).
- Computer-aided design (CAD) models: The digital twin is created using 3D models of the machinery, which accurately represent its components and physical dimensions.

- Software for modelling and simulation: These tools are used to build digital copies of the machinery and model how it would behave in various scenarios (Segovia & Garcia-Alfaro, 2022).
- Cloud computing and data storage: Cloud-based systems can supply the significant processing power and data storage capabilities needed for digital twins.
- Artificial intelligence and machine learning: can help with decision-making, optimization, and predictive maintenance using data from digital twins (Zhong et al., 2023).

Agricultural enterprises may reach greater levels of productivity, sustainability, and efficiency while lowering maintenance costs, downtime, and resource consumption by utilizing digital twins for farm machinery and equipment (Verdouw et al., 2021).

3.2.2 Farms and Croplands

Entire farms or crop fields, including the soil, weather, crop development phases, and other pertinent variables, can be represented by digital twins (Kim & Heo, 2024). Precision agricultural techniques like yield prediction, resource optimization, and variable rate input application (e.g., fertilizers, and pesticides) may all be implemented using these models. Agricultural operations today rely heavily on farm machinery and equipment, such as tractors, harvesters, irrigation systems, and speciality tools (Dhanaraju et al., 2022).

- Digital twin modelling: Using soil properties, weather patterns, crop growth phases, and other pertinent information, digital twins may be used to virtually recreate entire farms or agricultural fields. Accurate simulations and forecasts are made possible by these models, which support data-driven agricultural decision-making.
- Precision agriculture approaches: A number of approaches may be used in precision agriculture by utilizing digital twin models. These include a precise estimation of agricultural yields using simulations and real-time data known as yield prediction; resource optimization, the economical distribution and use of resources such as insecticides, fertilizers, and water; and variable rate input application, which minimizes waste and environmental effects by applying inputs (such as herbicides and fertilizers) precisely based on specific field conditions.
- Data-driven agriculture: Data-driven agriculture is made possible by the combination of innovative agricultural technology, precision agriculture techniques, and digital twins. Farmers may make well-informed decisions, improve operations, and maximize output while reducing environmental effects by gathering and evaluating data from a variety of sources (such as sensors, drones, and satellite imaging).
- A precision agricultural cycle is shown in Figure 3.2. To comprehend crop development patterns, data on environmental elements are first collected, and then they are analyzed and simulated. Decisions on changing inputs, like water and fertilizers, through input control systems, are made in light of this study. Following that, the crop's progress is tracked, and the outcomes are noted to provide farmers input. This feedback loop makes continuous improvement of crop management techniques and input application possible.

3.2.3 Livestock

It is possible to create digital twins for specific animals or entire livestock businesses. These models may be used to improve feeding schedules, housing arrangements, and breeding initiatives in addition to monitoring the well-being, productivity, and behaviour of animals (Dayoub et al., 2024). Livestock farming refers to the practice of raising and caring for animals for various purposes, such

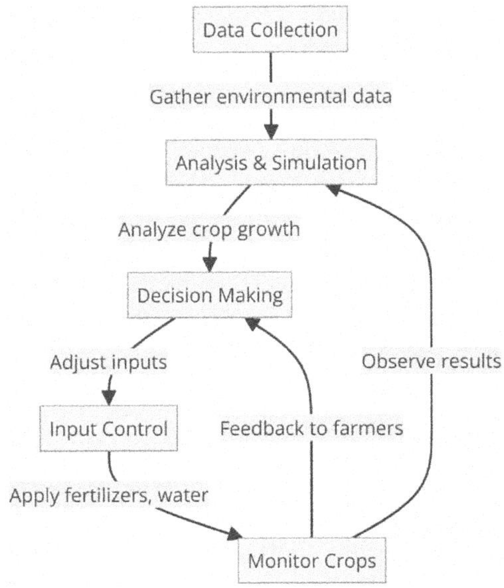

FIGURE 3.2 Intelligent crop management process.

as meat and dairy production. For cattle operations, the creation of digital twins can provide a variety of advantages and optimization opportunities:

- Monitoring of animal health and welfare: Wearable technology and sensors that measure different aspects of animal behaviour and health can be combined with data from digital twins (Mishra & Sharma, 2023). The body's temperature, heart rate, level of activity, eating habits, and breathing rates are just a few examples of these factors. The digital twin can identify early markers of disease, stress, or other health difficulties by continually monitoring these variables, which facilitates timely intervention and preventative actions (Vallée, 2023).
- Precision livestock farming: Digital twins can help with precision livestock farming by offering comprehensive information about individual animal performance and environmental variables (Haleem et al., 2023). Precision agriculture for crops operates similarly. By utilizing data, tailored breeding plans for each animal's specific needs can be developed, along with personalized feeding routines and living arrangements that consider factors such as humidity, temperature, and ventilation (Havelka et al., 2022).

  Digital twins can replicate the dynamics of a whole cattle herd, accounting for elements including genetics, nutrition, reproductive cycles, and environmental circumstances (Han et al., 2022). This allows for herd management and productivity improvement. Through the use of these simulations, farmers may improve production and profitability while maintaining animal welfare by refining herd structure, breeding plans, and management techniques (Van Arendonk, 2011).

  Digital twins can be used to simulate the transmission of infectious illnesses among a community of cattle, aiding in the monitoring and management of disease (Neethirajan & Kemp, 2021). Farmers may minimize the danger of outbreaks and the resulting financial losses by developing efficient disease management techniques, including vaccination programs, biosecurity measures, and quarantine protocols, by modelling a variety of scenarios and actions.
- Transparency and supply chain traceability: Digital twins can be combined with supply chain systems to offer livestock products complete traceability (Laubenbacher et al., 2021). The ability to trace animals from the farm to the customer's plate makes this information useful for quality control, food safety, and consumer confidence.

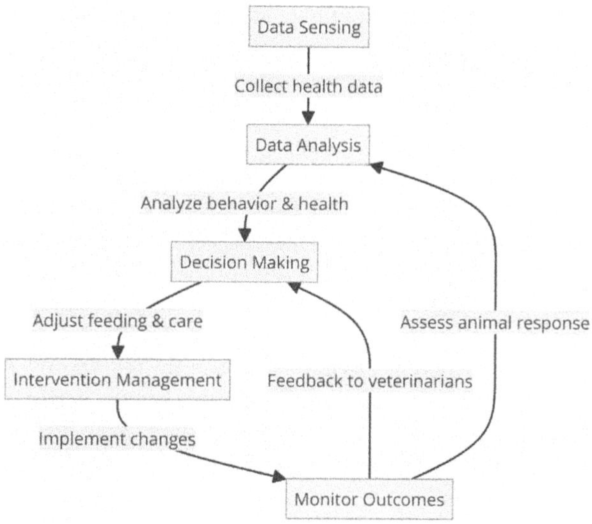

FIGURE 3.3 Precision livestock monitoring and management process.

A precision livestock management feedback loop is depicted in Figure 3.3. To begin gathering animal health data, data sensing is used. After that, these data are examined to learn more about their behaviour and state of health. Decisions on modifying feeding and care practices through intervention management are made in light of this analysis. The animal's reaction is evaluated, and the results of these therapies are tracked. Veterinarians get this input, which enables ongoing improvement and optimisation of animal management techniques.

There are various technologies utilized for creating and monitoring accurate digital replicas for cattle operations, including:

- Devices like wearable sensors and Internet of Things gadgets collect real-time data on the behaviour, health, and environment of animals.
- Biometric identification systems: Individual animals may be individually identified and linked to their digital twins using technologies such as RFID tagging and face recognition.
- Environmental monitoring systems: Sensors and monitoring systems track elements including temperature, humidity, air quality, and feed and water intake within livestock facilities (Aquilani et al., 2022).
- Genetic and genomic data: To improve breeding programs and forecast characteristics, data on an animal's genetic composition and genomic profiles may be integrated into digital twins.
- Data analytics and machine learning: To enable data-driven decision-making and predictive capabilities, massive amounts of data gathered from diverse sources are processed and analyzed using sophisticated algorithms and machine learning approaches (Fu et al., 2021).
- Cloud computing and data storage: Cloud-based systems can supply the substantial processing power and data storage capabilities needed for digital twins.

3.2.4 Controlled Environments and Greenhouses

The setup and functioning of greenhouses, vertical farms, and other controlled agricultural settings may be replicated using digital twins (Ariesen-Verschuur et al., 2022). With the help of these models, we can adjust environmental factors such as light, humidity, and temperature to create optimal conditions for crop growth.

By regulating a variety of environmental parameters, greenhouses and controlled environments—such as vertical farms and plant factories—are made to offer the best growth conditions for crops (Vatistas et al., 2022). For these systems, building digital twins can have the following advantages:

- Environmental monitoring and management: In a greenhouse or other controlled environment, digital twins can combine data from sensors and control systems to monitor and manage variables including temperature, humidity, illumination, and levels of carbon dioxide and nutrients (Bersani et al., 2022). The digital twin can offer real-time information and allow exact modifications to sustain ideal growth circumstances by continually monitoring key factors.
- Climatic and energy optimization: Digital twins can mimic various climatic conditions, crop varieties, and production schedules, as well as the energy consumption and temperature control needs of greenhouses and controlled settings (Slob & Hurst, 2022). By using these models, one may minimize the environmental effects of these systems, cut operating costs, and optimize energy use.
- Crop growth modelling and yield prediction: Digital twins can include crop growth models and simulations that account for plant genetics, nutritional profiles, and controlled environmental conditions (Mitsanis et al., 2024). By using these models to forecast crop growth rates, prospective yields, and the effects of different management techniques, producers may maximize their operations and make well-informed decisions.
- Management of pests and diseases: By integrating data on the dynamics of pests and diseases in controlled environments, digital twins enable the creation of focused and efficient control methods (Sun et al., 2023). Growers may avoid the use of hazardous pesticides and take preventative measures by modelling the spread and effect of pests and diseases.
- Facility design and optimization: Digital twins can be used to simulate and assess various design alternatives, layouts, and equipment combinations throughout the planning and building stages of new greenhouses or controlled environment facilities (Soori et al., 2023). This can aid in maximizing the facility's long-term sustainability, productivity, and effective use of its resources.

A cyclical procedure for controlled environment agriculture is shown in Figure 3.4. To identify the ideal growth circumstances, data on environmental elements are first gathered, and then analysis and simulation are performed. Based on this, the environment is managed by applying the proper settings for things like lighting, humidity, and temperature. The crop's reaction to these regulated circumstances is assessed, and the results are tracked. The operators get this feedback, which enables ongoing environmental condition optimization and correction for the best crop development.

Several technologies are used to build and manage precise digital twins for greenhouses and other controlled environments, such as:

- Environmental sensors: Within the controlled environment, a network of sensors is utilized to monitor temperature, humidity, light intensity, CO_2 levels, and other pertinent characteristics.
- Control systems: The digital twin is combined with automated control systems, such as lighting, climate control, and nutrition supply, to allow for the exact manipulation of environmental conditions (Knibbe et al., 2022).
- Plant growth models and simulations are utilized to represent how plants grow and develop in different environmental conditions and with various management strategies (Guo et al., 2011).
- Computer-aided design models are used to create a digital replica of the greenhouse or controlled environment facility, including detailed 3D representations of the physical layout and equipment configuration.

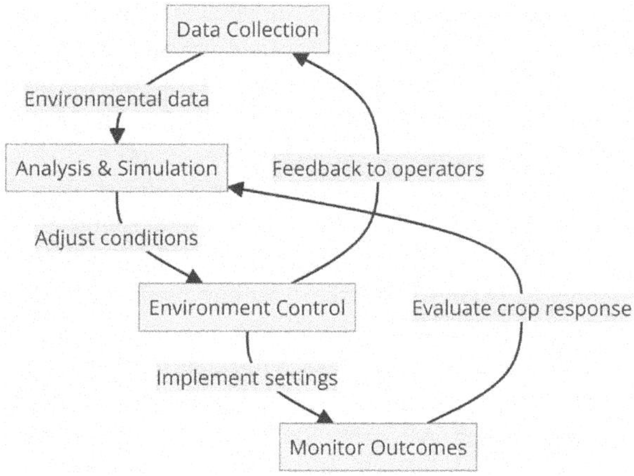

FIGURE 3.4 Intelligent greenhouse/growth chamber management process.

- Data analytics and machine learning techniques are employed to analyze extensive datasets from a variety of sources, enabling informed decision-making and predictive capabilities (Rahmani et al., 2021).
- Cloud computing and data storage: Cloud-based systems can supply the substantial processing power and data storage capabilities needed for digital twins.

3.2.5 Logistics and Supply Chains

The complete agricultural supply chain, including the activities involved in distribution, storage, and transportation, may be represented by digital twins (Guidani et al., 2024). These models can aid in waste reduction, traceability, and transparency improvements, as well as logistics optimization. The flow of goods, resources, and information from farm to consumer is a complicated network known as the agricultural supply chain (Ronaghi, 2021). For this supply chain, building digital twins can provide several advantages and optimization opportunities:

- End-to-end visibility and traceability: From the farm to the point of sale or consumption, digital twins may offer a thorough virtual depiction of the whole agricultural supply chain (Defraeye et al., 2021). According to Melesse and colleagues (2023), digital twins can track the movement of agricultural products, raw materials, and resources throughout the supply chain by integrating data from various sources such as sensors, logistics systems, and enterprise resource planning (ERP) systems. Traceability, quality assurance, and regulatory compliance are made possible by this visibility.
- Planning and optimization of the supply chain: Digital twins are capable of simulating various supply chain scenarios and configurations, accounting for variables like demand patterns, storage conditions, transit routes, and possible interruptions (Attaran & Celik, 2023). By pinpointing obstacles, minimizing inefficiencies, and streamlining the movement of goods and resources, these simulations can enhance productivity, cut expenses, and adapt more effectively to market demands.
- Logistics and transportation management: Real-time traffic and weather data, as well as data from vehicles like trucks, ships, and railcars, may all be integrated via digital twins (Sahal et al., 2021). To ensure the timely delivery of perishable agricultural goods while lowering transportation costs, fuel consumption, and emissions, this data may be utilized to optimize route, scheduling, and load planning.

- Digital twins can monitor and replicate the complete cold chain, from the farm to the consumer's refrigerator, for agricultural items that are sensitive to temperature (Dyck et al., 2023). Digital twins can help maintain product quality across the supply chain, minimizing spoiling, and ensuring optimal temperature management by integrating data from environmental monitoring systems and temperature sensors.
- Risk management and disruption mitigation: Natural catastrophes, equipment malfunctions, and other possible hazards and disruptions in the agricultural supply chain may all be modelled using digital twins (Huang et al., 2024). Through the process of modelling these situations, interested parties may create backup plans, put mitigation measures into action, and strengthen the supply chain's resilience.
- A cyclical supply chain management process in agricultural or food production is depicted in Figure 3.5. The first step is gathering product data, which is then used to optimise routes through transportation planning. After that, the logistics management distributes the items while conducting product inspections to ensure quality control. Logistics plans can be modified based on the input received from quality inspections and operations. Continuous improvement in product quality and supply chain efficiency is made possible by this feedback loop.

Several technologies are used to build and manage precise digital twins for the agricultural supply chain and logistics, including:

- Sensor networks and the Internet of Things: Throughout the supply chain, sensors and IoT devices are used to gather data in real time on product whereabouts, environmental conditions, and transportation asset performance (Taj et al., 2023).
- Geographic information systems (GISs) are used to analyze transportation routes, facility locations, and environmental conditions. This software allows for the integration and analysis of various geographical data.
- Supply chain management (SCM) and enterprise resource planning systems can be included in a digital twin to provide information on order fulfilment, manufacturing schedules, and inventory levels (Najafabadi et al., 2015). This allows for predictive analytics, optimization, and self-governing decision-making.

FIGURE 3.5 Supply chain management process.

- Software for modelling and simulation: These tools are used to generate digital copies of the supply chain and to model various configurations and situations.
- Cloud computing and data storage: Cloud-based systems can supply the substantial processing power and data storage capabilities needed for digital twins.

3.2.6 Water and Soil Resources

It is possible to construct digital twins for irrigation systems, water reservoirs, and soil profiles. Through the monitoring and optimization of soil moisture levels, nutrient content, water consumption, and drainage patterns, these models can support sustainable practices and more effective water management (Kassaye et al., 2020). Two essential elements that support agricultural sustainability and production are soil and water.

Digital twins for various resources can facilitate more effective and sustainable management techniques and provide insightful information:

- Mapping and tracking soil: High-resolution maps of the characteristics of soil may be produced across agricultural fields or regions by integrating data from several sources, including soil sampling, remote sensing, and geospatial technologies, using digital twins (Tsakiridis et al., 2023). The maps show various characteristics of the soil, including its texture, structure, nutrient levels, organic matter content, and pH levels. Digital twins can assist in identifying regions that need particular treatments, such as nutrient management, erosion control, or drainage improvements, by continually monitoring changes in soil parameters over time (Searle et al., 2021).
- Meticulous nutrition management: To improve fertilizer application rates and timing, digital twins can be used in conjunction with crop models and yield estimates (Kalyani et al., 2024). Digital twins have the potential to reduce the overapplication of fertilizer and minimize environmental consequences by guiding variable-rate fertilizer application by taking into account crop requirements, site-specific soil conditions, and weather patterns (Konfo et al., 2024). With less fertilizer runoff and groundwater pollution, this precise technique can increase crop yield, profitability, and efficiency in the use of nutrients.
- Management of water resources: Data from surface water bodies, irrigation systems, groundwater aquifers, rainfall patterns, and other sources may all be integrated via digital twins (Henriksen et al., 2022). "Using digital twins can help improve how water is allocated, irrigation schedules are planned, and drainage is managed by simulating the flow of water and maintaining a proper water balance in specific regions or watersheds." This can minimize water waste and its negative effects on the environment while increasing agricultural yields, decreasing water stress, and improving water efficiency.
- Optimization of irrigation systems: Irrigation system performance and efficiency may be represented by digital twins that account for several elements, including water sources, distribution networks, and application techniques (Alves et al., 2023). Digital twins can assist in identifying chances for system enhancements, leak detection, and optimum water distribution to crops by modelling various irrigation situations and techniques. This can lead to lower energy costs for water pumping and distribution, increased crop uniformity, and water savings.
- Erosion of soil and preservation: According to Silva et al. (2023), digital twins can replicate soil erosion patterns and the impact of various conservation methods such as contour planting, cover crops, and no-till farming. Digital twins have the potential to facilitate the adoption of successful soil conservation methods by modelling various scenarios.

Several technologies are used to build and manage precise digital twins for soil and water resources, including satellite imaging and remote sensing, in-situ sensors (weather stations, soil moisture

TABLE 3.2
Key Technologies for Building Digital Twins in Agriculture

Technology	Description
Sensors and the Internet of Things	Collecting real-time data from farm equipment, livestock, crops, and environmental conditions.
Remote Sensing and Satellite Imaging	Gathering data on crop health, soil characteristics, and environmental variables over large areas.
Geographic Information Systems	Integrating and analyzing spatial data for mapping and monitoring agricultural landscapes.
Simulation and Modelling Software	Creating virtual representations and simulating the behaviour of agricultural systems and processes.
Machine Learning and Data Analytics	Analyzing large datasets, identifying patterns, and enabling predictive capabilities.
Cloud Computing and Data Storage	Providing the necessary computing power and data storage for managing digital twin models.
Computer-Aided Design	Developing 3D models of farm machinery, facilities, and equipment for digital twin creation.
Bioinformatics and Genomics	Integrating genetic and genomic data for applications in agricultural biotechnology

probes, etc.), systems of geographic information, software for simulating soil and hydrology, machine learning and data analytics, data storage, and cloud computing.

Table 3.2 provides an overview of several technologies that are essential for facilitating digital twins and sophisticated analytics in the field of agriculture. The course encompasses various techniques for data collecting such as sensors, remote sensing, and GIS. It also explores advanced computational tools for simulation, modelling, machine learning, and data storage. Additionally, it covers design tools like CAD and new topics such as bioinformatics and genomics. These technologies work together to enable the development of virtual representations, data-driven decision-making, and the integration of various data sources to improve agricultural practices and systems.

3.2.7 Control of Diseases and Pests

The dynamics of disease outbreaks, insect populations, and their relationships with crops and the environment may all be represented using digital twins (Kamel Boulos & Zhang, 2021). By helping with early identification, forecasting, and the development of focused pest management measures, these models can lessen the need for the overuse of pesticides. According to Wilkinson et al. (2011), diseases and pests are significant threats to agricultural production and can result in farmers experiencing substantial financial losses. For the control of pests and diseases, digital twin creation can have the following advantages:

- Monitoring for diseases and pests: With the use of digital twins, agricultural fields and regions may be monitored for the presence, abundance, and dispersion of pests and illnesses through the integration of data from several sources, including meteorological data, field scouting, remote sensing, and pest traps (Rovira-Más et al., 2021). By identifying possible hotspots or early indicators of infestations, this real-time monitoring can provide prompt action and focused control efforts.
- Modelling diseases and pests: Predictive models that mimic the life cycles, behaviours, and population dynamics of various illnesses and pests can be incorporated into digital

twins (Mulder et al., 2022). These models allow for forecasts of possible outbreaks or disease propagation by accounting for variables including weather, crop phenology, and management techniques.

Digital twins can assist in assessing the efficacy of many control techniques, including pharmacological treatments, biological control agents, and cultural practices, by modelling numerous scenarios (Armeni et al., 2022).

- Accurate pest control: Based on site-specific pest or disease pressure, digital twins can direct the targeted use of pesticides or other management measures (Roberts et al., 2021). Digital twins can improve the timing, location, and dose of pest control methods, avoiding the use of unneeded pesticides and their negative effects on the environment (Deguine et al., 2021). This is achieved by combining data on insect/disease distribution, crop growth phases, and environmental variables.
- Handling resistance: Digital twins can simulate how populations of pathogens or pests develop and spread resistance to pesticides or other forms of management (Iost Filho et al., 2022). Digital twins can assist in the development of resistance management techniques by modelling various circumstances. These techniques may include the use of integrated pest management (IPM) techniques or the rotation of several forms of action.
- Systems for advice and decision support: Digital twins can be combined with advisory tools or decision support systems to give farmers real-time advice on managing pests and diseases (Purcell & Neubauer, 2023). These suggestions, which enable prompt and well-informed decision-making for efficient pest and disease control, can be based on the existing circumstances, past data, and forecast models.

Several technologies are used to produce and maintain precise digital twins for the management of pests and diseases, including tools for field observation (such as drones, sensors, and traps), satellite imaging and remote sensing, weather information and prediction systems, models of disease and pest populations, systems of geographic information, machine learning and data analytics, data storage and cloud computing etc.

3.2.8 Biotechnology in Agriculture

To produce and evaluate genetically modified crops, biofertilizers, and other biotechnological uses in agriculture, digital twins can be employed (Batlang et al., 2014). Research and development procedures may be sped up by using these virtual models to mimic the functionality and interactions of these technologies with their surroundings.

By utilizing scientific techniques and tools, agricultural biotechnology alters and improves plants, animals, and microorganisms for various purposes, such as increasing crop production, strengthening resistance to diseases and pests, improving nutritional quality, and developing biofuels and bioproducts (Srivastava, 2019). Within this field, the creation of digital twins can offer numerous benefits:

- Genetic engineering and the evolution of traits: The consequences of introducing certain genetic alterations or novel features into livestock or agricultural plants can be replicated using digital twins (Symeonaki et al., 2024). By studying how changes in genes impact an organism's physiology, growth, and development, scientists can predict outcomes and evaluate the success of targeted interventions. This can hasten the process of developing new varieties with the necessary traits, lessen the need for protracted field testing, and raise the likelihood of success (Gudi et al., 2022).
- Optimization of bioprocesses: "Digital twins can enhance the efficiency of bioprocesses for creating biofuels, biofertilizers, biopesticides, and other agricultural biotechnology

products" (Appl et al., 2020). Digital twins can be used to simulate the dynamics and circumstances of fermentation, enzyme reactions, and other biological processes to determine the ideal amounts of temperature, pH, nutrients, and feedstock composition (Appl et al., 2021). This may result in the manufacture of these biotechnology goods being more efficient, producing better yields, and generating less waste.

- Assessment of the environmental impact: The potential impacts on the environment from introducing genetically modified organisms (GMOs) or other biotech products into agricultural systems can be simulated using digital twin technology (Zilberman et al., 2018). Researchers can evaluate possible dangers, including gene flow, impacts on non-target animals, and long-term ecological ramifications, by modelling various scenarios (Bauer-Panskus et al., 2020). This knowledge can help ensure that agricultural biotechnology is used responsibly and sustainably by directing the creation of suitable biosafety measures.
- Risk management and compliance with regulations: Digital twins can help with risk evaluations and regulatory compliance for agricultural biotechnology goods (Purcell & Neubauer, 2023). "Digital twins play a crucial role in evaluating potential risks, creating strategies to reduce them, and showcasing adherence to safety and environmental regulations through realistic simulations and valuable analytical findings" (Björklund et al., 2023).
- Scaling and transmission of technology: Innovations in biotechnology can be more easily transferred from labs to real-world uses in the field because of digital twins (Holzinger et al., 2023). Digital twins may be used to simulate the behaviour and performance of biotechnology products in various agricultural and environmental settings (Peladarinos et al., 2023). This allows for the optimization of deployment methods for successful acceptance and scaling, as well as the identification of possible difficulties.

Several technologies are used to build and manage precise digital twins for agricultural biotechnology, including tools for bioinformatics and genomics, models of systems biology and computational biology, software for simulating biological processes, tools for risk assessment and environmental modelling, machine learning and data analytics, data storage, and cloud computing.

3.2.9 Packaging and Processing of Food

Food processing plants can have their equipment, manufacturing processes, and storage conditions digitally twined (Verboven et al., 2020). These models may aid in process optimization, guarantee food safety and quality, and cut waste through the supply chain. Before consuming food items, they go through processing and packaging to guarantee safety, quality, and longevity. These steps are crucial to the agricultural supply chain (Awuchi, 2023). For these processes, building digital twins can provide several advantages:

- Efficiency and process optimization: The complete food processing line, comprising different unit activities like cleaning, sorting, grinding, cooking, drying, and packing, may be simulated using digital twins (Krupitzer et al., 2022). "Digital twins can help identify bottlenecks, improve process conditions, and increase productivity by modelling the connections between materials, equipment, and process variables like temperature, pressure, and flow rates" (Lv, 2023). Reduced downtime, higher throughput, and cheaper operating expenses might result from this.
- Assurance of quality and uniformity of product: Throughout the production and packaging phases, food items' quality qualities may be tracked and predicted by digital twins (Maheshwari et al., 2023). Digital twins allow for the prompt correction of deviations from intended quality criteria (e.g., moisture content, texture, colour, and nutritional levels) by combining data from sensors, analytical tools, and product requirements (Lee & Lee, 2021).

This can lessen the possibility of product recalls or consumer complaints while also ensuring constant product quality.

- Traceability and food safety: "In the production and wrapping stages, digital replicas can oversee and track important factors that impact the safety of food, like temperature, humidity, and levels of microbiological contamination" (Valero et al., 2023). Through the integration of data from several sources (such as environmental monitoring systems, cleaning records, and sensors), digital twins provide quick incident response in the event of food safety concerns and offer end-to-end traceability (Henriksen et al., 2022). This can boost customer trust and ensure adherence to food safety rules.
- Asset management and equipment upkeep: Digital twins can monitor the performance, energy consumption, and maintenance needs of individual pieces of equipment or whole processing lines (Kalyani et al., 2023). Digital twins can optimize maintenance schedules, minimize unscheduled downtime, and increase equipment longevity by combining data from sensors and predictive maintenance models (Speicher & DeFranco, 2023). Better asset utilization, increased production, and cost savings are possible outcomes of this.
- Design and optimization of packaging: Scientists can use digital twins to simulate how different packaging materials and designs perform in different environmental conditions like temperature, humidity, shock, and vibration. This technology can help optimize packaging solutions to improve shelf-life, protect products better, and reduce waste by modelling how food products, packaging materials, and environmental factors interact.

Several technologies are used to build and manage precise digital twins for food processing and packaging, including systems for process monitoring and control, sensors and analytical tools, CAD models of facilities and equipment, fluid dynamics computation (CFD) and process modelling, machine learning and data analytics, data storage, and cloud computing.

3.2.10 Rules and Policies about Agriculture

The intricate relationships that exist between environmental elements, legislative frameworks, and agricultural practices may be represented using digital twins (Dihan et al., 2024). These models can aid in the evaluation of possible effects on productivity, sustainability, and economic considerations by policymakers and regulators (Sala et al., 2015).

The agricultural sector's practices, sustainability, and economic viability are significantly shaped by agricultural laws and regulations. In this area, building digital twins can provide stakeholders, regulatory agencies, and legislators with insightful information and assistance:

- Evaluation of the impact of policy: Digital twins can replicate the possible effects on several facets of the agricultural system, such as crop production, animal farming, resource management, and economic variables, of proposed agricultural policies or regulatory modifications (Ryan, 2023). Digital twins can offer data-driven assessments of the anticipated outcomes and unintended repercussions by modelling the intricate relationships between governmental initiatives, environmental variables, market dynamics, and farming methods (Føre et al., 2024). This can aid in the development of initiatives to reduce negative effects or increase good ones, as well as aid policymakers and regulatory bodies in making more informed choices and identifying possible trade-offs.
- Environmental compliance and sustainable agriculture: The impacts of different farming techniques on the environment, like greenhouse gas emissions, water contamination, soil damage, and loss of biodiversity, can be studied using digital twins (Kourmouli & Lesniewska, 2024). Digital twins can assist in assessing how well programs are working to support sustainable agriculture and ensure environmental standards are followed by modelling scenarios under different policy frameworks and regulatory constraints

(Delgado et al., 2019). This data can direct the creation of laws and other incentives that promote the use of ecologically friendly farming methods.

- Evaluation of risks and readiness: The possible hazards and vulnerabilities of agricultural systems to different threats, such as pest and disease outbreaks, natural catastrophes, climate change, and market disruptions, may be replicated using digital twins (Demestichas et al., 2020). Digital twins are useful for simulating various scenarios and stress-testing the robustness of agricultural systems. This allows for the creation of risk management plans, backup plans, and emergency response procedures. In the event of possible disruptions, this can assist regulatory agencies and legislators in strengthening the agriculture sector's readiness and resilience (Priyadarshini & Abhilash, 2021).
- Analysis of trade and markets: The intricate relationships between international supply networks, trade regulations, market dynamics, and agricultural output may all be modelled using digital twins (Badakhshan & Ball, 2023). Digital twins can offer insights into possible effects on local and global markets, pricing, and competitiveness by modelling various trade scenarios and governmental actions. The creation of market rules, tariff structures, and trade policies that encourage fair competition, food security, and economic stability can all be aided by the information provided (Zimmermann & Rapsomanikis, 2023).
- Supporting decision-making and involving stakeholders: With the use of digital twins, decision-makers, regulators, and other interested parties may examine many options, weigh trade-offs, and work together to create agricultural policies and regulations that work (Finger, 2023). Digital twins can help with evidence-based decision-making, consensus-building, and stakeholder involvement by offering a shared virtual environment for data visualization, simulation, and analysis (Davila Delgado & Oyedele, 2021).

Several technologies are used to build and maintain precise digital twins for agricultural laws and regulations, including GISs and the study of geographical data, tools for simulating the environment and climate, models for simulating the economy and markets, software for scenario planning and risk assessment, machine learning and data analytics, data storage and cloud computing, and platforms for decision support and visualization.

3.3 LIMITATIONS OF DIGITAL TWINS IN AGRICULTURE

Digital twin technology has a huge potential to benefit the agriculture sector, but it also has several limitations and challenges. The limitations and challenges need to be addressed to adopt the digital twin in increased numbers. Some of the key limitations are as follows:

- Availability and Quality of Data: The first and foremost important limitation is the availability and quality of required data. IoT technology has sensor devices that have enabled the agricultural data collection at various stages but the accuracy, consistency and accessibility of the data remains a hurdle. The construction of reliable digital twin models is further hampered by the extra difficulties associated with obtaining real-time data in isolated or rural locations with poor connectivity.
- Cost and Complexity: Building and maintaining digital twin models for agricultural systems can be a costly and complex endeavour. From the deployment of sensor networks and data acquisition infrastructure to the development of simulation algorithms and analytics platforms, the upfront investment required can be prohibitive for many farmers, especially smallholders and those in developing countries. Moreover, ongoing maintenance and updates to digital twin systems incur additional expenses, further limiting accessibility.
- Technical Expertise: Specialized technical knowledge and expertise in the field of data science, computer modelling and agricultural domain are essential for the effective utilization of

digital twin technology. Many farmers may lack the necessary skills and resources to develop and operate digital twin systems independently. Bridging the digital divide and providing training and support for farmers to leverage digital twin technology effectively is essential for democratizing access and maximizing its impact across diverse agricultural contexts.

- Scalability and Interoperability: In the context of deploying digital twin solutions across different agricultural landscapes, scalability and interoperability pose important issues. Development of standardized digital twin models that can be easily integrated and scaled across many contexts is difficult due to the considerable variations in size, complexity, and technical infrastructure among agricultural systems. The compatibility and interoperability of digital twin systems with current farm management software must be guaranteed by interoperability standards and frameworks.
- Privacy and Security Concerns: As digital twin technology spreads, questions concerning data ownership, security, and privacy are brought up. A valuable asset, agricultural data must be shielded from unwanted access, manipulation, and exploitation. This includes sensitive information regarding crop yields, livestock health, and farm operations. Establishing strong cybersecurity, encryption, and data governance policies is crucial to fostering trust and defending the rights to privacy and intellectual property held by farmers.
- Ethical Considerations: There are many intricate issues of equity, accountability, and autonomy raised by the ethical implications of digital twin technology in agriculture. Digital twins are having a more and greater impact on farm decision-making processes, raising questions about algorithmic bias, transparency, and the possibility of unforeseen repercussions. Furthermore, unequal access to digital twin technologies may worsen already-existing agricultural sector inequities by marginalizing smallholder farmers and escalating social and economic divides.
- Adoption Barriers: Despite digital twins' potential advantages for agriculture, adoption barriers still exist because of a lack of knowledge, a fear of change, and hostility to technology. Without strong proof of the technology's value proposition and demonstrable advantages, farmers could be reluctant to adopt new ones. Fostering an innovative culture and facilitating the widespread implementation of digital twin technologies in agriculture requires overcoming behavioural and cultural hurdles through education, demonstration projects, and stakeholder involvement.

Table 3.3 highlights some of the key challenges associated with the adoption of digital twin technology in the agriculture sector, along with potential solutions to address these challenges. Overcoming these obstacles will be crucial for realizing the full potential of digital twins in enhancing agricultural productivity, sustainability, and resilience.

3.4 CONCLUSION

The world is going to witness a significant shift in the way it thinks about farming methods because of the application of digital twin technology in agriculture. This chapter has explored how digital twins, with their unmatched sustainability, efficiency, and precision, are changing conventional techniques. Farmers are equipped to make data-driven choices, allocate resources as efficiently as possible, and reduce their negative effects on the environment. However, several obstacles remain: worries over data quality, cost, technical know-how, and privacy continue to exist. But there are also chances for cooperation, creativity, and advancement among these difficulties. Going forward, it will be important to tread carefully to fully realize the potential of digital twin technology, keeping in mind its ethical, societal, and environmental ramifications. To remove obstacles, promote innovation, and guarantee fair access to digital twin solutions, cooperation amongst stakeholders such as farmers, academics, legislators, and technology providers is essential.

TABLE 3.3
Adoption Challenges and Potential Solutions for Digital Twins in Agriculture

Challenge	Description	Potential Solutions
Data Quality and Availability	Lack of accurate, consistent, and accessible data can hinder the development of reliable digital twin models.	—Invest in sensor networks and IoT infrastructure —Improve data collection and management practices —Leverage remote sensing and satellite data
Technical Expertise	Specialized knowledge in data science, computer modelling, and agricultural domains is required to effectively develop and utilize digital twin systems.	—Provide training and capacity-building programs —Collaborate with academic and research institutions —Develop user-friendly interfaces and decision-support tools
Initial Investment Costs	The upfront costs of deploying sensor networks, computing infrastructure, and software platforms can be prohibitive, especially for smallholder farmers.	—Explore cost-sharing models and public-private partnerships —Offer incentives and subsidies for technology adoption —Develop scalable and modular digital twin solutions
Scalability and Interoperability	Ensuring digital twin systems can be scaled across diverse agricultural landscapes and integrated with existing farm management software.	—Establish interoperability standards and frameworks —Develop modular and flexible digital twin architectures —Encourage collaboration and knowledge sharing
Privacy and Security Concerns	Ensuring data privacy, security, and protection against cyber threats and unauthorized access to sensitive agricultural data.	—Implement robust cybersecurity measures and data governance policies —Establish clear data ownership and privacy guidelines —Promote transparency and build trust with stakeholders
Adoption Barriers	Cultural resistance, lack of awareness, and scepticism towards new technologies can impede the widespread adoption of digital twins.	—Conduct awareness campaigns and demonstration projects —Involve stakeholders in the development and deployment process —Highlight the tangible benefits and success stories

REFERENCES

Alves, R. G., Maia, R. F., & Lima, F. (2023). Development of a Digital Twin for smart farming: An irrigation management system for water saving. *Journal of Cleaner Production, 388*, 135920. https://doi.org/10.1016/j.jclepro.2023.135920

Appl, C., Baganz, F., & Hass, V. C. (2021). Development of a digital twin for enzymatic hydrolysis processes. *Processes, 9*(10), 1734. https://doi.org/10.3390/pr9101734

Appl, C., Moser, A., Baganz, F., & Hass, V. C. (2020). Digital Twins for bioprocess control strategy development and realisation. In C. Herwig, R. Pörtner, & J. Möller (Eds.), *Digital Twins* (Vol. 177, pp. 63–94). Springer International Publishing. https://doi.org/10.1007/10_2020_151

Aquilani, C., Confessore, A., Bozzi, R., Sirtori, F., & Pugliese, C. (2022). Review: Precision Livestock Farming technologies in pasture-based livestock systems. *Animal, 16*(1), 100429. https://doi.org/10.1016/j.animal.2021.100429

Ariesen-Verschuur, N., Verdouw, C., & Tekinerdogan, B. (2022). Digital Twins in greenhouse horticulture: A review. *Computers and Electronics in Agriculture, 199*, 107183. https://doi.org/10.1016/j.compag.2022.107183

Armeni, P., Polat, I., De Rossi, L. M., Diaferia, L., Meregalli, S., & Gatti, A. (2022). Digital Twins in healthcare: Is it the beginning of a new era of evidence-based medicine? A critical review. *Journal of Personalized Medicine, 12*(8), 1255. https://doi.org/10.3390/jpm12081255

Attaran, M., & Celik, B. G. (2023). Digital Twin: Benefits, use cases, challenges, and opportunities. *Decision Analytics Journal, 6*, 100165. https://doi.org/10.1016/j.dajour.2023.100165

Awuchi, C. G. (2023). HACCP, quality, and food safety management in food and agricultural systems. *Cogent Food & Agriculture, 9*(1), 2176280. https://doi.org/10.1080/23311932.2023.2176280

Badakhshan, E., & Ball, P. (2023). Applying digital twins for inventory and cash management in supply chains under physical and financial disruptions. *International Journal of Production Research, 61*(15), 5094–5116. https://doi.org/10.1080/00207543.2022.2093682

Batlang, U., Tsurupe, G., Segwagwe, A., & Obopile, M. (2014). Development and application of modern agricultural biotechnology in Botswana: The potentials, opportunities and challenges. *GM Crops & Food, 5*(3), 183–194. https://doi.org/10.4161/21645698.2014.945887

Bauer-Panskus, A., Miyazaki, J., Kawall, K., & Then, C. (2020). Risk assessment of genetically engineered plants that can persist and propagate in the environment. *Environmental Sciences Europe, 32*(1), 32. https://doi.org/10.1186/s12302-020-00301-0

Bersani, C., Ruggiero, C., Sacile, R., Soussi, A., & Zero, E. (2022). Internet of Things approaches for monitoring and control of smart greenhouses in Industry 4.0. *Energies, 15*(10), 3834. https://doi.org/10.3390/en15103834

Björklund, L., Schick, J., Lundteigen, M. A., & Glaser, M. (2023). Towards digital twins for safety demonstrations: Interfacing strategies for FPGA-targeted applications. *Computers & Industrial Engineering, 186*, 109733. https://doi.org/10.1016/j.cie.2023.109733

Cesco, S., Sambo, P., Borin, M., Basso, B., Orzes, G., & Mazzetto, F. (2023). Smart agriculture and digital twins: Applications and challenges in a vision of sustainability. *European Journal of Agronomy, 146*, 126809. https://doi.org/10.1016/j.eja.2023.126809

Davila Delgado, J. M., & Oyedele, L. (2021). Digital Twins for the built environment: Learning from conceptual and process models in manufacturing. *Advanced Engineering Informatics, 49*, 101332. https://doi.org/10.1016/j.aei.2021.101332

Dayoub, M., Shnaigat, S., Tarawneh, R., Al-Yacoub, A., Al-Barakeh, F., & Al-Najjar, K. (2024). Enhancing animal production through smart agriculture: Possibilities, hurdles, resolutions, and advantages. *Ruminants, 4*(1), 22–46. https://doi.org/10.3390/ruminants4010003

Defraeye, T., Shrivastava, C., Berry, T., Verboven, P., Onwude, D., Schudel, S., Bühlmann, A., Cronje, P., & Rossi, R. M. (2021). Digital twins are coming: Will we need them in supply chains of fresh horticultural produce? *Trends in Food Science & Technology, 109*, 245–258. https://doi.org/10.1016/j.tifs.2021.01.025

Deguine, J.-P., Aubertot, J.-N., Flor, R. J., Lescourret, F., Wyckhuys, K. A. G., & Ratnadass, A. (2021). Integrated pest management: Good intentions, hard realities. A review. *Agronomy for Sustainable Development, 41*(3), 38. https://doi.org/10.1007/s13593-021-00689-w

Delgado, J. A., Short, N. M., Roberts, D. P., & Vandenberg, B. (2019). Big data analysis for sustainable agriculture on a geospatial cloud framework. *Frontiers in Sustainable Food Systems, 3*, 54. https://doi.org/10.3389/fsufs.2019.00054

Demestichas, K., Peppes, N., & Alexakis, T. (2020). Survey on security threats in agricultural IoT and smart farming. *Sensors, 20*(22), 6458. https://doi.org/10.3390/s20226458

Dhanaraju, M., Chenniappan, P., Ramalingam, K., Pazhanivelan, S., & Kaliaperumal, R. (2022). Smart farming: Internet of Things (IoT)-based sustainable agriculture. *Agriculture*, *12*(10), 1745. https://doi.org/10.3390/agriculture12101745

Dihan, M. S., Akash, A. I., Tasneem, Z., Das, P., Das, S. K., Islam, M. R., Islam, M. M., Badal, F. R., Ali, M. F., Ahamed, M. H., Abhi, S. H., Sarker, S. K., & Hasan, M. M. (2024). Digital twin: Data exploration, architecture, implementation and future. *Heliyon*, *10*(5), e26503. https://doi.org/10.1016/j.heliyon.2024.e26503

Dyck, G., Hawley, E., Hildebrand, K., & Paliwal, J. (2023). Digital Twins: A novel traceability concept for post-harvest handling. *Smart Agricultural Technology*, *3*, 100079. https://doi.org/10.1016/j.atech.2022.100079

Finger, R. (2023). Digital innovations for sustainable and resilient agricultural systems. *European Review of Agricultural Economics*, *50*(4), 1277–1309. https://doi.org/10.1093/erae/jbad021

Føre, M., Alver, M. O., Alfredsen, J. A., Rasheed, A., Hukkelås, T., Bjelland, H. V., Su, B., Ohrem, S. J., Kelasidi, E., Norton, T., & Papandroulakis, N. (2024). Digital Twins in intensive aquaculture—Challenges, opportunities and future prospects. *Computers and Electronics in Agriculture*, *218*, 108676. https://doi.org/10.1016/j.compag.2024.108676

Fu, C., Xu, C., Xue, M., Liu, W., & Yang, S. (2021). Data-driven decision making based on evidential reasoning approach and machine learning algorithms. *Applied Soft Computing*, *110*, 107622. https://doi.org/10.1016/j.asoc.2021.107622

Gudi, S., Kumar, P., Singh, S., Tanin, M. J., & Sharma, A. (2022). Strategies for accelerating genetic gains in crop plants: Special focus on speed breeding. *Physiology and Molecular Biology of Plants: An International Journal of Functional Plant Biology*, *28*(10), 1921–1938. https://doi.org/10.1007/s12298-022-01247-8

Guidani, B., Ronzoni, M., & Accorsi, R. (2024). Virtual agri-food supply chains: A holistic digital twin for sustainable food ecosystem design, control and transparency. *Sustainable Production and Consumption*, *46*, 161–179. https://doi.org/10.1016/j.spc.2024.01.016

Guo, Y., Fourcaud, T., Jaeger, M., Zhang, X., & Li, B. (2011). Plant growth and architectural modelling and its applications. Preface. *Annals of Botany*, *107*(5), 723–727. https://doi.org/10.1093/aob/mcr073

Haleem, A., Javaid, M., Pratap Singh, R., & Suman, R. (2023). Exploring the revolution in healthcare systems through the applications of digital twin technology. *Biomedical Technology*, *4*, 28–38. https://doi.org/10.1016/j.bmt.2023.02.001

Han, X., Lin, Z., Clark, C., Vucetic, B., & Lomax, S. (2022). AI based digital twin model for cattle caring. *Sensors*, *22*(19), 7118. https://doi.org/10.3390/s22197118

Havelka, Z., Kunes, R., Kononets, Y., Stokes, J. E., Smutny, L., Olsan, P., Kresan, J., Stehlik, R., Bartos, P., Xiao, M., Kriz, P., Findura, P., & Roztocil, D. (2022). Technology of microclimate regulation in organic and energy-sustainable livestock production. *Agriculture*, *12*(10), 1563. https://doi.org/10.3390/agriculture12101563

Henriksen, H., Schneider, R., Koch, J., Ondracek, M., Troldborg, L., Seidenfaden, I., Kragh, S., Bøgh, E., & Stisen, S. (2022). A new Digital Twin for climate change adaptation, water management, and disaster risk reduction (HIP Digital Twin). *Water*, *15*(1), 25. https://doi.org/10.3390/w15010025

Holzinger, A., Keiblinger, K., Holub, P., Zatloukal, K., & Müller, H. (2023). AI for life: Trends in artificial intelligence for biotechnology. *New Biotechnology*, *74*, 16–24. https://doi.org/10.1016/j.nbt.2023.02.001

Huang, Y., Ghadge, A., & Yates, N. (2024). Implementation of digital twins in the food supply chain: A review and conceptual framework. *International Journal of Production Research*, 1–27. https://doi.org/10.1080/00207543.2024.2305804

Iost Filho, F. H., Pazini, J. D. B., Alves, T. M., Koch, R. L., & Yamamoto, P. T. (2022). How does the digital transformation of agriculture affect the implementation of Integrated Pest Management? *Frontiers in Sustainable Food Systems*, *6*, 972213. https://doi.org/10.3389/fsufs.2022.972213

Kalyani, Y., Bermeo, N. V., & Collier, R. (2023). Digital twin deployment for smart agriculture in Cloud-Fog-Edge infrastructure. *International Journal of Parallel, Emergent and Distributed Systems*, *38*(6), 461–476. https://doi.org/10.1080/17445760.2023.2235653

Kalyani, Y., Vorster, L., Whetton, R., & Collier, R. (2024). Application scenarios of Digital Twins for smart crop farming through cloud–fog–edge infrastructure. *Future Internet*, *16*(3), 100. https://doi.org/10.3390/fi16030100

Kamel Boulos, M. N., & Zhang, P. (2021). Digital Twins: From personalised medicine to precision public health. *Journal of Personalized Medicine*, *11*(8), 745. https://doi.org/10.3390/jpm11080745

Kassaye, K. T., Boulange, J., Lam, V. T., Saito, H., & Watanabe, H. (2020). Monitoring soil water content for decision supporting in agricultural water management based on critical threshold values adopted for Andosol in the temperate monsoon climate. *Agricultural Water Management*, *229*, 105930. https://doi.org/10.1016/j.agwat.2019.105930

Kheder, M. Q., & Mohammed, A. A. (2024). Real-time traffic monitoring system using IoT-aided robotics and deep learning techniques. *Kuwait Journal of Science*, *51*(1), 100153. https://doi.org/10.1016/j.kjs.2023.10.017

Kim, S., & Heo, S. (2024). An agricultural digital twin for mandarins demonstrates the potential for individualized agriculture. *Nature Communications*, *15*(1), 1561. https://doi.org/10.1038/s41467-024-45725-x

Knibbe, W. J., Afman, L., Boersma, S., Bogaardt, M.-J., Evers, J., Van Evert, F., Van Der Heide, J., Hoving, I., Van Mourik, S., De Ridder, D., & De Wit, A. (2022). Digital twins in the green life sciences. *NJAS: Impact in Agricultural and Life Sciences*, *94*(1), 249–279. https://doi.org/10.1080/27685241.2022.2150571

Konfo, T. R. C., Chabi, A. B. P., Amoussouga Gero, A., Lagnika, C., Avlessi, F., Biaou, G., & Sohounhloue, C. K. D. (2024). Recent climate-smart innovations in agrifood to enhance producer incomes through sustainable solutions. *Journal of Agriculture and Food Research*, *15*, 100985. https://doi.org/10.1016/j.jafr.2024.100985

Kourmouli, A., & Lesniewska, F. (2024). Losing ground: Targeting agricultural land take by enabling a circular economy in construction. *Circular Economy and Sustainability*, *4*(1), 459–473. https://doi.org/10.1007/s43615-023-00293-y

Krupitzer, C., Noack, T., & Borsum, C. (2022). Digital food twins combining data science and food science: System model, applications, and challenges. *Processes*, *10*(9), 1781. https://doi.org/10.3390/pr10091781

Laubenbacher, R., Sluka, J. P., & Glazier, J. A. (2021). Using digital twins in viral infection. *Science (New York, N.Y.)*, *371*(6534), 1105–1106. https://doi.org/10.1126/science.abf3370

Lee, D., & Lee, S. (2021). Digital twin for supply chain coordination in modular construction. *Applied Sciences*, *11*(13), 5909. https://doi.org/10.3390/app11135909

Lv, Z. (2023). Digital twins in Industry 5.0. *Research (Washington, D.C.)*, *6*, 0071. https://doi.org/10.34133/research.0071

Maheshwari, P., Kamble, S., Belhadi, A., Venkatesh, M., & Abedin, M. Z. (2023). Digital twin-driven real-time planning, monitoring, and controlling in food supply chains. *Technological Forecasting and Social Change*, *195*, 122799. https://doi.org/10.1016/j.techfore.2023.122799

Melesse, T. Y., Franciosi, C., Di Pasquale, V., & Riemma, S. (2023). Analyzing the implementation of digital twins in the agri-food supply chain. *Logistics*, *7*(2), 33. https://doi.org/10.3390/logistics7020033

Mishra, S., & Sharma, S. K. (2023). Advanced contribution of IoT in agricultural production for the development of smart livestock environments. *Internet of Things*, *22*, 100724. https://doi.org/10.1016/j.iot.2023.100724

Mitsanis, C., Hurst, W., & Tekinerdogan, B. (2024). A 3D functional plant modelling framework for agricultural digital twins. *Computers and Electronics in Agriculture*, *218*, 108733. https://doi.org/10.1016/j.compag.2024.108733

Mulder, S. T., Omidvari, A.-H., Rueten-Budde, A. J., Huang, P.-H., Kim, K.-H., Bais, B., Rousian, M., Hai, R., Akgun, C., van Lennep, J. R., Willemsen, S., Rijnbeek, P. R., Tax, D. M., Reinders, M., Boersma, E., Rizopoulos, D., Visch, V., & Steegers-Theunissen, R. (2022). Dynamic Digital Twin: Diagnosis, treatment, prediction, and prevention of disease during the life course. *Journal of Medical Internet Research*, *24*(9), e35675. https://doi.org/10.2196/35675

Najafabadi, M. M., Villanustre, F., Khoshgoftaar, T. M., Seliya, N., Wald, R., & Muharemagic, E. (2015). Deep learning applications and challenges in big data analytics. *Journal of Big Data*, *2*(1), 1. https://doi.org/10.1186/s40537-014-0007-7

Nasirahmadi, A., & Hensel, O. (2022). Toward the next generation of digitalization in agriculture based on digital twin paradigm. *Sensors*, *22*(2), 498. https://doi.org/10.3390/s22020498

Neethirajan, S., & Kemp, B. (2021). Digital Twins in livestock farming. *Animals*, *11*(4), 1008. https://doi.org/10.3390/ani11041008

Peladarinos, N., Piromalis, D., Cheimaras, V., Tserepas, E., Munteanu, R. A., & Papageorgas, P. (2023). Enhancing smart agriculture by implementing digital twins: A comprehensive review. *Sensors*, *23*(16), 7128. https://doi.org/10.3390/s23167128

Priyadarshini, P., & Abhilash, P. C. (2021). Agri-food systems in India: Concerns and policy recommendations for building resilience in post COVID-19 pandemic times. *Global Food Security*, *29*, 100537. https://doi.org/10.1016/j.gfs.2021.100537

Purcell, W., & Neubauer, T. (2023). Digital Twins in agriculture: A state-of-the-art review. *Smart Agricultural Technology*, *3*, 100094. https://doi.org/10.1016/j.atech.2022.100094

Pylianidis, C., Osinga, S., & Athanasiadis, I. N. (2021). Introducing digital twins to agriculture. *Computers and Electronics in Agriculture*, *184*, 105942. https://doi.org/10.1016/j.compag.2020.105942

Rahman, M. M., Ali, M. R., Oliver, M. M. H., Hanif, M. A., Uddin, M. Z., Tamim-Ul-Hasan, Saha, K. K., Islam, M. H., & Moniruzzaman, M. (2021). Farm mechanization in Bangladesh: A review of the status, roles, policy, and potentials. *Journal of Agriculture and Food Research*, *6*, 100225. https://doi.org/10.1016/j.jafr.2021.100225

Rahmani, A. M., Azhir, E., Ali, S., Mohammadi, M., Ahmed, O. H., Yassin Ghafour, M., Hasan Ahmed, S., & Hosseinzadeh, M. (2021). Artificial intelligence approaches and mechanisms for big data analytics: A systematic study. *PeerJ. Computer Science*, *7*, e488. https://doi.org/10.7717/peerj-cs.488

Relf-Eckstein, J. E., Ballantyne, A. T., & Phillips, P. W. B. (2019). Farming Reimagined: A case study of autonomous farm equipment and creating an innovation opportunity space for broadacre smart farming. *NJAS: Wageningen Journal of Life Sciences*, *90–91*(1), 1–23. https://doi.org/10.1016/j.njas.2019.100307

Roberts, D. P., Short, N. M., Sill, J., Lakshman, D. K., Hu, X., & Buser, M. (2021). Precision agriculture and geospatial techniques for sustainable disease control. *Indian Phytopathology*, *74*(2), 287–305. https://doi.org/10.1007/s42360-021-00334-2

Ronaghi, M. H. (2021). A blockchain maturity model in agricultural supply chain. *Information Processing in Agriculture*, *8*(3), 398–408. https://doi.org/10.1016/j.inpa.2020.10.004

Rovira-Más, F., Saiz-Rubio, V., & Cuenca-Cuenca, A. (2021). Sensing architecture for terrestrial crop monitoring: harvesting data as an asset. *Sensors (Basel, Switzerland)*, *21*(9), 3114. https://doi.org/10.3390/s21093114

Ryan, M. (2023). The social and ethical impacts of artificial intelligence in agriculture: Mapping the agricultural AI literature. *AI & Society*, *38*(6), 2473–2485. https://doi.org/10.1007/s00146-021-01377-9

Sahal, R., Alsamhi, S. H., Brown, K. N., O'Shea, D., McCarthy, C., & Guizani, M. (2021). Blockchain-empowered Digital Twins collaboration: smart transportation use case. *Machines*, *9*(9), 193. https://doi.org/10.3390/machines9090193

Sala, S., Ciuffo, B., & Nijkamp, P. (2015). A systemic framework for sustainability assessment. *Ecological Economics*, *119*, 314–325. https://doi.org/10.1016/j.ecolecon.2015.09.015

Searle, R., McBratney, A., Grundy, M., Kidd, D., Malone, B., Arrouays, D., Stockman, U., Zund, P., Wilson, P., Wilford, J., Van Gool, D., Triantafilis, J., Thomas, M., Stower, L., Slater, B., Robinson, N., Ringrose-Voase, A., Padarian, J., Payne, J., . . . Andrews, K. (2021). Digital soil mapping and assessment for Australia and beyond: A propitious future. *Geoderma Regional*, *24*, e00359. https://doi.org/10.1016/j.geodrs.2021.e00359

Segovia, M., & Garcia-Alfaro, J. (2022). Design, modeling and implementation of Digital Twins. *Sensors*, *22*(14), 5396. https://doi.org/10.3390/s22145396

Silva, L., Rodríguez-Sedano, F., Baptista, P., & Coelho, J. P. (2023). The Digital Twin paradigm applied to soil quality assessment: A systematic literature review. *Sensors (Basel, Switzerland)*, *23*(2), 1007. https://doi.org/10.3390/s23021007

Slob, N., & Hurst, W. (2022). Digital Twins and Industry 4.0 technologies for agricultural greenhouses. *Smart Cities*, *5*(3), 1179–1192. https://doi.org/10.3390/smartcities5030059

Soori, M., Arezoo, B., & Dastres, R. (2023). Digital twin for smart manufacturing, A review. *Sustainable Manufacturing and Service Economics*, *2*, 100017. https://doi.org/10.1016/j.smse.2023.100017

Speicher, T. L., & DeFranco, J. F. (2023). Industry 4.0 and Digital Twins. *Computer*, *56*(6), 84–88. https://doi.org/10.1109/MC.2023.3243450

Srivastava, R. K. (2019). Bio-energy production by contribution of effective and suitable microbial system. *Materials Science for Energy Technologies*, *2*(2), 308–318. https://doi.org/10.1016/j.mset.2018.12.007

Sun, T., He, X., & Li, Z. (2023). Digital twin in healthcare: Recent updates and challenges. *Digital Health*, *9*, 20552076221149651. https://doi.org/10.1177/20552076221149651

Symeonaki, E., Maraveas, C., & Arvanitis, K. G. (2024). Recent advances in Digital Twins for agriculture 5.0: Applications and open issues in livestock production systems. *Applied Sciences*, *14*(2), 686. https://doi.org/10.3390/app14020686

Taj, S., Imran, A. S., Kastrati, Z., Daudpota, S. M., Memon, R. A., & Ahmed, J. (2023). IoT-based supply chain management: A systematic literature review. *Internet of Things*, *24*, 100982. https://doi.org/10.1016/j.iot.2023.100982

Tsakiridis, N. L., Samarinas, N., Kalopesa, E., & Zalidis, G. C. (2023). Cognitive soil Digital Twin for monitoring the soil ecosystem: A conceptual framework. *Soil Systems*, *7*(4), 88. https://doi.org/10.3390/soilsystems7040088

Valero, M. R., Hicks, B. J., & Nassehi, A. (2023). A conceptual framework of a Digital-twin for a circular meat supply chain. In K.-Y. Kim, L. Monplaisir, & J. Rickli (Eds.), *Flexible Automation and Intelligent Manufacturing: The Human-Data-Technology Nexus* (pp. 188–196). Springer International Publishing. https://doi.org/10.1007/978-3-031-18326-3_19

Vallée, A. (2023). Digital twin for healthcare systems. *Frontiers in Digital Health, 5*, 1253050. https://doi.org/10.3389/fdgth.2023.1253050

Van Arendonk, J. A. M. (2011). The role of reproductive technologies in breeding schemes for livestock populations in developing countries. *Livestock Science, 136*(1), 29–37. https://doi.org/10.1016/j.livsci.2010.09.004

Van Dinter, R., Tekinerdogan, B., & Catal, C. (2022). Predictive maintenance using digital twins: A systematic literature review. *Information and Software Technology, 151*, 107008. https://doi.org/10.1016/j.infsof.2022.107008

Vatistas, C., Avgoustaki, D. D., & Bartzanas, T. (2022). A systematic literature review on controlled-environment agriculture: How vertical farms and greenhouses can influence the sustainability and footprint of urban microclimate with local food production. *Atmosphere, 13*(8), 1258. https://doi.org/10.3390/atmos13081258

Verboven, P., Defraeye, T., Datta, A. K., & Nicolai, B. (2020). Digital twins of food process operations: The next step for food process models? *Current Opinion in Food Science, 35*, 79–87. https://doi.org/10.1016/j.cofs.2020.03.002

Verdouw, C., Tekinerdogan, B., Beulens, A., & Wolfert, S. (2021). Digital twins in smart farming. *Agricultural Systems, 189*, 103046. https://doi.org/10.1016/j.agsy.2020.103046

Wilkinson, K., Grant, W. P., Green, L. E., Hunter, S., Jeger, M. J., Lowe, P., Medley, G. F., Mills, P., Phillipson, J., Poppy, G. M., & Waage, J. (2011). Infectious diseases of animals and plants: An interdisciplinary approach. *Philosophical Transactions of the Royal Society of London. Series B, Biological Sciences, 366*(1573), 1933–1942. https://doi.org/10.1098/rstb.2010.0415

Zhong, D., Xia, Z., Zhu, Y., & Duan, J. (2023). Overview of predictive maintenance based on digital twin technology. *Heliyon, 9*(4), e14534. https://doi.org/10.1016/j.heliyon.2023.e14534

Zilberman, D., Holland, T., & Trilnick, I. (2018). Agricultural GMOs—What we know and where scientists disagree. *Sustainability, 10*(5), 1514. https://doi.org/10.3390/su10051514

Zimmermann, A., & Rapsomanikis, G. (2023). Trade and sustainable food systems. In J. Von Braun, K. Afsana, L. O. Fresco, & M. H. A. Hassan (Eds.), *Science and Innovations for Food Systems Transformation* (pp. 685–709). Springer International Publishing. https://doi.org/10.1007/978-3-031-15703-5_36

4 AI-Enabled Blockchain for Livestock Tracking and Health Management in Smart Agriculture

Ruchi Jakhmola Mani, Naman Kapoor, Harsh Kashyap, Abhishake Shukla, and Deepshikha Pande Katare

4.1 OVERVIEW OF ARTIFICIAL INTELLIGENCE

Artificial intelligence (AI) represents a rapidly evolving domain within computer science, focused on developing systems that emulate human-like intelligence. This multifaceted field encompasses an array of techniques and methodologies aimed at enabling machines to perform tasks traditionally associated with human cognition, including learning from experience, reasoning, problem-solving, and decision-making. At its core, AI seeks to replicate and automate cognitive functions, empowering machines to analyze data, recognize patterns, and make informed decisions autonomously. Machine learning, a prominent subset of AI, is instrumental in developing algorithms that enable computers to learn from data and enhance their performance over time. Through iterative training on extensive datasets, machine learning algorithms gain proficiency in recognizing patterns, making predictions, and generating insights. Deep learning, an advanced form of machine learning, involves training artificial neural networks with multiple layers to process vast amounts of data and extract complex features (Russell & Norvig, 2016). This approach, inspired by the interconnected structure of neurons in the human brain, excels in tasks such as image recognition, natural language processing, and speech recognition. Furthermore, reinforcement learning algorithms, another key aspect of AI, refine their actions through trial and error, ultimately discovering optimal strategies for achieving predefined objectives. This adaptability makes reinforcement learning well-suited for diverse applications, including robotics, game playing, and autonomous decision-making. In the agricultural sector, AI technologies, including precision livestock farming (PLF) and automated monitoring and control systems, optimize livestock management by leveraging sensors and predictive analytics. These systems enhance productivity, minimize risks, and enable early disease detection. Additionally, AI-driven genetic improvement and breeding programs utilize genomic analysis to enhance livestock quality and productivity through informed selective breeding decisions (Johnson et al., 2021).

4.2 UNDERSTANDING LIVESTOCK TRACKING

Livestock tracking is a critical component of modern agriculture, enabling farmers to monitor the movement, health, and behaviour of their animals effectively. We delve into the importance of livestock tracking in agriculture, examine traditional methods used for tracking livestock, and explore the advantages offered by artificial intelligence in revolutionizing livestock tracking systems.

Tracking plays a pivotal role in ensuring the efficient management and welfare of animals in agricultural settings. One of the primary reasons for tracking livestock is to monitor their location and

 DOI: 10.1201/9781003507390-4

movement, thereby preventing loss or theft and facilitating timely intervention in case of emergencies such as natural disasters or disease outbreaks. By tracking individual animals or herds, farmers can also monitor grazing patterns, optimize pasture management, and prevent overgrazing, thereby promoting sustainable land use practices. Moreover, livestock tracking enables farmers to monitor the health and well-being of their animals closely. By collecting data on parameters such as body temperature, heart rate, and rumination patterns, farmers can detect signs of illness or distress early, allowing for prompt veterinary intervention and treatment. Additionally, tracking animal behaviour can provide insights into social dynamics within herds, helping farmers identify dominant individuals, detect mating behaviours, and manage breeding programs effectively. Furthermore, livestock tracking facilitates traceability and quality assurance throughout the supply chain. By maintaining detailed records of individual animals, including their origin, vaccination history, and feed intake, farmers can ensure compliance with regulatory standards, enhance food safety, and meet consumer demands for transparency and accountability (Ji-Hoon, 2022).

Farmers have utilized a variety of traditional ways, including mechanical and electrical equipment in addition to manual procedures, to track their cattle. One of the simplest forms of livestock tracking involves visual observation, where farmers visually inspect their animals regularly to monitor their whereabouts and condition. While this method can be effective for small-scale operations, it is labour intensive and thus does not provide real-time data on animal location or health status. Another traditional method of livestock tracking is the use of physical markers or identifiers, such as ear tags, collars, or brands (Figure 4.1). These markers contain unique identifiers that enable farmers to identify individual animals and track their movements over time. while the physical markers help identify animals and track their movements but can be lost or damaged, relying on accurate manual record-keeping (TheTagFactory).

In recent years, electronic tracking technologies have gained popularity, offering more advanced solutions for livestock monitoring and management. *Radio frequency identification (RFID)* tags use radio waves to transmit unique identification codes, allowing farmers to track animals remotely using readers or scanners. GPS tracking systems, on the other hand, utilize satellite technology to provide real-time location data, enabling farmers to monitor the movement of animals across large grazing areas (Figure 4.2) (TheTagFactory).

FIGURE 4.1 This image depicts a cow adorned with an ear tag, symbolizing the traditional method of livestock tracking employed in agricultural practices (https://creativecommons.org/publicdomain/zero/1.0/).

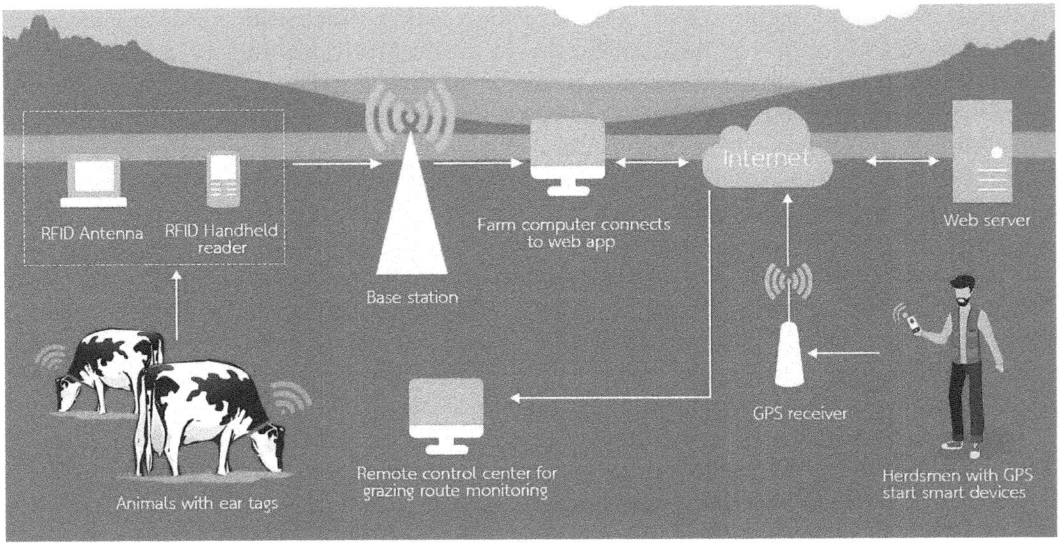

FIGURE 4.2 This image illustrates a cow equipped with an RFID ear tag, representing the modern approach to livestock management through radio frequency identification technology.

4.2.1 Advantages of AI in Livestock Tracking

The integration of AI with sensor arrays and machine learning has revolutionized livestock monitoring, offering automated precision and efficacy.

1. **Enhanced Data Collection and Analysis**: Utilizing AI-driven technology, livestock monitoring systems are able to collect and analyze the extensive datasets regarding animal behaviours, health metrics, and movement patterns instantly. Through the utilization of sophisticated machine learning algorithms, such systems can discern trends, pinpoint irregularities, and produce valuable insights essential for informed decision-making processes.
2. **Improved Accuracy and Precision:** AI algorithms can process complex sensory data from various sources, such as GPS trackers, accelerometers, and environmental sensors, with high accuracy and precision. This enables farmers to monitor individual animals or herds more effectively, identify potential health risk, and optimize management practices accordingly (Sivakumar et al., 2023).
3. **Predictive Analytics for Disease Prevention:** Livestock tracking systems powered by AI leverage predictive analytics to anticipate disease outbreaks by analyzing historical data, environmental conditions, and indicators of animal health. This enables farmers to identify early signs of illness, allowing for the timely implementation of preventive measures like vaccination initiatives and quarantine procedures.
4. **Autonomous Monitoring and Decision-Making**: AI-enabled livestock tracking systems can operate autonomously, continuously monitoring animal behaviour and environmental conditions without human intervention. By leveraging advanced sensors, data analytics, and decision-making algorithms, these systems can adapt to changing circumstances in real time, optimizing resource allocation and improving overall farm efficiency (Garcia & Kim, 2021).
5. **Integration with Precision Livestock Farming**: AI technologies can integrate seamlessly with other precision livestock farming techniques, such as automated feeding systems,

environmental controls, and robotic milking machines. By combining data from multiple sources, AI-driven livestock tracking systems can provide holistic insights into animal welfare, productivity, and sustainability, enabling farmers to make informed decisions and optimize farm performance (Reham et al., 2023).

4.3 EVOLUTION OF LIVESTOCK TRACKING WITH AI

AI techniques have revolutionized livestock tracking and health management, offering innovative solutions to monitor and optimize the well-being of animals in agricultural settings. Several AI techniques employed in livestock tracking and health management like computer vision algorithms analyze visual data from cameras in livestock facilities to track animals based on unique features, detecting abnormalities for prompt intervention. Lameness in cattle can be identified early through gait or posture analysis early detection and treatment. Following AI-driven sensors, data fusion integrates information from multiple sensors to comprehensively monitor livestock health. Wearable sensor devices with AI can detect signs of stress, illness, and reproductive activity, empowering informed animal health management. Other predictive analytics use AI to forecast disease outbreaks, optimize feeding, and plan breeding programs in livestock tracking. Machine learning models trained on historical health records predict disease occurrence based on various factors, enabling proactive preventive measures for farmers. GPS and RFID with AI track livestock in outdoor grazing, offering real-time location data to optimize grazing and resource allocation. AI analyzes GPS data for sustainable land management and predicts optimal grazing locations for improved pasture utilization. Last, natural language processing (NLP) helps analyze text data from veterinary records, research papers, and farmer diaries for insights on animal health and management. Algorithms can extract key information on diseases and treatment recommendations from veterinary reports (Szczepan, 2022). Now the design principles for AI-Blockchain integration in agriculture can be found in figure 4.3.

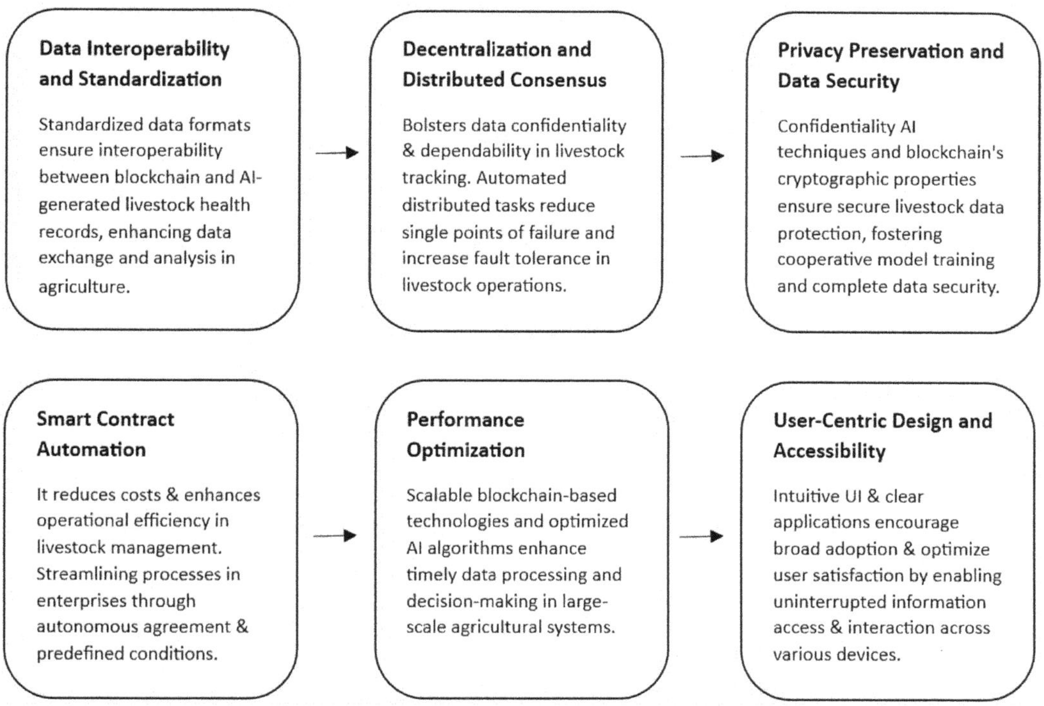

FIGURE 4.3 The design principles for AI–blockchain integration in agriculture.

4.3.1 Benefits and Use Cases of AI-Enabled Livestock Tracking

The agriculture sector as a whole and farmers both stand to gain greatly from AI-enabled livestock tracking. AI has numerous and broad benefits when it comes to cattle tracking systems. A few primary advantages are:

1. **Enhanced Animal Welfare**: AI-enabled livestock tracking systems enable farmers to monitor the health and behaviour of individual animals more effectively, thereby promoting better animal welfare practices. By detecting signs of illness or distress early, farmers can provide timely intervention and treatment, reducing suffering and improving overall animal well-being. For example, real-time monitoring of behavioural patterns and physiological parameters allows farmers to identify potential health issues such as lameness, respiratory distress, or reproductive problems (Gao et al., 2023).
2. **Increased Productivity:** AI-driven livestock tracking systems optimize management practices, such as feeding, breeding, and health management, leading to increased productivity and efficiency in agricultural operations. By leveraging real-time data and predictive analytics, farmers can make informed decisions to maximize the yield and profitability of their livestock enterprises. For instance, they can examine past output data, environmental factors, and genetic data to enhance breeding initiatives and identify animals possessing exceptional qualities.
3. **Improved Resource Management:** AI technologies help farmers optimize resource allocation by monitoring environmental conditions, animal behaviour, and feed intake in real time. By identifying inefficiencies and optimizing inputs such as water, feed, and energy, AI-enabled livestock tracking systems enable farmers to minimize waste and reduce environmental impact, for example, analyze data on grazing patterns and forage availability to optimize pasture management practices, reducing overgrazing and soil erosion while maximizing nutrient utilization.
4. **Disease Detection and Management:** By analyzing data from various sensors, AI algorithms can detect early signs of illness, monitor disease progression, and facilitate timely intervention and treatment. For instance, machine learning models can analyze physiological data to predict the onset of diseases such as mastitis or respiratory infections, enabling farmers to implement preventive measures and reduce the spread and hence minimize economic losses for farmers (Sharma et al., 2023).
5. **Data-Driven Decision Making**: AI-enabled livestock tracking systems generate vast amounts of data on animal health, behaviour, and performance, providing farmers with valuable insights to inform decision-making. By analyzing data trends and patterns, farmers can identify areas for improvement, optimize management practices, and adapt strategies to changing conditions. For example, they can analyze historical production data to identify factors influencing feed conversion ratios and optimize feed formulations accordingly, maximizing efficiency and profitability (Rekiek & Reklaoui, 2023).

Artificial intelligence has transformed livestock monitoring in a number of agricultural domains by providing creative ways to track, manage, and maximize livestock operations. Several important use cases of AI for tracking cattle are outlined; for example, precision livestock farming (PLF) involves the use of advanced technologies, including AI, to monitor and manage individual animals or herds with precision. AI-driven monitoring systems can track animal behaviour, detect deviations from normal patterns, and provide early warnings for potential health issues. By automating routine tasks and providing real-time insights, these systems enable farmers to enhance productivity and minimize resource wastage. A dairy farm can utilize AI-powered sensors and cameras to monitor the behaviour and health of its dairy cows. Computer vision algorithms analyze video footage to detect abnormalities in gait or posture, while sensors measure physiological parameters such as rumination, body temperature, and activity levels, enabling farmers to take timely action to prevent illness and optimize milk production (Nigade et al., 2023).

AI technologies play a crucial role in disease detection and management within livestock farming. By analyzing data from various sensors, AI algorithms can detect early signs of illness, monitor disease progression, and facilitate timely intervention and treatment. Predictive analytics models leverage historical data to forecast disease outbreaks and identify risk factors, enabling farmers to implement preventive measures and minimize economic losses. A poultry farm can use AI-powered sensors to monitor the respiratory health of its chickens. Algorithms analyze temperature, humidity, and airflow sensor data to identify variations that may be related to disease outbreaks or respiratory distress. Early detection minimizes production losses and stops the spread of illness by enabling producers to isolate afflicted birds, put biosecurity measures in place, and provide the necessary treatment (Kaur et al., 2023).

AI technologies also facilitate genetic improvement and breeding programs in livestock farming by analyzing genomic data to identify desirable traits and genetic markers. Predictive modelling AI enables the estimation of breeding values and genetic parameters, facilitating informed decision-making in breeding programs. By incorporating genomic information into breeding strategies, farmers can selectively breed animals with superior traits, stimulating the genetic potential of their livestock population. A pig breeding company can use AI algorithms to analyze genomic data from thousands of pigs to identify genetic markers associated with desirable traits such as growth rate, feed efficiency, and disease resistance. By selecting animals with superior genetic profiles for breeding, the company accelerates the genetic improvement of its pig population, leading to offspring with enhanced performance and productivity characteristics. AI-driven predictive modelling enables the estimation of breeding values and genetic parameters, thereby enhancing the genetic potential of their livestock population and ensuring the long-term sustainability and profitability of their breeding operations.

AI-enabled livestock tracking systems operate by autonomously and continuously monitoring animal behaviour and environmental conditions without human intervention. By using advanced sensors, data analytics, and decision-making algorithms, these systems adapt to changing circumstances in real time, optimizing resource allocation and improving farm efficiency. A sheep farm can employ AI-powered drones equipped with thermal cameras and GPS trackers to monitor the health and location of its sheep flock. These algorithms analyze thermal images to detect signs of illness or distress, while GPS trackers provide real-time location data. If anomalies are detected, the drones autonomously redirect their flight path for further investigation. By autonomously monitoring and managing the flock, the farm improves productivity, reduces labour costs, and enhances animal welfare (Kaur et al., 2023).

4.4 BLOCKCHAIN TECHNOLOGY IN LIVESTOCK TRACKING

Blockchain is a decentralized and distributed ledger system that records transactions in a secure, transparent, and immutable manner. Unlike traditional centralized databases controlled by a single authority, blockchain operates on a network of interconnected nodes, where each node maintains a copy of the entire blockchain. Transactions are validated and added to the blockchain through a consensus mechanism agreed upon by the network participants. Blockchain technology has emerged as a transformative tool and offers numerous use cases in livestock tracking and management, revolutionizing traditional practices with its immutable, transparent, and decentralized nature (Shen et al., 2023).

It enables the creation of tamper-evident records of livestock movement, health data, and supply chain transactions. Each transaction recorded on the blockchain provides a tamper-proof audit trail, enhancing traceability from farm to fork. Utilizing blockchain traceability, consumers can scan a QR code on beef packaging to uncover comprehensive insights into the livestock's farm-to-fork journey. This data encompasses breed specifics, farm origins, feed details, medication history, and organic farming certifications. Every supply chain stage is meticulously documented on the blockchain, guaranteeing transparency and empowering consumers to make informed and confident purchasing choices. This transparency fosters trust among consumers and stakeholders by allowing them to verify the authenticity and quality of livestock products. It also enables real-time monitoring of livestock movements, optimizing supply chain logistics and reducing processing delays. Smart contracts deployed on the blockchain automate contractual agreements and transactions, streamlining processes, reducing administrative overheads, and ensuring timely delivery of livestock products while upholding quality standards and regulatory compliance. It facilitates real-time monitoring of cattle transportation to the processing facility. IoT devices securely track movements and record data on the blockchain, while smart contracts automate agreements among stakeholders, ensuring timely delivery and upheld quality standards (Wang & Yu, 2023).

Real-time monitoring of livestock health is done by the integration of data from IoT sensors, veterinary records, and diagnostic tests. Automated alerts for deviations from normal parameters streamline intervention and disease control. This results in enhanced disease surveillance, swift outbreak responses, improved animal welfare, and reduced economic losses. At a dairy farm, a blockchain-based health monitoring system is used to oversee the well-being of its cows. Real-time data from IoT sensors, including body temperature, heart rate, and milk production, is securely recorded on the blockchain. At any signs of illness, such as a sudden increase in body temperature, the system automatically alerts the farm manager. Immediate alerts allow for timely intervention and prevent the spread of disease (Techane & Gebresenbet, 2023).

Blockchain-based marketplaces and trading platforms connect livestock producers with buyers and sellers in a secure, transparent manner and also provide access to new markets by demonstrating compliance with rigorous standards and certifications. Smart contracts executed on the blockchain automate the negotiation, execution, and settlement of transactions, eliminating the need for intermediaries and reducing transaction costs (Xiong et al., 2020). A blockchain-based livestock trading platform connects cattle farmers with buyers. The platform ensures transparent and authentic information, listing cattle details such as breed, age, and health status. Smart contracts automate transactions, eliminating the need for intermediaries, reducing costs, and ensuring fair prices and timely payments for both parties, thus enhancing market efficiency and profitability for the farmer (Zhang et al., 2020).

4.5 INTEGRATION OF AI AND BLOCKCHAIN FOR LIVESTOCK TRACKING AND HEALTH MANAGEMENT

The combination of blockchain technology and artificial intelligence signals a significant shift in contemporary agriculture, especially in the areas of health management and livestock tracking. A number of industries could be revolutionized by this alliance, with agriculture having the potential to gain tremendously. With this integration, a new era of effectiveness and long-term viability is anticipated to be ushered in by confronting important obstacles, strengthening data security, expanding traceability measures, and optimizing management procedures in the livestock sector.

However, in order to capitalize effectively on this integration, strong principles of design that balance the complementary effects of blockchain and AI must be established. These guiding principles will guarantee the smooth functioning and optimal performance of these state-of-the-art livestock management technology. Establishing explicit design principles is crucial because of many advantages that the integration deals, such as improved data security, increased traceability, and efficient management procedures. These regulations will not only facilitate the seamless integration of blockchain and AI technologies but also pave the way for an agriculturally resilient ecosystem poised for the future, leveraging innovation to propel sustainable growth (Neethirajan, 2023). These principles are the following.

The use of Internet of Things (IoT) devices in conjunction with blockchain technology for intelligent livestock management within the Indian agricultural landscape is an excellent example of how artificial intelligence and blockchain technology can be combined to track and manage the health of livestock. Real-time monitoring and tracking of vital livestock health indicators, position data, and other metrics are achieved through the integration of wearable AI devices and IoT ear tags with radio frequency identification technology. Indian farmers can increase the accuracy and effectiveness of livestock management techniques by strategically utilizing AI algorithms and the safe, transparent ledger features of blockchain technology. The agriculture sector in India is more sustainable and more productive because of this innovative method (Farooq et al., 2020).

4.5.1 Development of Smart Contracts for Livestock Management

Smart contracts, essential to blockchain networks, automate predefined terms and conditions, removing the need for intermediaries. These programs ensure contract negotiations and performance occur smoothly, guaranteeing accuracy, security, and transparency in transactions. In livestock management, smart contracts revolutionize processes like tracking, insurance, and handling, leveraging blockchain technology for automation and efficiency. The use of smart contracts in livestock management is a step toward more transparency and efficiency, which reinforces the critical role that technology plays in improving agricultural operations (Sinha et al., 2020).

Smart contracts create clear and automatic agreement protocols between the parties concerned and are essential to streamlining the livestock insurance negotiating process. Therefore, it lowers the possibility of disagreements and shortens the time it takes to process claims. The precise architecture of smart contracts allows for the real-time collection of vital information on the location, health, and breeding of livestock. This important data is stored safely on blockchain networks, providing stakeholders with clear, unchangeable records. By automating the fulfilment of predefined agreements about transactions and payments, smart contracts transform the livestock sector. These contracts remove the need for intermediaries, streamlining transactional procedures and lowering related expenses in the livestock trade. Last, it can be used to proactively generate alerts and responses that are in line with preset health indicators by connecting with AI-driven monitoring systems and IoT devices. The use of a proactive approach guarantees prompt resolution of livestock health concerns, hence mitigating the likelihood of disease outbreaks or health consequences among the livestock population (Shen et al., 2023).

Major steps are involved in the development of smart contracts for the livestock management start from the identification of requirement till to security and compliance. These steps are as follows:

1. **Identification of Requirements:** Determine the particular conditions, such as contract terms, transaction kinds, participant responsibilities, and data inputs for automation, that must be met for smart contracts for livestock management (Pranto et al., 2021).
2. **Design and Architecture:** Outline the code structure, conditional statements, event triggers, and data storage techniques in the design and architecture of the smart contract. Think about IoT integration, scalability, and interoperability.
3. **Coding and Testing:** Use Chaincode for Hyperledger Fabric or Solidity for Ethereum to create smart contracts. Ensure that functionality, security, and dependability are met by doing thorough testing using vulnerability assessments and scenario simulations (Buterin, 2014).
4. **Deployment and Integration:** Install tested smart contracts on the blockchain network of your choice. Integrate for smooth data flow and transaction automation with current systems, Internet of Things devices, and data sources.
5. **Monitoring and Maintenance:** Deployed smart contracts should be maintained and monitored to guarantee best practices and adherence to changing legal requirements. Manage complaints, handle exceptions, and apply updates or improvements as needed (Al-Turjman & Khalil, 2020).

6. **Security and Compliance:** To protect data and adhere to regulations, implement security measures including audit trails, access control, and encryption. Incorporate ethical issues to guarantee that animals are treated fairly and avoid algorithmic prejudices.

4.5.2 Data Privacy and Security Considerations in AI–Blockchain Integration

Blockchain technology and artificial intelligence have enormous potential for use in industries like agriculture, but data security and privacy must be seriously considered. Effective safety safeguards must be implemented to prevent illegal accessibility, data theft, and privacy violations, especially in agriculture, where private data such as supply chain details and livestock health records are significant. The integration of health management and livestock tracking highlights benefits as well as obstacles in managing sensitive data.

1. **Encryption and Data Anonymization:** In AI–blockchain integrated systems, the use of strong encryption algorithms for data in transit and at rest is essential to protecting sensitive data from manipulation and unwanted access. The highest level of security and secrecy for data stored on the blockchain is guaranteed. Moreover, anonymizing sensitive data protects individual security and that of livestock owners by preventing the release of private information from transaction records (Li et al., 2020).
2. **Identity Management and Access Control**: Managing access to data pertaining to livestock requires incorporating secure identity management and access control methods within AI–blockchain ecosystems. These mechanisms dictate who can read, alter, and interact with the data stored on the blockchain. Sensitive data exposure is significantly reduced by limiting illegal access by the use of cryptographic keys, permissions-based access protocols, role-based access control (RBAC), and digital identity verification. By establishing only those with explicit authorization can access particular data according to their responsibilities and permissions.
3. **Secure Data Sharing Protocols:** Authorized entities can maintain data privacy while accessing relevant livestock data by implementing secure data sharing protocols within AI–blockchain networks. Secure data sharing is made possible while maintaining the privacy of sensitive data by utilizing strategies like zero-knowledge proofs and decentralized storage methods (Assiri & Humayun, 2023).
4. **Audit Trails and Transparency:** The blockchain maintains comprehensive audit trails, which significantly improve accountability and transparency. Assuring accountability and traceability of data exchanges within AI–blockchain integrated systems, the immutability of data transactions permits the establishment of visible and auditable records. As a result, interested parties are able to track the origins of data, spot illegal changes, and penalize those who commit privacy or security violations (Zhang et al., 2020).
5. **Consensus Mechanisms:** Maintaining data security and integrity in blockchain networks requires careful consideration of the consensus methods to be used. In order to successfully combat malicious efforts at data manipulation, consensus algorithms such as Proof of Work (PoW) or Proof of Stake (PoS) are essential for confirming and adding authorized transactions to the blockchain. By establishing a strong framework within which network users must agree on the blockchain's current state, these techniques guarantee that only legitimate transactions are approved, boosting the system's general dependability.
6. **Regulatory Compliance:** Ensuring adherence to rigorous data protection rules and industry standards, such as the General Data Protection Regulation (GDPR) and Health Insurance Portability and Accountability Act (HIPAA) is crucial for the smooth operation of AI and blockchain in cattle management. Respecting data protection laws and agriculture rules fosters confidence and responsibility in data management while reducing legal and reputational concerns (Zhou et al., 2023).

4.6 REAL-WORLD APPLICATIONS AND PRACTICAL CASE STUDIES

CASE STUDY 1: AI–BLOCKCHAIN INTEGRATION IN LIVESTOCK MONITORING SYSTEMS

INTRODUCTION AND BACKGROUND:

Carlos Callejero (CEO), Ignacio Gomez-Maqueda (CTO), and Rubén Blanco-Carrera (commercial director) founded DigitalAnimal in 2016 in response to the findings of inefficiencies in the conventional livestock management methods that are common in India. Driven by a goal to bring these approaches up to date and maximize their effectiveness, the organization set out on a bold quest to transform livestock management in the agricultural sector by introducing state-of-the-art technology. DigitalAnimal became a groundbreaking force in agricultural innovation with the primary goal of improving the welfare, production, and general health of livestock. A progressive strategy to change livestock management methods in India was demonstrated by their initiative, which was built on the integration of blockchain technology and artificial intelligence (Smith & Patel, 2022).

IMPLEMENTATION:

DigitalAnimal introduced an advanced monitoring solution tailored for beef cattle, enhanced by a sophisticated "Powered by FIWARE" architecture. This cutting-edge solution transforms cattle management techniques by flawlessly fusing blockchain technology, AI algorithms, sentinel satellite imaging, and IoT wearables.

Internet of Things devices are closely connected to the cloud, where a machine learning algorithm analyzes real-time sensor data to reliably anticipate the health state of livestock. Farmers can gain detailed insights into the health and well-being of their livestock through ongoing data collecting. In order to make data visualization easier, a user-friendly mobile app was also developed, affording farmers an easy way to track and assess the information (Smith & Patel, 2022).

Blockchain technology was strategically integrated into DigitalAnimal's monitoring systems to guarantee data security and transparency. The company has built a decentralized platform for securely capturing and validating livestock data by putting in place a permissioned blockchain network. Every data transaction, ranging from health records to location updates, undergoes cryptographic hashing and is stored on the blockchain, guaranteeing tamper-proof and immutable records. This integration not only improves data integrity but also builds confidence through open and safe data management procedures (Kumar & Singh, 2021).

CHALLENGES FACED BY THE ORGANIZATION:

They ran into a number of challenges throughout the implementation stage, leading to innovations being required. The interoperability between AI and blockchain systems posed a significant hurdle, demanding meticulous integration efforts to enable smooth data flow. Additional challenges included low profitability, a slow uptake of digital solutions, and limited connectivity in remote locations. The company's commitment to addressing increasing concerns regarding animal welfare and the environmental impact of cattle also surfaced as a substantial challenge.

SOLUTION:

To get over barriers, they worked together with blockchain and AI specialists to create unique APIs and data interfaces. These customized solutions enabled smooth connection between AI-enabled devices and the blockchain network, ensuring safe, instantaneous data transfers and effectively resolving interoperability issues. The business refined consensus methods inside its blockchain network and used creative sharding strategies to address scalability issues. Their commitment to improving farm production, sustainability, and animal welfare was evident when they used partitioning algorithms to generate manageable blockchain fragments and effective Proof of Stake consensus processes. A comprehensive strategy for improving livestock management methods is demonstrated by the development of IoT wearables, such as collars, ear tags, and gateways, which have increased data collecting possibilities across animal monitoring, farm operations, and even transportation (Siham & Reklaoui, 2023).

CASE STUDY 2: REAL-TIME HEALTH MANAGEMENT USING AI AND BLOCKCHAIN

INTRODUCTION AND BACKGROUND:

Forward-thinking businesses in India's dynamic agriculture sector are adopting cutting-edge technology like blockchain and artificial intelligence to manage livestock health in a new way. As a pioneer in the application of AI and blockchain technology for real-time health management, AgriData Solutions Pvt. Ltd. is leading the way in this revolutionary endeavour. In their groundbreaking efforts to transform real-time livestock health monitoring, AgriData Solutions has employed unique approaches and implementation tactics. This case study illuminates these strategies, difficulties, and solutions.

It stands as a pioneering force in agricultural technology solutions, recognizing the imperative need for advanced health management solutions to address the multifaceted challenges encountered by livestock farmers in India. With the unwavering goal of improving animal welfare and production, they started a trailblazing adventure to create innovative solutions that paved the way for previously unheard-of developments in the field of livestock health management (Oliveira & Silva, 2023).

IMPLEMENTATION:

Through the seamless integration of blockchain technology and AI-powered health monitoring, AgriData Solution revolutionized the management of livestock health. They developed an innovative real-time health management system that tracks the physical attributes and behavioural patterns of livestock by utilizing cutting-edge computer vision and machine learning techniques. The company created a sophisticated framework for real-time monitoring of critical health metrics in livestock enclosures by strategically placing cameras and sensors.

In order to ensure data integrity and transparency, the gathered data was safely saved and encrypted on a blockchain network. AgriData Solution, a permissioned blockchain network operator, created a decentralized platform for safely capturing and authenticating health data. All health-related transactions, including food evaluations, anomaly detection, and behavioural analytics, were cryptographically hashed and recorded on the blockchain to provide unchangeable and untouchable data. This all-encompassing strategy demonstrated the company's dedication to improving animal welfare by utilizing cutting-edge technological solutions (Dahiphale et al., 2023).

CHALLENGES FACED BY THE ORGANIZATION:

They overcame several obstacles throughout the installation process. These included things such as ensuring data accuracy, making the process of integrating blockchain and AI systems more efficient, and navigating the complex landscape of regulatory standards governing data privacy and security. In addition, addressing scalability became a critical issue as the business aimed to extend its livestock monitoring capabilities to bigger population. Creating novel approaches to maximize resource usage and increase system throughput while maintaining performance and efficiency was necessary to meet this challenge.

SOLUTION:

They partnered with blockchain and artificial intelligence expertise. The creation of unique algorithms and protocols suited for smooth integration as a result of this strategic partnership improved the synergy between blockchain and AI technology. Also, the company improved the quality and dependability of their data gathering procedures, which strengthened the efficacy of their real-time health monitoring system.

Also, they developed cutting-edge encryption technologies that adhere to rigorous regulatory requirements for data privacy and security, augmenting the security of data kept on the blockchain network. By integrating extensive consensus processes and dynamic resource allocation strategies including partitioning and parallel processing into their blockchain architecture, AgriData Solutions managed to increase scalability and efficiency. This improvement serves as an expression of their dedication to leading the way in agricultural technological developments (Morales & Villalobos, 2023).

4.7 FUTURE DIRECTIONS AND CHALLENGES

4.7.1 Emerging Trends in AI–Blockchain Integration for Smart Agriculture

Smart agriculture, which combines blockchain technology with artificial intelligence, has sparked a revolution in the field of agriculture by introducing innovative solutions that improve data-driven decision-making, sustainability, and efficiency. With its potential to transform the agricultural industry by utilizing cutting-edge technology to provide meaningful results, this integration represents a substantial development in farming techniques. Prominent trends in this convergence consist of the following.

4.7.1.1 Predictive Analytics for Crop Management

An innovative method for streamlining agricultural operations is crop management using AI-driven predictive analytics. These analytics enable farmers significant insights into factors influencing agricultural production by examining past patterns in crop performance and projecting the future. Predictive analytics facilitates proactive decision-making and accurate resource allocation by utilizing an abundance of data related to weather patterns, soil conditions, and past crop performance. Farmers have the capability to make data-driven decisions in real time, which eventually results in improved crop management techniques and higher yields by predicting future trends based on past crop performance.

The cornerstone of predictive analytics in agriculture lies in analyzing historical crop performance trends. Integrating these trends into predictive analytics frameworks offers a holistic view of the factors that have historically impacted crop outcomes. Advanced machine learning algorithms use historical crop performance data to power predictive analytics models that estimate crop yields and anticipate insect outbreaks. These models provide important insights into the factors that determine agricultural production. These forecasts provide farmers with accurate and timely projections by taking into account a wide range of variables, including market dynamics, crop types, insect occurrences, weather patterns, soil moisture levels, and input prices (Karimah et al., 2023).

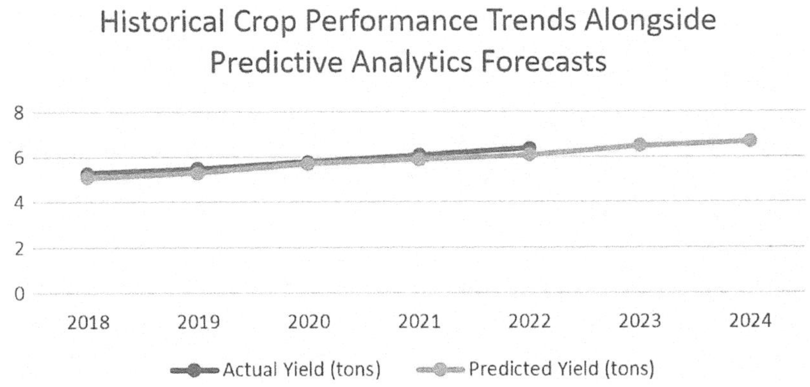

FIGURE 4.4 Crop yield trends and predictive analytics analysis.

By using this information, farmers may proactively modify their management tactics to take advantage of opportunities, control risks, and maximize yields while minimizing losses. Thus, real-time informed decisions are facilitated, enhancing productivity and resource allocation. Moreover, preserving data integrity in predictive analytics models requires the incorporation of blockchain technology. Reliability and trust are ensured in decision-making processes by safely storing agricultural data on an immutable ledger. Blockchain facilitates transparent data exchange between agricultural stakeholders, hence promoting cooperation and confidence within the agricultural ecosystem (Chataut et al., 2023).

Figure 4.4 shows the difference between the projected and actual crop yields in tons for the years 2018 through 2024. It contains projections for 2023 and 2024 in addition to historical actual yields from 2018 to 2022. This data shows the performance trend of agricultural yields over the designated years as well as the precision of predictive analytics projections in predicting crop yields in the future.

Based on prior statistics, agricultural output has consistently increased over the last years, with a minor dip in 2020. With predicted yields of 6.5 tons in 2023 and 6.7 tons in 2024, the predictive analytics projections point to the continuance of this rising trend. Remember that this is sample data and that actual crop performance and predictive analytics estimates might differ based on a range of variables, including weather, crop management techniques, and technology developments (Dhanaraju et al., 2022; Rane et al., 2023).

4.7.1.2 AIoT Applications for Precision Agriculture

A new era of sophisticated monitoring and management systems is being ushered in by the convergence of artificial intelligence and the Internet of Things into the AIoT, which is transforming precision agriculture. By providing farmers with immediate, real-time insights on critical factors like plant health, irrigation requirements, and pest management, these innovative programs are revolutionizing the way agricultural operations are carried out. Through the use of this data-driven methodology, AIoT plays a critical role in maximizing crop growth and putting effective resource conservation plans into place, therefore promoting efficient and sustainable practices in the agricultural industry (Pranto et al., 2021).

The incorporation of AIoT solutions is a key factor in the conservation of agricultural resources, as it provides farmers with up-to-date information on crop health and facilitates the early identification and prompt resolution of issues. For instance, AIoT-based irrigation systems use weather predictions and soil moisture level analysis to provide targeted water applications, reducing water waste and encouraging effective water use. These solutions, which make use of IoT sensors and AI algorithms, guarantee ongoing plant condition monitoring. This allows for proactive pest management

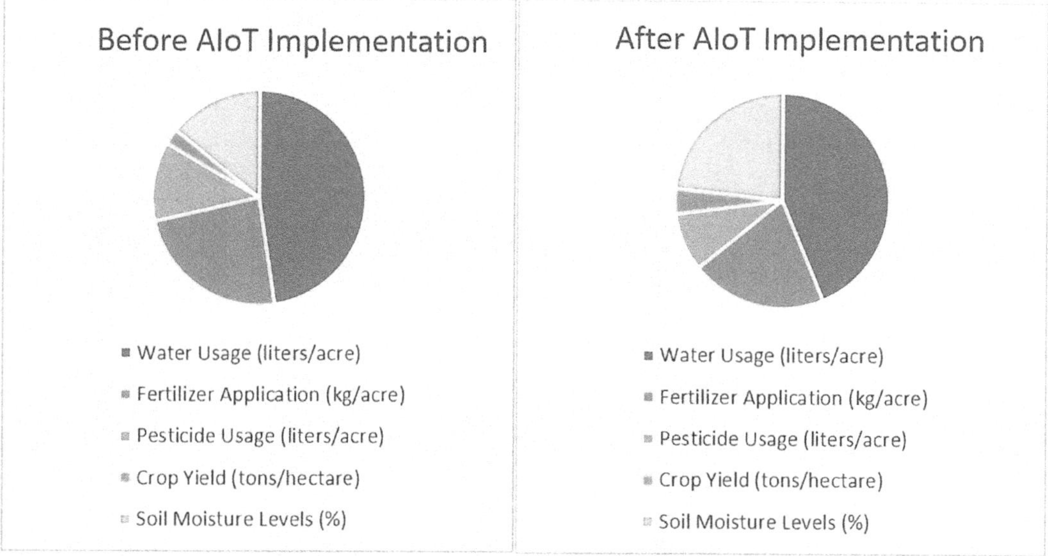

FIGURE 4.5 Visualizing the evolution of precision irrigation system: before and after AIoT implementation.

systems that use AI to fight possible pest threats and preserve the vibrancy of plants. In addition to encouraging crop development through timely care, the proactive aspect of AIoT solutions powered by simultaneous insights also helps to improved yields, sustainable agriculture practices, and environmental conservation (Ankur et al., 2023).

The use of AIoT in agriculture has led to significant achievements in resource efficiency and crop yield enhancement. By integrating precision irrigation systems, AIoT contributed to a 25% reduction in water consumption, optimized fertilizer application by reduction of 15 units of fertilizer usage, and reduced pesticide usage by 10 units. AIoT-enabled soil moisture levels increased by 10%, leading to a 1.5-ton increase in crop yield. Although the exact results may differ based on variables like crop variety, weather, and the particular AIoT solutions used, it's vital to remember that these statistics show how AIoT may support effective and sustainable agricultural methods (Figure 4.5). (Demestichas et al., 2020; Adil et al., 2022).

4.7.1.3 Smart Contract Applications for Agricultural Transactions

Smart contracts driven by blockchain technology have completely changed the agriculture commerce, including crop sales, land leasing agreements, and insurance payouts. It automates and protects these transactions through the use of decentralized ledger technology, greatly reducing operational barriers and increasing stakeholder confidence.

Agricultural transactions were complicated before smart contracts became widely used. They involved a lot of intermediaries, lengthy paper-based procedures, and manual verifications. These cumbersome procedures frequently caused miscommunications, disagreements, delays, and a lack of transparency, which in turn caused a significant degree of mistrust between the parties. As an example, tension between landowners and renters may arise from arguments over conditions and missed payments in land leasing agreements. Contractual conditions were difficult to enforce in the absence of automatic execution methods, which opened up the possibilities of fraud and misunderstanding. They have changed as a result of the incorporation of blockchain-based smart contracts, which incorporate predefined rules and conditions into self-executing code, eliminating intermediary reliance and decreasing the likelihood of human errors or manipulation. Blockchain technology's intrinsic transparency and immutability ensure that each party has a single, unchangeable

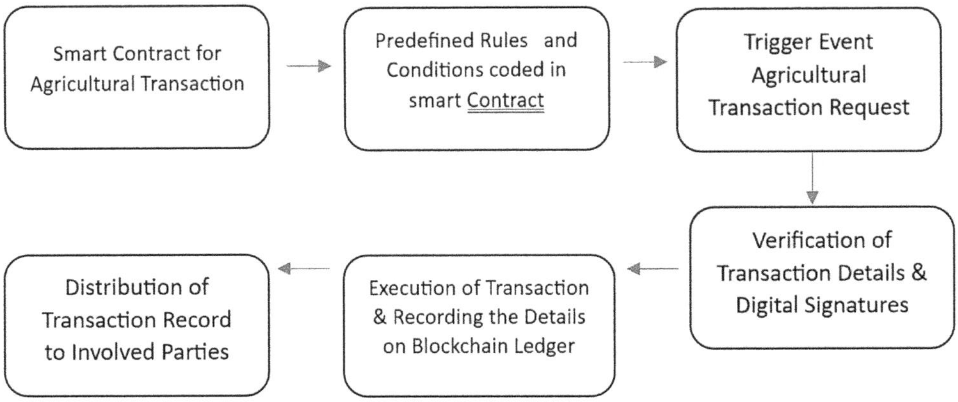

FIGURE 4.6 Automated process flow of a smart contract for agricultural transactions, showing reduction of operational friction.

copy of the contract, significantly reducing the likelihood of disagreements and misinterpretations (Zwitter & Hazenberg, 2020).

Through the assurance that transactions take place out precisely as agreed upon, free from the risk of fraud or unauthorized changes, this automated procedure fosters confidence between parties.

Illustrating a seamless smart contract execution for agricultural transactions, pre-set rules trigger the process upon request. Verification authenticates digital signatures, ensuring accuracy. The smart contract executes the transaction, recording it securely on a blockchain ledger. Parties gain access to the transaction data, streamlining operations and enhancing efficiency in agricultural transactions (Figure 4.6) (Bibri et al., 2024).

4.8 CONCLUSION

In conclusion, the fusion of blockchain technology with artificial intelligence has revolutionized traditional agricultural methods and brought in a new era of sustainability, efficiency, and accuracy in livestock management. It demonstrates how AI is essential to tracking cattle and managing their health, transforming operations with automated monitoring, illness prevention, and real-time decision-making. Notably, the proactive application of AI-driven natural language processing, computer vision, and predictive analytics improves productivity, welfare, and disease surveillance in livestock management, and blockchain technology offers safe and transparent data management by guaranteeing data integrity, transparency, and automation through smart contracts.

Modern organizations have effectively improved real-time agricultural decision-making and animal health monitoring through the use of blockchain and AI. Precision agriculture is undergoing a transformation thanks to the use of AI-driven predictive analytics and Internet of Things technologies, which enhance resource management and boost crop yields. Agricultural transactions are becoming more efficient and transparent thanks to smart contracts. With an emphasis on improving productivity, resource economy, and transparency using cutting-edge technologies like artificial intelligence, the Internet of Things, and smart contracts, the sector is moving toward decision-making based on data and sustainable practices. The potential for tackling the many complex issues facing the agriculture industry, such as supply chain management, crop cultivation, animal management, and sustainable resource management, is enormous when AI and blockchain join. It is clear that this combination of cutting-edge technologies has the potential to transform livestock management, support resilient agricultural ecosystems, foster profitable development, benefit our livestock's well-being, and increase agricultural productivity. We just need to responsibly navigate the challenges and take advantage of the many opportunities presented by blockchain and artificial intelligence.

REFERENCES

Adil, El, Mane, Younes Chihab, Khalid Tatane, and Redouan Korchiyne. "Agriculture Supply Chain Management Based on Blockchain Architecture and Smart Contracts." *Applied Computational Intelligence and Soft Computing* (2022). https://doi.org/10.1155/2022/8011525.

Al-Turjman, Fadi, and Ali Khalil. "Blockchain and Smart Contracts for Data Management in Agriculture: A Survey." *Sustainability* 12, no. 16 (2020): 6585.

Ankur, M. Goyal, Hoshiyar Singh, Kanyal, and Bhasker Sharma. "Analysis of IoT and Blockchain Technology for Agricultural Food Supply Chain Transactions." *International Journal on Recent and Innovation Trends in Computing and Communication* (2023). https://doi.org/10.17762/ijritcc.v11i3.6342.

Assiri, Mohammed, and M. Humayun. "A Blockchain-Enabled Framework for Improving the Software Audit Process." *Applied Sciences* (2023). https://doi.org/10.3390/app13063437.

Bibri, Simon Elias, John Krogstie, A. Alahi, and Ali Kaboli. "Smarter Eco-Cities and Their Leading-Edge Artificial Intelligence of Things Solutions for Environmental Sustainability: A Comprehensive Systematic Review." *Environmental Science and Ecotechnology* 19 (2024): 100330. https://doi.org/10.1016/j.ese.2023.100330.

Buterin, Vitalik. "A Next-Generation Smart Contract and Decentralized Application Platform." *White Paper*. Ethereum, 2014.

Chataut, Ramesh, Anousone Phoummalayvane, and Robert Akl. "Unleashing the Power of IoT: A Comprehensive Review of IoT Applications and Future Prospects in Healthcare, Agriculture, Smart Homes, Smart Cities, and Industry 4.0." *Sensors (Basel)* 23, no. 16 (2023): 7194. https://doi.org/10.3390/s23167194.

Dahiphale, Devendra, Pratik Shinde, Koninika Patil, and Vijay Dahiphale. "Smart Farming: Crop Recommendation Using Machine Learning with Challenges and Future Ideas." *Authorea Preprints* (2023). https://doi.org/10.36227/techrxiv.23504496.v1.

Demestichas, Konstantinos, Nikolaos Peppes, Theodora Alexakis, and Eleni Adamopoulou. "Blockchain in Agriculture Traceability Systems: A Review." *Applied Sciences* 10, no. 12 (2020): 4113. https://doi.org/10.3390/app10124113.

Dhanaraju, M., P. Chenniappan, K. Ramalingam, S. Pazhanivelan, and R. Kaliaperumal. "Smart Farming: Internet of Things (IoT)-Based Sustainable Agriculture." *Agriculture* 12, no. 10 (2022): 1745. https://doi.org/10.3390/agriculture12101745.

Farooq, M. S., S. Riaz, A. Abid, T. Umer, and Y. B. Zikria. "Role of IoT Technology in Agriculture: A Systematic Literature Review." *Electronics* 9, no. 2 (2020): 319. https://doi.org/10.3390/electronics9020319.

Gao, Jerry, Charanjit Kaur Bambrah, Nidhi Parihar, Sharvaree Kshirsagar, Sruthi Mallarapu, Hailong Yu, Jane Wu, and Yunyun Yang. "Building Smart Surveillance Solutions for Livestock Farms." (2023). https://doi.org/10.21203/rs.3.rs-2685214/v1.

Garcia, Luis, and Yuna Kim. "Application of AI and Data Analytics in Precision Livestock Farming: A Comprehensive Review." *Livestock Science* 7, no. 3 (2021): 112–125.

Ji-Hoon, Ahn. "History of Wildlife Tracking Technologies." *Oxford Research Encyclopedia of Environmental Science* (2022). https://doi.org/10.1093/acrefore/9780199389414.013.725.

Johnson, Robert, Laura White, and Maria Garcia. "Innovative Applications of AI in Livestock Management." *Agricultural Innovations Journal* 9, no. 2 (2021): 88–101.

Karimah, Hasyiya, Adli Muhammad, Akmal Remli, Khairul Nizar Syazwan, Wan Salihin Wong, Alfonso González-Briones, Juan M. Corchado, and Mohd Saberi Mohamad. "Recent Advancements and Challenges of AIoT Application in Smart Agriculture: A Review." *Sensors* (2023). https://doi.org/10.3390/s23073752.

Kaur, Upinder, Victor Marco Rocha Malacco, Huiwen Bai, Tanner P. Price, Arunashish Datta, Lei Xin, Shreyas Sen, Robert A. Nawrocki, George T.-C. Chiu, Shreyas Sundaram, Byung-Cheol Min, K. M. Daniels, Robin R. White, Shawn S. Donkin, Luiz F. Brito, and Richard M. Voyles. "Invited Review: Integration of Technologies and Systems for Precision Animal Agriculture—A Case Study on Precision Dairy Farming." *Journal of Animal Science* (2023). https://doi.org/10.1093/jas/skad206.

Kumar, Amit, and Sandeep Singh. "Blockchain-based Solutions for Agricultural Data Management: Opportunities and Challenges." *International Journal of Information Management* 45 (2021): 102–118.

Li, Changyou, Lin Xu, and Zhaoxiang Zhang. "Blockchain-Enabled Secure Data Sharing with Fine-Grained Access Control in Agriculture." *IEEE Internet of Things Journal* 7, no. 11 (2020): 11627–11635.

"Livestock Management—TheTagFactory." *TheTagFactory*. Accessed April 26, 2024. https://thetagfactory.com/livestock-management.

Morales, Alejandro, and Francisco J. Villalobos. "Using Machine Learning for Crop Yield Prediction in the Past or the Future." *Frontiers in Plant Science* (2023). https://doi.org/10.3389/fpls.2023.1128388.

Neethirajan, Suresh. "Artificial Intelligence and Sensor Technologies in Dairy Livestock Export: Charting a Digital Transformation." *Sensors* 23, no. 16 (2023): 7045. https://doi.org/10.20944/preprints202307.0473.v1.

Nigade, Akshay Suresh, Madhuri Shelke, Swapnil Gavhane, Shweta Mashale, and Bhakti Ratnaparkhi. "Review Paper on IoT based Cattle Health Monitoring System." In *2023 IEEE 8th International Conference for Convergence in Technology (I2CT)*, pp. 1–4. IEEE, 2023. https://doi.org/10.1109/i2ct57861.2023.10126158.

Oliveira, R. C. d., and R. D. d. S. e. Silva. "Artificial Intelligence in Agriculture: Benefits, Challenges, and Trends." *Applied Sciences* 13, no. 7405 (2023). https://doi.org/10.3390/app13137405.

Pranto, Tasnim H., A. A. Noman, A. Mahmud, and A. B. Haque. "Blockchain and Smart Contract for IoT Enabled Smart Agriculture." *PeerJ Computer Science* 7 (2021): e407. https://doi.org/10.7717/peerj-cs.407.

Rane, Kantilal, K. A. Narendra Kumar, and Nitesh Behare. "An IoT-Based System for Managing and Monitoring Smart Irrigation through Mobile Integration." *Journal of Machine and Computing* (2023). https://doi.org/10.53759/7669/jmc202303018.

Reham, A., Hosny, Nayerah AlAtfeehy, May F. Abdelaty. "Application Of Artificial Intelligence In The Management Of Poultry Farms And Combating Antimicrobial Resistance." *Egyptian Journal of Animal Health* (Print), (2023). https://doi.org/10.21608/ejah.2023.302769.

Rekiek, Siham, and Kamal Reklaoui. "Poultry-Edge-AI-IoT System for Real-Time Monitoring and Predicting by Using Artificial Intelligence." *International Journal of Interactive Mobile Technologies* (2023). https://doi.org/10.3991/ijim.v17i12.38095.

Russell, Stuart, and Peter Norvig. *Artificial Intelligence: A Modern Approach.* 3rd ed. Pearson, 2016.

Sharma, Sandhya, Kazuhiko Sato, and Bishnu Prasad Gautam. "A Methodological Literature Review of Acoustic Wildlife Monitoring Using Artificial Intelligence Tools and Techniques." *Sustainability* (2023). https://doi.org/10.3390/su15097128.

Shen, Li, Zhenyu Zhang, Yiming Zhou, and Yu Xu. "Applying Blockchain Technology and the Internet of Things to Improve the Data Reliability for Livestock Insurance." *Sensors (Basel)* 23, no. 14 (2023): 6290. https://doi.org/10.3390/s23146290.

Sinha, Pankaj, Surender Saini, and Jaswinder Pal Singh. "Blockchain and Internet of Things (IoT) Integration in Agriculture: A Survey." *IEEE Internet of Things Journal* 7, no. 12 (2020): 11172–11187.

Sivakumar, S. A., B. Maruthi Shankar, M. Mahaboob, N. Adhish, R. Dineshkumar, and N. Rahul. "Sustainable Farming and Customized Livestock Management Using Internet of Things." In *Proceedings of Third International Conference on Sustainable Expert Systems: ICSES 2022*, pp. 543–552. Singapore: Springer Nature Singapore, 2023. https://doi.org/10.1007/978-981-19-7874-6_40.

Smith, John, and Rakesh Patel. "Advances in Livestock Monitoring Technologies: A Review." *Journal of Agricultural Engineering* 23, no. 1 (2022): 45–60.

Szczepan, Figiel. "Development of Artificial Intelligence and Potential Impact of Its Applications in Agriculture on Labor Use and Productivity." *Zagadnienia Ekonomiki Rolnej* (2022). https://doi.org/10.30858/zer/153583.

Techane, Bosona, and Girma Gebresenbet. "The Role of Blockchain Technology in Promoting Traceability Systems in Agri-Food Production and Supply Chains." *Sensors* (2023). https://doi.org/10.3390/s23115342.

Wang, Dong, and Ao Yu. "Supply Chain Resources and Economic Security Based on Artificial Intelligence and Blockchain Multi-Channel Technology." *International Journal of Information Technologies and Systems Approach* (2023). https://doi.org/10.4018/ijitsa.322385.

Xiong, Hu, Thomas Dalhaus, Peter Wang, and Jianguo Huang. "Blockchain Technology for Agriculture: Applications and Rationale." *Frontiers in Blockchain* 3 (2020): 7. https://doi.org/10.3389/fbloc.2020.00007.

Zhang, Yanbo, Lihua Hu, and Qi Chen. "Blockchain-Based Traceability System for Food Safety in Agriculture." *IEEE Transactions on Industrial Informatics* 16, no. 6 (2020): 4121–4130.

Zhang, Yanbo, Shuzhi Wang, and Xiaowei Wang. "A Blockchain-Based Traceability System for Livestock Supply Chain: Design Principles and Implementation." *Computers and Electronics in Agriculture* 175 (2020): 105551.

Zhou, Sisi, Kuan Li, Lijun Xiao, Jiahong Cai, Wei Liang, and Arcangelo Castiglione. "A Systematic Review of Consensus Mechanisms in Blockchain." *Mathematics* (2023). https://doi.org/10.3390/math11102248.

Zwitter, Andrej, and Jelte Hazenberg. "Decentralized Network Governance: Blockchain Technology and the Future of Regulation." *Frontiers in Blockchain* 3 (2020): 12. https://doi.org/10.3389/fbloc.2020.00012.

5 A Blockchain-Enabled Framework to Enhance Cybersecurity Challenges in Smart Agriculture

Joseph Bamidele Awotunde, Abidemi Emmanue Adeniyi, Peace Busola Falola, and Temitayo Caroline Adeniran

5.1 INTRODUCTION

By 2050, there will be 9 billion people on the planet, meaning that 70% more food would need to be produced in order to feed them (Fróna et al. 2019; Forslund et al. 2023). In addition to growing resilience to crop pests and diseases, farmers are dealing with a number of business difficulties that have an impact on farm productivity, including significant shifts brought on by climate change (Zulfikri et al. 2024; Chakauya et al. 2023). As a result of growth in food prices, over 40 million people fell into poverty in 2010 (Ivanic et al. 2012; Headey & Hirvonen 2023; Bedasa & Bedemo 2023), which emphasizes the need for more efficient agricultural initiatives (Adisa et al. 2024). Researchers studying agri-food in this setting are developing strategies to increase crop productivity while lowering yield risk (Chams et al. 2020). The ICT-based revolution has significantly enhanced agricultural productivity through its benefits. Nonetheless, it is evident that a new revolution is required to support smart agriculture to reduce or aid the aforementioned challenges. Services that are fact based, supported by scientific knowledge, and that provide farmers with affordable yet insightful guidance are desperately needed. Therefore, it is anticipated that smart agriculture will decrease hazards for farmers, protect the environment, raise production (both quantitatively and qualitatively), and lower production costs (Rogotis & Marianos 2021).

Precision agriculture, frequently referred to as smart agriculture, combines traditional farming methods with advanced technologies like the Internet of Things (IoT), drones, artificial intelligence (AI), and data analytics to maximize efficiency, sustainability, and productivity. Farmers can make data-driven decisions about irrigation, fertilization and crop management by using sensors to monitor soil moisture, nutrition levels, and climate conditions, thereby reducing resource wastage and increasing yields. In addition, drones that are fitted with cameras plus sensors can offer real-time airborne pictures which help in determining the health status of crops and discovering pests, leading to early actions. Through the application of smart agriculture techniques, farmers can achieve more precise resource allocation, mitigate environmental impacts, and enhance overall farm management practices.

Smart agriculture is a burgeoning field that leverages big data, AI, and connected data to bring the advantages of digitalization to the agricultural sector. Smart agriculture is more about leveraging technology to create and use knowledge than just integrating technological advances into agriculture. Information technology should be utilized to enable agricultural machines and devices to process, analyze, and make decisions semi-automatically (Márta et al. 2022). The rise of big data technologies, the IoT, satellite observation, linked data, and AI is revolutionizing all stages of smart agriculture especially supply chain (Kumar et al. 2022; Misra et al. 2020). The global market value of smart farming, including precision farming, is projected to reach 23.1 billion dollars by 2050 (Karnaushenko et al. 2023).

DOI: 10.1201/9781003507390-5

Smart agriculture, while promising numerous benefits like increased efficiency and yield, faces several security challenges. One significant concern is data privacy and confidentiality, as the extensive collection of sensitive information about crops, weather patterns, and farming practices could be vulnerable to breaches, leading to potential misuse or exploitation (Gupta et al. 2020; Yazdinejad et al. 2021). Moreover, the interconnected nature of smart agriculture systems introduces risks of cyber-attacks, such as unauthorized access to automated machinery or manipulation of sensor data, which could disrupt operations or compromise crop health (West 2020; Ferrag et al. 2021). Furthermore, the reliance on internet connectivity leaves these systems susceptible to disruptions from network outages or malicious interference, emphasizing the need for strong cybersecurity measures to protect the reliability and integrity of smart agriculture technologies.

To ensure the security of smart agriculture and to safeguard against cyber threats, various security measures have been employed to ensure that intruders do not gain unauthorized entry into the systems and guarantee their security by putting in place resilient authentication mechanisms, for instance, biometric authentication or multi-factor authentication. Encryption, regular security audits, software updates, intrusion detection systems, firewalls, and education about cybersecurity best practices are crucial for preventing data breaches and ensuring data transmission and storage security. Hence, this chapter suggests a blockchain-incorporated security pattern for smart agriculture to protect the technology from numerous security vulnerabilities that could arise while deploying smart agriculture.

An end-to-end interaction between the entities that make up the smart agriculture system is displayed in Figure 5.1. Data is generated by field sensors and livestock, and user applications are used to command and execute actions. Connected to edge nodes supported by the gateway, these electronic devices on the farm are enabling real-time agronomy analytics, filtering sensor data, and

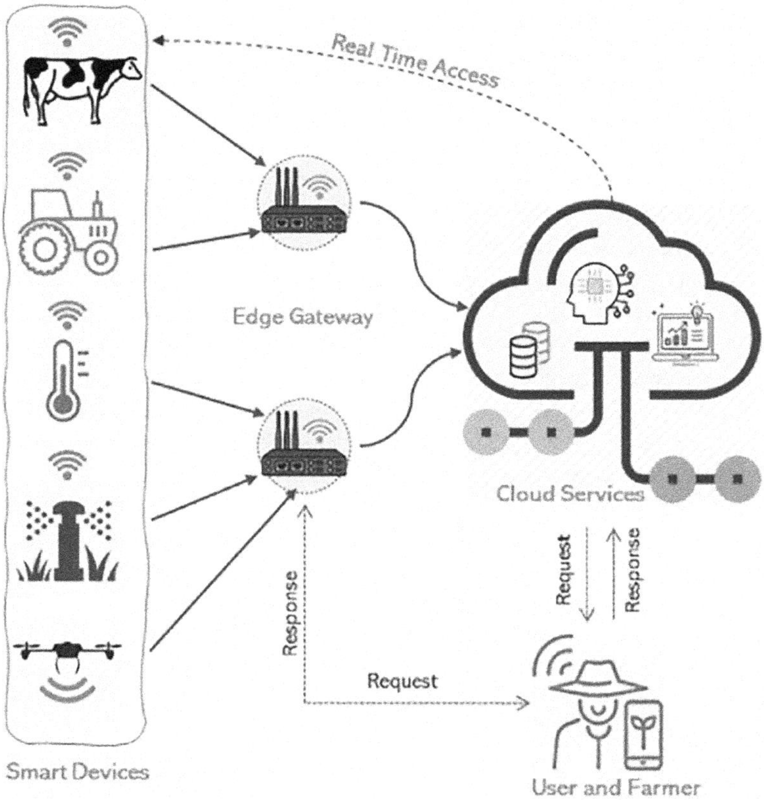

FIGURE 5.1 Various entities interact end-to-end in smart agriculture.

in-farm device communication, and much data and information, including but not limited to environmental data, are stored in cloud data lakes, such as fertility status, moisture level, monitoring information such as smart machinery status, data on energy management, and other private details. Modification and disclosure of such data, together with damage to hardware and software, can have detrimental effects on security and privacy. The figure displays various objects interrelated end-to-end in smart agriculture.

A blockchain-enabled security framework for smart agriculture offers a decentralized and tamper-resistant solution to safeguard sensitive data and transactions within agricultural systems. Utilizing blockchain technology, each transaction or data entry is cryptographically linked and stored among a collection of nodes across a number of networks, verifying transparency and immutability. The blockchain has smart contracts hardcoded in it, easing agreements between individuals, hence popularizing procedures such as asset tracking, payments, and supply chain management and reducing instances of fraud in operations. Furthermore, blockchain enhances the trustworthiness of data by eliminating reliance on a single centralized authority, but offering a decentralized and tamper-resistant system for data preservation. This framework empowers stakeholders in the agricultural ecosystem to securely exchange information, track the provenance of products, and establish trust among participants, ultimately enhancing efficiency and traceability throughout the agricultural value chain. Therefore, this chapter reviews the potential security and privacy issues in smart agriculture and the various countermeasures against these challenges. The following are the key contributions of this chapter:

1. the chapter reviews smart agriculture, which focuses on security and privacy, multi-layered architecture, and blockchain-based security framework;
2. the chapter highlights potential cybersecurity threats in smart agriculture, categorizing them into data, networks, supply chains, and other common attacks;
3. this chapter evaluates current cybersecurity research and countermeasures in smart agriculture and focuses on the contributions and weaknesses of these studies; and
4. the chapter highlights the ongoing security research challenges in the area of smart agriculture.

5.1.1 General Overview and Concepts of Smart Agriculture

Smart agriculture, which is also referred to as precision agriculture or digital farming, involves the use of modern technologies to improve productivity, efficiency, and sustainability in agricultural processes (Falola et al. 2024). It leverages data-driven insights and automation to optimize farming processes, from planting to harvesting, ensuring better management of resources and improved crop yields (Koshariya et al. 2024; Shakoor et al. 2019). Smart agriculture refers to the incorporation of superior technologies in enhancing agricultural efficiency, productivity, and sustainability of farming processes. IoT devices like sensors and drones collect real-time data on different aspects of agriculture like soil moisture, crop health, and weather conditions (Boursianis et al. 2022; Rehman et al. 2022; Dhal et al. 2024). Farmers benefit from utilizing big data analytics and machine learning algorithms on this data, thus enabling them to make sound choices with regard to water supply, chemicals applied, or getting rid of pests. By precisely targeting the needs of crops, smart agriculture minimizes resource wastage and maximizes yields, leading to more efficient and sustainable farming operations (Vishnoi & Goel 2024; Júnior et al. 2024).

One of the important smart farming components is geographic information system (GIS) technology, since it helps in field mapping as well as monitoring at micro levels. The introduction of GIS tools like variable rate technology (VRT) to farmers has helps them to precisely apply water, fertilizers, and pesticides at varying rates across various farming setting based on specific environmental location conditions. This approach not only optimized resource use but also enhanced the efficiency and sustainability of agricultural practices, hence, addressing the variability and complexity of farm conditions (Lanucara et al. 2024). This precision approach not only optimizes the use of resources but also reduces environmental impact by preventing over-application and runoff. Additionally, autonomous machinery, including tractors and harvesters equipped with Global Positioning System (GPS) and Artificial Intelligence (AI) systems, can perform agricultural tasks with high precision and minimal human intervention, further enhancing efficiency and productivity (Gupta & Gupta 2024; Rampalli et al. 2024).

Smart agriculture, despite its numerous advantages, faces a number of problems in terms of adoption. High initial costs of technology implementation may prove a stumbling block, particularly for small and medium-sized farms and households (Fragomeli et al. 2024; Friedman et al. 2024). Moreover, the effective use of smart agriculture requires robust data management systems and skilled personnel to analyze and interpret data (Wanyama et al. 2024; Bazargani & Deemyad 2024). Connectivity issues, especially in remote rural areas, can hinder the seamless operation of IoT devices and other technologies (Cabrera-Castellanos et al. 2021). Furthermore, ensuring data privacy and security is crucial to protecting sensitive agricultural information from cyber threats (Gupta et al. 2020; Demestichas et al. 2020; de Araujo Zanella et al. 2020). Addressing these challenges through supportive policies, training programs, and infrastructure development is essential for the widespread adoption and success of smart agriculture (Raile et al. 2021; Azadi et al. 2021; Gaber et al. 2023).

5.2 KEY CONCEPTS OF SMART AGRICULTURE

The use of modern technology in the farming process is referred to as smart agriculture. It aims to improve farm practices by enhancing their efficiency, sustainability, and profitability. A number of ICT instruments being utilized include IoT devices, big data, and artificial intelligence internet-enabled applications. Internet of Things devices, soil moisture sensors like Soil salinity, and transmission equipped with a global positioning system provide information instantly concerning the agricultural environment. This data is then analyzed using AI and big data analytics to provide actionable insights, such as optimal planting times, pest control measures, and irrigation schedules. By utilizing these technologies, farmers can make informed decisions that maximize crop yields while minimizing resource use and environmental impact.

Another key concept of smart agriculture is the integration of automation and robotics in farming operations. Drones and autonomous tractors, for instance, can do tasks such as sowing seed, spraying pesticides, and reaping crops accurately. That means the machine does not need much human input when doing these jobs. This way, it cuts down on the cost of human power and enhances the accuracy and efficiency of these tasks. Furthermore, smart agriculture includes sustainable practices such as precision irrigation, which ensures that water is used efficiently and only when necessary, reducing waste. The combination of these technologies and practices helps create a more resilient agricultural system that can adapt to changing environmental conditions and meet the growing global demand for food. The following are some of the key concepts of smart agriculture. Figure 5.2 shows the significant concepts of smart agriculture systems.

Sensors and IoT Devices: These are placed in different fields in order to gather online information on soil moisture, temperature, air humidity, and plant conditions (Lloret et al. 2021; Thilakarathne et al. 2023). Soil moisture, temperature, air humidity, and plant conditions are some of the information collected by these sensors, which is significant in enabling smart farming by the use of IoT devices that support precision farming, hence increasing effectiveness and productivity (Shafi et al. 2020; Jaiswal et al. 2020). These devices collect real-time data on various environmental and crop conditions, such as soil moisture, temperature, humidity, light intensity, and nutrient levels (Karuna et al. 2024; Supriya et al. 2024). Soil sensors, for example, provide precise information about the moisture content and pH level of the soil, helping farmers to optimize irrigation and fertilization processes (Venkatachalam et al. 2024). Weather sensors track local climate conditions, allowing for more accurate predictions and adjustments to farming practices. The data collection and analysis that is done in real time helps in making decisions that are well informed in addition to reducing wastage. It also enhances an increase in crop yields (Medina et al. 2024).

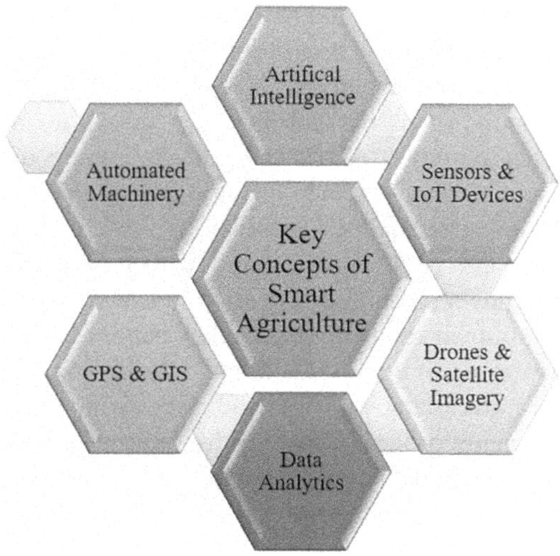

FIGURE 5.2 Key concepts of smart agriculture.

Moreover, IoT devices facilitate automation and remote monitoring, which are essential for modern agricultural operations. Drones can monitor crop health over large areas, detecting pest infestations or nutrient deficiencies early with the help of multispectral sensors (Pungavi & Praveenkumar 2024). Automated irrigation systems, controlled via IoT networks, ensure that crops receive the right amount of water at the right time, reducing water usage and preventing over-irrigation (Zhang & Wang 2024). Livestock monitoring systems use wearable sensors to track the health and activity of animals, alerting farmers to any signs of illness or distress (Et-taibi et al. 2024). This integration of IoT devices in agriculture not only improves efficiency and sustainability but also reduces the labor burden on farmers, allowing for more scalable and resilient farming practices (Shrivastav et al. 2024).

Artificial Intelligence: In the context of smart agriculture, AI algorithms comb through huge troves of data derived from distinct sources like sensors, drones, satellites, and weather stations to generate useful information (Rao et al. 2021; Patil et al. 2024). Artificial intelligence has become instrumental in transforming intelligent agriculture via steering data-oriented decision-making and prophetic analytics (Awotunde et al. 2022a). These insights range from predicting weather patterns and identifying pest outbreaks to optimizing crop management practices (Kumari et al. 2024). Artificial intelligence-driven predictive models help farmers prepare for and mitigate threats as they occur, thus aiding them in making timely decisions (Mishra & Mishra 2023). For instance, AI is capable of studying past weather records so as to predict future rainfall trends, a move that enables farmers to plan for watering periods better, thereby conserving water and reducing crop stress (Ajagbe et al. 2022).

Moreover, AI facilitates the development of intelligent decision-support systems tailored to the specific needs of farmers. These systems provide personalized recommendations and actionable insights based on the analysis of data related to soil conditions, crop health, and market trends (Salehi 2024). AI can be used by farmers to make sure that they allocate their resources more effectively, get the most out of crops, or even simply run things more smoothly (Kose et al. 2022). Furthermore, AI enables the automation of repetitive tasks, such as weed detection and sorting of harvested produce, freeing up farmers' time and labor for more strategic activities (Patil et al. 2024). As AI continues to advance, its integration into smart agriculture holds the potential to tackle global food insecurity, increase agricultural productivity and sustainability (Kumari et al. 2024).

Drones and Satellite Imagery: Unmanned aerial vehicles (UAVs) and satellites provide high-resolution images of crops, enabling precise monitoring of plant health, growth patterns, and potential issues such as disease outbreaks or nutrient deficiencies (Abdullah et al. 2023). Drones and satellite imagery have become indispensable tools in smart agriculture, offering advanced capabilities for monitoring and managing crops (Abbas et al. 2023). Equipped with high resolution cameras and multispectral sensors, drones can snap detailed images of various fields, offering useful information on crop health, soil state, and pest presence (Ivošević et al. 2023). By flying over large areas, drones can quickly and efficiently. Omia et al. (2023) gather data that would be very time consuming and laborious to collect manually. This allows farmers to catch early manifestations of sickness or lacking nutrients in their crops in time so they can intervene. Furthermore, drones can be employed to do precise spraying of both pesticides and fertilizers, ensuring that these inputs are applied exactly where needed, thus reducing waste and minimizing environmental impact.

Satellite imagery complements drone technology by offering a broader, macro-level view of agricultural landscapes over time (Brockmann & Iskander 2023). Satellites equipped with advanced imaging sensors can monitor crop growth, land use, and environmental conditions across entire regions. This data is crucial for large-scale agricultural planning and management, as it provides information on crop health, yield predictions, and the impact of climatic changes. Satellites can capture images at regular intervals, allowing for continuous monitoring and trend analysis (Niu et al. 2024). This long-term perspective is valuable for assessing the effectiveness of agricultural practices and making strategic decisions. Together, drones and satellite imagery provide a comprehensive, multi-scale approach to managing agricultural production, enhancing efficiency, sustainability, and resilience in farming operations.

Global Positioning Systems and Geographic Information Systems: GPS and GIS technologies facilitate accurate mapping and analysis of farmland. These tools help in planning and executing field operations with precision, reducing waste, and optimizing resource use. GPS is a fundamental technology in smart agriculture, providing precise location data that is essential for various precision farming applications (Obi Reddy et al. 2023). GPS enables farmers to map their fields accurately, facilitating the use of automated equipment and machinery. For instance, GPS-guided tractors can follow precise paths to plant seeds, apply fertilizers, and harvest crops with minimal overlap and waste (Yadav & Sidana 2023). This precision reduces input costs, enhances efficiency, and minimizes environmental impact (Yadav & Sidana 2023). Furthermore, GPS is instrumental in enabling variable rate technology, where inputs such as water, seeds, and fertilizers are applied at varying rates across a field based on specific soil and crop needs, optimizing resource use and improving crop yields (Ahmad & Sharma 2023).

GIS complements GPS by providing a framework for gathering, managing, and analyzing spatial and geographic data. In smart agriculture, GIS is used to create detailed maps and models of agricultural fields, integrating data from various sources such as soil sensors, drones, and satellite imagery (Obi Reddy et al. 2023). This integration allows for comprehensive spatial analysis and decision-making. For example, GIS can be used to analyze soil properties, identify variations in crop performance, and determine areas that may require different management practices (Ahmad & Sharma 2023). By overlaying different types of data, farmers can gain a deeper understanding of their fields and make informed decisions about planting, irrigation, pest control, and harvesting. Together, GPS and GIS enhance the ability to implement precision agriculture practices, leading to more efficient, productive, and sustainable farming operations (Yadav & Sidana 2023).

Automated Machinery: Tractors and harvesters that have been equipped with robot and automation technologies can carry out activities like planting crops, weeding them, or harvesting them, all without much help from human beings, making them more effective and reducing the expenses of farming work (Rai et al. 2023). There is a technological breakthrough through automated farming equipment which has improved operational efficiency levels in smart farming. These machines, which include autonomous tractors, robotic harvesters, and automated planters, can perform a wide range of agricultural tasks with minimal human intervention (Gupta & Gupta 2024). At the end of the day, autonomous tractors can optimize fields for crop planting, applying fertilizers, and harvesting (Yadav & Sidana 2023). In the process, it minimizes human labor requirements and improves the precision of field operations and allows farmers to manage larger areas more effectively. Robotic harvesters, designed to pick fruits and vegetables, can operate continuously and handle delicate produce with care, reducing damage and increasing productivity (Kumar et al. 2023).

In addition to performing basic farming tasks, automated machinery in smart agriculture is often integrated with sophisticated data analytics and IoT technologies. These machines can collect and transmit real-time data on crop conditions, soil health, and environmental factors, enabling more informed decision-making (Mail et al. 2023). Automated irrigation systems ensure optimal water delivery by changing it using real-time soil moisture data, for example, water use and reducing waste. Similarly, automated machinery can be programmed to apply Kumar et al. (2023) fertilizers, and pesticides may be applied only in places they are required. Such a focused strategy enhances productivity in addition to cutting down on potential dangers posed by agricultural activities (Gupta & Gupta 2024). Overall, the integration of automated machinery in smart agriculture supports sustainable and productive farming by leveraging technology to optimize every aspect of the agricultural process (Sarfraz et al. 2023).

Data Analytics and Artificial Intelligence: Big data analytics and AI algorithms process vast amounts of data collected from various sources (Awotunde et al. 2022a). This analysis provides actionable insights, such as predicting weather patterns, optimizing crop yields, and managing supply chains effectively. Data analytics and artificial intelligence are central to the advancements in smart agriculture, transforming how farmers manage their operations by turning vast amounts of data into actionable insights (Javaid et al. 2022). Data analytics is the process of bringing together different kinds of information and data, including but not limited to sensors on farmland (these measure how moist or dry it is), meteorological data, and images taken from above by unmanned flying vehicles (drones) and others like satellites. Such data offers facts that are more detailed concerning the fertility status of soils as well as the way plants develop over time during different seasons, as opposed to environmental conditions at that particular point in time, which determine if conditions are conducive for pest multiplication and activities; for example, their migration pattern data enables farmers and other users to make sense of these aspects when they analyze them (Gebresenbet et al. 2023). For instance, predictive analytics can forecast crop yields and detect potential issues so farmers can take proactive measures to enhance productivity and prevent losses before they become significant problems (Gebresenbet et al. 2023).

Artificial intelligence further enhances smart agriculture by providing advanced tools for automation, optimization, and decision support. Systems that are AI powered have the capability to study elaborate datasets so as to anticipate results, refine utilization of resources, and bolster agronomic practices. For instance, by studying weather changes alongside soil composition, these systems are able to give advice on the best periods when farmers should plant their crops and irrigate them. Furthermore, it can be applied to recognize pests or diseases from images of plants from drone or satellite images, enabling timely and precise interventions. Additionally, AI can optimize supply chain logistics, ensuring that crops are harvested, stored, and transported efficiently. The integration of AI and data analytics in smart agriculture leads to more efficient, sustainable, and resilient farming practices, ultimately increasing food production and reducing environmental impact.

Blockchain Technology: To predict outcomes, optimize resource use, and improve crop management practices, machine learning algorithms can analyze complex datasets. For instance, based on AI, weather patterns and soil conditions may be analyzed for purposes of giving advice about when one should plant their crops and irrigation schedules. Additionally, image recognition to identify pests or diseases attacking plants can also be done by artificial intelligence (Mukherjee et al. 2022). Blockchain is essentially a decentralized, immutable ledger that records transactions securely and verifiably. On the farm-to-table journey in agriculture, blockchain can track products (Bosona & Gebresenbet 2023). Therefore, each phase of the production cycle, starting from planting and picking through refining and distribution, is recorded. Consumers can scan the QR code on a product for more details on quality and handling, fostering trust and confidence in the food they consume (Bosona & Gebresenbet 2023). This transparency also helps in ensuring compliance with safety standards and certifications, which is crucial for food safety and quality assurance (Bosona & Gebresenbet 2023).

Moreover, blockchain technology facilitates efficient and secure transactions within the agricultural sector. Trade among farmers can be streamlined and automated through the use of smart contracts to facilitate transactions seamlessly between suppliers and buyers (Torky & Hassanein 2020). These contracts ensure that payments are made promptly once predefined; the conditions have to be met in order to avoid intermediaries as well as to cut down on fraud risks. For example, when a shipment is delivered to a customer, it's possible that a farmer will receive payments automatically because there is an intelligent agreement involved if produce is delivered and verified (Torky & Hassanein 2020). Additionally, blockchain can support fair trade practices by providing a transparent record of transactions, ensuring that farmers receive fair compensation for their products. Overall, the integration of blockchain technology in smart agriculture enhances the efficiency, reliability, and sustainability of agricultural operations, benefiting both producers and consumers (Bosona & Gebresenbet 2023).

5.3 SECURITY AND PRIVACY CHALLENGES IN SMART AGRICULTURE SYSTEMS

Understanding the security issues in smart agriculture systems requires a comprehensive examination of the technological landscape, the intricacies of agricultural operations, and the potential vulnerabilities inherent in integrating advanced technology into traditional farming practices (Awotunde et al. 2023). In this chapter, we delve into the multifaceted aspects of security challenges in smart agriculture systems. Smart agriculture leverages various technologies to optimize farming practices, enhance crop yield, and minimize resource wastage. While these innovations promise significant in terms of enhancing efficiency, productivity, and sustainability, and gains have recently been made, new security risks have been introduced which may compromise the integrity, confidentiality, and availability of agricultural operations (Awotunde et al. 2023).

IoT devices and sensor usage for smart agriculture systems have been encouraged for collection of real-time data like soil moisture levels, temperature humidity, and crop status, among others. The devices are seamlessly connected through wireless networks, enabling uninterrupted communication and efficient information exchange across various components of the agricultural ecosystem. However, this interconnectedness also creates a larger attack surface, exposing vulnerabilities that malicious actors can exploit to infiltrate the system.

Moreover, the adoption of cloud computing and edge computing technologies in smart agriculture facilitates data storage, processing, and analysis on remote servers or decentralized edge devices. While cloud computing offers scalability and flexibility, it introduces concerns related to data sovereignty, privacy, and the security of data transmission over the Internet. Similarly, edge computing raises issues of the security of distributed computing resources and the management of edge devices deployed in remote agricultural locations. Security challenges in smart agriculture systems can be categorized into several key areas.

5.4 ATTACKS ON HARDWARE IN SMART AGRICULTURE

In smart agriculture, hardware attacks pose significant threats to the integrity, confidentiality, and availability of data and operations. Here are several types of hardware attacks that could target smart agriculture systems:

Tampering with Sensors: Attackers may physically tamper with sensors deployed in the field to manipulate data collection. For example, they might cover or damage sensors to disrupt environmental monitoring or input false data, leading to incorrect decisions by the automated systems relying on that data.

Side-Channel Attacks: Side-channel attacks in smart agriculture target hardware by exploiting unintended data leaks, such as electromagnetic emissions, power consumption, or timing information, from IoT devices and sensors. These attacks bypass cryptographic defenses, extracting sensitive information like encryption keys or operational data. For example, a side-channel attack could compromise a sensor monitoring crop conditions, manipulating or stealing data. This jeopardizes system reliability, leading to potential crop losses, financial damage, and weakened trust in smart agriculture technologies.

Supply Chain Attacks: Malicious actors can compromise hardware components during the manufacturing or distribution process. For instance, they might implant malicious firmware or hardware backdoors into sensors, controllers, or other devices, allowing unauthorized access or control over the system (de Araujo Zanella et al. 2020).

Denial-of-Service (DoS) Attacks: In smart agriculture, DoS attacks can target hardware components such as irrigation controllers or communication gateways, disrupting essential operations (de Araujo Zanella et al. 2020). By overwhelming these devices with excessive requests or malicious traffic, attackers can render them unavailable, causing significant financial losses or crop damage.

Interference with Communication Systems: Attackers may jam or intercept wireless communication between sensors, actuators, and control systems. By disrupting communication links, they can prevent data transmission, manipulate commands, or eavesdrop on sensitive information, compromising the reliability and security of the entire system (Jaiswal et al. 2020).

Physical Destruction: Sabotaging or destroying hardware components such as drones, robots, or autonomous vehicles can disrupt farming operations and cause financial losses (Yaacoub et al. 2022). Attackers might employ various methods, including vandalism, explosives, or targeted attacks using firearms, to damage or disable critical infrastructure (Yaacoub et al. 2022).

Hardware Trojan: In this type of attack, malicious circuitry is inserted into the hardware design during the manufacturing process. These Trojans can remain dormant until triggered by specific conditions, such as a particular command or a specific date, allowing attackers to gain unauthorized access or manipulate the system's behavior (Venugopalan & Patterson 2018).

GPS Spoofing: Smart agriculture systems often rely on GPS for precise positioning and navigation. Attackers can spoof GPS signals to deceive drones, tractors, or other autonomous vehicles, leading to inaccurate mapping, navigation errors, or even collisions with obstacles, endangering personnel and equipment (Naseer et al. 2024).

5.5 PRIVACY PROTECTION IN SMART AGRICULTURE SYSTEMS

Protecting against these hardware attacks requires a combination of physical security measures, cryptographic protections, supply chain integrity verification, intrusion detection systems, and resilient communication protocols (Abosata et al. 2021; AbdulRaheem et al. 2023). Regular security audits and updates are essential to mitigate emerging threats and ensure the robustness of smart agriculture systems (Awotunde et al. 2022a). Data privacy on hardware in smart agriculture is crucial for protecting sensitive information collected by various devices and systems (Adeniyi et al. 2024). Here are several aspects of data privacy in this context:

Data Encryption: Implementing encryption mechanisms on hardware devices helps to keep data safe. Data should be encrypted not only when it is being transferred but also while it is still. This means that if someone gets inside your sensors, controllers, or whatever else, they still cannot read what has been stored there because everything stays unreadable without an appropriate key. Similarly, encrypting communication channels between devices and backend servers ensures that data transmitted over the network cannot be intercepted and deciphered by unauthorized parties.

Access Control: Access control in smart agriculture is vital to prevent unauthorized access to hardware systems like IoT devices, sensors, and controllers. Without robust authentication and authorization, attackers can manipulate or disrupt critical operations, such as irrigation, pest control, or data collection, compromising productivity and safety. Techniques like multi-factor authentication, encryption, and role-based access help safeguard hardware. Strong access control minimizes risks of tampering, ensuring secure and efficient operation in smart agriculture environments.

Anonymization and Pseudonymization: To minimize the risk of data breaches and protect the privacy of individuals, sensitive data collected by hardware devices should be anonymized or pseudonymized whenever possible. By removing or replacing personally identifiable information (PII) with pseudonyms or random identifiers, such as in the case of farm workers or landowners, data can be aggregated and analyzed without compromising individuals' privacy.

Data Minimization: Smart agriculture systems should adhere to the principle of data minimization, collecting only the necessary information required for specific tasks or operations. By reducing the amount of data stored on hardware devices, the risk of unauthorized access or misuse is mitigated. Additionally, regularly purging unnecessary data helps prevent the accumulation of obsolete or redundant information that could be exploited by attackers.

Secure Data Storage: Hardware devices used in smart agriculture should employ secure storage mechanisms to protect data from unauthorized access or tampering. This includes using hardware-based security modules, such as trusted platform modules (TPMs) or secure elements, to store cryptographic keys and sensitive data securely. Additionally, implementing secure file systems with access controls and integrity checks helps safeguard data stored on local storage media, such as flash memory or hard drives.

Secure Firmware Updates: Ensuring the integrity and authenticity of firmware updates is essential for preventing malicious actors from installing unauthorized or compromised software on hardware devices. Implementing secure boot mechanisms and digital signatures allows devices to verify the authenticity and integrity of firmware updates before applying them, reducing the risk of firmware-level attacks.

Auditing and Logging: Hardware devices should maintain detailed audit logs of all interactions and data access attempts to facilitate forensic analysis and accountability. By recording information such as device activities, user authentication events, and data access timestamps, suspicious or unauthorized activities can be detected and investigated promptly, helping to identify and mitigate potential security breaches.

By addressing these aspects of data privacy on hardware devices in smart agriculture, organizations can enhance the security and confidentiality of sensitive information collected and processed by their systems, mitigating the risk of data breaches and ensuring compliance with privacy regulations.

5.6 ATTACKS ON NETWORKS AND EQUIPMENT IN SMART AGRICULTURE

Smart agriculture, like any other interconnected system, is vulnerable to a variety of attacks that can compromise the integrity, confidentiality, and availability of data and resources. Here are some common types of attacks on networks and equipment in smart agriculture:

Denial-of-Service Attacks: These attacks aim to make resources unavailable to legitimate users by overwhelming the target system with a flood of illegitimate service requests. In smart agriculture, this could disrupt critical services like irrigation systems or data collection sensors.

Man-in-the-Middle (MitM) Attacks: In a MitM attack, the attacker intercepts and possibly alters communication between two parties without their knowledge. In smart agriculture, this could involve intercepting data from sensors or control systems, leading to incorrect decisions being made based on manipulated data.

Data Tampering: Attackers may alter sensor data or control signals to manipulate the operation of agricultural equipment, leading to suboptimal or harmful outcomes. For example, changing temperature readings could result in incorrect climate control decisions in a greenhouse.

Eavesdropping: Attackers may intercept wireless communications between IoT devices, such as sensors or actuators, to gain access to sensitive information about crop conditions or farm operations. This information could be used for various malicious purposes, such as insider trading or sabotage.

Physical Attacks: Physical access to agricultural equipment can also pose a security risk. Attackers may tamper with or sabotage equipment directly, leading to production losses or safety hazards. Additionally, physical theft of equipment or data storage devices could result in data breaches or operational disruptions.

Malware and Ransomware: Malicious software can infect smart agricultural systems, compromising their functionality or stealing sensitive information. Ransomware attacks, in particular, can encrypt data or lock users out of their systems until a ransom is paid, causing significant financial losses and operational downtime.

Social Engineering: Attackers may exploit human vulnerabilities through techniques such as phishing or pretexting to gain unauthorized access to smart agricultural systems. This could involve tricking employees into revealing login credentials or other sensitive information, enabling the attacker to compromise the system remotely.

To mitigate these risks, smart agriculture stakeholders need to implement robust cybersecurity measures, including encryption, authentication, access control, intrusion detection systems, and regular security updates. Additionally, training personnel on cybersecurity best practices and maintaining physical security measures can help prevent and mitigate the impact of attacks on networks and equipment in smart agriculture.

5.7 ATTACKS ON SUPPLY CHAIN IN SMART AGRICULTURE

Smart agriculture, relying heavily on interconnected technologies, is susceptible to various attacks on its supply and support chain. These attacks can disrupt operations, compromise data integrity, or even cause physical damage. Each of these attacks poses unique risks to the supply and support chains in smart agriculture and requires specific security measures to mitigate them effectively. To mitigate these risks, stakeholders in smart agriculture need to implement robust security measures, including encryption, authentication, access controls, intrusion detection systems, and regular security audits. Additionally, raising awareness among employees and partners about cybersecurity best practices can help prevent successful attacks on the supply and support chain. Here are some common attacks:

Attack Type	Description
Data Interception	Attackers intercept data flowing between sensors, controllers, and servers, compromising sensitive information like crop data, weather forecasts, or logistics plans. Attackers may intercept data flowing between sensors, controllers, and servers in the supply chain. By eavesdropping on communications, they can gain access to sensitive information such as crop data, weather forecasts, or logistics plans. This interception can lead to data manipulation or theft.

(Continued)

(Continued)

Attack Type	Description
Denial-of-Service	Attackers overwhelm critical infrastructure such as irrigation systems or weather monitoring stations with excessive traffic or requests, disrupting operations. These attacks aim to disrupt the availability of services or systems. In smart agriculture, DoS attacks can target critical infrastructure like irrigation systems, weather monitoring stations, or crop monitoring platforms. By overwhelming these systems with excessive traffic or requests, attackers can render them unusable, causing significant operational disruptions.
Sensor Data Manipulation	Manipulating sensor data to provide false readings, leading to inaccurate analyses and decision-making. Smart agriculture relies on data from various sensors to make informed decisions. Attackers can manipulate sensor data to provide false readings, leading to inaccurate analyses and decision-making. For example, altering soil moisture or temperature readings can mislead farmers into incorrect irrigation or fertilization practices, impacting crop health and yield.
Supply Chain Compromise	Infiltrating suppliers or distributors to introduce counterfeit or compromised components into smart agriculture systems, compromising system integrity. Attackers may target the supply chain itself, infiltrating suppliers or distributors to introduce counterfeit or compromised components into smart agriculture systems. For instance, compromised sensors or controllers could be used to inject malicious code into the network, compromising system integrity or stealing sensitive information.
Physical Attacks	Sabotaging equipment, tampering with sensors, or destroying critical infrastructure components, leading to downtime and financial losses. Physical attacks on infrastructure components like sensors, drones, or machinery can disrupt operations and cause damage. Attackers may sabotage equipment, tamper with sensors, or destroy critical infrastructure components, leading to downtime and financial losses.
Malware and Ransomware	Infecting devices and systems within the supply chain, causing data breaches, operational disruptions, or financial losses. Malicious software can infect devices and systems within the smart agriculture supply chain, leading to data breaches, operational disruptions, or financial losses. Ransomware attacks, in particular, can encrypt data or systems, demanding ransom payments for decryption keys, further disrupting operations and causing financial harm.
Social Engineering	Manipulating individuals within the supply chain through techniques like phishing, pretexting, or impersonation to divulge sensitive information or perform malicious actions. Attackers may employ social engineering techniques to manipulate individuals within the supply chain, such as employees or vendors, into divulging sensitive information or performing actions that compromise security. Phishing emails, pretexting, or impersonation tactics can be used to trick unsuspecting targets into revealing credentials or executing malicious actions.

5.8 ATTACKS ON SENSOR DEVICES IN SMART AGRICULTURE

Smart agriculture relies heavily on sensors for various tasks such as monitoring soil moisture, temperature, and humidity, and detecting pests. These sensors are vulnerable to several types of attacks, which can disrupt agricultural operations and compromise data integrity. Here are some common attacks on sensor devices in smart agriculture:

Physical Tampering: Attackers may physically tamper with sensors by damaging or altering them. This could involve cutting wires, placing physical barriers over sensors, or even stealing the sensors themselves. Physical tampering can lead to inaccurate data collection or complete failure of sensor functionality. Physical tampering in smart agriculture involves direct interference with sensor devices to disrupt their functionality or accuracy. This can include actions such as cutting wires, blocking sensor inputs, altering sensor placement, or outright theft of the sensors. Such tampering can lead to erroneous data collection or complete sensor failure, resulting in incorrect decisions regarding irrigation, fertilization, or pest control. To mitigate this risk, it's crucial to implement protective measures like robust enclosures, tamper-evident seals, and surveillance systems to monitor and secure sensor installations.

Jamming: Jamming attacks involve emitting interference signals to disrupt the communication between sensors and the central control system. This can prevent sensors from transmitting data accurately or at all, leading to gaps in data collection and analysis. Jamming in smart agriculture refers to deliberate interference with communication signals between sensor devices and their central control systems. By emitting strong, disruptive signals on the same frequency bands used by the sensors, attackers can prevent the transmission of crucial data related to soil moisture, temperature, humidity, or pest presence. This disruption can lead to significant gaps in data collection, impairing the ability to make timely and accurate agricultural decisions, such as when to irrigate or apply pesticides. To counteract jamming attacks, employing frequency hopping, spread spectrum technologies, and robust encryption can enhance the resilience of sensor networks against such disruptions.

Spoofing: In spoofing attacks, attackers impersonate legitimate sensors by transmitting false data to the central control system. By doing so, attackers can manipulate the decision-making process based on erroneous information, leading to incorrect actions such as over-watering or under-fertilizing crops.

Eavesdropping: Eavesdropping attacks involve intercepting and listening to communication between sensors and the central control system. Attackers can gain valuable information about the operation of the system, including sensor data and control commands. This information can be used to plan further attacks or exploit vulnerabilities in the system.

Replay Attacks: Replay attacks on sensor devices in smart agriculture involve capturing legitimate sensor data and retransmitting it at a later time to deceive the central control system. This manipulation can cause the system to make incorrect decisions based on outdated or irrelevant information. For example, an attacker could record data indicating optimal soil moisture during a wet period and replay it during a dry spell, leading the irrigation system to remain off when watering is critically needed. Such attacks can disrupt crop management and harm agricultural productivity. To mitigate replay attacks, incorporating time stamps, sequence numbers, and robust encryption methods in sensor communications can help ensure the timeliness and authenticity of the data being processed.

Denial of Service: DoS attacks aim to disrupt the availability of sensor devices by overwhelming them with a high volume of requests or traffic. This can prevent sensors from performing their intended functions, leading to delays or failures in data collection and analysis. A DoS attack on sensor devices in smart agriculture involves overwhelming the sensors or their communication networks with excessive traffic or requests, rendering them unable to function properly. This flood of data can cause sensors to become unresponsive, leading to gaps in monitoring critical parameters such as soil moisture, temperature, and humidity. Consequently, the agricultural management system may fail to receive timely information, disrupting automated processes like irrigation and pest control, which can severely affect crop health and yield. To protect against DoS attacks, implementing network security measures such as traffic filtering, rate limiting, and anomaly detection systems is essential to maintain the availability and reliability of sensor networks.

Firmware Attacks: Attackers may exploit vulnerabilities in sensor firmware to gain unauthorized access or control over devices. This could involve installing malicious firmware that alters sensor behavior, exfiltrates sensitive data, or provides backdoor access to the system. Firmware attacks on sensor devices in smart agriculture involve exploiting vulnerabilities in the sensor's firmware to gain unauthorized access or control. Attackers might install malicious firmware to alter the sensor's behavior, extract sensitive data, or create a backdoor for future exploits. Such attacks can lead to the transmission of false data, malfunctioning of the sensors, or complete system disruption, jeopardizing the accuracy of agricultural monitoring and decision-making processes. For instance, compromised firmware might send incorrect soil moisture levels, causing over- or under-irrigation. To prevent firmware attacks, it is crucial to regularly update firmware, implement strong authentication mechanisms, and use secure boot processes to ensure only verified firmware is installed on sensor devices.

Battery Drainage: Attackers may employ techniques to drain the batteries of sensor devices quickly, rendering them inoperable. This could involve continuously sending requests to the sensors or activating power-intensive features, reducing the lifespan of the sensors and disrupting agricultural operations. Battery drainage attacks on sensor devices in smart agriculture involve deliberately exhausting the battery power of these devices, rendering them inoperative. Attackers may achieve this by continuously sending signals or requests that force the sensors to remain active and consume power rapidly. This can result in premature battery depletion, causing sensors to fail and disrupt the collection of critical agricultural data such as soil moisture, temperature, and pest presence. Consequently, automated systems may lack the necessary information to perform tasks like irrigation or pest control, negatively impacting crop health and yield. To mitigate battery drainage attacks, implementing energy-efficient protocols, setting limits on request frequencies, and incorporating low-power modes can help preserve battery life and ensure continuous sensor operation.

Protecting sensor devices in smart agriculture requires implementing robust security measures such as encryption, authentication, intrusion detection, and regular firmware updates to mitigate these and other potential threats.

5.8.1 General Framework Using Blockchain-Based Infrastructure in Smart Agriculture

Blockchain-based architecture in smart agriculture leverages the decentralized, transparent, and immutable nature of blockchain technology to enhance various aspects of the agricultural supply chain. One of the primary applications is in supply chain traceability. All players in the agrifood chain, including producers, distribution channels, and consumers, would have one single source of information about movement and every transaction concerning agricultural commodities through the use of blockchain technology. This transparency helps ensure food safety, reduce fraud, and increase consumer trust by verifying the authenticity and origin of products. Additionally, blockchain can facilitate more efficient and secure transactions by enabling smart contracts, which are self-executing contracts with the terms of the agreement directly written into code. These clever contracts can handle payments automatically, make sure quality standards are met, and make transport more efficient. Consequently, there are no intermediaries involved, who usually lead to increased prices.

Moreover, blockchain technology can significantly improve data management and sharing in smart agriculture. Farmers can securely record and share data related to crop yields, soil health, weather conditions, and pest management on a blockchain (Awotunde et al. 2022b). This data can be accessed by researchers, agricultural extension services, and other farmers to develop better farming practices and make informed decisions. Furthermore, blockchain can enhance the transparency and efficiency of agricultural subsidies and insurance programs. By using blockchain to verify and record farmers' eligibility and claims, governments and insurance companies can reduce fraud and ensure timely disbursement of funds. Additionally, blockchain can support the implementation of decentralized autonomous organizations (DAOs) in agriculture, where farmers can collectively make decisions and manage resources more democratically and transparently. Overall, blockchain-based architecture offers a robust framework for enhancing the efficiency, transparency, and sustainability of smart agriculture. Figure 5.3 displays a general framework for blockchain technology for smart agriculture.

The proposed framework can be used for efficient commodity price prediction, information scheduling, and management, using regression-based linear and logistic methods for efficient product classification. The framework significantly impacts production and consumption, ensuring security and privacy. It optimizes cost-efficient scheduling and forecasting in agriculture commodity distribution applications. The collaborative architecture showcases various aspects, including efficient forecasting, which necessitates more effective resource distribution and classification. The architecture optimizes data from food and safety departments, storing commodity production details through network layers for efficient forecasting and defining the overall scenario of data received in the process operation environment.

The framework uses bio-inspired forecasting and intelligent statements to automate payments, ensure quality standards, simplify logistics, eliminate intermediaries, and reduce transaction expenses. The system calculates the cost of delay and generates commodity forecasts, allowing stakeholders to request services dynamically. The main goal is to ensure that each product forecast is optimized for itself, while we strive to stay on track with the large volume of work currently we have.

The collaborative approach entails analyzing schedules, managing and forecasting programs, and enhancing monitoring efforts by efficiently capturing and organizing network layers. It effectively evaluates delays, prediction accuracy, and efficient network transmission. Smart contracts are deployed for secure, distributed forecasting and management. Internal resources like node interconnectivity, the states of hyperledgers, and REST API are calculated. The Hyperledger Sawtooth engineer handles transaction records and addresses, thus, addressing redundancy and confirming node delivery.

5.9 RESEARCH ISSUES AND FUTURE WORK

Research issues and future work in the context of a blockchain-enabled framework to enhance cybersecurity in smart agriculture involve addressing several technical, operational, and regulatory challenges. The following are some of the key research issues and potential areas for future work.

5.9.1 Research Issues

5.9.1.1 Scalability

Data Volume: Smart agriculture generates large volumes of data from various IoT devices and sensors. Ensuring that the blockchain can handle this data without compromising performance is a significant challenge.

Transaction Throughput: Traditional blockchains like Bitcoin and Ethereum have limited transaction throughput. Research is needed to develop or adopt blockchain technologies that can handle the high transaction rates typical in smart agriculture.

5.10 LATENCY

Real-Time Data Processing: Many applications in smart agriculture require real-time or near-real-time data processing. The inherent latency in blockchain operations (due to consensus mechanisms) can be a bottleneck.

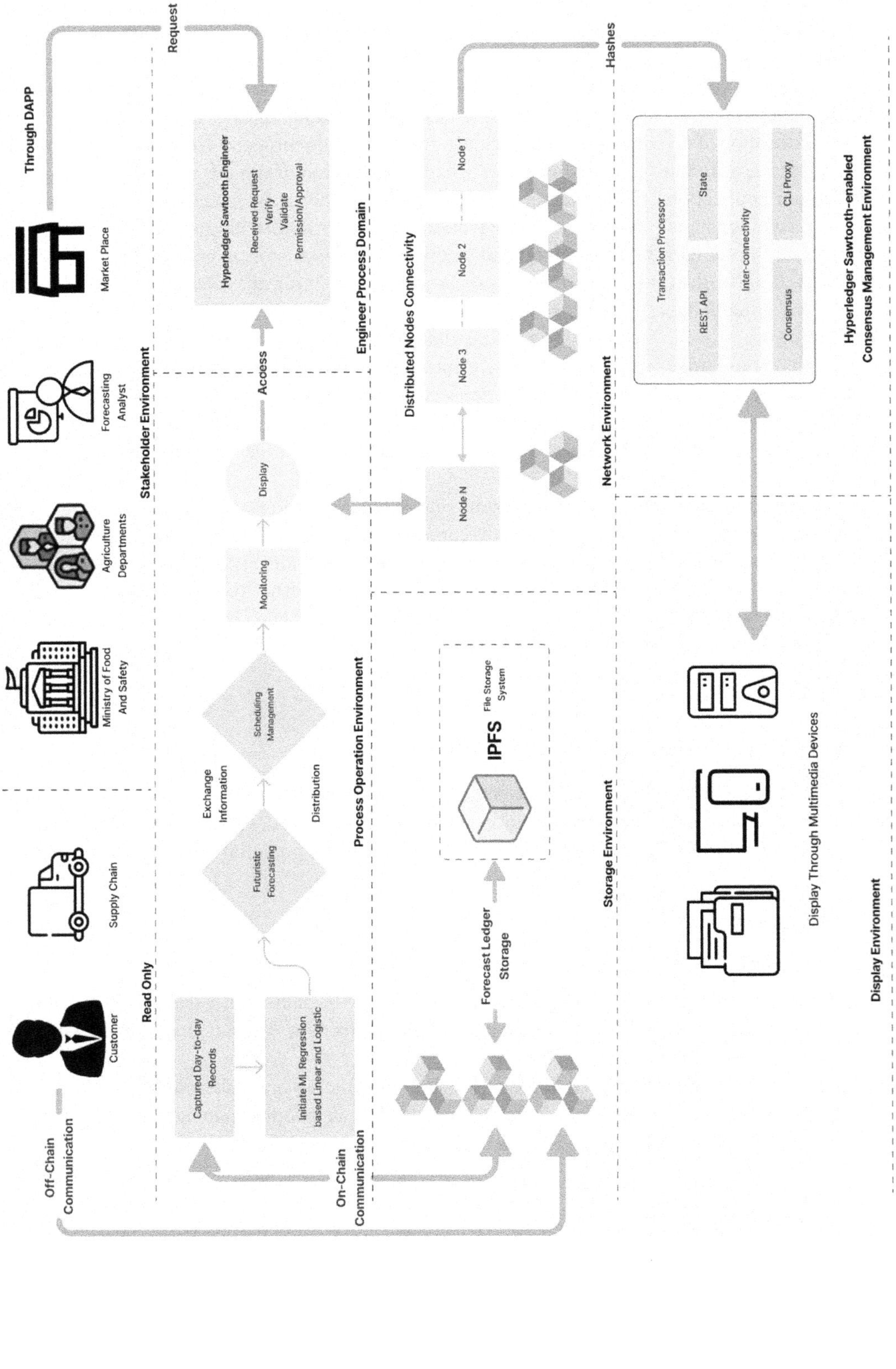

FIGURE 5.3 The general architecture of blockchain technology for smart agriculture.

Consensus Mechanisms: Exploring and developing faster and more efficient consensus mechanisms suitable for real-time applications in agriculture.

5.11 ENERGY CONSUMPTION

IoT Device Constraints: IoT devices used in agriculture often have limited battery life and computational power. Integrating these devices with blockchain solutions without excessive energy consumption is crucial.

Green Blockchain Technologies: Developing and implementing blockchain technologies that are energy-efficient and suitable for low-power IoT devices.

5.12 DATA PRIVACY AND SECURITY

Sensitive Information: Agricultural data can be sensitive, including proprietary farming techniques and personal data of farm owners. Ensuring data privacy while using a public or consortium blockchain is challenging.

Access Control: Designing robust access control mechanisms that allow only authorized entities to read/write data on the blockchain.

5.13 INTEROPERABILITY

Heterogeneous Devices: Smart agriculture involves a wide variety of devices and sensors. Ensuring that these heterogeneous systems can effectively communicate and operate within a blockchain framework.

Integration with Existing Systems: Seamlessly integrating blockchain solutions with existing agricultural management systems and IoT platforms.

5.14 COST

Implementation and Maintenance: Blockchain solutions can be costly to implement and maintain. Cost-effective solutions that provide a good balance between security and performance should be developed.

Transaction Fees: Minimizing transaction fees to ensure that the system is economically viable for farmers and agribusinesses.

5.15 REGULATORY AND LEGAL ISSUES

Compliance: Ensuring that blockchain solutions comply with relevant agricultural and data protection regulations.

Legal Frameworks: Developing legal frameworks that support the use of blockchain in agriculture, including issues related to smart contracts and liability.

5.16 FUTURE WORK

Advanced Consensus Mechanisms: Researching and developing new consensus algorithms that are energy efficient and capable of supporting the high transaction throughput required in smart agriculture.

Hybrid Blockchain Solutions: Exploring the use of hybrid blockchains that combine the benefits of public and private blockchains to achieve better scalability, privacy, and security.

Edge Computing Integration: Integrating edge computing with blockchain to process data closer to the source (i.e., on the farm) to reduce latency and improve real-time decision-making capabilities.

AI and Machine Learning (ML): Utilizing AI and ML to enhance the analysis of blockchain-stored agricultural data, improving predictive analytics and optimizing farming practices.

IoT Device Improvements: Developing more efficient and secure IoT devices tailored for blockchain integration, with enhanced power management and secure communication protocols.

Decentralized Applications: Creating decentralized applications specifically designed for smart agriculture, facilitating various use cases such as supply chain management, traceability, and resource management.

Policy Development: Collaborating with policymakers to develop regulatory frameworks that support blockchain adoption in agriculture, addressing issues related to data ownership, privacy, and liability.

In addressing these research issues and focusing on the outlined future work, the potential of blockchain technology to enhance cybersecurity and overall efficiency in smart agriculture can be fully realized.

5.17 CONCLUSION

Smart agriculture is a burgeoning field that leverages big data, AI, and connected data to bring the advantages of digitalization to the agricultural sector. Smart farming is integrating new technologies (including IoT, drones, AI, and data analytics) in the agricultural sector into traditional farming practices to optimize efficiency, sustainability, and productivity. However, smart agriculture faces several security challenges, such as data, networks, supply chains, and other common attacks. The chapter provides an overview of the current security and privacy concerns in smart agriculture and how a blockchain-enabled approach can mitigate these issues. Furthermore, the chapter discusses the pros and cons of using this method, while suggesting areas for future investigation. Prospects for smart agriculture using blockchain technology are promising in addressing the numerous security and privacy challenges currently facing the industry. By leveraging the decentralized, immutable, and transparent nature of blockchain, agricultural data can be securely managed, ensuring the integrity and confidentiality of sensitive information. This approach not only enhances trust among stakeholders but also promotes efficient and traceable supply chain operations. As smart agriculture continues to evolve, further research and development into blockchain applications will be essential to address emerging threats and optimize the benefits of this technology. Finally, a blockchain-enabled approach offers a robust framework to safeguard the future of smart agriculture, fostering sustainable and resilient agricultural practices.

REFERENCES

Abbas, A., Zhang, Z., Zheng, H., Alami, M. M., Alrefaei, A. F., Abbas, Q., . . . & Zhou, L. (2023). Drones in plant disease assessment, efficient monitoring, and detection: a way forward to smart agriculture. *Agronomy*, *13*(6), 1524.

Abdullah, H. M., Mohana, N. T., Khan, B. M., Ahmed, S. M., Hossain, M., Islam, K. S., . . . & Ahamed, T. (2023). Present and future scopes and challenges of plant pest and disease (P&D) monitoring: Remote sensing, image processing, and artificial intelligence perspectives. *Remote Sensing Applications: Society and Environment*, 100996.

AbdulRaheem, M., Awotunde, J. B., Chakraborty, C., Adeniyi, E. A., Oladipo, I. D., & Bhoi, A. K. (2023). Security and privacy concerns in smart healthcare system. In *Implementation of Smart Healthcare Systems using AI, IoT, and Blockchain* (pp. 243–273). Academic Press.

Adeniyi, A. E., Jimoh, R. G., & Awotunde, J. B. (2024), A systematic review on elliptic curve cryptography algorithm for internet of things: Categorization, application areas, and security. *Computers and Electrical Engineering*, *118*(Part A), 109330. ISSN: 0045-7906, https://doi.org/10.1016/j.compeleceng.2024.109330.

Abosata, N., Al-Rubaye, S., Inalhan, G., & Emmanouilidis, C. (2021). Internet of things for system integrity: A comprehensive survey on security, attacks and countermeasures for industrial applications. *Sensors*, *21*(11), 3654.

Adisa, O., Ilugbusi, B. S., Adewunmi, O., Franca, O., & Ndubuisi, L. (2024). A comprehensive review of redefining agricultural economics for sustainable development: Overcoming challenges and seizing opportunities in a changing world. *World Journal Of Advanced Research and Reviews*, *21*(1), 2329–2341.

Ahmad, U., & Sharma, L. (2023). A review of best management practices for potato crop using precision agricultural technologies. *Smart Agricultural Technology*, 100220.

Ajagbe, S. A., Awotunde, J. B., Oladipupo, M. A., & Oye, O. E. (2022). Prediction and forecasting of coronavirus cases using artificial intelligence algorithm. In *Machine Learning for Critical Internet of Medical Things: Applications and Use Cases* (pp. 31–54). Cham: Springer International Publishing.

Awotunde, J. B., Ajagbe, S. A., & Florez, H. (2022a, October). Internet of things with wearable devices and artificial intelligence for elderly uninterrupted healthcare monitoring systems. In *International Conference on Applied Informatics* (pp. 278–291). Cham: Springer International Publishing.

Awotunde, J. B., Chakraborty, C., & Folorunso, S. O. (2022b). A secured transaction based on blockchain architecture in mobile banking platform. *International Journal of Internet Technology and Secured Transactions*, *12*(4), 287–303.

Awotunde, J. B., Gaber, T., Prasad, L. N., Folorunso, S. O., & Lalitha, V. L. (2023). Privacy and security enhancement of smart cities using hybrid deep learning-enabled blockchain. *Scalable Computing: Practice and Experience*, *24*(3), 561–584.

Awotunde, J. B., Jimoh, R. G., Ogundokun, R. O., Misra, S., & Abikoye, O. C. (2022). Big data analytics of iot-based cloud system framework: Smart healthcare monitoring systems. In *Artificial intelligence for cloud and edge computing* (pp. 181–208). Cham: Springer International Publishing.

Awotunde, J. B., Misra, S., & Pham, Q. T. (2022, November). A secure framework for Internet of Medical Things security-based system using lightweight cryptography-enabled blockchain. In *International Conference on Future Data and Security Engineering* (pp. 258–272). Singapore: Springer Nature Singapore.

Awotunde, J. B., Imoize, A. L., Adeniyi, E. A., Jimoh, R. G., & Falola, P. B. (2023). Artificial intelligence-enabled security Systems for 6G wireless networks: Algorithms, strategies, and applications. In *Security and Privacy Schemes for Dense 6G Wireless Communication Networks* (pp. 61–87). IET Publisher.

Azadi, H., Moghaddam, S. M., Burkart, S., Mahmoudi, H., Van Passel, S., Kurban, A., & Lopez-Carr, D. (2021). Rethinking resilient agriculture: From climate-smart agriculture to vulnerable-smart agriculture. *Journal of Cleaner Production*, *319*, 128602.

Bazargani, K., & Deemyad, T. (2024). Automation's Impact on Agriculture: Opportunities, Challenges, and Economic Effects. *Robotics*, *13*(2), 33.

Bedasa, Y., & Bedemo, A. (2023). The effect of climate change on food insecurity in the Horn of Africa. *Geojournal*, *88*(2), 1829–1839.

Bosona, T., & Gebresenbet, G. (2023). The role of blockchain technology in promoting traceability systems in agri-food production and supply chains. *Sensors*, *23*(11), 5342.

Boursianis, A. D., Papadopoulou, M. S., Diamantoulakis, P., Liopa-Tsakalidi, A., Barouchas, P., Salahas, G., . . . & Goudos, S. K. (2022). Internet of things (IoT) and agricultural unmanned aerial vehicles (UAVs) in smart farming: A comprehensive review. *Internet of Things*, *18*, 100187.

Brockmann, J., & Iskander, D. (2023). Mapping patchy malaria: The role of drone technologies in depicting particular environments and contingent risk. In *Planetary health approaches to understand and control vector-borne diseases* (pp. 183–205). Wageningen Academic.

Cabrera-Castellanos, D. F., Aragón-Zavala, A., & Castañón-Ávila, G. (2021). Closing connectivity gap: An overview of mobile coverage solutions for not-spots in rural zones. *Sensors*, *21*(23), 8037.

Chakauya, R., Materechera, S. A., Jiri, O., Chakauya, E., & Machete, M. (2023). Climate change impacts on agriculture, adaptation and resilience: insights from local farmers in South-East Zimbabwe. In *Routledge Handbook of Climate Change Impacts on Indigenous Peoples and Local Communities* (pp. 273–286). Routledge.

Chams, N., Guesmi, B., & Gil, J. M. (2020). Beyond scientific contribution: Assessment of the societal impact of research and innovation to build a sustainable agri-food sector. *Journal of environmental management*, *264*, 110455.

de Araujo Zanella, A. R., da Silva, E., & Albini, L. C. P. (2020). Security challenges to smart agriculture: Current state, key issues, and future directions. *Array*, *8*, 100048.

Demestichas, K., Peppes, N., & Alexakis, T. (2020). Survey on security threats in agricultural IoT and smart farming. *Sensors*, *20*(22), 6458.

Dhal, S., Wyatt, B. M., Mahanta, S., Bhattarai, N., Sharma, S., Rout, T., . . . & Acharya, B. S. (2024). Internet of Things (IoT) in digital agriculture: An overview. *Agronomy Journal*, *116*(3), 1144–1163.

Et-taibi, B., Abid, M. R., Boufounas, E. M., Morchid, A., Bourhnane, S., Hamed, T. A., & Benhaddou, D. (2024). Enhancing Water Management in Smart Agriculture: A Cloud and IoT-Based Smart Irrigation System. *Results in Engineering*, 102283.

Falola, P. B., Adeniyi, A. E., Madamidola, O. A., Awotunde, J. B., Olukiran, O. A., & Akinola, S. O. (2024). Artificial Intelligence in Agriculture: The Potential for Efficiency and Sustainability, With Ethical Considerations. In *Exploring Ethical Dimensions of Environmental Sustainability and Use of AI* (pp. 307–329). IGI Global.

Ferrag, M. A., Shu, L., Friha, O., & Yang, X. (2021). Cyber security intrusion detection for agriculture 4.0: Machine learning-based solutions, datasets, and future directions. *IEEE/CAA Journal of Automatica Sinica*, *9*(3), 407–436.

Forslund, A., Tibi, A., Schmitt, B., Marajo-Petitzon, E., Debaeke, P., Durand, J. L., . . . & Guyomard, H. (2023). Can healthy diets be achieved worldwide in 2050 without farmland expansion?. *Global Food Security*, *39*, 100711.

Fragomeli, R., Annunziata, A., & Punzo, G. (2024). Promoting the Transition towards Agriculture 4.0: A Systematic Literature Review on Drivers and Barriers. *Sustainability*, *16*(6), 2425.

Friedman, N., Tan, Z., Haskins, M. N., Ju, W., Bailey, D., & Longchamps, L. (2024). Understanding Farmers' Data Collection Practices on Small-to-Medium Farms for the Design of Future Farm Management Information Systems. *Proceedings of the ACM on Human-Computer Interaction*, *8*(CSCW1), 1–28.

Fróna, D., Szenderák, J., & Harangi-Rákos, M. (2019). The challenge of feeding the world. *Sustainability*, *11*(20), 5816.

Gaber, T., Awotunde, J. B., Torky, M., Ajagbe, S. A., Hammoudeh, M., & Li, W. (2023). Metaverse-IDS: Deep learning-based intrusion detection system for Metaverse-IoT networks. *Internet of Things*, *24*, 100977.

Gebresenbet, G., Bosona, T., Patterson, D., Persson, H., Fischer, B., Mandaluniz, N., . . . & Nasirahmadi, A. (2023). A concept for application of integrated digital technologies to enhance future smart agricultural systems. *Smart agricultural technology*, *5*, 100255.

Gupta, M., Abdelsalam, M., Khorsandroo, S., & Mittal, S. (2020). Security and privacy in smart farming: Challenges and opportunities. *IEEE access*, *8*, 34564–34584.

Gupta, N., & Gupta, P. K. (2024). Robotics and Artificial Intelligence (AI) in Agriculture with Major Emphasis on Food Crops. *Digital Agriculture: A Solution for Sustainable Food and Nutritional Security*, 577–605.

Headey, D., & Hirvonen, K. (2023). Higher food prices can reduce poverty and stimulate growth in food production. *Nature Food*, *4*(8), 699–706.

Ivanic, M., Martin, W., & Zaman, H. (2012). Estimating the short-run poverty impacts of the 2010–11 surge in food prices. *World Development*, *40*(11), 2302–2317.

Ivošević, B., Kostić, M., Ljubičić, N., Grbović, Ž., & Panić, M. (2023). A drone view for agriculture. In *Unmanned Aerial Systems in Agriculture* (pp. 25–47). Academic Press.

Jaiswal, A., Jindal, R., & Verma, A. K. (2020). Crop health monitoring system using IoT. *Int. Res. J. Eng. Technol*, 2485–2489.

Javaid, M., Haleem, A., Singh, R. P., & Suman, R. (2022). Enhancing smart farming through the applications of Agriculture 4.0 technologies. *International Journal of Intelligent Networks*, *3*, 150–164.

Júnior, M. R. B., de Almeida Moreira, B. R., dos Santos Carreira, V., de Brito Filho, A. L., Trentin, C., de Souza, F. L. P., . . . & Shiratsuchi, L. S. (2024). Precision agriculture in the United States: A comprehensive meta-review inspiring further research, innovation, and adoption. *Computers and Electronics in Agriculture*, *221*, 108993.

Karnaushenko, A., Tanklevska, N., Povod, T., Kononenko, L., & Savchenko, V. (2023). Implementation of blockchain technology in agriculture: fashionable trends or requirements of the modern economy.

Karuna, G., RP Ram Kumar, Polepaka Sanjeeva, Palakurthy Deepthi, Hazim Y. Saeed, V. Asha, Lavish Kansal, and Praveen Praveen. "Crop recommendation system and crop monitoring using IoT." In *E3S Web of Conferences*, vol. 507, p. 01063. EDP Sciences, 2024.

Kose, U., Prasath, V. B., Mondal, M., Podder, P., & Bharati, S. (Eds.). (2022). *Artificial intelligence and smart agriculture technology*. CRC Press.

Koshariya, A. K., Rameshkumar, P. M., Balaji, P., Cavaliere, L. P. L., Dornadula, V. H. R., & Singh, B. (2024). Data-Driven Insights for Agricultural Management: Leveraging Industry 4.0 Technologies for Improved Crop Yields and Resource Optimization. In *Robotics and Automation in Industry 4.0* (pp. 260–274). CRC Press.

Kumar, P., Singh, A., Rajput, V. D., Yadav, A. K. S., Kumar, P., Singh, A. K., & Minkina, T. (2022). Role of artificial intelligence, sensor technology, big data in agriculture: next-generation farming. In *Bioinformatics in Agriculture* (pp. 625–639). Academic Press.

Kumar, S., Mohan, S., & Skitova, V. (2023). Designing and Implementing a Versatile Agricultural Robot: A Vehicle Manipulator System for Efficient Multitasking in Farming Operations. *Machines*, *11*(8), 776.

Kumari, A., Khan, M. N., & Sinha, A. K. (2024). Harvesting Intelligence: AI and ML Revolutionizing Agriculture. In *Data-Driven Farming* (pp. 126–141). Auerbach Publications.

Lanucara, S., Praticò, S., Pioggia, G., Di Fazio, S., & Modica, G. (2024). Web-based spatial decision support system for precision agriculture: A tool for delineating dynamic management unit zones (MUZs). *Smart Agricultural Technology*, *8*, 100444.

Lloret, J., Sendra, S., Garcia, L., & Jimenez, J. M. (2021). A wireless sensor network deployment for soil moisture monitoring in precision agriculture. *Sensors*, *21*(21), 7243.

Mail, M. F., Maja, J. M., Marshall, M., Cutulle, M., Miller, G., & Barnes, E. (2023). Agricultural harvesting robot concept design and system components: A review. *AgriEngineering*, *5*(2), 777–800.

Márta, A., Jung, A., & Molnár, B. (2022). Information Technology Drivers in Smart Farming Management Systems. In *Smart Farming-Integrating Conservation Agriculture, Information Technology, and Advanced Techniques for Sustainable Crop Production*. IntechOpen.

Medina, D. C., Delgado, M. G., Ramos, J. S., Amores, T. P., Rodríguez, L. R., & Domínguez, S. Á. (2024). Empowering urban climate resilience and adaptation: Crowdsourcing weather citizen stations-enhanced temperature prediction. *Sustainable Cities and Society*, *101*, 105208.

Mishra, H., & Mishra, D. (2023). Artificial intelligence and machine learning in agriculture: Transforming farming systems. *Res. Trends Agric. Sci*, *1*, 1–16.

Misra, N. N., Dixit, Y., Al-Mallahi, A., Bhullar, M. S., Upadhyay, R., & Martynenko, A. (2020). IoT, big data, and artificial intelligence in agriculture and food industry. *IEEE Internet of things Journal*, *9*(9), 6305–6324.

Mukherjee, A. A., Singh, R. K., Mishra, R., & Bag, S. (2022). Application of blockchain technology for sustainability development in agricultural supply chain: Justification framework. *Operations Management Research*, *15*(1), 46–61.

Naseer, A., Shmoon, M., Shakeel, T., Ur Rehman, S., Ahmad, A., & Gruhn, V. (2024). A Systematic Literature Review of the IoT in Agriculture-Global Adoption, Innovations, Security Privacy Challenges. *IEEE Access*.

Niu, H., Peddagudreddygari, J. R., Bhandari, M., Landivar, J. A., Bednarz, C. W., & Duffield, N. (2024). In-Season Cotton Yield Prediction with Scale-Aware Convolutional Neural Network Models and Unmanned Aerial Vehicle RGB Imagery. *Sensors*, *24*(8), 2432.

Obi Reddy, G. P., Dwivedi, B. S., & Ravindra Chary, G. (2023). Applications of geospatial and big data technologies in smart farming. In *Smart Agriculture for Developing Nations: Status, Perspectives and Challenges* (pp. 15–31). Singapore: Springer Nature Singapore.

Omia, E., Bae, H., Park, E., Kim, M. S., Baek, I., Kabenge, I., & Cho, B. K. (2023). Remote sensing in field crop monitoring: A comprehensive review of sensor systems, data analyses and recent advances. *Remote Sensing*, *15*(2), 354.

Patil, R., Sable, N. P., Mahalle, P. N., Shinde, G. R., Dhotre, P., & Chandre, P. (2024). Role of data-centric artificial intelligence in agriculture. In *Data-Centric Artificial Intelligence for Multidisciplinary Applications* (pp. 159–176). Chapman and Hall/CRC.

Pungavi, R., & Praveenkumar, C. (2024). Unmanned Aerial Vehicles (UAV) for smart agriculture. In *Artificial Intelligence and Smart Agriculture: Technology and Applications* (pp. 251–269). Singapore: Springer Nature Singapore.

Rai, A. K., Kumar, N., Katiyar, D., Singh, O., Sreekumar, G., & Verma, P. (2023). Unlocking productivity potential: The promising role of agricultural robots in enhancing farming efficiency. *International Journal of Plant & Soil Science*, *35*(18), 624–633.

Raile, E. D., Young, L. M., Kirinya, J., Bonabana-Wabbi, J., & Raile, A. N. (2021). Building public will for climate-smart agriculture in Uganda: Prescriptions for industry and policy. *Journal of Agricultural & Food Industrial Organization*, *19*(1), 39–50.

Rampalli, N. S., Sri, Y. G., & Bhuvaneshwari, K. S. (2024). Autonomous Agriculture and Food Production: Agritech Revolution. In *The Convergence of Self-Sustaining Systems with AI and IoT* (pp. 40–63). IGI Global.

Rao, E. P., Rakesh, V., & Ramesh, K. V. (2021). Big Data analytics and Artificial Intelligence methods for decision making in agriculture. *Indian J. Agron, 66*, 279–287.

Rehman, A., Saba, T., Kashif, M., Fati, S. M., Bahaj, S. A., & Chaudhry, H. (2022). A revisit of internet of things technologies for monitoring and control strategies in smart agriculture. *Agronomy, 12*(1), 127.

Rogotis, S., & Marianos, N. (2021). Smart farming for sustainable agricultural production. *Big Data in Bioeconomy: Results from the European DataBio Project*, 191–205.

Salehi, F. (2024). The role of artificial intelligence in revolutionizing the agriculture industry in Canada. *Asian Journal of Research and Review in Agriculture, 6*(1), 70–78.

Sarfraz, S., Ali, F., Hameed, A., Ahmad, Z., & Riaz, K. (2023). Sustainable agriculture through technological innovations. In *Sustainable agriculture in the era of the OMICs revolution* (pp. 223–239). Cham: Springer International Publishing.

Shafi, U., Mumtaz, R., Iqbal, N., Zaidi, S. M. H., Zaidi, S. A. R., Hussain, I., & Mahmood, Z. (2020). A multimodal approach for crop health mapping using low altitude remote sensing, internet of things (IoT) and machine learning. *IEEE Access, 8*, 112708–112724.

Shakoor, N., Northrup, D., Murray, S., & Mockler, T. C. (2019). Big data driven agriculture: big data analytics in plant breeding, genomics, and the use of remote sensing technologies to advance crop productivity. *The Plant Phenome Journal, 2*(1), 1–8.

Shrivastav, V., Yadav, M., Sharma, A., Kumar, D., Sharma, S., & Chauhan, A. S. (2024, February). IoT and IoE transformations in precision farming agriculture: Sensor based monitoring, Automated irrigation and Livestock monitoring. In *2024 IEEE International Students' Conference on Electrical, Electronics and Computer Science (SCEECS)* (pp. 1–14). IEEE.

Supriya, M., Tyagi, A. K., & Tiwari, S. (2024). Sensor-Based Intelligent Recommender Systems for Agricultural Activities. In *AI Applications for Business, Medical, and Agricultural Sustainability* (pp. 197–235). IGI Global.

Thilakarathne, N. N., Bakar, M. S. A., Abas, P. E., & Yassin, H. (2023). Towards making the fields talks: A real-time cloud enabled iot crop management platform for smart agriculture. *Frontiers in Plant Science, 13*, 1030168.

Torky, M., & Hassanein, A. E. (2020). Integrating blockchain and the internet of things in precision agriculture: Analysis, opportunities, and challenges. *Computers and Electronics in Agriculture, 178*, 105476.

Venkatachalam, S., Kavitha, P., Ingle, P. K., Ramachandran, G., Sasikala, R., & Muthumanickam, T. (2024, January). Analysis of Internet of Things Based Agriculture Fertilizer Nutrient Management Soil Health Irrigation System and its Applications. In *2024 2nd International Conference on Intelligent Data Communication Technologies and Internet of Things (IDCIoT)* (pp. 64–69). IEEE.

Venugopalan, V., & Patterson, C. D. (2018). Surveying the hardware trojan threat landscape for the internet-of-things. *Journal of Hardware and Systems Security, 2*(2), 131–141.

Vishnoi, S., & Goel, R. K. (2024). Climate smart agriculture for sustainable productivity and healthy landscapes. *Environmental Science & Policy, 151*, 103600.

Wanyama, J., Bwambale, E., Kiraga, S., Katimbo, A., Nakawuka, P., Kabenge, I., & Oluk, I. (2024). A Systematic Review of Fourth Industrial Revolution Technologies in Smart Irrigation: Constraints, Opportunities, and Future Prospects for sub-Saharan Africa. *Smart Agricultural Technology*, 100412.

West, J. (2020). Advances in data security for more effective decision-making in agriculture. In *Improving data management and decision support systems in agriculture* (pp. 59–94). Burleigh Dodds Science Publishing.

Yaacoub, J. P. A., Noura, H. N., Salman, O., & Chehab, A. (2022). Robotics cyber security: Vulnerabilities, attacks, countermeasures, and recommendations. *International Journal of Information Security, 21*(1), 115–158.

Yadav, N., & Sidana, N. (2023). Precision Agriculture Technologies: Analysing the Use of Advanced Technologies, Such as Drones, Sensors, and GPS, In Precision Agriculture for Optimizing Resource Management, Crop Monitoring, and Yield Prediction. *Journal of Advanced Zoology, 44*.

Yazdinejad, A., Zolfaghari, B., Azmoodeh, A., Dehghantanha, A., Karimipour, H., Fraser, E., . . . & Duncan, E. (2021). A review on security of smart farming and precision agriculture: Security aspects, attacks, threats and countermeasures. *Applied Sciences, 11*(16), 7518.

Zhang, W., & Wang, L. (2024). Advancing Agricultural Practices through IoT-Driven Crop Field Monitoring and Automated Irrigation Systems for Seamless Farm Management. *Advances in Urban Resilience and Sustainable City Design*, *16*(02), 1–17.

Zulfikri, A., Ningsih, E. M. N., Harsono, I., & Susanto, H. (2024). Agricultural Adaptation Strategies to Weather Fluctuations for Improved Agribusiness Sustainability in West Java. *West Science Nature and Technology*, *2*(01), 17–23.

6 Transformative Potential
IoT, Cloud, Big Data, and AI-Driven Blockchain Digital Twins in Agriculture

Garima Jain, Ankush Jain, Anuj Jaiswal, Shakeel Ahmad, Vinod M. Kapse, and Sovers Singh Bisht

6.1 INTRODUCTION

6.1.1 The Significance of Smart Farming

The importance of smart farming extends far beyond conventional agricultural practices. By incorporating sensors and automated irrigation systems, smart farming enables the monitoring of soil quality, temperature, and moisture levels. Agriculture serves as the cornerstone for essential resources such as food, fabrics, construction wood, and paper products. In a world increasingly focused on balancing economic growth with environmental protection and social equity, aligning industry practices with Sustainable Development Goals (SDGs) has become essential (Bendre et al., 2016). The emergence of Industry 5.0 presents a promising opportunity to adopt cleaner production methods and achieve sustainable competitive advantages. However, realizing this vision requires a thorough understanding and strategic implementation. All these essential resources stem from agricultural produce. Implementing smart farming primarily aims to leverage data collected through specialized tools and convert it into actionable insights.

The synergy of IoT, cloud computing, and big data is pivotal in advancing sustainable agricultural practices, as depicted in Figure 6.1.

Intelligent farming aims to offer the agricultural industry an extensive framework for harnessing cutting-edge technologies such as big data, cloud computing, and IoT. This integration elevates operational efficiency, fosters sustainability, and boosts overall productivity, leading the way toward a future in agriculture that is both sustainable and technologically advanced.

6.1.2 IoT, Cloud Computing, and Big Data in Agriculture

Big data analytics is pivotal in the development of IoT solutions for automation and optimization. IoT serves as the primary data source, while big data analyzes the data, and cloud computing provides storage, scalability, and rapid accessibility.

The integration of AI and IoT in agriculture encompasses several key areas, including precision farming, agricultural drones, hydroponic systems, livestock monitoring, climate monitoring, intelligent greenhouses, and AI and IoT-driven computer imagery (Klerkx et al., 2019). The primary objective of IoT is to establish self-reporting systems that enhance efficiency and expedite the emergence of critical information, surpassing the speed of human-dependent processes (Stosch et al., 2017).

Cloud computing plays a pivotal role in collecting, processing, and storing agricultural data. Cloud-connected wireless sensors gather field data, which is then processed in real time using

DOI: 10.1201/9781003507390-6

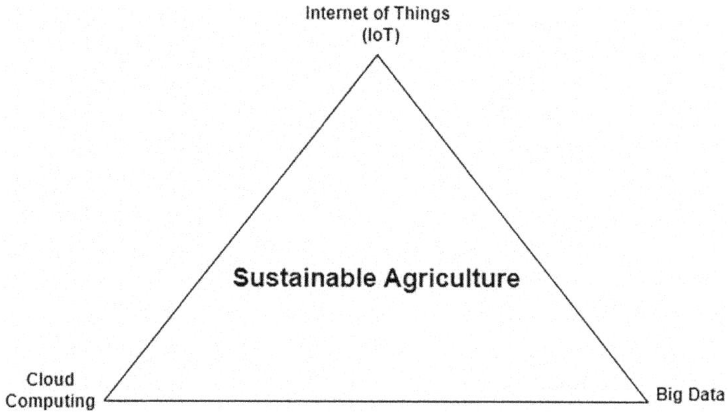

FIGURE 6.1 Sustainable agriculture using IoT, cloud computing, and big data.

machine learning algorithms. The recent transformation of the agriculture sector has been propelled by the integration of IoT, cloud computing, and big data technologies (Pivoto et al., 2018). This convergence is reshaping traditional farming practices, ushering in Industry 5.0, and contributing to the attainment of Sustainable Development Goals (SDGs).

IoT facilitates the deployment of sensors and devices across agricultural landscapes to gather real-time data on soil moisture, temperature, crop health, and other critical parameters. This information is transmitted to centralized cloud computing platforms, where it is processed and analyzed in near real time. Cloud computing provides the infrastructure for efficiently storing, managing, and analyzing vast amounts of agricultural data. Subsequently, big data analytics algorithms are applied to derive actionable insights from the data collected by IoT sensors, empowering farmers and agricultural stakeholders to make informed decisions concerning crop management, resource allocation, and operational optimization through data-driven insights.

By leveraging IoT, cloud computing, and big data, farmers can enhance productivity, minimize resource wastage, and improve agricultural sustainability. Furthermore, this integration of technologies aligns with the principles of Industry 5.0, emphasizing the fusion of human intelligence with advanced technologies to create smarter and more sustainable production systems. Industry 5.0 aims to foster innovation, increase productivity, and drive economic growth while minimizing environmental impact and promoting social inclusivity.

In conclusion, the fusion of IoT, cloud computing, and big data in agriculture underscores the transformative potential of these technologies in advancing Industry 5.0 and contributing to achieving the Sustainable Development Goals. The agriculture sector can lay the groundwork for a deeply sustainable and resilient environment by embracing digitalization and data-driven approaches.

6.1.3 Objectives of the Chapter

This chapter explores the concept of smart farming, which endeavors to improve production capacity and quality by leveraging advanced agriculture and animal husbandry technologies. The central focus lies in optimizing resource utilization while minimizing environmental impact. Sustainable agriculture, a fundamental tenet of smart farming, underscores the importance of reducing dependence on external energy sources and transitioning from non-renewable to renewable energy sources like solar and wind power, biofuels sourced from agricultural waste, and, where applicable, human or animal labor (AGRI Committee, 2019; Rijswijk et al., 2019; Řezník et al., 2017). Smart agriculture leverages IoT technology to gather data on weather, humidity, temperature, and soil fertility. This data is used to develop agricultural products designed to monitor fields using sensors and automate irrigation systems, including products for monitoring plants, water levels, precise location, field interruption, field development, and horticulture. With IoT-based systems, farmers and related entities can easily monitor field conditions from anywhere. For instance, when the water level drops, sensors detect it and automatically activate the water pump, while the fan starts when the temperature exceeds a predefined level. Cloud-based APIs and services, such as weather forecasting, help farmers plan their crop selection and cultivation and harvest schedules. IoT-based smart agriculture employs numerous sensors to monitor environmental climate conditions. These sensors collect data across fields and transmit it to the cloud for analysis. The realm of big data analytics presents a beacon of hope for farmers, offering them detailed insights into crucial aspects like rainfall patterns, water cycles, fertilizer availability, and beyond. Armed with this wealth of information, farmers gain the capacity to make well-informed decisions, ranging from selecting the most suitable crops for planting to pinpointing the ideal times for harvesting, ultimately resulting in amplified yields.

6.2 IOT IN AGRICULTURE

6.2.1 Understanding IoT and Its Application in Farming

This chapter explores the concept of the Internet of Things (IoT) in agriculture, which focuses on Internet-based connectivity and coordination of everyday equipment (Uprety et al., 2023). In IoT, the term "things" refers to various common items that are connected to or interfaced with the Internet. IoT represents an advanced amalgamation of automation and analytics, incorporating artificial intelligence (AI), sensors, networking, electronics, and cloud messaging. This integration facilitates the comprehensive development of systems for various products and services.

IoT encompasses a network of physical objects integrated with sensors, software, and advanced technologies that facilitate connectivity and data exchange over the Internet.

Figure 6.2 exemplifies the application of IoT technology in agricultural environments, illustrating its profound influence on revolutionizing farming methodologies (Wolfert et al., 2017).

6.2.2 Elevating Agriculture through IoT: Exploring Its Application in Farming

The agricultural sector integrates robots, drones, remote sensors, and computer imaging, combined with dynamic machine learning tools and analytics (Dankan Gowda et al., 2020; Ferrández-Pastor et al., 2016; Tekinerdogan, 2018). These technologies monitor crops, survey fields, and provide

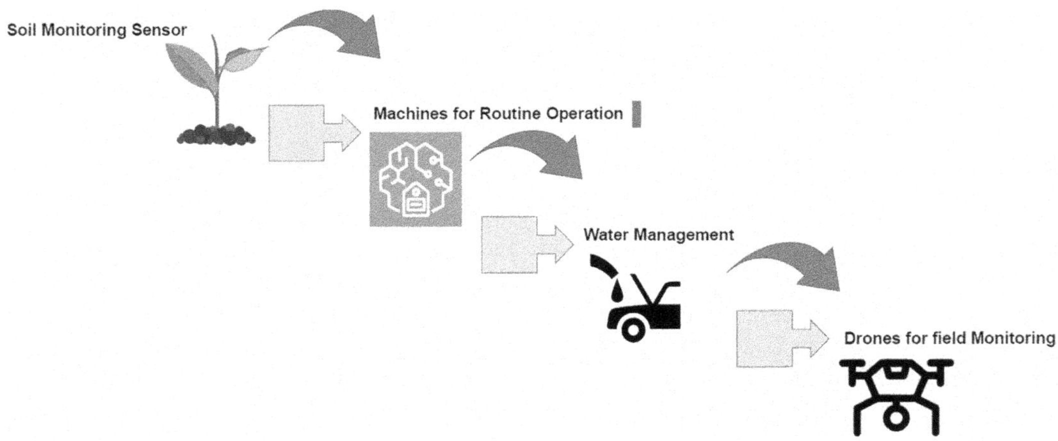

FIGURE 6.2 IoT in agriculture.

actionable data to farmers, streamlining farm management strategies and optimizing resource usage. Various sensing technologies underpin these sensors and devices.

- Optical sensors: Assess soil properties like clay content, organic matter, and moisture.
- Electrochemical sensors: These devices offer vital information for precision farming, including soil pH and nutrient levels.
- Mechanical sensors measure the soil's mechanical resistance or compaction.
- Dielectric soil moisture sensors determine moisture levels by measuring the soil's dielectric constant.
- Airflow sensors: They may estimate the soil's air permeability by measuring the air pressure required to break through the ground at particular depths.

6.2.3 Integration of IoT with Farming Technology

The incorporation of artificial intelligence enriches the harnessing of sensor data through vigilant monitoring and intelligent management. Challenges in smart agriculture encompass sensor integration, data linkage with analytics, dynamic automation, and responsive actions. The Internet of Things holds immense potential to revolutionize agriculture by rationalizing processes and systems for enhanced efficiency and efficacy. Flow sensors, light sensors, and pressure sensors, among others, permit the automation and standardization of various farming operations (Ramesha et al., 2020). Smart farming utilizes the internet to automate irrigation systems, enable remote field evaluations, and monitor fields with sensors that measure light, humidity, temperature, and soil moisture.

6.2.4 Challenges and Benefits of Implementing IoT in Agriculture

When considering IoT adoption in agriculture, it becomes evident that while it offers numerous benefits, it also faces several challenges. These challenges include information gaps, high implementation costs, and security concerns. Educating farmers about the benefits of IoT through practical demonstrations involving drones, sensors, and related technologies is essential in addressing these issues (Ramesha et al., 2020). Furthermore, addressing issues such as hardware compatibility, networking, data security, and governance is pivotal to ensuring the seamless integration of IoT in agriculture. By implementing these solutions, the agricultural sector can mitigate risks, promptly identify anomalies and inconsistencies in crop production, reduce waste, control costs, and ultimately boost overall output.

In the realm of smart agriculture, several pressing issues demand immediate attention. These include problems related to networking, infrastructure, hardware, signal interference, data security, and administrative challenges (Dankan Gowda et al., 2020). Additionally, IoT implementation in agriculture faces significant sectoral and technological obstacles, such as interoperability, connectivity, data processing power, and explicit data governance, as well as data security and privacy concerns. Nevertheless, despite these challenges, IoT agricultural solutions have the potential to mitigate some of the risks associated with farming. By swiftly identifying abnormalities and discrepancies, farmers can effectively reduce waste, manage costs, and enhance crop productivity.

6.3 CLOUD COMPUTING FOR AGRICULTURE DATA MANAGEMENT

6.3.1 Introduction: Harnessing Cloud Computing for Enhanced Agricultural Practices

Cloud computing is a valuable tool for gathering, analyzing, and storing agricultural data efficiently. Wireless sensors connected to the cloud enable real-time data collection from fields, facilitating enhanced insights into crop development through machine learning algorithms. This approach offers elasticity and reliability and enhances performance and efficiency while reducing IT costs. In agricultural management, cloud computing revolutionizes traditional practices by providing farmers with accessible computing resources and robust data storage capabilities via the Internet (Naveena Pai et al., 2020). Utilizing cloud-based platforms allows farmers to streamline operations, optimize resource allocation, and boost overall productivity. Whether it's monitoring crop conditions or managing inventory and logistics, cloud computing enables remote access to vital information and the execution of essential tasks through internet connectivity. This transformative potential highlights the diverse advantages of cloud computing in agriculture and its ability to revolutionize traditional farming practices. Additionally, the analytical capabilities of cloud technology help farmers better understand their production environment and improve farming outcomes. There are six main deliverables in general:

- Information on crops: Farmers can use data processing to gather and evaluate data on crops cultivated in the past and present, assisting in decision-making from now on.
- Info on the weather: With the ability to get forecasts for certain seasons, edge computing can deliver real-time, regional weather information essential for farming.
- Details on the soil: A farmer must make decisions about their land based on more than just the soil profile; they must also be able to forecast the future course of the soil's behavior. For example, understanding if soil has a propensity to become acidic or alkaline may be vital for future farming activities.
- Tracking of growth: Controlling the growth of diverse crops in various places at regular intervals provides farmers with a wealth of knowledge about growth intervals, which is crucial for profitable and sustainable farming.
- Farmer data: Agricultural policy and strategy can be developed step by step using data from farmers that can be collected across an entire region.
- E-commerce: Farming operations in remote locations might find direct routes to the market, avoiding exploitation. Farmers may benefit from a cloud-based information management system for agriculture that enables direct sales of their goods.

6.3.2 Challenges and Considerations in Cloud Computing for Agriculture

While cloud computing holds immense promise for enhancing agricultural practices, it also presents diverse challenges and considerations that must be resolved. One of the primary challenges is the availability and reliability of internet connectivity, particularly in rural areas where access to high-speed internet may be limited (Batte, 2020; Choudhury et al., 2022). Data privacy, security, and ownership concerns may also arise when entrusting sensitive agricultural data to cloud-based platforms. Furthermore, implementing and maintaining cloud computing infrastructure can be prohibitive for small-scale farmers. Understanding these challenges and considering potential solutions is crucial for maximizing the benefits of cloud computing while mitigating risks. The possibilities offered by cloud computing in the agricultural sector seem intriguing. However, three ominous clouds are gathering: low security, poor speed, and huge expenses. Despite these challenges, industry leaders have turned to edge computing as a solution.

- First problem: Farming security with cloud technologies There are hazards everywhere there is data. Since IoT systems that communicate data to the cloud are made up of numerous devices, there are many possible points for leaks when considering cloud computing in the agriculture sector. The likelihood of data being compromised, stolen, or misused increases dramatically with each back-and-forth transfer as your system for digital farming expands.
- Second problem: In the agricultural sector, cloud computing moves quickly. In today's digital agriculture, slow internet connections are one of the most common problems. Slow connection rates, for instance, might seriously hinder agriculture that depends on cloud

computing. Acquiring, transferring, and analyzing data takes much time and isn't always "forgivable" in agribusinesses, where immediate responses are required. Therefore, certain businesses must choose between processing speed and depth of insights in a cloud computing-based agriculture system.

- Third problem: Cloud farming costs The cost of cloud computing for farming is based on how much data is created by the IoT system, transported to the cloud, and then transferred back (Chen et al., 2004). The price of cloud computing will continue to rise as the typical digitally driven farm needs more sensors. More data points are required as farms expand, which raises the cost because a better "brain" is needed to process the data.

6.3.3 Addressing Security and Privacy Concerns in Cloud-Based Agricultural Systems

Security and privacy are crucial considerations in the adoption of cloud-based agricultural systems. Safeguarding agricultural data is paramount for maintaining the integrity and confidentiality of critical information. Implementing strong security measures such as encryption, access controls, and routine security audits is vital in preventing unauthorized access, breaches, and cyber threats. These measures ensure the protection of agricultural data stored in the cloud, providing peace of mind for farmers and stakeholders. Additionally, farmers should meticulously review and negotiate service-level agreements (SLAs) with cloud service providers to clarify data ownership, usage rights, and liability during security incidents. Farmers can confidently embrace cloud-based agricultural systems while safeguarding sensitive information and upholding data integrity by proactively addressing security and privacy concerns.

The integration of Agriculture 4.0 systems introduces vulnerabilities to various environmental factors, climatic changes, and human activities (Ehlers, 2008). The dependability and trustworthiness of these systems can be compromised by inadequate security measures, leaving them susceptible to physical damage, remote manipulation, and unintentional usage of compromised data. Moreover, while the rapid advancements in cloud technology have brought about increased product elasticity, cost savings, and scalability, they have also introduced numerous privacy and security challenges. The evolving nature of cloud computing necessitates the adoption of unforeseen safety precautions. It is crucial to delve into the top seven privacy issues cloud computing users face, emphasizing the importance of vigilance and proactive measures in safeguarding agricultural data and systems.

- Data Privacy Concerns: When outsourcing minimal and sensitive data to a cloud service provider, data privacy is a crucial concern. Implementing strong access control policies and conventions is essential for protecting privacy. Ensuring that private data is inaccessible to unauthorized individuals is a key measure to safeguard sensitive information.
- Problems with Data Thrashing: One of the leading security challenges cloud providers face is data thrashing or data theft. Over 60% of consumers are reluctant to use the services of a cloud provider if there have been previous incidents of data thrashing or theft involving significant or sensitive data. Businesses like Dropbox, Microsoft, and Amazon frequently use cloud services.
- Data Storage Concerns: Since cloud communications are dispersed across various geographical locations worldwide, users often worry about their data being stored outside their legal jurisdiction. This raises concerns about the applicability and enforcement of local laws and policies on data stored in other regions.
- Multi-Tenancy Safety Issues: The multi-tenancy model allows different tenants to share computing resources, data storage, applications, and services, all hosted on the same physical or logical platform at the cloud service provider's location. While this strategy maximizes the provider's earnings, it poses risks to the client due to shared resources.

- Lucidity Concerns: In cloud computing security, lucidity refers to a cloud service provider's willingness to disclose various information and characteristics related to their security measures. This includes details about safety, isolation, and service-level standards and rules for compliance. Evaluating lucidity also involves considering factors like security vigilance and provider transparency.
- Hypervisor-Related Problems: The hypervisor, which controls multiple virtual machines, is a prime target for attackers. If the hypervisor is compromised, multiple virtual machines are at risk. Advancements in hypervisor technology, such as improved isolation, security hardening, and access control, have introduced new ways for adversaries to exploit the system.
- Managerial Challenges: Cloud privacy issues encompass technological, managerial, and non-technical components. Examples include the lack of control over virtualization safety and privacy management, the creation of detailed SLAs, and the interaction between cloud service providers and users (Ojo et al., 2018).

6.4 BIG DATA ANALYTICS FOR PRECISION FARMING

6.4.1 Introduction

A fundamental tool in precision agriculture, a geographic information system (GIS) is essential computer software designed to gather, assess, and visually represent spatial data on maps. Utilizing GIS empowers farmers to efficiently handle diverse sets of data and analytics, significantly enhancing decision-making processes and farm planning strategies.

Based upon the data outlined previously, farmers gain access to vital information regarding key aspects such as crop status, weather predictions, and ecological shifts (Adamchuk et al., 2004; Dhanaraju et al., 2022; Srisruthi et al., 2017). Moreover, farmers can tap into extensive datasets sourced from big data analytics encompassing rainfall patterns, water cycles, and compost availability. Armed with this in-depth understanding, farmers are equipped to make informed decisions with precision and reliability, thereby optimizing the productivity of their agricultural endeavors.

6.4.2 Smart Agricultural Importance of Big Data in the Industry

Big data serves as a valuable resource for farmers, offering comprehensive insights into fertilizer availability, water cycle patterns, and rainfall trends. Armed with this data, farmers can make informed decisions regarding crop selection to enhance fertility and determine the optimal timing of harvests, among other strategic choices. Additionally, agricultural data plays a pivotal role in estimating, planning, and forecasting agricultural operations for specific area units at precise points in time, enabling efficient and effective management of farming activities.

- Big data significantly enhances farming productivity by facilitating crop production forecasting and the monitoring of crop yields. This empowers farmers to make informed business decisions, ensuring that they remain as perceptive to market price fluctuations as any other industry.
- A significant proportion of food, estimated at around 30%, is lost or wasted throughout different supply chain stages. Big data plays a critical role in combating food waste by gathering insights on stores and other supply chain components to pinpoint areas and reasons for cost reductions.
- Enabling farm equipment and machinery with sensors to detect and record critical data, such as fuel levels and service due dates, provides farmers with valuable insights. These data applications can effectively process and analyze the collected information, supporting essential decisions for the optimization of agricultural equipment.

- Improves Supply Chain Organization: Tracking the number of food delivery trucks and their routes helps bridge the supply-and-demand gap between producers and the market. This helps prevent food waste that may occur due to a lack of big data.
- Utilizing big data to track the fleet of food delivery trucks and their optimal routes facilitates the alignment of supply with market demand, effectively reducing food waste attributable to a lack of big data insights.
- Big data provides farmers with essential guidance on the safe quantities and appropriate timings for pesticide application, contributing to cost reduction through the avoidance of excessive pesticide usage and potential health risks.
- Big data in agriculture offers critical insights into climatic trends, fertilizer availability, and soil moisture levels, ensuring a sustainable food supply by exploring methods to enhance per-acre agricultural output.

6.4.3 Big Data Analytics for Crop Management in Precision Farming

Precision agriculture, also known as satellite crop management or site-specific crop management, is an agricultural management approach that focuses on monitoring crop variations within individual fields and across multiple fields.

Figure 6.3 illustrates the pivotal role of digital agriculture in optimizing crop production processes, showcasing its transformative impact on modern farming.

The primary goal of precision agriculture is to optimize inputs and outputs across various agricultural systems, creating a data-driven support system for comprehensive farm management. Key technologies utilized in precision agriculture include global positioning systems (GPSs), drones, and

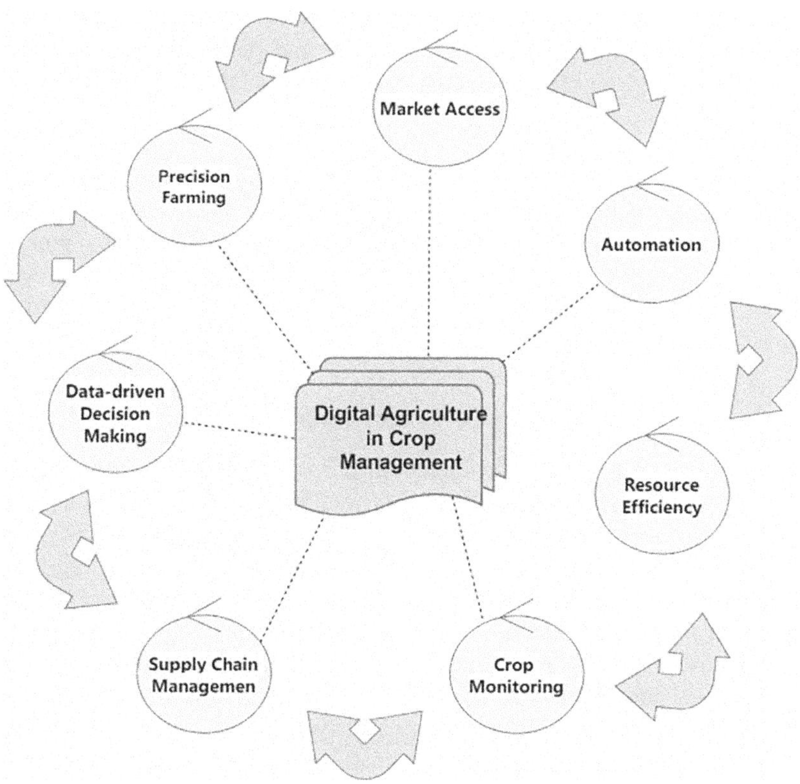

FIGURE 6.3 Role of digital agriculture in crop production.

advanced irrigation technologies. Digital farming, alternatively termed smart farming, represents a contemporary evolution of precision agriculture (Jain & Jain, 2022). This methodology leverages digital and smart devices such as sensors, cameras, satellites, drones, and GPS to enhance the efficiency of precision farming practices. By harnessing these technologies, digital farming aims to maximize resource utilization, improve crop yields, and optimize agricultural operations for greater sustainability and profitability.

6.5 SMART SUSTAINABLE FARMING FRAMEWORK

6.5.1 Introduction to Smart Sustainable Farming

Agriculture has evolved through four main eras:

- Traditional Agriculture (1.0): This era, which began in the Neolithic era around 10,000 years ago, relied on manual labor and animal power.
- Mechanized Agriculture (2.0): Starting in the 19th century with steam-powered machinery, this era brought increased productivity and efficiency, along with environmental issues such as pollution and soil degradation.
- Intelligent Agriculture (3.0): Beginning in the 20th century with the advent of computers and electronics, this era enabled more precise farming practices like precision irrigation and pest control.
- The convergence of cutting-edge technologies like the IoT, big data, and AI heralds a new phase in agriculture, emphasizing advancements in productivity, sustainability, and efficiency. This transformative era signifies a transition towards intelligent, sustainable farming methodologies that harness sophisticated technologies to streamline resource allocation, boost yields, and champion ecological conservation. With the agricultural sector progressing, these technological innovations stand poised to reshape farming practices significantly and tackle the complexities of sustenance amid a burgeoning global population.

The highlighted technologies represent a fraction of the digital tools integrated into modern agriculture. Their transformative impact has the potential to revolutionize the agricultural landscape and offer solutions to the pressing issue of sustaining a burgeoning global population as they progress and evolve (Jain et al., 2024).

Figure 6.4 portrays the array of smart farming tools, showcasing their innovative role in modern agricultural practices.

6.5.2 Exploring Technologies Used in Smart Farming

Technology plays a pivotal role as a vital pillar in precision agriculture, facilitating the collection and analysis of crucial data on crops, soil, and weather. This information empowers farmers to make well-informed decisions regarding crop management practices such as optimal planting times, precise fertilizer application, and efficient irrigation strategies. The integration of various cutting-edge technologies like global positioning systems, sensors, remote sensing, and geographic information systems forms the backbone of precision agriculture, enabling farmers to boost crop yields, lower input costs, and conserve valuable resources. Smart farming emerges as a transformative approach that amalgamates diverse technologies to revolutionize agricultural methodologies. From leveraging IoT sensors and drones for comprehensive data acquisition to harnessing the power of AI and big data analytics for strategic decision-making, these technologies require farmers with the tools to make informed choices and streamline farming operations for peak efficiency and productivity. Enhancing the environmental footprint of farming practices is pivotal for ensuring

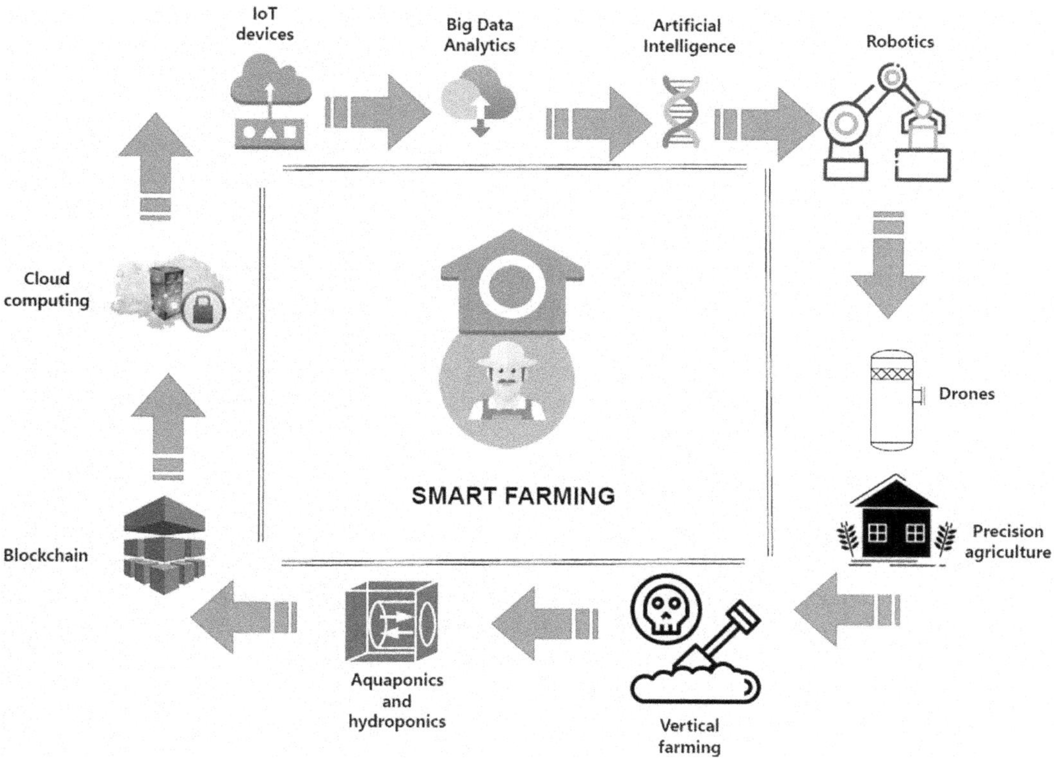

FIGURE 6.4 Smart farming tools.

agriculture's long-term sustainability. The adoption of precision farming practices offers numerous benefits for sustainability, including minimized water, fertilizer, and pesticide usage, enhanced soil health, and amplified crop yields. Precision agriculture holds immense potential in advancing agricultural sustainability, promising a more effective and responsible approach to food production as technology continues to progress. While challenges like initial costs, data management complexities, and varying levels of farmer acceptance exist, the trajectory of precision agriculture toward sustainable farming practices remains promising. As technology evolves and becomes more accessible, it is anticipated that farmers will increasingly embrace these innovative technologies, heralding a new era of sustainable farming practices that optimize productivity while minimizing environmental impact.

6.5.3 Embracing Sustainability in Agricultural Practices

Sustainable agriculture practices are anticipated to yield a host of benefits, including environmental preservation, enhanced availability of natural resources, and improved soil fertility (Baroda, 2021). The primary goal of sustainable agriculture is to boost farm profitability by pursuing a multitude of objectives aimed at fostering sound environmental stewardship. These initiatives encourage effective environmental management while optimizing resource utilization for the long-term prosperity of agricultural systems. Sustainability is at the core of smart farming, encompassing environmental stewardship, economic viability, and social responsibility. By adopting precision agriculture, organic farming, and regenerative agriculture practices, farmers can minimize their environmental footprint, conserve natural resources, and promote biodiversity while ensuring long-term profitability and resilience. Data accessibility is crucial for sustainable agriculture. Smart farming utilizes navigation satellites and earth observation data to support sustainable and cost-effective agriculture, enabling farmers to make informed agricultural decisions easily. Sustainability in agriculture can be categorized into three main forms:

- Ecological Sustainability: It aims to prevent nature from being exploited as an infinite resource and ensures its common and uninterrupted use, considering various factors. Practical adaptations include interests in sustainable energy sources, environmental protection, and water conservation.
- Social sustainability lays the groundwork for the development of individuals, communities, and societies to support fair and equitable individual well-being, healthcare, and education on a global scale.
- Economic Sustainability: Economic sustainability ensures that business operations and the flow of financial resources support alternative pillars of viability for overall development.

6.5.4 Developing a Sustainable Farming Framework

Sustainable farms foster environments conducive to the natural breakdown of waste materials or employ natural predators and competitors to mitigate pest populations. A sustainable farming framework provides a structured approach for integrating sustainability principles into agricultural operations. This framework encompasses soil health management, water conservation, energy efficiency, waste reduction, and community engagement (Jain & Rani, 2020).

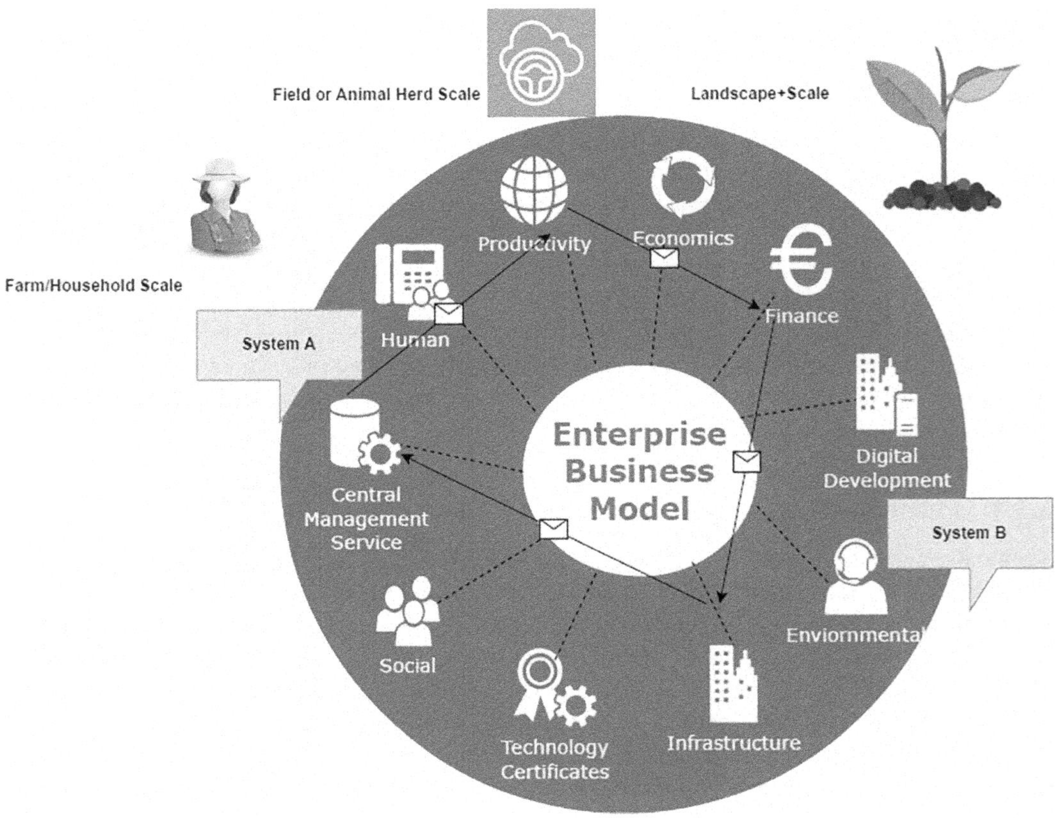

FIGURE 6.5 Sustainable farming framework across the five domains of sustainable growth.

Figure 6.5 demonstrates a comprehensive sustainable farming framework spanning five domains, illustrating its holistic approach towards sustainable growth. By establishing and executing a sustainable farming framework, farmers can adeptly harmonize economic prosperity with environmental and social well-being, laying the foundation for a more sustainable future in agriculture.

6.6 SMART FARMING: IOT, CLOUD, BIG DATA, AI, AND BLOCKCHAIN INTEGRATION

6.6.1 Cultivating Connectivity: Unleashing the Power of IoT in Agricultural Practices

In the contemporary context, with humanity grappling with the critical issues of food security, environmental sustainability, and climate change, the agricultural sector stands as a vanguard of innovation and adaptation. Against this backdrop, the integration of IoT technologies into agriculture has emerged as a pivotal catalyst, poised to transform conventional farming methods and inaugurate a new era characterized by enhanced efficiency, productivity, and sustainability.

Figure 6.6, denoting the interconnected network of devices, sensors, and software applications that gather, interpret, and exchange data, shows changing agricultural practices. Across diverse farming operations, IoT technologies are being implemented to monitor, control, and optimize essential parameters such as soil moisture, temperature, humidity, crop health, and irrigation systems in real time.

6.6.2 Benefits of IoT in Agriculture

The incorporation of IoT in agriculture offers a diverse range of advantages that hold the potential to revolutionize the cultivation, harvesting, and marketing of crops. Precision farming, facilitated by IoT sensors and monitoring devices, equips farmers with the capability to collect accurate and real-time data on soil conditions, weather patterns, and crop health. This data forms the cornerstone for informed decision-making concerning precise irrigation, fertilization, and pest control, thereby resulting in optimized resource utilization and improved crop yields. Moreover, IoT solutions grant farmers the ability to remotely monitor and oversee their agricultural operations from any location, using smartphones or computers, thereby elevating operational efficiency and reducing labor costs.

IoT platforms harness the power of predictive analytics to analyze extensive datasets, generating informed projections concerning crop yields, pest infestations, and market demand. This capability facilitates proactive decision-making and risk management. Furthermore, IoT aids in curbing

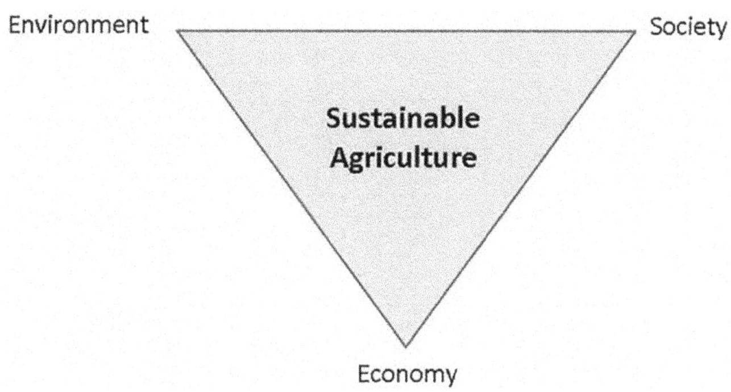

FIGURE 6.6 Sustainable agriculture gives equal weight to ecological, communal, and financial concerns in agriculture.

resource wastage and environmental impact by optimizing inputs based on real-time data, fostering sustainable farming practices, and conserving water resources.

By meticulously monitoring crop parameters such as moisture levels, nutrient content, and ripeness, farmers can produce high-quality crops that align with market specifications and consumer preferences, bolstering market competitiveness and profitability. Additionally, IoT technologies enable comprehensive traceability and transparency throughout the agricultural supply chain, ensuring food safety, minimizing losses, and refining logistics and distribution processes.

6.6.3 Harnessing the Power of Case Studies and Success Stories

Figure 6.7 shows a multitude of real-world instances that illustrate the transformative impact of IoT on agriculture across the globe. In India, IoT-based weather stations and soil moisture sensors are helping farmers in drought-prone regions optimize water usage and increase crop resilience. In the United States, IoT-enabled precision agriculture techniques have enabled farmers to reduce fertilizer usage, improve soil health, and achieve higher yields while minimizing environmental impact.

6.6.4 Challenges and Opportunities

To harness the full benefits of IoT in agriculture, several challenges must be addressed. These include concerns related to data privacy and security, interoperability of IoT devices and platforms, high upfront costs of implementation, and limited technical expertise among farmers. However, these challenges also present opportunities for collaboration and innovation among

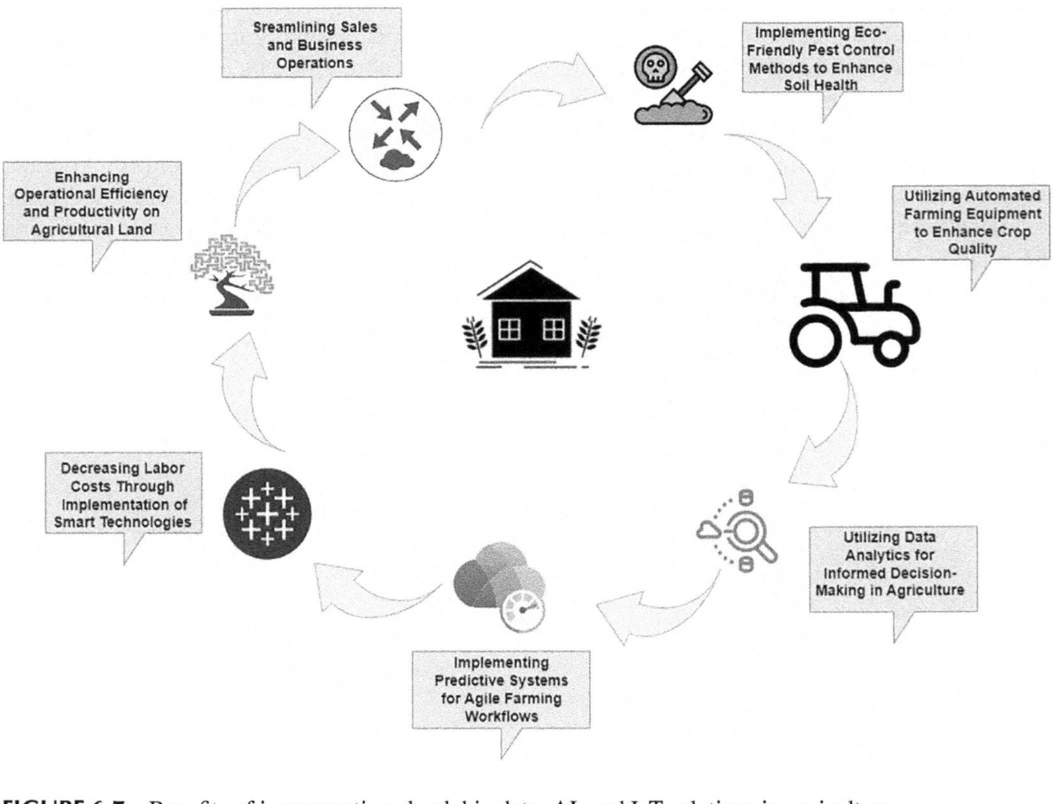

FIGURE 6.7 Benefits of incorporating cloud, big data, AI, and IoT solutions in agriculture.

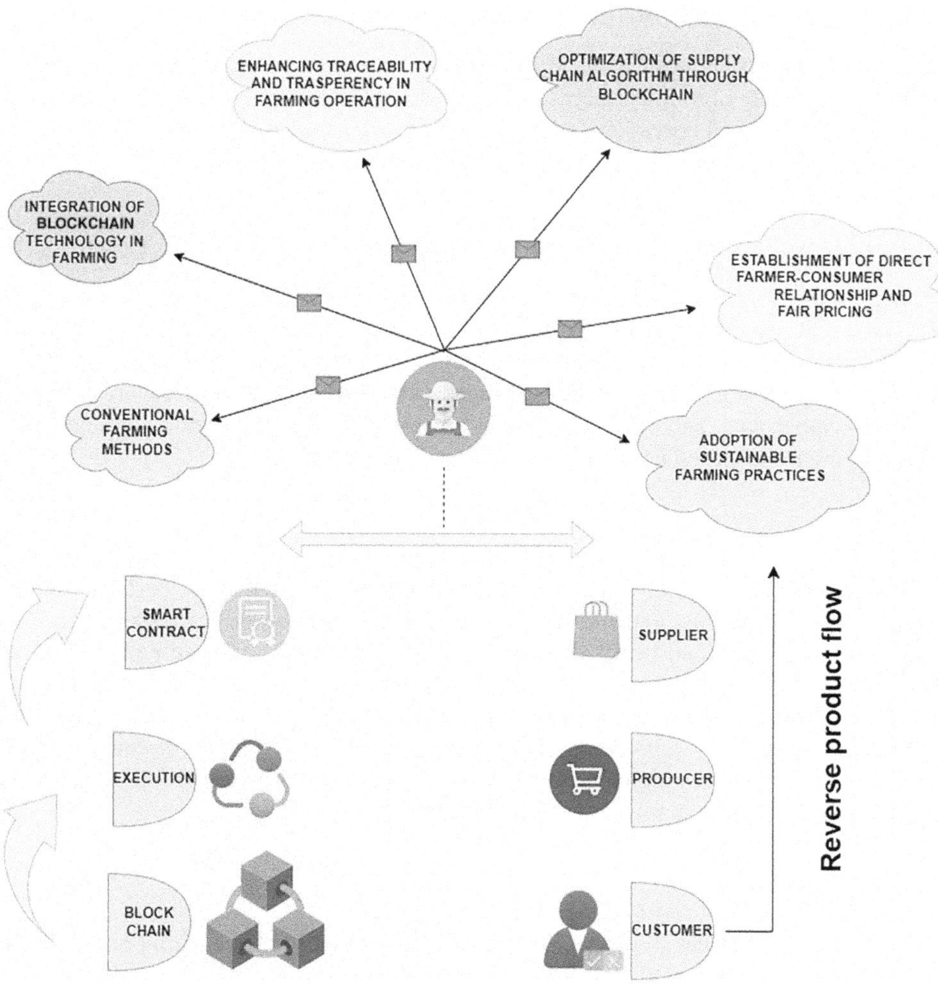

FIGURE 6.8 Blockchain-driven agricultural transformation.

stakeholders in the agriculture ecosystem. Governments, industry players, research institutions, and technology providers can collaborate to develop standards, regulations, and best practices for IoT adoption in agriculture. Embracing the Fourth Industrial Revolution, IoT in agriculture emerges as a potent tool for tackling global challenges related to food security, environmental sustainability, and rural development, paving the way for a brighter future for farmers and communities worldwide.

Figure 6.8 showcases the profound impact and multi-faceted benefits of integrating cloud, big data, AI, and IoT solutions in agricultural operations.

6.6.5 Cloud Cultivation: Optimizing Farming Operations through Cloud Computing

Cloud computing revolutionizes farming operations by offering scalable and flexible data storage, processing, and analysis solutions. With cloud-based platforms, farmers can securely access and manage vast amounts of agricultural data from any location with an internet connection. This enables real-time decision-making, enhances stakeholder collaboration, and facilitates the integration of IoT devices and other technologies into agricultural workflows.

6.6.6 Benefits of Cloud Computing in Agriculture

Cloud computing platforms offer scalability, allowing farmers to manage seasonal fluctuations efficiently. They provide convenient access to agricultural data, tools, and applications from any device, reducing capital and operational costs. Robust security measures protect sensitive data, and collaboration is enhanced through seamless data sharing among stakeholders.

6.6.7 Exemplary Case Studies and Success Narratives

Several case studies demonstrate the transformative impact of cloud computing on farming operations worldwide. For example, a dairy farm in New Zealand uses cloud-based software to monitor cow health, track milk production, and optimize feeding schedules, leading to improved productivity and profitability. In Brazil, soybean farmers utilize cloud computing platforms to analyze satellite imagery, weather data, and soil moisture levels, optimizing planting and irrigation strategies, ultimately leading to increased crop yields and resource efficiency.

6.6.8 Challenges and Considerations

Despite the numerous benefits of cloud computing in agriculture, challenges such as connectivity issues, data privacy concerns, and integration complexities need to be addressed. Farmers may struggle to fully leverage cloud-based solutions in remote rural areas with limited internet access. Additionally, ensuring the security and privacy of agricultural data stored in the cloud requires robust cybersecurity measures and compliance with regulations such as GDPR (General Data Protection Regulation) and CCPA (California Consumer Privacy Act). Furthermore, integrating cloud-based systems with existing farm management software and IoT devices may require technical expertise and investment in training and infrastructure.

6.6.9 Future Outlook

As cloud computing continues to evolve, the future of agriculture holds exciting possibilities. The progress in edge computing, artificial intelligence, and machine learning is poised to elevate the capabilities of cloud-based agricultural solutions, facilitating predictive analytics, autonomous decision-making, and tailored farming recommendations. By embracing cloud cultivation, farmers can unlock new opportunities for innovation, efficiency, and sustainability in agriculture.

6.6.10 Data Harvest: Maximizing Yield with Advanced Big Data Analytics

In today's data-driven agriculture landscape, big data analytics is crucial in maximizing crop yield, optimizing resource allocation, and mitigating risks. Farmers can gain valuable insights from large and complex datasets by leveraging advanced analytics techniques, enabling informed decision-making, proactive management, and continuous improvement in farming practices.

6.6.11 Benefits of Big Data Analytics in Agriculture

Big data analytics is transforming agriculture by providing farmers with invaluable insights and recommendations. By analyzing historical data, current conditions, and predictive models, farmers can make well-informed decisions regarding planting, irrigation, fertilization, pest control, and harvesting. This data-driven approach contributes to optimized crop yields, reduced losses, and enhanced profitability. Moreover, big data analytics facilitates precision agriculture techniques such as variable rate application and site-specific management, resulting in resource efficiency, cost reduction, and minimal environmental impact.

Additionally, big data analytics plays a pivotal role in risk management by identifying and mitigating risks such as pests, diseases, extreme weather events, market fluctuations, and supply chain disruptions, contributing to enhanced resilience and sustainability in agriculture. Furthermore, it fosters innovation and research in agriculture by fostering collaboration among farmers, scientists, agronomists, and technology providers, driving continuous improvement and adaptation to emerging challenges and opportunities.

6.6.12 Exemplary Case Studies and Success Narratives

Multiple case studies illustrate the transitive impact of big data analytics on agricultural productivity and sustainability. For example, a vineyard in California uses sensor data, satellite imagery, and weather forecasts to optimize irrigation schedules, resulting in water savings of up to 30% and increased grape quality. In Australia, a wheat farm applies machine learning algorithms to analyze historical yield data, soil samples, and climate data to predict crop performance and inform planting decisions, leading to higher yields and profits.

6.6.13 Challenges and Considerations

Despite its potential benefits, adopting big data analytics in agriculture requires overcoming several challenges, including data quality issues, interoperability constraints, privacy concerns, and skill shortages. Integrating data from disparate sources, such as sensors, drones, satellites, and farm management systems, into a unified analytics platform can become intricate and time consuming. Additionally, ensuring data privacy, security, and regulatory compliance is critical to building trust and confidence among farmers, data providers, and consumers. Moreover, training farmers and agricultural professionals in data literacy, analytics tools, and best practices is essential for realizing the full potential of big data analytics in agriculture.

6.6.14 Future Outlook

As the volume, velocity, and variety of agricultural data continue to expand, the future of big data analytics in agriculture appears promising. Anticipated advances in artificial intelligence, machine learning, and predictive modeling are set to elevate the capabilities of big data analytics, enabling personalized farming recommendations, autonomous decision-making, and real-time monitoring and control. With the harnessing of data's potential, farmers can unlock new opportunities for innovation, efficiency, and sustainability in agriculture, ultimately contributing to global food security and environmental stewardship.

6.6.15 Cultivating Intelligence: Revolutionizing Agricultural Systems with AI Integration

Artificial intelligence revolutionizes agricultural systems by offering advanced data analysis, decision-making, and automation capabilities. By integrating AI technologies such as machine learning, computer vision, and natural language processing into farming operations, farmers can improve productivity, optimize resource allocation, and enhance sustainability across the agricultural value chain.

6.6.16 Benefits of AI Integration in Agriculture

- Predictive Analytics: AI algorithms analyze historical data, current conditions, and external factors to predict crop yields, pest outbreaks, market trends, and weather patterns, enabling proactive decision-making and risk management.

- Precision Farming: AI-powered systems collect, process, and interpret data from sensors, drones, satellites, and other sources to create detailed maps of fields, identify areas for improvement, and optimize inputs such as water, fertilizers, and pesticides, resulting in heightened efficiency and yield.
- Autonomous Machinery: AI-driven robots, drones, and autonomous vehicles that perform tasks like planting, spraying, harvesting, and monitoring with precision and efficiency can reduce labor costs, minimize environmental impact, and enhance safety in agriculture.
- Disease Detection: AI algorithms analyze images of plants, animals, and soil to detect signs of diseases, pests, nutrient deficiencies, and other abnormalities, enabling early intervention and targeted treatments to prevent crop losses and ensure food safety.
- Market Intelligence: AI-powered systems analyze market data, consumer preferences, supply chain dynamics, and competitive trends to identify opportunities for product differentiation, pricing optimization, and marketing strategies, enhancing market competitiveness and profitability.

6.6.17 Exemplary Case Studies and Success Narratives

The Transformative Impact of AI Integration in Agricultural Productivity and Sustainability: For example, a dairy farm in the Netherlands uses AI-powered sensors and algorithms to monitor cow behavior, health, and productivity, enabling early detection of health issues, optimized feeding schedules, and improved milk quality. In Japan, a rice farmer applies AI-based image analysis to assess crop growth, detect pests, and optimize irrigation, resulting in higher yields and improved resource efficiency.

6.6.18 Blockchain Blossoms: Ensuring Transparency and Security in Agriculture through Blockchain Technology

The Technological Revolution in Agriculture: The Emergence of Blockchain Innovation: Initially, associated with cryptocurrencies, blockchain technology has found transformative applications in agriculture. Figure 6.8 highlights the evolutionary shift toward blockchain-driven agricultural transformation, showcasing its potential to enhance transparency, efficiency, and security across the agricultural landscape. This section explores how blockchain reshapes smart farming practices by ensuring transparency, enabling traceability, and revolutionizing the production, distribution, and consumption of agricultural products.

6.6.19 Understanding Blockchain in Agriculture

The integration of blockchain technology into agriculture encompasses various aspects of farming and food production. Fundamentally, blockchain is a decentralized, immutable digital ledger that records transactions across a network of computers. Each transaction, or "block," links to the previous one, establishing a transparent, tamper-resistant system ideal for ensuring supply chain transparency and traceability.

6.6.20 Advantages of Blockchain in Agriculture

Blockchain technology offers numerous benefits for revolutionizing the agricultural industry:

Enhanced Traceability: Real-time tracking of agricultural products from farm to fork provides consumers with detailed information about origin, quality, and journey.

Improved Food Safety: Quick identification of contamination sources enables targeted recalls, minimizing consumer and industry impacts.

Supply Chain Efficiency: Blockchain streamlines supply chain management by reducing paperwork, manual records, and intermediaries through automated tasks.

Fairer Prices for Farmers: Direct farmer-consumer links empower fair pricing, fostering equitable agricultural systems and sustainable practices.

Sustainable Agriculture: Encouraging eco-friendly product choices incentivizes sustainable farming practices, promoting biodiversity and conservation.

6.6.21 Applications of Blockchain in Smart Farming

Blockchain's integration into smart farming introduces numerous possibilities:

Smart Contracts for Automated Agreements: Self-executing contracts automate farmer–distributor agreements, ensuring timely payments and quality adherence.

IoT and Data Management: Blockchain securely manages and shares IoT-generated farm data, facilitating data-driven decision-making for optimal outcomes.

Crop Monitoring and Quality Control: Combining blockchain with IoT sensors allows real-time crop monitoring, aiding informed decisions and maintaining quality.

Fair Trade and Ethical Labeling: Blockchain verifies fair trade and ethical sourcing, with QR codes providing product journey and authenticity.

Certification and Compliance: Streamlined certification processes for organic and sustainable practices enhance transparency for regulators and consumers.

6.6.22 Ensuring Data Privacy and Security

While blockchain is secure, protecting stored data from unauthorized access is vital. Encryption and permissioned blockchains address these concerns, ensuring sensitive information is accessible only to authorized parties. Several real-world implementations demonstrate blockchain's positive impact on agriculture:

Walmart's Food Traceability Initiative: Using blockchain to trace produce origins ensures quick contamination identification and food safety.

Crop Insurance on the Blockchain: Farmers secure blockchain-based crop insurance, triggering payouts during predefined conditions like extreme weather.

Coffee Supply Chain Transparency: QR codes on coffee packages provide transparency, supporting sustainable and ethically sourced products.

Blockchain-Backed Food Safety Platform: Seafood companies use blockchain to track seafood, ensuring freshness and authenticity while combating illegal fishing.

Farm-to-Table Vegetable Tracking: Blockchain tracks vegetables from farm to table, enabling consumers to verify origin and quality, boosting food confidence.

6.7 CHALLENGES AND FUTURE DIRECTIONS OF SMART FARMING

Integrating smart agriculture encounters various obstacles, including challenges in disease prediction, implementing automated irrigation systems, ensuring continuous monitoring of crops and environmental conditions, and optimizing energy harvesting for sustainable operations (Ramesha et al., 2020). Crop infestation and disease pose significant threats to agricultural productivity. The prevailing challenges, including small farm sizes, inadequate infrastructure, limited adoption of advanced farming technologies and best practices, soil degradation due to excessive fertilization, and prolonged pesticide use, contribute to the low agricultural yields in India.

Farmers globally grapple with an array of daunting challenges that directly and indirectly impact their livelihoods. Agricultural activities and practices often demand significant time and financial resources. Addressing the issues confronted by farmers in the food sector is imperative. The primary issues faced by farmers include:

- Demand for infrastructure
- Demand for insurance
- Provisional water rights
- Requirement for a living wage

Farmers engaging in smart farming possess a comprehensive understanding of crucial elements such as water, terrain, plant species, and soil types. This knowledge empowers them to make well-informed decisions regarding the management of limited resources within their production areas in an environmentally and economically sustainable manner. Contemporary farming techniques like precision farming, hydroponics, aquaponics, and vertical farming are gaining traction among farmers, promoting efficiency; waste reduction; and the supply of fresh, sustainable produce to consumers. An array of technologies, including sensors, smart irrigation systems, drones, automation, and satellite/GPS systems, enable precise agriculture, facilitating efficient resource utilization.

6.8 CONCLUSION

Smart farming significantly revolutionizes the agricultural sector by leveraging advanced technologies for tracking, monitoring, automating, and analyzing operations, incorporating big data, cloud computing, and the Internet of Things. Sustainable agriculture is guided by the paramount goals of ecological well-being, economic productivity, and social equality, shaped by diverse principles, regulations, and methodologies. The integration of state-of-the-art technologies such as IoT sensors, drones, and AI-driven analytics progressively reshapes agriculture, promoting intelligent and sustainable farming practices. These tools enable real-time monitoring of crucial parameters like temperature, moisture levels, and soil conditions, providing farmers with actionable insights for optimizing resource allocation and decision-making processes. By leveraging predictive analytics, farmers can anticipate and mitigate potential challenges, ensuring efficient water usage and maximizing production rates for crops and livestock. Moreover, intelligent farming systems empower users to remotely monitor and control field operations based on precise data insights, including irrigation and pest management, enhancing operational efficiency while reducing labor costs and environmental impacts. Additionally, smart agriculture eliminates soil needs by utilizing nutrient solutions tailored to plant requirements, promoting rapid nutrient absorption and healthy crop growth. The key to the success of intelligent farming lies in our ability to collect, analyze, and utilize data effectively. By harnessing real-time data, farmers can make informed decisions across all operations, leading to higher productivity, reduced expenses, and improved ecological stewardship. Furthermore, adopting innovative farming technologies enables farmers to adapt to changing environmental conditions and market demands, ensuring long-term sustainability and resilience in agriculture. In summary, intelligent, sustainable farming represents a transformative shift towards more efficient, productive, and environmentally sustainable agricultural practices. By embracing advanced technologies and data-driven decision-making, farmers can enhance productivity, reduce costs, and promote ecological stewardship, ensuring the continued success and resilience of the agricultural sector in the face of evolving challenges.

REFERENCES

Adamchuk, V. I., Hummel, J. W., Morgan, M. T., and Upadhyaya, S. K. (2004). On-the-go soil sensors for precision agriculture. *Computers and Electronics in Agriculture*, 44(1):71–91.

AGRI Committee (2019). *Impacts of the Digital Economy on the Food Chain and the CAP*. AGRI Committee.

Batte, M. (2020). *Precision Farming—Factors Influencing Profitability*. Fred N. VanBuren Professor of Farm Management Department of Agricultural, Environmental, and Development Economics, April.

Bendre, M. R., Thool, R. C., and Thool, V. R. (2016). Big Data in Precision Agriculture Through ICT: Rainfall Prediction Using Neural Network Approach. *Advances in Intelligent Systems and Computing*, 438:165–175.

Chen, F., Kissel, D. E., West, L. T., Adkins, W., Clark, R., Rickman, D., and Luvall, J. C. (2004). Field Scale Mapping of Surface Soil Clay Concentration. *Precision Agriculture*, 5(1):7–26.

Choudhury, A., Biswas, A., Singh T. P., and Santanu, G. K. (2022). *Smart Agriculture Automation Using Advanced Technologies*. Technical Report. https://www.researchgate.net/publication/373554499_Smart_Agriculture_Automation_using_Advanced_Technologies

Dankan Gowda, V., Kumar, P. M., Pai, N. G., Muralidhar, K., and Kumar, V. B. (2020). Dynamic Analysis and Control Strategies of an Anti-lock Braking System. *Proceedings of the 4th International Conference on Electronics, Communication and Aerospace Technology, ICECA 2020*, 1677–1682.

Dhanaraju, M., Chenniappan, P., Ramalingam, K., Pazhanivelan, S., and Kaliaperumal, R. (2022). Smart Farming: Internet of Things (IoT)-Based Sustainable Agriculture. *Agriculture (Switzerland)*, 12(10):1–26.

Ehlers, M. (2008). Geoinformatics and Digital Earth Initiatives: A German Perspective. *International Journal of Digital Earth*, 1(1):17–30.

Ferrández-Pastor, F. J., García-Chamizo, J. M., Nieto-Hidalgo, M., Mora-Pascual, J., and Mora-Martínez, J. (2016). Developing Ubiquitous Sensor Network Platform Using Internet of Things: Application in Precision Agriculture. *Sensors (Switzerland)*, 16(7).

Jain, A., Jain, G., and Tewari, D. (2024). KNetwork: Advancing Cross-Lingual Sentiment Analysis for Enhanced Decision-Making in Linguistically Diverse Environments. *Knowledge and Information Systems*, 66:2925–2943. https://doi.org/10.1007/s10115-023-02051-w

Jain, G., and Jain, A. (2022). Blockchain for 5G-Enabled Networks in Healthcare Service Based on Several Aspects. *Blockchain Applications for Healthcare Informatics*, 471.

Jain, G., and Rani, N. (2020). Awareness Learning Analysis of Malware and Ransomware in Bitcoin. *Lecture Notes in Networks and Systems*, 121:765–776.

Klerkx, L., Jakku, E., and Labarthe, P. (2019). A Review of Social Science on Digital Agriculture, Smart Farming and Agriculture 4.0: New Contributions and a Future Research Agenda. *NJAS – Wageningen Journal of Life Sciences*, 90–91:100315.

Naveena Pai, G., Swathi Pai, M., Dankan Gowd, V., Shruthi, M., and Kumar, N. B. (2020). Internet of Things: A Survey on Devices, Ecosystem, Components and Communication Protocols. In *Proceedings of the 4th International Conference on Electronics, Communication and Aerospace Technology, ICECA 2020*, 611–616.

Ojo, O. I., Ilunga, F., Ojo, O. I., and Ilunga, F. (2018). Geospatial Analysis for Irrigated Land Assessment, Modeling and Mapping. *Multi-purposeful Application of Geospatial Data*, 9:65–84.

Pivoto, D., Waquil, P. D., Talamini, E., Finocchio, C. P. S., Dalla Corte, V. F., and de Vargas Mores, G. (2018). Scientific Development of Smart Farming Technologies and Their Application in Brazil. *Information Processing in Agriculture*, 5(1):21–32.

Ramesha, M., Dankangowda, V., Jeevan, K. M., and Sathisha, B. M. (2020). Gowda Electronic Stethoscope. *MPCIT 2020—Proceedings: IEEE 3rd International Conference on Multimedia Processing, Communication and Information Technology*, 103–106.

Rijswijk, K., Klerkx, L., and Turner, J. A. (2019). Digitalisation in the New Zealand Agricultural Knowledge and Innovation System: Initial Understandings and Emerging Organisational Responses to Digital Agriculture. *NJAS—Wageningen Journal of Life Sciences*, 90–91:100313.

Srisruthi, S., Swarna, N., Susmitha Ros, G. M., and Elizabeth, E. (2017). Sustainable Agriculture Using Eco-Friendly and Energy Efficient Sensor Technology. In *2016 IEEE International Conference on Recent Trends in Electronics, Information and Communication Technology, RTEICT 2016—Proceedings*, 1442–1446. IEEE.

Stosch, K. C., Quilliam, R. S., Bunnefeld, N., and Oliver, D. M. (2017). Managing Multiple Catchment Demands for Sustainable Water Use and Ecosystem Service Provision. *Water (Switzerland)*, 9(9).

Tekinerdogan, B. (2018). *Strategies for Technological Innovation in Agriculture 4.0*. Wageningen University.

Uprety, D., Banarjee, D., and Kumar, N. (2023). *Easy Chair Preprint Smart Sustainable Farming Using IoT, Cloud Computing and Big Data*. https://easychair.org/publications/preprint/SXjm/download

Wolfert, S., Ge, L., Verdouw, C., and Bogaardt, M. J. (2017). Big Data in Smart Farming—A review. *Agricultural Systems*, 153:69–80.

Řezník, T., Lukas, V., Charvát, K., Charvat Jr., K., Křivánek, Z., Kepka, M., Herman, L., and Řezníková, H. (2017). Disaster Risk Reduction in Agriculture Through Geospatial (Big) Data processing. *ISPRS International Journal of Geo-Information*, 6(8).

University of Baroda. (2021). Crops Protection System Using IoT and UAV in 5G. *Journal of the Maharaja Sayajirao*. https://papers.ssrn.com/sol3/papers.cfm?abstract_id=4113040

7 Blockchain Implementation in IoT-Digital Twin-Based Crop Production

Ushaa Eswaran, Vivek Eswaran, Keerthna Murali, and Vishal Eswaran

7.1 INTRODUCTION

7.1.1 Overview of the Importance of Agriculture and Crop Production

For billions of people worldwide, agriculture provides food, a means of subsistence, and economic stability. It is the foundation of human civilization. The United Nations forecasted in 2019 that there will be 9.7 billion people on the planet by the year 2050, meaning that agriculture will play a crucial role in feeding this expanding population.

Economically, agriculture plays a vital role, contributing significantly to national GDPs, employment rates, and exports in many countries. It forms the foundation of rural economies, providing income and livelihoods to millions of farmers and workers globally. Moreover, agriculture fosters cultural and social ties, deeply rooted in traditions and practices passed down through generations. It shapes the identity of rural communities and contributes to their social cohesion and resilience.

7.1.2 Challenges Faced in Traditional Crop Production Practices

1. **Resource Constraints:** Traditional crop production faces numerous resource constraints, including limited arable land, water scarcity, and the inefficient use of fertilizers and pesticides. The growing population exerts pressure on available land resources, leading to deforestation and soil degradation. Additionally, water scarcity exacerbates the challenge, as agriculture accounts for a significant portion of global water consumption.
2. **Effects of Climate Change:** The threat posed by climate change is substantial to crop production, with its effects manifesting in extreme weather events, temperature fluctuations, and shifting precipitation patterns. These climatic variations disrupt traditional farming practices, leading to reduced crop yields, decreased productivity, and increased vulnerability to natural disasters.
3. **Pest and Disease Infestations:** Pests and diseases represent major threats to crop health and productivity. Weeds, infections, and insects can all lead to extensive harm to crops, resulting in yield losses and compromised food quality. The indiscriminate use of chemical pesticides contributes to environmental pollution and poses health risks to humans and ecosystems.
4. **Supply Chain Inefficiencies:** Inefficient supply chain management in agriculture results in poor traceability, food waste, and lack of transparency. Challenges such as inadequate infrastructure, inadequate storage facilities, and limited access to markets hinder the efficient distribution of agricultural produce. As a result, farmers frequently struggle to get their products into markets and get paid fairly for them.

DOI: 10.1201/9781003507390-7

5. **Absence of Data-Driven Inference**: Traditional agricultural methods rely on outdated methods and often lack availability of real-time data and analytics. This limits farmers' ability to make well-informed choices on risk reduction, resource allocation, and crop management. The absence of data-driven insights hampers productivity and inhibits the adoption of innovative agricultural practices.

7.1.3 The Need for Digital Transformation in Agriculture

Recognizing the challenges faced by traditional crop production practices, there is an urgent need for digital transformation in agriculture. Digital technologies have the power to completely transform the agricultural sector, facilitating farmers to overcome existing limitations and achieve sustainable growth.

Digital change encompasses the integration of emerging technologies like the Internet of Things (IoT), digital twins, and blockchain into agricultural systems. These technologies enable precision agriculture, wherein farmers are able to observe and manage soil quality, environmental factors, and crop health parameters in real time. By leveraging data analytics and artificial intelligence (AI), farmers can make data-driven decisions, optimize resource use, and raise agricultural yields (Eswaran et al., 2023a)

The adoption of digital transformation in agriculture offers numerous advantages, such as better food security, decreased environmental effect, and enhanced production. By utilizing technology's potential, farmers can mitigate the effects of climate change, minimize resource wastage, and ensure the sustainability of agricultural production for future generations.

7.1.4 Objectives of the Chapter

1. To explore the integration of blockchain technology, Internet of Things, and digital twins in crop production and supply chain management.
2. To analyze the benefits and applications of these technologies in addressing challenges faced in traditional agricultural practices.
3. To examine the architecture, data flow, and security considerations of blockchain-enabled IoT–digital twin systems.
4. To investigate the role of smart contracts in automating processes, resource allocation, and crop insurance schemes.

7.1.5 Organization of the Chapter

The chapter starts with an introduction giving a summary of the importance of agriculture and crop production, as well as the challenges faced in traditional crop production practices. It then delves into the Internet of Things in agriculture, explaining its principles, applications, and benefits in precision farming, livestock monitoring, greenhouse automation, and irrigation management. The chapter continues by exploring digital twins in agriculture, including their definition, components, and applications in crop modeling, predictive analytics, and decision support systems. Next, it introduces blockchain technology, its key features, and use cases in agriculture, such as supply chain traceability, smart contracts, and data integrity. The chapter then focuses on integrating blockchain with IoT and digital twins for crop production, discussing the architecture, data flow, and the role of smart contracts in automation. Data management and security concerns, as well as blockchain-based solutions for secure data storage and sharing, are also addressed. The chapter then examines smart contracts and automation, covering their applications in crop production, supply chain automation, and resource allocation. Case studies from the real world and best practices are presented to illustrate successful implementations and lessons learned. Subsequently, the chapter discusses challenges and future directions, including technical challenges, regulatory and legal concerns, and potential advancements in areas such as AI integration, edge computing, quantum cryptography, and sustainable agriculture practices. Finally, the chapter ends with summarizing the benefits and potential of blockchain-enabled IoT–digital twin systems in crop production and providing final remarks and recommendations for adoption and implementation.

7.2 INTERNET OF AGRICULTURE THINGS

7.2.1 Definition and Principles of IoT

The Internet of Things is a network of sensors and devices that gather, share, and process data without the need for human interaction. It depends on automation and connection to enable smooth communication between digital systems and real-world items.

- **Device Connectivity**: IoT devices, including sensors, actuators, and gateways, are interconnected through wireless or wired communication networks, allowing them to transmit data to centralized data processing platforms (Eswaran et al., 2023b).
- **Data Exchange Protocols**: IoT devices exchange data safely and effectively over networks using standardized communication protocols like HTTP (hypertext transfer protocol) and MQTT (message queuing telemetry transport).
- **Interoperability**: IoT systems prioritize interoperability, which guarantees smooth communication and collaboration amongst gadgets from various vendors and manufacturers. This promotes flexibility and scalability in IoT deployments.
- **Automation:** Automation is a key principle of IoT, enabling devices to perform tasks and make decisions autonomously based on predefined rules or algorithms. This automation reduces the need for human intervention and streamlines operations.

7.2.2 IoT Applications in Agriculture

7.2.2.1 Precision Farming

- **Soil Monitoring:** IoT sensors embedded in the soil can continuously monitor soil nitrogen levels, temperature, and moisture content. With the use of this data, farmers can optimize crop growth and resource utilization by putting accurate irrigation and fertilizer application methods into practice.
- **Variable Rate Technology:** By leveraging IoT data, farmers can apply inputs like seeds, fertilizer, and insecticides at different rates depending on the conditions unique to their particular field thanks to variable rate technology.

This targeted approach maximizes crop yields while minimizing input costs.

- **Yield Mapping:** IoT-enabled yield monitors installed on harvesting equipment collect data on crop yields as they are harvested. Using this data, they can create yield maps, helping farmers identify areas of high and low productivity within their fields.

7.2.2.2 Livestock Monitoring

- **Health Monitoring:** IoT sensors attached to livestock can monitor essential indicators and degree of activity, and behavior patterns. This real-time data enables early identification of health issues and allows farmers to provide timely interventions, improving animal welfare and productivity.
- **Environmental Monitoring:** IoT systems installed in barns or grazing areas can keep an eye on the ventilation, humidity, and temperature in the surroundings. By guaranteeing ideal circumstances for livestock, farmers can minimize stress and disease outbreaks, enhancing overall herd health (Eswaran and Eswaran, 2023a).

7.2.2.3 Greenhouse Automation

- **Climate Control:** IoT systems in greenhouses can regulate temperature, humidity, and CO_2 levels to create an optimal growing environment for crops. Automated climate control systems ensure consistent conditions, promoting healthy plant growth and maximizing yields (Sadek and Shehata, 2024).
- **Irrigation and Nutrient Management:** IoT sensors are able to track soil moisture content. and nutrient concentrations in greenhouse substrates. Automated irrigation and nutrient dosing systems adjust water and nutrient delivery based on real-time data, ensuring plants receive the necessary resources for optimal growth.

7.2.2.4 Irrigation Management

- **Efficient Water Use:** IoT sensors deployed in fields can monitor soil moisture levels and weather conditions to optimize irrigation scheduling. Automated irrigation systems adjust water application rates and timing based on crop needs and environmental factors, minimizing water waste and promoting water conservation.

7.2.2.5 Weather Monitoring

- **Real-Time Data:** Internet of Things weather stations gather data in real time on solar radiation, wind speed, temperature, humidity, and precipitation. Making educated decisions about crop management techniques, such as planting, harvesting, and pest control, depends on this data.

7.2.3 Benefits of IoT in Crop Production

7.2.3.1 Increased Yield

IoT-enabled precision farming practices leverage real-time data from sensors to optimize crop inputs and management, ultimately leading to higher yields and increased productivity. By precisely monitoring and controlling factors such as insect activity, nutritional levels, and soil moisture infestations, farmers can guarantee that crops receive the best conditions for expansion and development. This targeted approach minimizes crop stress and maximizes yield potential, resulting in improved harvests and profitability.

7.2.3.2 Resource Optimization

IoT solutions are essential for crop production resource optimization. Sensors provide useful insights by continuously monitoring soil moisture, nutrients, and weather. Farmers can minimize waste and their influence on the environment by modifying pest management, fertilization, and irrigation. The optimization of resources encourages agriculture that is sustainable.

7.2.3.3 Real-Time Monitoring

The environment, soil, and crop health can all be continuously and real-time monitored thanks to IoT. Farmers are able to monitor changes and trends since sensors gather data on a regular basis. It is possible to identify problems like pests or nutritional shortages early on. Prompt notifications enable farmers to take preventative action, reducing losses and increasing yields (Eswaran and Eswaran, 2023b).

7.2.3.4 Automated Processes

Climate control and irrigation are two farming tasks that are automated by IoT technology. Temperature controls are made by climate systems, while soil moisture sensors modify watering regimens. Automation boosts productivity, lowers labor expenses, and frees up farmers to concentrate on other farm management duties (Eswaran et al., 2024).

TABLE 7.1
IoT Applications in Agriculture

Application	Description
Precision Farming	Utilizes AI for soil monitoring, variable rate technology, and yield mapping.
Livestock Monitoring	AI-based health monitoring and environmental monitoring for livestock.
Greenhouse Automation	AI-driven climate control, irrigation, and nutrient management in greenhouse environments.
Irrigation Management	AI-driven irrigation systems that use soil moisture and meteorological data to optimize water use.
Weather Monitoring	Real-time data collection and analysis of temperature, humidity, precipitation, and wind conditions.

7.2.3.5 Data-Driven Decision-Making

IoT provides real-time data on crop performance, soil conditions, and weather patterns, which helps farmers make data-driven decisions. This data is processed by sophisticated analytics systems, which find patterns and correlations to help guide decisions. Planting dates and fertilizer applications can be optimized by farmers to increase output, profitability, and sustainability (Eswaran and Khang, 2024).

An overview of IoT applications in agriculture is shown in the accompanying Table 7.1, which also shows how artificial intelligence is used to improve farming techniques. These IoT-powered applications help to increase agricultural sustainability, efficiency, and productivity through real-time monitoring and management of important aspects of crop production, livestock health, and environmental conditions.

7.3 DIGITAL TWINS IN AGRICULTURE

7.3.1 Definition and Concept of Digital Twins

Digital twins in agriculture combine simulations and real-time data to function as virtual copies of actual operations and offer insights. They allow farmers to monitor and optimize crop production systems in real time by bridging the gap between the physical and digital worlds. These models capture the intricacies of agriculture, enabling in-depth examination and forecasting abilities to boost production and efficiency (Christos Mitsanis et al., 2024).

Key components of a digital twin system include the physical entity (such as a crop field or farm equipment), a virtual model that replicates its behavior, data integration mechanisms to gather real-time information, and sophisticated analytics tools to examine the data and draw conclusions.

7.3.2 Components of a Digital Twin System

- **Sensor Data Acquisition:**
  An essential component of digital twin systems in agriculture is real-time data gathering. Data on the environment, soil moisture content, crop health, and other topics are gathered from a variety of sources, such as IoT devices, drones, satellites, and weather stations.
- **Data Processing:**
  In order to preserve the quality and utility of the collected data, data processing procedures make sure that it is integrated, cleaned, and preprocessed. To prepare the data for virtual modelling and analysis, this process entails data fusion, transformation, and aggregation.

- **Virtual Modeling:**
  Virtual modeling involves creating digital replicas of physical agricultural systems, incorporating crop growth models, soil models, and environmental models. These virtual representations accurately simulate the behavior of real-world components, enabling detailed analysis and predictive capabilities.
- **Simulation and Analysis:**
  Simulation techniques and sophisticated analytics, encompassing AI and machine learning, are applied to the virtual models to predict crop growth, simulate different scenarios, and optimize management practices. These analyses provide valuable insights into crop performance, resource utilization, and potential risks.
- **Visualization and Decision Support:**
  User-friendly interfaces and visualization tools are essential for presenting insights and decision support to farmers and stakeholders. visualization methods like dashboards, charts, and maps enable users to interpret complex data and make well-informed choices about crop management techniques, resource distribution, and risk reduction tactics.

7.3.3 Applications of Digital Twins in Crop Production

- **Crop Modeling and Simulation:**
  Digital twins simulate crop growth, yield, and response to environmental and management factors, facilitating scenario analysis and optimization of farming practices. By accurately modeling crop behavior, farmers can experiment with different strategies and predict outcomes to maximize productivity.
- **Predictive Analytics:**
  Leveraging both current and historical data, digital twins employ forecasting analytics to forecast crop yields, detect potential issues, and provide early warning systems for pest infestations, diseases, or adverse weather conditions. These insights enable proactive decision-making and risk management.
- **Decision Support Systems:**
  Digital twins serve as decision support systems, offering recommendations for crop management practices, resource allocation, and risk mitigation strategies. By integrating data-driven insights with farmer expertise, these systems empower farmers to make informed decisions aligned with their goals and objectives.
- **Scenario Analysis:**
  Digital twins facilitate scenario analysis by simulating different conditions and events, such as climate change impacts, pest outbreaks, or changes in management practices. Farmers can explore various scenarios, assess their potential impact, and develop adaptive strategies to mitigate risks and optimize outcomes.
- **Precision Agriculture:**
  Supporting precision agriculture practices, digital twins provide site-specific recommendations and enable variable-rate application of inputs, including irrigation water, herbicides, and fertilizers. Farmers may maximize crop output, reduce waste, and optimize resource use by customizing interventions to each field's unique needs.

Applications of digital twins in agricultural production are shown in Figure 7.1. The purpose of these virtual representations is to monitor, simulate, and optimize crop production systems using real-time data, analytics, and simulations. They are used by farmers to control crop growth, yield, environmental factors, and soil health. Agricultural production and sustainability are increased through the use of digital twins, which facilitate data-driven decision-making, result prediction, and management optimization.

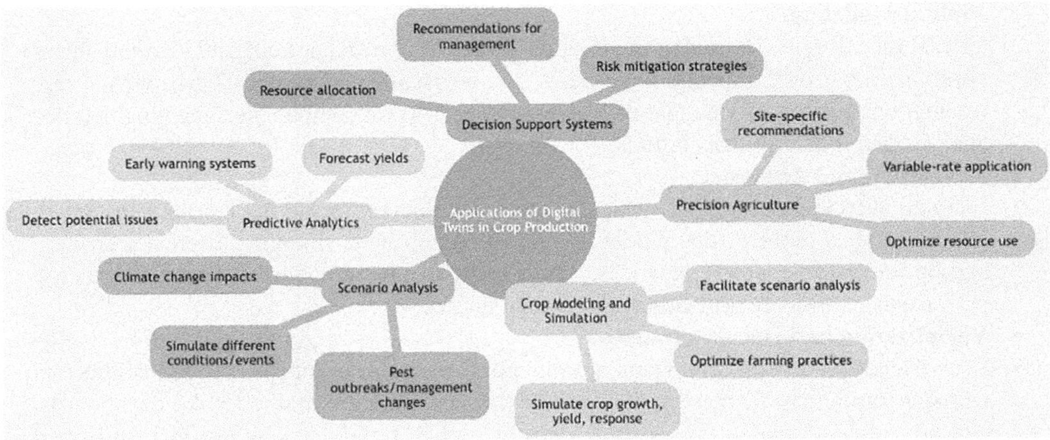

FIGURE 7.1 Digital twin applications in crop production.

7.4 BLOCKCHAIN METHODS

7.4.1 Introduction to Blockchain Technology

Blockchain technology, originating from the Bitcoin whitepaper by Satoshi Nakamoto in 2008, revolutionized the concept of decentralized systems. Fundamentally, blockchain is a distributed ledger that securely and openly records transactions via a network of computers. Blockchain functions on the ideas of decentralization, consensus procedures, and cryptographic security, in contrast to conventional centralized systems.

7.4.2 Key Features of Blockchain

Blockchain ensures that no single entity controls the system by operating on a decentralized network. A network of nodes records and verifies transactions, promoting resilience against single points of failure and transparency.

One essential component is transparency, which promotes accountability and fosters trust by enabling all participants to see the complete transaction history recorded on the ledger.

Transactions are tamper-proof and unchangeable once they are recorded, guaranteeing that data is reliable and consistent throughout time.

Cryptographic techniques, such as digital signatures and hashing, improve security by thwarting unauthorized access. Proof-of-work and proof-of-stake are two more consensus techniques that guarantee only legitimate transactions are added to the blockchain (Ramana Kadiyala et al., 2023).

7.4.3 Blockchain Use Cases in Agriculture

Blockchain can revolutionize supply chain management in agriculture by providing end-to-end traceability of products. Smart contracts automate and enforce agreements, streamlining processes like crop insurance claims and payments. Blockchain ensures data integrity, preventing tampering with agricultural data collected from IoT devices (Eswaran and Khang, 2023). It also offers secure digital identity management for stakeholders and enables provenance tracking of agricultural inputs (Bansal et al., 2024).

7.5 INTEGRATING BLOCKCHAIN WITH IOT AND DIGITAL TWINS FOR CROP PRODUCTION

7.5.1 Challenges in Traditional Crop Production Systems

Data Silos: Traditional crop production systems often suffer from data silos, where different stakeholders maintain separate and isolated data repositories. This fragmentation hinders collaboration and data sharing among farmers, suppliers, processors, and retailers, leading to inefficiencies and missed opportunities for optimization (Sundararajan et al., 2024).

Lack of Transparency: The lack of transparency in traditional supply chains poses significant challenges for stakeholders and consumers. Without visibility into the origin and journey of agricultural products, it becomes difficult to verify sustainable practices, ensure food safety, and trace the source of contamination or adulteration incidents.

Inefficient Processes: Manual record-keeping, paper-based contracts, and delays in payments and settlements contribute to inefficiencies in traditional crop production processes. These outdated practices not only slow down operations but also increase the risk of errors, disputes, and financial losses for stakeholders along the supply chain.

Vulnerabilities to Fraud and Counterfeit Products: Traditional crop production systems are vulnerable to fraud, counterfeiting, and adulteration, posing risks to food safety and customer confidence. Without robust mechanisms for verifying the genuineness and uprightness of agricultural products, consumers may fall victim to fraudulent practices, undermining the credibility of the complete chain of supplies.

7.5.2 The Role of Blockchain in Addressing These Challenges

Breaking Down Data Silos: Blockchain's decentralized and transparent nature offers a solution to the challenge by offering a common, unchangeable ledger, data silos for recording transactions and data exchanges. By eliminating intermediaries and enabling direct peer-to-peer interactions, blockchain promotes trust, transparency, and collaboration among stakeholders.

End-to-End Transparency: Blockchain's immutability and traceability features empower stakeholders and consumers with real-time visibility into the origin, journey, and handling of agricultural products. By recording every transaction on the blockchain, from the field to the fork, stakeholders can confirm the authenticity, quality, and sustainability of products, building reliance and accountability across the supply chain.

Streamlining Processes with Smart Contracts: Smart contracts automate procedures and transactions in supply chain management and agricultural production. They are self-executing contracts with established rules and conditions.

By encoding business logic and contractual agreements on the blockchain, smart contracts enable seamless execution of tasks such as payments, deliveries, and quality inspections, reducing delays and disputes.

Enhancing Security and Integrity: Blockchain's cryptographic security and consensus mechanisms protect against fraud, counterfeiting, and data tampering in crop production systems. By maintaining the integrity of the data and authenticity, blockchain mitigates the risks of unauthorized entry, manipulation, and corruption, safeguarding the integrity of agricultural products and supply chain data.

7.5.3 Architecture and Components of a Blockchain-Enabled IoT-Digital Twin System for Crop Production

Architectural Overview: A blockchain-enabled IoT–digital twin system for crop production comprises interconnected components, including the blockchain network, IoT devices and sensors, digital twin models, data storage, and analytics platforms. These components work synergistically to collect, analyze, and leverage data for optimizing crop production and supply chain operations.

Data Flow and Integration: Data from IoT sensors, such as soil moisture sensors, weather stations, and crop monitoring devices, is collected and transmitted to the blockchain network, where it is recorded in a transparent and immutable ledger. This data is then integrated with digital twin models, which create virtual representations of crop production systems, enabling simulations, predictions, and optimizations (Brandín and Abrishami, 2024).

Role of Smart Contracts: Smart contracts are essential to automating processes and transactions based on predefined conditions and triggers. For example, smart contracts have the ability to run automatically provide payments to farmers based on crop yield data collected from IoT sensors, initiate supply chain actions in response to quality assurance checks, or trigger alerts for adverse weather conditions affecting crop growth.

Integration with AI and ML: The blockchain-enabled IoT–digital twin system can leverage artificial intelligence and machine learning techniques for advanced analytics and decision support. Through analysis, data collected from IoT sensors and digital twins, AI models can provide insights, predictions, and recommendations in order to maximize supply chain logistics, resource allocation, and crop management techniques.

Scalability and Interoperability: To ensure scalability and interoperability, the blockchain-enabled IoT–digital twin system should be designed to accommodate diverse data sources, standards, and protocols. Interoperability between blockchain networks, IoT devices, and digital twin platforms enables seamless data exchange and collaboration across different stakeholders and systems in the agricultural ecosystem.

In the complex landscape of modern agriculture, the integration of blockchain technology, Internet of Things, and digital twin systems has emerged as a transformative solution. The mind map shown in Figure 7.2 visualizes the interconnected components and flow of data within this innovative

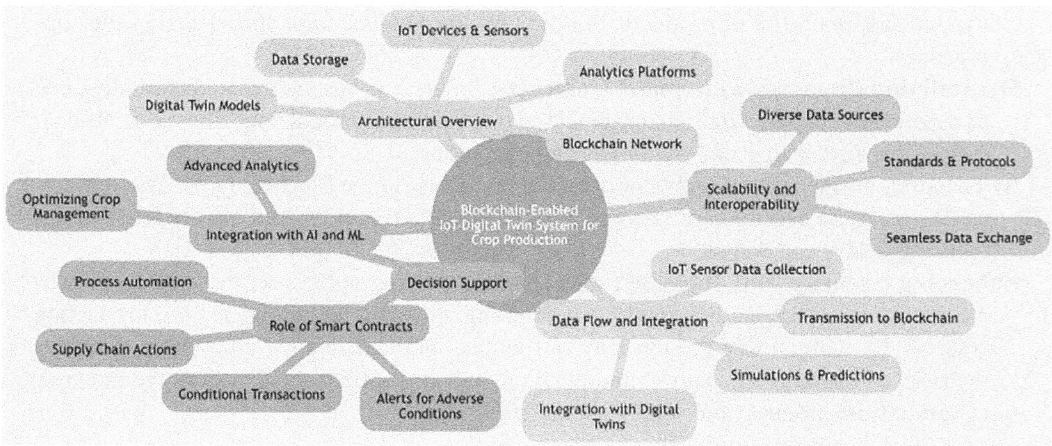

FIGURE 7.2 Blockchain-enabled IoT–digital twin system for crop production.

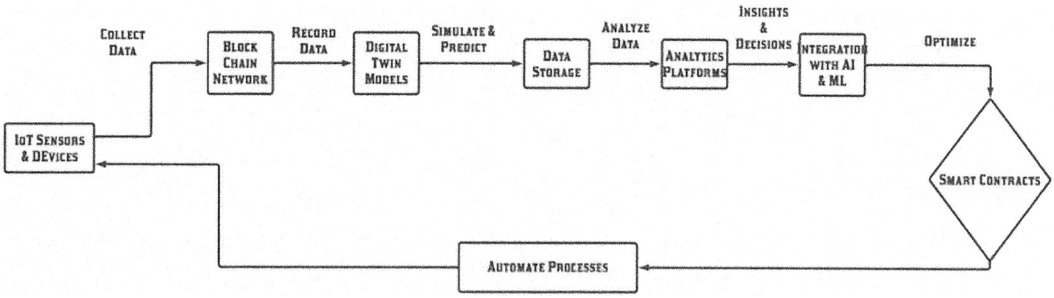

FIGURE 7.3 Data flow in a blockchain-enabled IoT–digital twin system.

system. By leveraging IoT sensors, blockchain networks, digital twin models, smart contracts, and AI integration, this system revolutionizes crop production and supply chain management. It ensures transparency, traceability, automation, and data-driven judgement, ultimately resulting in increased yield, source optimization, real-time monitoring, and scalability in agricultural operations. This comprehensive approach addresses the evolving challenges of modern farming while paving the way for sustainable, efficient, and resilient crop production practices.

Deep insights into the operating dynamics of an Internet of Things digital twin system can be gained by visualizing the data flow within it. We are able to obtain a thorough grasp of the methods used to gather, interpret, and utilize real-time data in order to optimize crop output by following the data's path from Internet of Things sensors to the blockchain network and integrating it with digital twin models. This visualization shown in Figure 7.3 highlights the crucial role that data plays in facilitating automated processes, efficient agricultural practices, and well-informed decision-making. It also emphasizes the smooth collaboration between interrelated components.

7.6 DATA MANAGEMENT AND SECURITY

7.6.1 Types of Data Collected in IoT-Digital Twin Crop Production Systems

Sensor Data: Different kinds of sensor data are gathered in IoT–digital twin agricultural production systems to keep an eye on crop health and environmental conditions. This contains information about plant development characteristics like height, canopy cover, and biomass as well as temperature, humidity, leaf wetness, and solar radiation.

Aerial and Satellite Imagery: To obtain high-resolution aerial photography and remote sensing data, drones and satellites are used. Insights regarding crop health, nutrient shortages, pest infestations, and other variables influencing crop development and yield can be gained from this imagery.

Yield Data: Yield data, including yield maps and historical yield records, are collected to assess the performance of various areas within a field and track changes in productivity over time. Farmers can use this information to guide their decisions on resource allocation and crop management techniques.

Supply Chain Data: Data related to the movement of agricultural products through the supply chain is collected to track product flow, transportation routes, storage conditions, and logistics information. This data enables stakeholders to optimize supply chain operations, improve traceability, and ensure product quality and safety.

Environmental Data: Environmental data, such as weather conditions, precipitation, and climate patterns, are collected to understand the impact of weather variability on crop growth and yield. This data informs decision-making regarding irrigation scheduling, pest management, and other agronomic practices.

Operational Data: Operational data, including equipment usage, labor inputs, and resource consumption (e.g., water, fertilizers, pesticides), are collected to monitor farm operations and optimize resource efficiency. This data helps identify inefficiencies, reduce costs, and improve overall farm productivity.

7.6.2 Data Privacy and Security Concerns

Privacy Concerns: Collecting and handling sensitive data, such as personal information of farmers, proprietary business data, and location data, raises privacy concerns. Farmers may be reluctant to share their data if they fear it could be misused or compromised, leading to challenges in data sharing and collaboration.

Security Risks: Data breaches, unauthorized access, and potential misuse of data pose significant security risks in IoT–digital twin crop production systems. Malicious actors may attempt to steal or manipulate data for financial gain or sabotage, jeopardizing the integrity of crop production processes and supply chain operations.

Data Integrity: Ensuring data integrity and preventing data tampering are critical to maintaining trust and confidence in crop production systems. Any unauthorized changes to data could lead to incorrect decisions, financial losses, and reputational damage. Therefore, strong security protocols must be established to protect data integrity throughout its lifecycle.

7.6.3 Blockchain-Based Solutions for Secure Data Storage and Sharing

Safe Data Storage: By utilizing its decentralized and immutable characteristics, blockchain technology offers a safe and impenetrable option for data storage. The blockchain ensures data integrity and authenticity by preventing data tampering or deletion. This makes blockchain an ideal platform for storing sensitive agricultural data securely (Verma and Kanrar, 2024).

Cryptographic Techniques: Blockchain uses digital signatures and hashing, two cryptographic techniques, to guarantee the validity and integrity of data. Every data transaction is cryptographically hashed and connected to the one before it, forming a chain of blocks that cannot be changed without the network's members agreeing.

Access Control Mechanisms: Blockchain allows for the putting in place access restriction measures and permission oversight systems to regulate data access and sharing among authorized parties. Smart contracts can be used to enforce access policies and automate data sharing based on predefined conditions, ensuring that sensitive data can only be accessed by authorized people.

Decentralized Data Storage: Decentralized data storage options, including distributed file systems and the InterPlanetary File System (IPFS), offer alternatives to centralized data storage architectures. These solutions distribute data across multiple nodes in a network, enhancing data resilience, scalability, and availability while maintaining data integrity and security.

Combining Blockchain with Other Security Technologies: Blockchain can be combined with other security technologies, such as to improve data privacy, use safe multi-party computation, homomorphic encryption and secure data analysis. These cryptographic techniques allow for computations to be performed on encrypted data without revealing sensitive information, ensuring privacy and confidentiality.

The line graph shown in Figure 7.4 illustrates the projected growth of agricultural data volume over the years, showcasing a steady upward trend. As depicted, the volume of agricultural data should

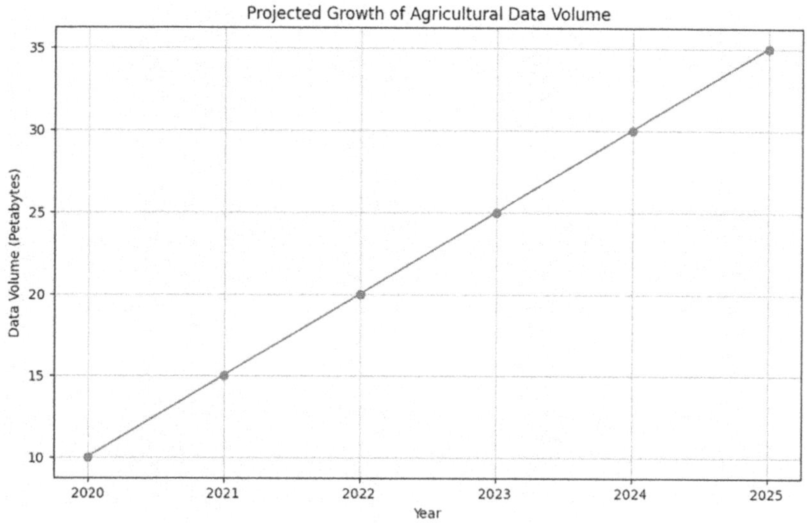

FIGURE 7.4 Projected growth of agricultural data volume (2020–2025).

rise from 10 petabytes in 2020 to 35 petabytes in 2025. This upward trajectory underscores the significance of data in modern agricultural practices, highlighting the need for robust data management and analysis solutions to harness the potential insights for optimized crop production and supply chain management.

7.7 SMART CONTRACTS AND AUTOMATION

7.7.1 Introduction to Smart Contracts

Definition: Smart contracts are self-executing programs that are implemented on a blockchain network with predetermined conditions and regulations expressed in code. Agreements are automatically executed and enforced via these contracts. without the need for intermediaries, relying on blockchain technology for transparency, security, and immutability.

Key Characteristics: Smart contracts exhibit deterministic execution, meaning their outcomes are predetermined by the code and cannot be altered. They cannot be changed once they are put on the blockchain, since they are immutable. Furthermore, all users on the blockchain network may view the code and outcomes of smart contract execution, making them transparent.

Role in Automation: Smart contracts are essential in automating processes by enabling trustless and tamper-proof execution of agreements. They get rid of the necessity for intermediaries, streamline transactions, and ensure transparency and accountability in business dealings.

7.7.2 Applications of Smart Contracts in Crop Production

Supply Chain Automation: Smart contracts can automate various supply chain processes, including tracking shipments, verifying product authenticity, and executing payments. For example, smart contracts have the ability to release payment automatically to suppliers after the products are authenticated and delivered.

Crop Insurance Schemes: Smart contracts can streamline the implementation of crop insurance schemes by automating the payout process. Payouts can be triggered automatically based on predefined conditions, such as adverse weather events, pest infestations, or crop yield fluctuations.

Farmer-Buyer Agreements: Smart contracts can facilitate agreements between farmers and buyers, ensuring fair pricing, timely payments, and adherence to contractual terms. For instance, smart contracts can automatically execute payments to farmers upon delivery of agricultural products.

Resource Allocation and Management: Smart contracts can optimize the allocation and management of agricultural resources, such as using data from IoT sensors and digital twin models in real time for water, fertilizer, and pesticide applications. For example, smart contracts can automatically adjust irrigation schedules based on soil moisture levels detected by IoT sensors.

7.7.3 Integration of Smart Contracts with IoT and Digital Twins

Integration with IoT Devices: To automate processes based on real-time data, smart contracts can be connected with sensors and Internet of Things devices. For example, smart contracts can be used to instruct irrigation systems to water crops when IoT sensors detect a drop in soil moisture levels below a predetermined threshold (Iakovos Pittaras et al., 2024).

Integration with Digital Twins: Smart contracts can leverage data from digital twin models to inform decision-making and automate processes. For example, smart contracts can execute actions based on predictions generated by digital twin simulations, such as adjusting fertilizer application rates based on crop growth forecasts.

Combination with Artificial Intelligence and Machine Learning: Smart contracts can be enhanced by combining them with ML and artificial intelligence algorithms for adaptive and dynamic contract execution. For instance, smart contracts will be able to evaluate past data and forecast future events crop yields, automatically adjusting contract terms and conditions accordingly.

7.8 CASE STUDIES AND REAL-WORLD EXAMPLES

7.8.1 Successful Implementations of Blockchain-Enabled IoT-Digital Twin Systems

CASE STUDY 1: PRECISION FARMING IN THE UNITED STATES

- **Challenges:** Traditional farming practices in the United States faced several challenges, including inefficient resource utilization, limited insights into crop health, and a lack of transparency in the supply chain. Farmers struggled in order to maximize their inputs like water and fertilizers, leading to decreased yields and increased production costs. Moreover, the lack of real-time data on crop health made it difficult for farmers to detect and deal with problems promptly, leading to lower-quality produce and reduced market competitiveness (Jaramillo-Hernández et al., 2024)
- **Solution:** To address these challenges, a consortium of agricultural technology companies collaborated to develop a blockchain-enabled IoT–digital twin system tailored for precision farming. The system integrated various IoT sensors for soil monitoring, drones equipped with multispectral cameras for aerial imaging, and blockchain technology for supply chain traceability. Soil sensors provided real-time data on soil health by continuously monitoring temperature, moisture content, and nutrient

content. High-resolution aerial photos of crop fields were taken by drones, giving farmers the ability to keep an eye on crop health, identify pests, and apply fertilizer and irrigation more efficiently. The blockchain technology recorded every transaction and data point in the supply chain, from seed planting to product distribution, ensuring transparency and traceability.

- **Benefits:** The implementation of the precision farming system resulted in significant benefits for farmers and stakeholders. Farmers experienced a 15% increase in yield due to optimized resource utilization and improved crop management practices. Water consumption was reduced by 20%, contributing to water conservation efforts and environmental sustainability. The enhanced traceability provided by blockchain technology enabled farmers to get their produce into exclusive marketplaces and fetch greater prices, leading to increased profitability and market competitiveness.
- **Applications:** The precision farming system facilitated various applications, including precision irrigation, variable rate fertilization, and predictive analytics for crop management decisions. Farmers could precisely monitor and control irrigation and fertilizer application grounded in real time soil data, cutting back on waste and maximizing crop yield. In order to give farmers information on crop growth, yield estimates, and the best times to harvest their crops, predictive analytics algorithms examined both historical and current data. This allowed farmers to make well-informed decisions and maximize their operations.

CASE STUDY 2: SMART CONTRACT–BASED CROP INSURANCE IN INDIA

- **Challenges:** The Indian agricultural sector faced numerous challenges, including inefficient and manual crop insurance processes, delays in claims processing, and a lack of transparency in insurance transactions. Farmers often encountered difficulties in accessing crop insurance due to cumbersome paperwork, complex eligibility criteria, and lengthy claims processing times. Moreover, the traditional insurance model relied heavily on manual processes, leading to inefficiencies, delays, and potential errors in claims settlement (Sugandh et al., 2024).
- **Solution:** To address these challenges, a government-led initiative introduced a blockchain-enabled smart contract system for crop insurance in India. The system leveraged blockchain technology to automate the execution of insurance contracts and claims processing, reducing administrative overhead and streamlining the insurance process. Smart contracts, deployed on a blockchain network, encoded the terms and conditions of insurance policies and automatically executed claims payments based on predefined triggers, such as weather conditions, yield losses, or pest infestations. Farmers were registered on the blockchain network, and their insurance coverage and claims history were recorded transparently and immutably.
- **Benefits:** The adoption of smart contract-based crop insurance resulted in significant benefits for farmers, insurers, and government agencies. Claims processing times were reduced from months to days, enabling farmers to receive timely compensation for crop losses and minimize financial hardships. The automation of insurance processes reduced administrative costs by 30% and eliminated manual errors and discrepancies in claims settlement. Moreover, the transparency and immutability provided by blockchain technology increased trust and confidence among farmers

and insurers, encouraging greater participation in crop insurance schemes and promoting financial inclusion in rural communities.

- **Applications:** The smart contract-based crop insurance system facilitated various applications, including automated claims processing, transparent insurance transactions, and risk mitigation for farmers. Smart contracts automatically triggered claims payments based on predefined triggers, such as adverse weather events, pest infestations, or yield losses, ensuring prompt and accurate compensation for crop losses. The transparent and immutable nature of blockchain technology increased trust and transparency in insurance transactions, reducing fraud and enhancing accountability in the insurance industry.

CASE STUDY 3: SUPPLY CHAIN TRACEABILITY IN AUSTRALIA

- **Challenges:** The Australian agricultural sector grappled with supply chain transparency, product provenance tracking, and food safety assurance issues. Consumers demanded transparency and traceability in food products, but traditional supply chains lacked visibility, hindering verification of product authenticity and quality. Concerns about food safety and fraud underscored the need for enhanced traceability. (Charlebois et al., 2024)
- **Solution:** Industry stakeholders collaborated to develop a blockchain-enabled supply chain traceability system for Australian agricultural products. Blockchain recorded every supply chain step, creating an immutable ledger of product transactions. QR codes or NFC tags linked physical products to digital records, enabling consumers to access detailed information.
- **Benefits:** The system provided consumers with accurate product information, fostering informed decisions. Producers gained consumer trust, leading to increased market demand and premium pricing. Blockchain technology reduced fraud risk, ensuring product authenticity and safety.
- **Applications:** The system facilitated transparent product tracking, consumer engagement, and fraud prevention. Each product received a unique digital identity, enabling stakeholders to trace its journey accurately. Consumers could access detailed product information by scanning packaging, fostering trust. Blockchain's tamper-proof nature ensured secure transaction recording, reducing fraud risk.

7.8.2 Lessons Learned and Best Practices

Stakeholder Collaboration: Successful implementations of blockchain-enabled IoT–digital twin systems require close collaboration among farmers, tech providers, and government agencies. Engaging diverse stakeholders fosters partnerships, ensuring solutions meet everyone's needs. Collaboration facilitates knowledge sharing and problem-solving, leading to more robust solutions.

Change Management and Training: Adoption of new technologies like blockchain-enabled IoT–digital twin systems requires changes in mindset and behavior. Comprehensive training programs educate stakeholders about the benefits and implications of the new system. Hands-on workshops and ongoing support empower users to leverage the technology effectively.

Data Governance and Standardization: Good data governance makes sure the interoperability, reliability, and security of blockchain-enabled IoT–digital twin systems. Clear policies and standards define roles, enforce data integrity, and mitigate risks. Standardization efforts focus on common data formats and interoperability standards, ensuring seamless data exchange.

Scalability and Infrastructure: Scalable infrastructure is crucial for supporting the growing demands of blockchain-enabled IoT–digital twin systems. Investments in resilient infrastructure, including cloud-based solutions and edge computing technologies, ensure scalability and responsiveness. Cybersecurity measures protect sensitive data from threats and attacks.

Regulatory Support: Clear regulatory frameworks support the implementation of blockchain-enabled IoT–digital twin systems in crop production. Governments should collaborate to develop regulations addressing legal, ethical, and security considerations. Incentives like grants and tax incentives encourage widespread adoption of these transformative technologies.

It is crucial to include Table 7.2 that summarizes the most important best practices for deploying blockchain-enabled IoT–digital twin systems in agricultural production so that readers may quickly refer to and understand it. An overview of the best practices that were discussed is given in Table 7.2.

7.9 CHALLENGES AND FUTURE DIRECTIONS

7.9.1 Technical Challenges

- **Integration with Legacy Systems:**

The integration of blockchain-enabled IoT–digital twin systems with existing legacy systems in the agricultural sector poses significant technical challenges. Legacy systems may have disparate data formats, communication protocols, and infrastructure, making seamless integration or migration complex.

The creation of middleware solutions and standardized application programming interfaces (APIs) to enable data interchange and interoperability between legacy and contemporary systems are viable approaches for overcoming these obstacles. Additionally, phased migration approaches and pilot projects can help mitigate risks and validate integration strategies before full-scale deployment.

TABLE 7.2
Best Practices for Implementing Blockchain-Enabled IoT–Digital Twin Systems in Agriculture

Best Practice	Description
Stakeholder Collaboration	Engage diverse stakeholders and foster partnerships to ensure alignment and shared goals.
Change Management and Training	Educate users and provide support for technology adoption through change management strategies and comprehensive training programs.
Data Governance and Standardization	Define data policies, protocols, and interoperability standards to ensure consistent and reliable data for AI systems.
Scalability and Infrastructure	Invest in scalable and resilient infrastructure to accommodate the growing demands of AI applications and ensure efficient operations.
Regulatory Support	Develop clear regulatory frameworks and incentives to facilitate the responsible and ethical use of AI technologies while ensuring compliance with legal requirements.

- **Data Quality and Integrity:**

Ensuring high-quality and accurate data inputs from IoT sensors and other sources is crucial to maintaining the integrity and reliability of digital twin models and blockchain-based systems. However, data quality issues such as noise, bias, and incomplete data can affect the accuracy and effectiveness of predictive analytics and decision-making algorithms.

Addressing these challenges requires robust data validation and cleansing processes, as well as continuous monitoring and quality assurance mechanisms. Implementing data validation rules, anomaly detection algorithms, and data fusion techniques can help identify and mitigate data quality issues in real time.

- **User Experience and Adoption:**

User-friendly interfaces and training programs are crucial for facilitating the adoption of blockchain-enabled IoT–digital twin systems. Complex interfaces and lack of training can hinder usability and adoption.

Designing intuitive interfaces, providing comprehensive training, and fostering digital literacy within the agricultural community can improve user experience and promote adoption.

7.9.2 Regulatory and Legal Challenges

- **Data Privacy and Security rules:** To safeguard sensitive data and maintain confidence in blockchain-enabled IoT–digital twin systems, adherence to data privacy and security rules, such as the General Data Protection Regulation (GDPR) and industry-specific laws, is crucial. Ensuring compliance and managing complicated regulatory obligations, however, can be difficult.

In order to preserve user privacy and adhere to regulatory requirements, organizations need to put strong data protection mechanisms in place, like encryption, access limits, and data anonymization. In order to make sure that data handling procedures adhere to regulatory requirements, conducting routine audits and assessments can assist in locating and addressing compliance gaps.

- **Intellectual Property Rights:**

Intellectual property rights, particularly related to digital twin models, algorithms, and data ownership, present legal challenges in the implementation of blockchain-enabled IoT–digital twin systems. Questions around data ownership, licensing, and intellectual property infringement require careful consideration and legal guidance.

Establishing clear ownership rights and licensing agreements for digital twin models and algorithms can help protect intellectual property and prevent unauthorized use or exploitation. Collaboration agreements and contractual arrangements should define rights, responsibilities, and limitations concerning intellectual property ownership and usage rights.

- **Legal and Governance Frameworks:**

Clear legal and governance frameworks are essential to regulate the use of blockchain technology, IoT devices, and digital twins in crop production. Legal issues such as liability, accountability, and dispute resolution mechanisms require careful consideration to ensure fair and transparent governance of blockchain-enabled systems.

Governments, regulatory bodies, and industry associations should collaborate to develop and enforce clear legal frameworks that address emerging issues in agricultural technology. Establishing regulatory sandboxes, pilot projects, and industry standards can facilitate experimentation and innovation while ensuring compliance with legal and ethical standards.

- **Cross-Border Data Sharing and Harmonization:**

Cross-border data sharing presents challenges related to data sovereignty, jurisdictional differences, and regulatory harmonization. Inconsistent regulations and data protection laws across different jurisdictions can hinder the seamless exchange of data and information, particularly in global supply chains.

International cooperation and standardization efforts are needed to harmonize regulations and facilitate cross-border data sharing in agricultural systems. Establishing bilateral or multilateral agreements, data-sharing protocols, and interoperability standards can promote collaboration and facilitate data exchange while respecting privacy and sovereignty concerns.

7.9.3 Future Research Directions and Potential Advancements

7.9.3.1 Artificial Intelligence and Machine Learning Integration

Integrating AI and ML with blockchain-enabled IoT–digital twin systems promises advanced analytics, predictive modeling, and autonomous decision-making in crop production. AI algorithms analyze data from IoT sensors and digital twins to identify patterns, predict yields, and optimize farming practices in real time. Collaborative research is essential for developing tailored AI models and accelerating advancements in AI-driven agriculture.

7.9.3.2 Edge Computing and Decentralized Processing

Edge computing addresses scalability and latency challenges by moving processing closer to IoT devices and digital twins. Research should focus on developing edge computing frameworks for real-time analytics in agricultural systems.

7.9.3.3 Quantum Computing and Post-Quantum Cryptography

Quantum computing poses security challenges for blockchain systems. Research should prioritize developing post-quantum cryptographic solutions to protect blockchain from quantum attacks.

7.9.3.4 Hybrid Architectures and Interoperability

Hybrid blockchain architectures combine public and private blockchains, offering flexibility and interoperability. Research should focus on developing standards and protocols for seamless data exchange (Sugarbroad, 1990).

7.9.3.5 Sustainable Agriculture

Blockchain-enabled IoT systems can promote sustainable farming practices. Research should develop blockchain-based solutions for monitoring and verifying sustainability initiatives.

7.9.3.6 Decentralized Autonomous Organizations

Decentralized autonomous organizations (DAOs) can revolutionize governance in agriculture, enabling collective decision-making. Research should focus on developing DAO frameworks tailored to agricultural contexts.

7.10 CONCLUSION

Blockchain-based benefits of IoT–digital twin systems include improved crop production efficiency, traceability, and transparency. They minimize the effects of climate change, solve problems with conventional methods, and streamline supply chain operations. They facilitate well-informed decision-making to improve agriculture's sustainability and profitability by offering real-time data and analytics. All things considered, these integrated systems present a viable way to raise food security and transparency in the agriculture industry.

7.10.1 Final Remarks and Recommendations for Adoption and Implementation

Stressing collaboration among stakeholders is crucial for successful adoption of blockchain-enabled IoT–digital twin systems in agriculture. Comprehensive strategies considering technical, regulatory, and socio-economic factors are recommended for smooth transition. Continuous research, innovation, and pilot projects are essential to drive evolution and demonstrate real-world benefits. Education and training programs are vital to increase awareness and adoption. Ultimately, collaborative efforts are needed for a sustainable, efficient, and resilient agricultural ecosystem empowered by these technologies.

REFERENCES

Bansal, A., Tewari, A., Sharma, A., and Bansal, A. (2024). Implementations and Rationale for Blockchain Technique in Agriculture. In S. S. Chouhan, U. P. Singh, and S. Jain (Eds.), *Applications of Computer Vision and Drone Technology in Agriculture 4.0.* Springer, Singapore. https://doi.org/10.1007/978-981-99-8684-2_6

Brandín, R., and Abrishami, S. (2024). IoT-BIM and Blockchain Integration for Enhanced Data Traceability in Offsite Manufacturing. *Automation in Construction*, 159, article 105266. ISSN: 0926-5805. https://doi.org/10.1016/j.autcon.2024.105266.

Charlebois, S., Latif, N., Ilahi, I., Sarker, B., Music, J., and Vezeau, J. (2024). Digital traceability in agri-food supply chains: A comparative analysis of OECD member countries. *Foods*, 13, 1075. https://doi.org/10.3390/foods13071075

Eswaran, U., and Eswaran, V. (2023a). Real-Time Monitoring of Environmental Parameters in Agriculture using WSN. *i-manager's Journal on Embedded Systems*, 11(2), 13–20. https://doi.org/10.26634/jes.11.2.19970

Eswaran, U., and Eswaran, V. (2023b). Integrating IoT, ML and Cloud Computing for Sustainable Agriculture: Opportunities and Challenges. *i-manager's Journal on IoT & Smart Automation*, 1(1), 34–47. Retrieved from https://www.imanagerpublications.com/article/19933/?utm_source=TrendMD&utm_medium=cpc&utm_campaign=JIoT__TrendMD_0

Eswaran, U., Eswaran, V., Murali, K., and Eswaran, V. (2023a). Blockchain-based Privacy-preserving Data Sharing in Healthcare Systems. *Journal of Advanced Database Management & Systems*, 10(3), 28–34p https://stmcomputers.stmjournals.com/index.php/JoADMS/article/view/714

Eswaran, U., Eswaran, V., Murali, K., and Eswaran, V. (2023b). Precision Agriculture Revolution: A Low-cost Smart Irrigation System Empowered by IoT for Small-scale Farmers. *Journal of Mechatronics and Automation*, 10(3), 21–29p. https://engineeringjournals.stmjournals.in/index.php/JoMA/article/view/7744

Eswaran, U., Eswaran, V., Murali, K., and Eswaran, V. (2024). Application of Sensors for Smart Farming. In A. Khang (Ed.), *Agriculture and Aquaculture Applications of Biosensors and Bioelectronics* (pp. 18–44). IGI Global. https://doi.org/10.4018/979-8-3693-2069-3.ch002

Eswaran, U., and Khang, A. (2023). Advanced IoT Technologies and Applications in the Industry 4.0 Digital Economy. In J. Smith (Ed.), *Advances in IoT Technologies* (pp. 150–170). CRC Press, Taylor & Francis Group. https://doi.org/10.1201/978-1-003-43426-9

Eswaran, U., and Khang, A. (2024). Application of Internet of Things (IoT) and Sensors Technologies for Agriculture. In *Advanced IoT Technologies and Applications in the Industry 4.0 Digital Economy* (1st Ed.). CRC Press. https://doi.org/10.1201/9781003434269-8

Jaramillo-Hernández, J. F., Julian, V., Marco-Detchart, C., and Rincón, J. A. (2024). Application of Machine Vision Techniques in Low-Cost Devices to Improve Efficiency in Precision Farming. *Sensors*, 24, 937. https://doi.org/10.3390/s24030937

Kadiyala, R., Mohana, M., Reddy, K. K., and G., T. R. (2023). Block Chain Based Data Sharing System for Cloud Based Internet of Things Systems with Efficient Smart Contracts. In *IEEE ICC 2023*, Rome, Italy. https://doi.org/10.1109/ICCWorkshops57953.2023.10283747

Mitsanis, C., Hurst, W., and Tekinerdogan, B. (2024). A 3D Functional Plant Modelling Framework for Agricultural Digital Twins. *Computers and Electronics in Agriculture*, 218, article 108733. ISSN: 0168-1699. https://doi.org/10.1016/j.compag.2024.108733.

Pittaras, I., Fotiou, N., Karapapas, C., Siris, V. A., and Polyzos, G. C. (2024). Secure Smart Contract-based Digital Twins for the Internet of Things. *Blockchain: Research and Applications*, 5(1), article 100168. ISSN: 2096-7209. https://doi.org/10.1016/j.bcra.2023.100168.

Sadek, N., and Shehata, D. (2024). Internet of Things Based Smart Automated Indoor Hydroponics and Aeroponics Greenhouse in Egypt. *Ain Shams Engineering Journal*, 15(2), 102341.

Sugandh, U., Nigam, S., and Khari, M. (2024). Integrated Approach using Blockchain, Sensors and Smart Contracts for Weather Alert. In *2024 11th International Conference on Computing for Sustainable Global Development (INDIACom)*, New Delhi, India (pp. 1699–1705). https://doi.org/10.23919/INDIACom61295.2024.10498872.

Sugarbroad, I. (1990). An OSI-based Interoperability Architecture for Managing Hybrid Networks. *IEEE Communications Magazine*, 28(3), 61–69, March. https://doi.org/10.1109/35.52893.

Sundararajan, N. et al. (2024). Mitigating Global Challenges: Harnessing Green Synthesized Nanomaterials for Sustainable Crop Production Systems. *Global Challenges*, 8(1), 2300187.

Verma, G., and Kanrar, S. (2024). Secure Document Sharing Model Based on Blockchain Technology and Attribute-based Encryption. *Multimedia Tools and Applications*, 83, 16377–16394. https://doi.org/10.1007/s11042-023-16186-z

8 Digital Twin Applications in the Agriculture and Farming Domain

Sivaretinamohan R

8.1 INTRODUCTION: BACKGROUND

Farmers may better understand plant development patterns, assess resource requirements, and respond to crop diseases by closely monitoring the Internet of Things (IoT), soil, and weather using digital twins. These produce a real-time mirror of a field's current condition (Van Der Burg et al., 2021). An IoT-based smart farming system automatically triggers an irrigation system that monitors the agricultural land. Farmers may monitor field conditions from anywhere using temperature, light, humidity, and soil moisture sensors (Poornima & Gowda, 2024). Smart farming based on the Internet of Things is far more efficient than traditional farming methods (Jayaraman et al., 2016). As of right now, digital agriculture (DA) is generally considered to mean employing contemporary tools, data analytics, and monitoring to enhance and/or optimise farming systems, boost crop quality and production, cut waste, and control pest and disease pressure (Purcell & Neubauer, 2023). The growing unpopularity of soil-based agriculture, especially hydroponics, can be attributed to problems with soil-based agriculture, such as soil erosion, labour-intensiveness, reusability of water, and overall productivity (Niu & Masabni, 2022). Digital twins can be used in hydroponics in various ways, including developing, operating, monitoring, optimising, and maintaining systems.

A novel method in agriculture is using hanging robots with livestock farming help. These robots monitor temperature, moisture, and other factors, and if a disease is found, the system will automatically spray pesticides and fertilisers on the crops (Shamia et al., 2023). IoT-Architecture (IoT-A), a reference architecture for IoT systems, enables cloud services, protocols, and networked devices to work together to form an IoT ecosystem.

Digital twin applications like FarmLogs, Granular, Trimble Agriculture, Agrian, and Iteris ClearAg are made explicitly for within-field management zoning in agriculture. Precision livestock farming applications such as BouMatic's SmartDairy, DeLaval DelPro Farm Manager, Cowlar, Moocall, and SCR Dairy offer real-time monitoring and management solutions to enhance the productivity and well-being of cows.

Technologies such as Priva, HortiMaX, Hoogendoorn, Argus Controls, and Primalogik offer comprehensive solutions for real-time monitoring, controlling, and optimising supply chain processes and greenhouse environments. At the same time, there may be a dedicated digital twin platform for chain-integrated greenhouse production. Advanced sensors, data analytics, and machine learning algorithms are used by Blue River Technology's See & Spray, ecoRobotix's WeedVision, Xarvio's Weed Manager, Resson's WeedScout, and Trimble's WeedSeeker 2 to give farmers precise tools and actionable insights for managing weeds, thereby promoting more sustainable and effective agricultural practices. Comprehensive systems for managing pig farms are provided by Porcitec, PigKnows, PigCHAMP, FarmSight by John Deere, Topigs Norsvin's Total systems, and PigKnows. These solutions integrate sensor technology and data analytics to maximise animal welfare and production efficiency. Crop management and farm operations are optimised by integrating data analytics and virtual modelling in agricultural digital twin solutions such as Climate

DOI: 10.1201/9781003507390-8

FieldView, Taranis, Granular Insights, Iteris ClearAg, Trimble Ag Software, and John Deere Operations Center. Thus, digital twin applications in agriculture enhance farm operations and crop management by utilising data analytics and virtual modelling, boosting sustainability and production (Peladarinos et al., 2023).

8.2 PREREQUISITES FOR AGRICULTURAL DIGITAL TWINS

An agricultural digital twin is the real-time digital coupling that combines a virtual representation with a functional output with the condition of a physical asset or process. Physical assets, sensors, and actuators are found in the physical (real-world) environment, whereas multiphysics, multiscale, or probabilistic simulation models are found in the virtual area. Data analysis with real-time information gathered from previous studies is supported by machine learning (ML) models and simulations. Subsequently, the development of intelligent digital technologies and their integration into production has occurred. These technologies include industrial Internet of Things networks, artificial intelligence (AI), big data, robots, and automation (Escribà-Gelonch et al., 2024). This helps farmers avoid external risks from uncontrollable variables like weather, increase profitability, reduce expenses associated with logistics, and shorten the time it takes for agricultural products to reach markets (Attaran et al., 2024). Therefore, the following prerequisites are considered necessary for agricultural digital twins

(i). Uniqueness: The digital twin must be unique and have a designated name in agriculture.
(ii). Almost Real Time: This implies that the twin should persist for as long as the real-time version exists.
(iii). Information Based: The sample must be digitally measured using the same instrument.
(iv). Practical and Attainable: The digital twin must closely resemble the genuine item in shape and characteristics in agriculture. It needs to be grounded in reality.
(v). Doable: Information about the actual object needs to be action oriented.

8.3 HOW DO DIGITAL TWINS WORK IN SMART AGRICULTURE?

Digital twins in smart agriculture refer to creating virtual replicas of real agricultural assets, systems, or operations. These replicas are produced using many data sources, including historical data, satellite imagery, Internet of Things sensors, and weather forecasts (Verdouw et al., 2021). Typically, they function like this:

(i). Data gathering: Sensors installed in the agricultural location capture real-time data on temperature, humidity, crop growth stages, soil moisture, and weather conditions. Drones and satellites may be used to capture additional images for more comprehensive data.
(ii). Data integration: The information gathered from many sources is integrated into a centralised platform or system. It makes the creation and maintenance of the digital twin easier.
(iii). Digital Twin Creation: State-of-the-art modelling and simulation techniques are applied to generate a virtual twin, or replica, of the agricultural setting. The virtual twin simulates the mechanics of classifying and tracking agricultural and livestock production systems and their respective trajectories, as well as the physical characteristics of soil, which include texture, colour, structure, and porosity.
(iv). Analysis and Prediction: Farmers and agricultural authorities analyse past data, monitor the existing, and estimate impending consequences with the help of the digital twin. They predict weather patterns and how they impact crop harvest.
(v). Decision Support: Crop production can be increased by adopting an irrigation system, applying fertiliser and pesticide at a lower rate, and scheduling when to sow seeds and harvest.

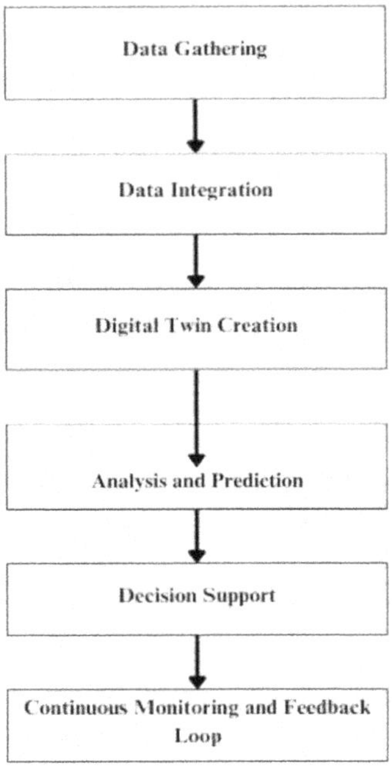

FIGURE 8.1 Modus operandi of the digital twin in agriculture and farming.

(vi). Continuous Monitoring and Feedback Loop: Real-time data updates are uninterruptedly supplied to the digital counterpart when the agricultural system changes. Therefore, unending evaluation and adjustment of processes guarantee the best farming practices

8.4 POTENTIAL DIGITAL TWINS IN SMART AGRICULTURE

8.4.1 Digital Twins in Within-Field Management Zoning

As part of precision agriculture, within-field management zoning divides agricultural fields into smaller zones according to topography, soil qualities, and other variables. Farmers may adjust their management strategies to meet the unique requirements of each zone in their field by acknowledging the variety within the field. It makes it easier to apply resources like pesticides, fertiliser, and water efficiently. Additional inputs can be added based on what each location needs.

Weather sensors are often positioned across the field to gather data on soil moisture, crop, temperature, nitrogen levels, and soil composition as part of the within-field management zone implementation procedure. GPS trackers precisely identify the boundaries of each zone and control the input application. Actuators such as variable rate application systems then adjust the input rates based on the sensor data that has been gathered. Figure 8.2 displays the key mechanisms of digital twins in within-field management zoning (implementation model).

Numerous apps and software platforms can now be accessible for digital twin solutions developed for agricultural within-field management zoning. A few of them are:

(i). FarmLogs: It provides a digital farming platform with tools for crop health monitoring, yield analysis, soil sampling, and field mapping. It facilitates farmers' use of within-field management zoning measures and the construction of digital twins in their agriculture zone to increase production and resource productivity.

(ii). Granular: A farm management and analytics app called Granular is used to increase farming productivity, profitability, and yield. Farm managers may use this software to set up work orders and planned chores that are sent straight to the farm operator or staff.

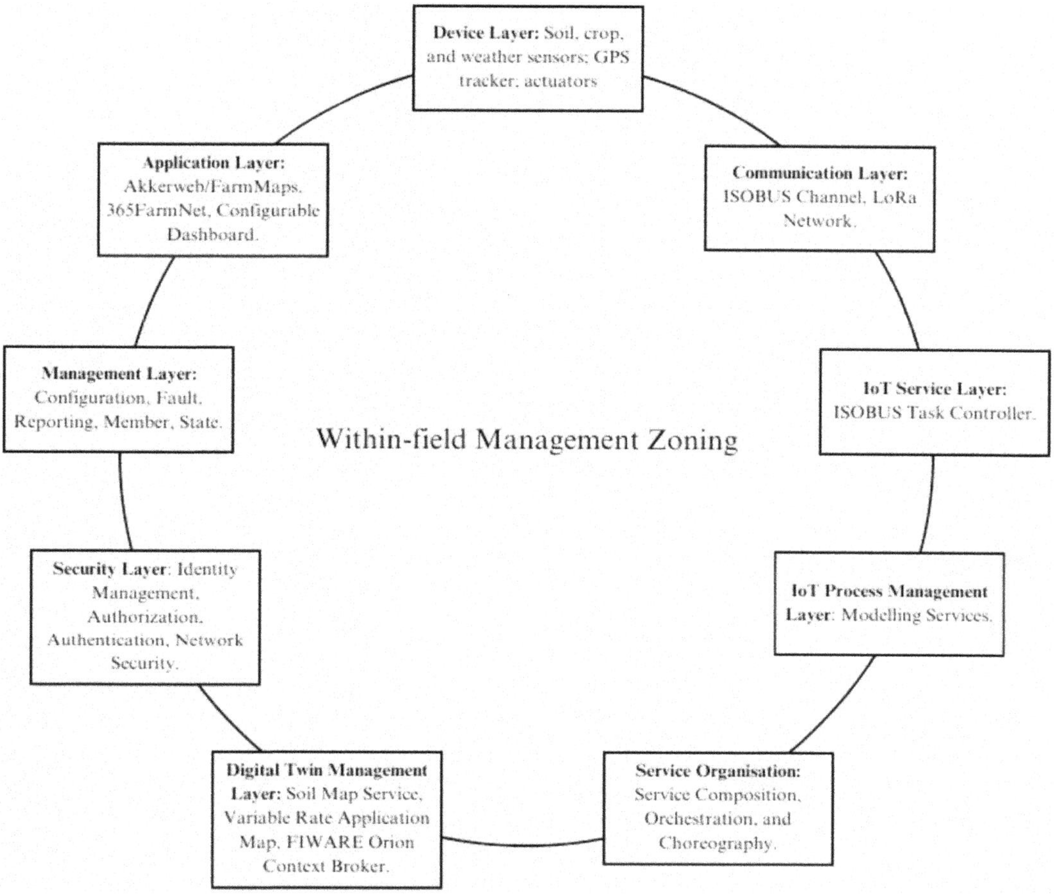

FIGURE 8.2 Key mechanisms of digital twins in within-field management zoning.

The work order provides the worker with comprehensive instructions on the tasks that must be completed, along with field directions, comments, product images, and pictures. Farm managers can access real-time updates, photographs, and notes taken by their field workers to enhance operational efficiency.

(iii). Trimble Agriculture: To satisfy the demands of farms, Trimble provides a broad range of precision agricultural solutions, including hardware, software, cloud-based applications, and positioning services. Farmers and advisors may spend limited resources to generate a safe and consistent food supply in a lucrative and ecologically friendly way with the help of Trimble Agricultural Solutions. Trimble solutions are compatible with most agricultural equipment and span all seasons, crops, terrains, and farm sizes. All facets of farming, including equipment and inventory management and crop and animal management, are readily tracked and managed using Trimble Ag Software (TAS). Farmers may create digital twins of their agricultural fields using this technology, and they can get precise and current data on inventory levels, equipment usage, crop growth, and livestock health. This allows farmers to optimise their operations and react swiftly to changing circumstances.

(iv). Agrian: The Agrian platform offers an accessible, user-friendly solution for capturing, storing, and sharing data. It now offers services in reporting, record keeping, and agronomy. By combining data from several sources, such as yield monitors, satellites, soil maps, and tissue and soil testing labs, into a single centralised data platform that is meant to function as a whole, Agrian's software does away with the requirement for "point products" that only do one function. Because Agrian's software is scalable, agribusinesses of all sizes may consolidate their workflows into a single, unified platform. Farmers can develop digital twins of their fields and use zoning schemes for within-field management with the help of Agrian's software.

(v). Iteris ClearAg: ClearAg offers agricultural data to help farmers make better decisions, produce higher-quality products, and increase income. Users may make better decisions by using the pertinent agronomic weather information that ClearAg material offers. Crop canopy and field-level weather insights are included in ClearAg content, which is accessible worldwide. To continuously enhance decision support analytics, the ClearAg Service uses strong agronomic and proprietary land surface models, machine learning, and artificial intelligence algorithms that dynamically learn about particular farm operations. Iteris EMPowerTM Platform, an adaptive forecasting engine for precision farming that integrates machine learning technology with meteorological and agronomic data, powers the ClearAg Service. Digital twins for within-field management zoning can be created using their platform

Besides, within-field management zoning can lessen the negative effects on soil and water quality using these applications.

8.4.2 Digital Twins in Happy Cow

A concept in animal husbandry called "Happy Cow" aims to ensure dairy cows are happy and healthy. In animal husbandry, the use of digital twin technology enhances animal welfare management and, therefore, animal product production. Farmers may create virtual, digital twins of their livestock in real time by employing generative adversarial networks and machine learning algorithms to create 3D faces of the animals. Farmers may more accurately forecast animal behaviour, foresee stress in their cattle, and see early symptoms of illness and discomfort by looking at details like ear position and eye areas on the virtual model (Han & Lin, 2022). The owner of a cow may locate a specific cow in a vast grazing area and analyse grazing habits and behaviours by employing wireless sensor networks (WSNs) and global positioning system (GPS) tracking technologies. Furthermore, the cow owner can readily identify which cows have been in close vicinity to ill animals if a disease is discovered early on thanks to monitoring technology, which helps stop the spread of disease and the death of livestock.

Dairy cow digital twins are useful to learn about the stages of animal growth and development from calf to adult. Sensor and longitudinal data are combined by multi-agent technology systems to create animal phenotypic characteristics. It is possible to record an animal's form, behaviour, and physiological processes to manage life phases and production cycles and plan the multi-agent growth of animals. When paired with cutting-edge AI technology, the camera system converts this real-time visual data into actionable insights that the farmer may evaluate on a desktop computer, mobile device, or phone. When a dairy cow is in estrus and ready for breeding, digital twins can use sophisticated pedometers to detect the movements of the cows. Figure 8.3 shows the key mechanisms of digital twins in Happy Cow (implementation model).

Obviously, modern livestock farming uses technology and applications to improve cow production and well-being. Here are a few instances:

(i). SCR Dairy:. Dairy farmers now have cutting-edge technology to monitor their cows in real time, around the clock: the Neck Collar Heat Detection and Health Monitoring Device for Cow. It combines animal activity and rumination for heat detection functionality. Cow monitoring systems with wearable sensors are provided by SCR Dairy. Additionally, the SCR system assesses the reproductive, health, nutritional, and general well-being of cows promptly and accurately.

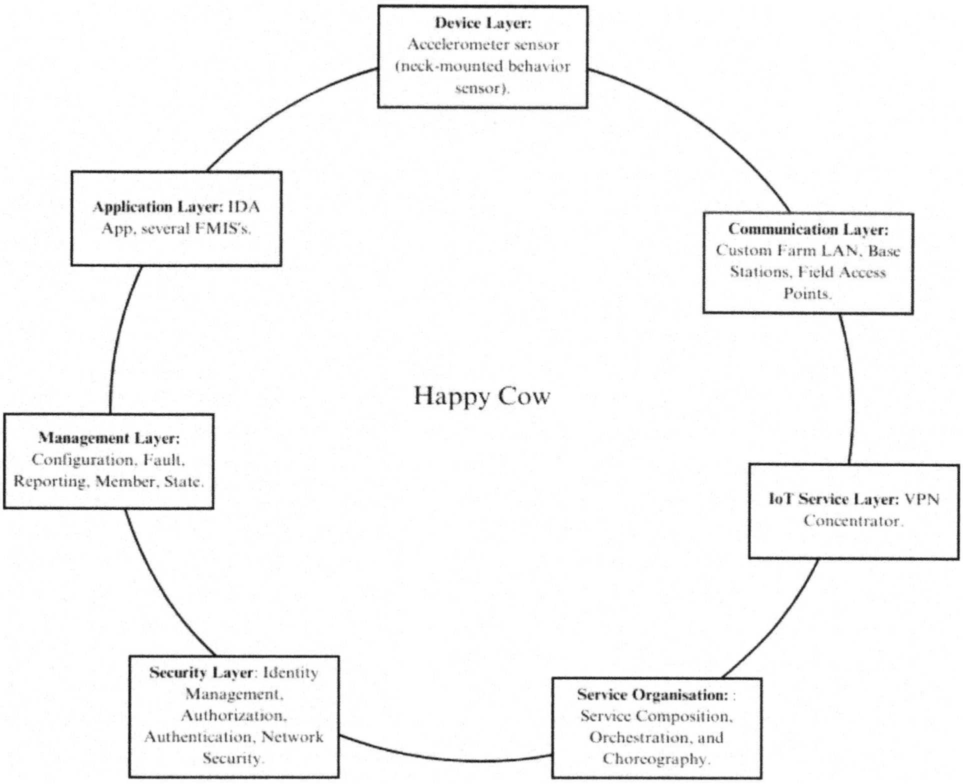

FIGURE 8.3 Key mechanisms of digital twins in Happy Cow.

(ii). DelPro Farm Manager by DeLaval: Farmers may monitor the performance of their dairy farm from a single, user-friendly application by using DelPro Farm Manager, which provides the data that DelPro has collected, processed, and analysed. DelPro Farm Manager transforms all of the farm's recorded data into insightful, comprehensible, and useful information. Monitor Board is a dashboard that presents a concise overview of all the information farmers want in a comprehensible and assessable visual manner.

(iii). Cowlar: Wearables from Cowlar monitor essential health metrics, allowing farmers to identify and treat diseases early on and maintain a healthy herd. Cowlar assists with determining the best times to breed, resulting in increased milk production, by tracking activity levels and health information.

(iv). Moocall: A motion sensor put on the tail called Moocall detects tail movement to forecast when calving will occur precisely. The gadget is fastened to the cow's tail by simply folding it over with one hand and tightening it with one or two clicks on the ratchet. Further, the pregnant cow is kept under observation by the Moocall calving sensor. To precisely predict when she will most likely give birth, Moocall team keep an eye on her contractions. Following that, when calving happens—typically an hour beforehand—an SMS notice will be issued to up to two phones of farmers/users

(v). BouMatic's SmartDairy: Dairy management may assess cows, milk output, and worker productivity by using the smart diary, which controls pulsation, meters, and detachers. SmartDairy manages sort gates and monitors stall operations, including entrance and exit gates, to ensure smooth cow traffic flow. Specialised feeding and hoof care systems are used to maintain the comfort and health of cows.

Therefore, farmers can increase animal health and production using digital twin approaches for Happy cows

8.4.3 Digital Twins in Chain-Integrated Greenhouse Production

Chain-integrated greenhouse production requires integrating several elements within the greenhouse environment to maximise crop productivity. This method covers every manufacturing process step, from planting and seeding to harvesting and packing. Farmers may establish ideal crop growth conditions by closely synchronising these elements, leading to increased yields and superior-quality products (Ariesen-Verschuur et al., 2022).

In chain-integrated greenhouse production, sensors and actuators are used to monitor and manage environmental parameters, including temperature, humidity, light levels, and watering. Farmers can understand and manage their crops using real-time data. Light sensors measure the total humidity in the greenhouse to ensure it stays within ideal bounds and offer data on various lighting circumstances so the greenhouse can adjust to specific lighting requirements. Actuators then alter the environmental conditions required for crop growth. The digital twins created for the chain-integrated greenhouse production additionally include fertigation, ventilation, and climate control systems. Figure 8.4 shows the key mechanisms of digital twins in chain-integrated greenhouse production (Implementation model).

Greenhouse farming uses numerous technologies and software programs in the digital twin platform to optimise supply chain management and production trials. Let's see a few of them:

(i). Priva: The most advanced process computer for ideal cultivation and optimal production is called Priva. Its user-friendly operating programs enable it to be fully self-managing and functioning around the clock. The web-based application Priva Connext Operator is

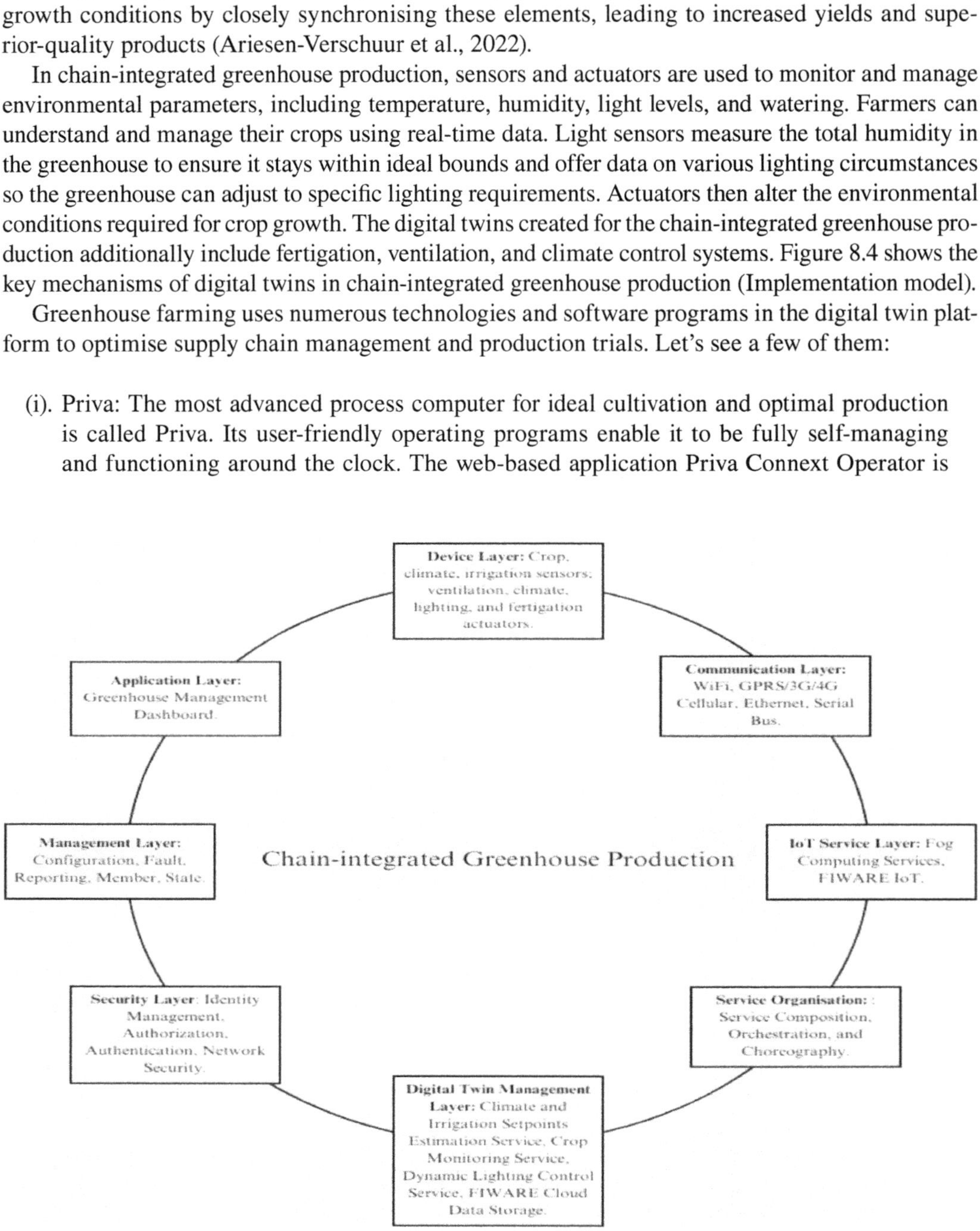

FIGURE 8.4 Key mechanisms of digital twins in chain-integrated greenhouse production.

available for tablets and smartphones. It offers insight into required procedures and makes the most significant system modifications consistently and globally. The Priva Office Direct software monitors and thoroughly examines all of the settings and controls of an automated greenhouse.

(ii). HortiMaX: The mobile remote control and simple touchscreen controls make the Ridder HortiMaX-Go user-friendly and accessible. This control system is among the best for automating the greenhouse's irrigation and climate. Together, they can improve greenhouse management and production.

(iii). Hoogendoorn: Using data analytics and predictive modelling, Hoogendoorn creates greenhouse temperature control and management systems that optimise growth conditions and increase crop yields.

(iv). Argus Controls: Argus Controls provides greenhouse farmers with real-time environmental monitoring and control, irrigation, and fertigation systems that maximise output and resource efficiency.

(v). Primalogik: They offer supply chain traceability and management software. It enables greenhouse growers to monitor their output from the point of growing to the point of distribution, guaranteeing quality and compliance through the supply chain.

Chain-integrated greenhouse production relies heavily on communication networks to share data between sensors, actuators, and central management systems. Farmers can quickly modify production procedures by ensuring information flows fluidly throughout the greenhouse environment through technologies like Ethernet, Serial Bus, GPRS/3G/4G Cellular, Wi-Fi, and Ethernet.

8.4.4 Digital Twins in Added-Value Weeding Data

Using cutting-edge technology and data analytics to maximise weed control methods in agriculture is known as "added-value weeding data." For farmers, weeds pose a serious problem because they may compete with crops for resources like sunshine, water, and nutrients, resulting in inferior crop quality and lower yields. Farmers may increase overall output and reduce the detrimental effects of weeds on their crops by implementing creative weed management strategies (Verdouw et al., 2019). Using sensors and actuators to detect and target weeds more accurately adds value to weeding data. These sensors may be placed all over the field to track the growth and spread of weeds and provide farmers with up-to-date information on infestations. Afterwards, weeds can be treated or removed carefully while causing the least amount of harm to crops by using actuators, such as robotic weeders or precision herbicide applicators. Communication networks provide data flow between field sensors, actuators, and central farm management systems, a crucial component of added value weeding data systems. Farmers may guarantee that information flows smoothly throughout the agricultural operation using Wi-Fi, GPRS, and XMPP technologies. This will allow them to make quick judgments on weed management measures. Figure 8.5 shows the key mechanisms of digital twins in added value weeding data (Implementation model).

Although a specialised application may be known as a digital twin for enhanced-value weeding data in agriculture, weed control and data-driven decision-making are at the forefront of various new precision agricultural applications and technology. These are a few instances:

(i). The See & Spray technology: It employs computer vision and machine learning to target weeds with herbicide accurately, hence lowering herbicide consumption and boosting productivity. It is one of the smart agriculture technologies developed and implemented by Blue River Technology (formerly known as John Deere AI).

(ii). The ecoRobotix WeedVision: With the use of cameras and artificial intelligence algorithms installed on autonomous robotic weeders, ecoRobotix's WeedVision intelligent weed identification technology can accurately identify and eradicate weeds, saving money

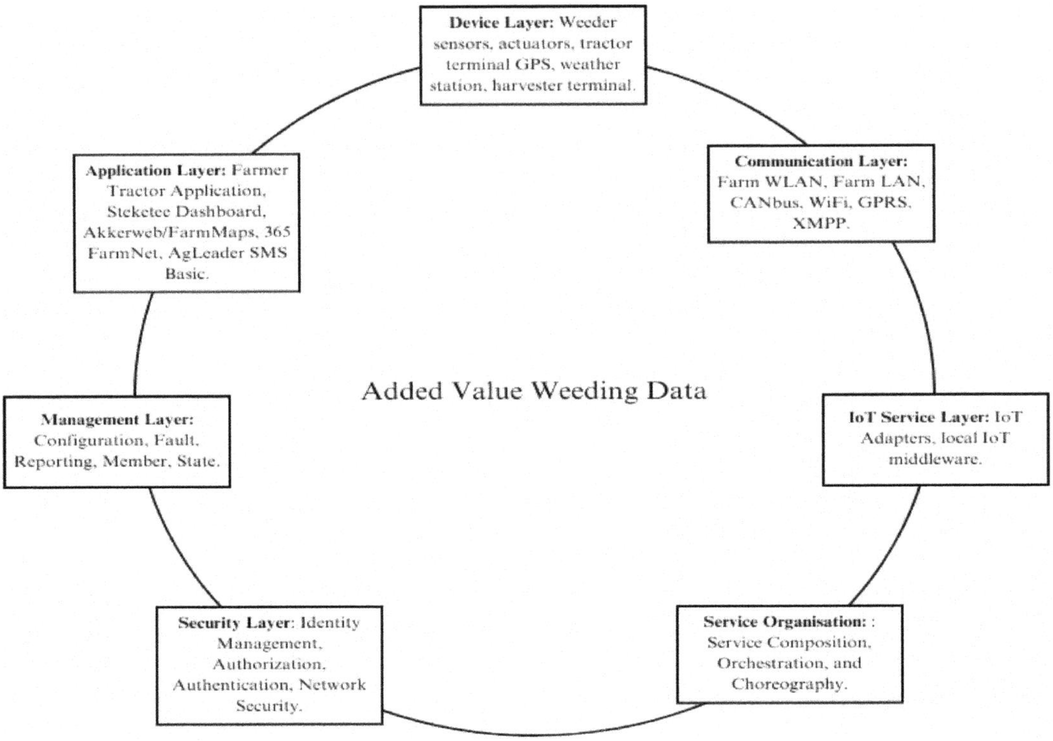

FIGURE 8.5 Key components of digital twins in added value weeding data.

on labour and pesticide use. Because there are weeds, less herbicide is used, which saves money.

(iii). Farmers may reap several advantages by implementing added value weeding data approaches. These include better environmental sustainability, lower pesticide dependency, and higher agricultural yields. Farmers may improve their production operations and lessen their environmental impact by using chemical inputs sparingly and targeting weeds more efficiently. Farmers may also save labour expenses and increase overall operational efficiency in weed control by implementing precision weed management strategies.

(iv). Resson's WeedScout: It is an AI-driven platform for mapping and detecting weed infestations in fields. It does this by using machine learning algorithms and aerial photography. This allows for focused weed management measures.

(v). Trimble WeedSeeker 2: Trimble WeedSeeker 2 is a precision herbicide application system that lowers herbicide consumption and costs by using optical sensors to detect weeds in real time as the sprayer travels over the field. This allows for targeted herbicide delivery just where weeds are present.

With the help of these apps and technology, farmers can manage weeds more effectively and sustainably by using precise tools and actionable information.

8.4.5 Digital Twins in Pig Farm Management

Pig farm management consists of pig farming activities, from shelter and food to breeding and health care. Modern pig farming operations depend on technology to protect animal health, welfare, and production efficiency (Jeong et al., 2023). RFID tags and readers administer pig farms because they

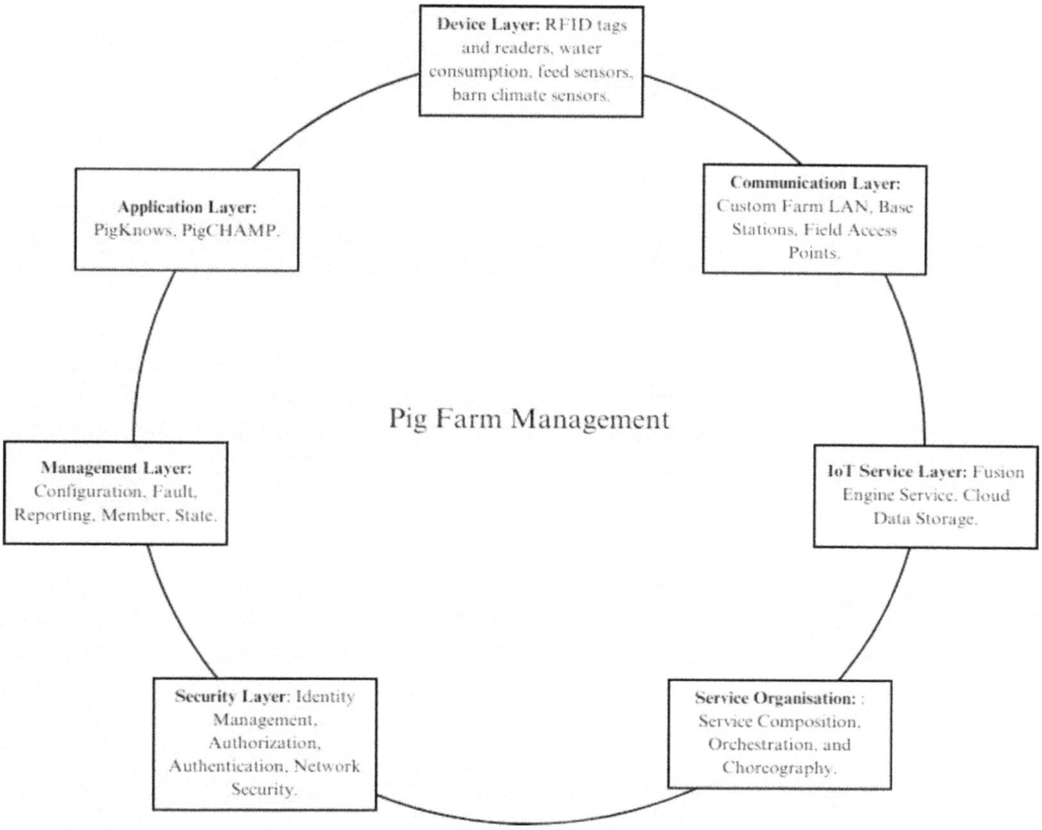

FIGURE 8.6 Key components of digital twins in pig farm management.

allow for tracking individual pigs and observing their behaviour and movements. Sensors are used to measure water consumption, feed, and barn climate. It monitors the environmental conditions within the pig housing facilities and ensures the safety of animals. Communication networks help pig operations. It facilitates data interchange between sensors, actuators, and central farm management systems. Farmers understand information flows across the farm using emerging technologies like base stations and bespoke farm LANs. Pig farm management systems monitor the pig's health and welfare. They guarantee legal compliance. Before a pigsty is built, farmers may test how well windows, fans, and heaters provide the ideal environment by creating a digital twin of the structure.

Figure 8.6 demonstrates the key mechanisms of digital twins in pig farm management (implementation model).

A digital twin platform for managing pig farms may exist, although the most recent precision animal farming apps and technology concentrate on streamlining pig farm operations. Here are a few instances:

(i). PigKnows: To assist pig farmers in maximising herd performance and profitability, PigKnows provides an all-inclusive swine management software platform that combines data from several sources, such as productivity, health, and genetics.
(ii). PigCHAMP: PigCHAMP offers swine producers herd management software solutions specifically designed to help them track and evaluate key performance indicators, control sow fertility, and keep an eye on the health and growth of their pigs.

(iii). FarmSight: John Deere's precision livestock farming solutions include tracking and monitoring individual pigs, monitoring their behaviour, and optimising feeding and management techniques. These systems make use of RFID tags and sensors.
(iv). Topigs Norsvin's Total Solutions: By utilising data analytics and digital technologies, Topigs Norsvin provides integrated pig genetics, health, and management solutions intended to enhance pig performance and efficiency throughout the production cycle.
(v). Porcitec: Porcitec offers swine farm management software that helps pig farmers make data-driven choices to maximise productivity and profitability by allowing them to track and control every element of their operations, including production, health, and finances.

Pig farmers may enhance productivity, health, and management practices in their operations by utilising the tools and insights offered by these apps and technology. Pig farm management techniques are essential to maintaining the health and welfare of the animals and maximising productivity. Farmers may monitor and manage their operations more efficiently by utilising new technology and data analytics, leading to increased yields and improved profitability

8.4.6 Digital Twins in Soil and Irrigation

To maintain plant production, soil quality must be monitored, assessed, and considered an essential component of crop health and yield (Knibbe et al., 2022). By lowering the possible usage of chemical pesticides and fertilisers in agricultural soil, digital technologies contribute to preserving human health, the environment and the quality of subsurface water. In digital agriculture, soil monitoring sensors such as those for organic matter, temperature, moisture, and pollutants are essential. By detecting the zones that might result in low crop output, digital soil assessment techniques directly influence crop performance and production. Digital twin models for soil in agriculture might result from using wireless sensors, IoT, and AI in digital agriculture.

Digital technologies based on smart irrigation systems include wireless networks, the Internet of Things, edge computing, local weather-based controllers, and soil sensors (Alves et al., 2023). To display satellite and drone photos, a digital twin system design incorporating monitoring devices (such as soil probes, weather data, irrigation systems, machinery, and other equipment) in a physical system (farm) can bridge the virtual and physical systems. Digital soil mapping uses data from complicated soil variation at a particular depth, time, and distinctive places to represent soil variation in digital twins.

A field avatar, crop agent, and soil agent are digital representations of the field, including meteorological and geological data. In particular, IoT, cloud, fog, and edge computing technologies transfer information from the soil as a physical system to a virtual system in the digital twin

8.4.7 Digital Twins in Crop Production and Post-Harvesting

Tractors combine harvesters, fertilisers, sprayers, and other equipment that use digital and ICT technologies that may save fuel, fertiliser, and labour costs while improving production efficiency and sustainability. With the use of sensors and ICT tools, a digital farm machine may transmit and receive data, enabling the devices to operate automatically (Chaudhary et al., 2021). Artificial intelligence, fuzzy logic, and big data analysis can monitor machine behaviour, operation, and maintenance status. Machine failures can be identified before or in the early stages of a breakdown through the integration of big data analytics with AI models. Digital twin techniques enable the integration of digital tools with autonomous machines and robotics and the real-time viewing of smart farm equipment conditions (Rogachev et al., 2022).

By introducing a remote operating system, the notions of digital twins can virtualise the robot environment. A gaming platform (as a virtual system) and a robot (as a real item) may monitor the robot's movement, simplify complicated environment programming, and provide a remote operating system for platform-to-platform communication. Sensor-based autonomous navigation can be optimised by field robots.

Digital twins in post-harvest can really be viewed as intelligent, statistical, and mechanical models. Mango fruit may be qualified and simulated via a post-harvest supply chain by creating a digital twin of the fruit. Agricultural items' shelf life across the cold chain can be predicted using digital twins (Nasirahmadi & Hensel, 2022).

Thus, farmers may be given more accurate information to help them make better decisions using advanced statistical models, machine learning, and data analytics.

8.5 EMERGING DIGITAL TWIN APPLICATIONS IN AGRICULTURE AND DIGITAL FARMING

Digital twin applications in agriculture and digital farming are emerging to maximise many areas of agricultural operations. Following are a few instances:

(i). Climate FieldView: It is the leading software platform for digital farming, analysing farm data down to the acre. It gathers, stores, and displays vital field data with ease. To maximise profit and maximum production, track and quantify the effects of the agronomic actions on the land under cultivation. Further, Climate FieldView provides prescription mapping, data analysis, and field monitoring (Karydas et al., 2023).

(ii). Taranis: Crop and farm management software is provided by Taranis. Using AI and machine learning, it offers farmers precise and creative precision agricultural solutions that maximise crop yields (Bacco et al., 2019). Therefore, practical insights from Taranis may empower farmers.

(iii). Granular Insights: The simplest method for farmers and their advisers to work together to plan, grow, and assess each field is with Granular Insights. It explains crop production, field conditions, and resource usage. It integrates data from field sensors and equipment telematics (Monteiro et al., 2021).

(iv). Iteris ClearAg: FluroSat incorporates Iteris ClearAg environmental intelligence to improve crop health decision support. ClearAg data will comprise historical, forecast, and current location-specific meteorological data for the United States, Australia, New Zealand, and Latin America. Agronomists and progressive farmers will be able to maximise fertiliser and water consumption while increasing crop output with the use of ClearAg data, which is accessible through FluroSat's platform, and crop protection. It also provides meteorological and agronomic information services.

(v). Trimble Ag Software: With the use of Trimble software, farmers can more effectively plan their crops, carry out farm tasks, monitor field activities, and keep an accurate record of the whole crop year.

(vi). John Deere Operations Center: This digital farming platform combines information from sensors, John Deere machinery, and outside sources to give farmers insights into field conditions, equipment performance, and agronomic techniques.

These apps use the principles of digital twins by creating virtual models of farms, crops, and agricultural processes. Farmers can examine data, simulate situations, and make well-informed choices.

8.6 CONCLUSION

The human capacity is insufficient for activities like high-precision environmental control, real-time decision-making, and continuous environmental monitoring. Significantly, a digital twin's capacity to carry out its intended role is intrinsically linked to its environment. The progress of farming and agriculture shows that digital twins have a generally positive impact in enhancing the sustainability of agriculture. Digital technologies like the Internet of Things, AI in agriculture, smart farming, and precision agriculture are necessary for agricultural digital twins to improve sustainable agriculture. The discussion in this chapter showed that digital twins support farmers in within-field management zoning, soil and irrigation, crop production, post-harvesting, chain-integrated greenhouse production, added value weeding data, and pig farm management in real time situations (Adade & De Vries, 2023). Therefore, future agricultural and farming practices will be more profitable, efficient, safe, and environmentally friendly because of sophisticated technologies like robotics, precision agriculture, aerial imagery, temperature and moisture sensors, GPS, and sophisticated devices.

REFERENCES

Adade, D., & De Vries, W. T. Digital Twin for active stakeholder participation in Land-Use planning. *Land*, 12(3), 538(2023). https://doi.org/10.3390/land12030538

Alves, R. G., Maia, R. F., & Lima, F. R. Development of a Digital Twin for smart farming: Irrigation management system for water saving. *Journal of Cleaner Production*, 388, 135920(2023). https://doi.org/10.1016/j.jclepro.2023.135920

Ariesen-Verschuur, N., Verdouw, C., & TekiNerdoğan, B. Digital Twins in greenhouse horticulture: A review. *Computers and Electronics in Agriculture*, 199, 107183(2022). https://doi.org/10.1016/j.compag.2022.107183

Attaran, M., Attaran, S., & Çelik, B. G. Revolutionizing agriculture through digital twins. In *Advances in information quality and management*, 1–14(2024). https://doi.org/10.4018/978-1-6684-7366-5.ch053

Bacco, M., Barsocchi, P., Ferro, E., Gotta, A., & Ruggeri, M. The digitisation of agriculture: A survey of research activities on smart farming. *Array*, 3–4, 100009(2019). https://doi.org/10.1016/j.array.2019.100009

Chaudhary, G., Khari, M., & Elhoseny, M. (Eds.). (2021). *Digital twin technology* (1st ed.). CRC Press. https://doi.org/10.1201/9781003132868

Escribà-Gelonch, M., Escribà-Gelonch, M., Escribà-Gelonch, M., Escribà-Gelonch, M., Escribà-Gelonch, M., & Escribà-Gelonch, M. Digital twins in agriculture: orchestration and applications. *Journal of Agricultural and Food Chemistry* (2024). https://doi.org/10.1021/acs.jafc.4c01934

Han, X., & Lin, Z. AI based digital twin model for cattle caring. *Sensors*, 22(19), 7118(2022). https://doi.org/10.3390/s22197118

Jayaraman, P. P., Yavari, A., Georgakopoulos, D., Morshed, A., & Zaslavsky, A. Internet of Things Platform for smart Farming: Experiences and lessons learnt. *Sensors*, 16(11), 1884(2016). https://doi.org/10.3390/s16111884

Jeong, D., Jo, S., Lee, I., Shin, H., & Kim, J. Digital Twin Application: Making a virtual pig house toward digital livestock farming. *IEEE Access*, 11, 121592–121602(2023). https://doi.org/10.1109/access.2023.3313618

Karydas, C., Chatziantoniou, M., Tremma, O., Milios, A., Stamkopoulos, K., Vassiliadis, V., & Mourelatos, S. Profitability assessment of precision agriculture applications—A step forward in Farm Management. *Applied Sciences*, 13(17), 9640(2023). https://doi.org/10.3390/app13179640

Knibbe, W. J., Afman, L., Boersma, S., Bogaardt, M-J., Evers, J., van Evert, F., van der Heide, J. et al. Digital twins in the green life sciences. *NJAS: Impact in Agricultural and Life Sciences*, 94(1), 249–79(2022). https://doi.org/10.1080/27685241.2022.2150571.

Monteiro, A., Santos, S. R. D., & Gonçalves, P. Precision agriculture for crop and livestock farming—Brief review. *Animals*, 11(8), 2345(2021). https://doi.org/10.3390/ani11082345

Nasirahmadi, A., & Hensel, O. Toward the next generation of digitalization in agriculture based on digital twin paradigm. *Sensors*, 22(2), 498(2022). https://doi.org/10.3390/s22020498

Niu, G., & Masabni, J. Hydroponics. In *Elsevier eBooks*, 153–166(2022). https://doi.org/10.1016/b978-0-323-85152-7.00023-9

Peladarinos, N., Piromalis, D., Cheimaras, V., Tserepas, E., Munteanu, R., & Papageorgas, P. Enhancing smart agriculture by implementing Digital Twins: A comprehensive review. *Sensors*, 23(16), 7128(2023). https://doi.org/10.3390/s23167128

Poornima, G., & Gowda, S. Digital twin for smart farming. In *Bentham Science Publishers eBooks*, 1–16(2024). https://doi.org/10.2174/9789815196177124010004

Purcell, W., & Neubauer, T. Digital Twins in Agriculture: A State-of-the-art review. *Smart Agricultural Technology*, 3, 100094(2023). https://doi.org/10.1016/j.atech.2022.100094

Rogachev, A. F., Skiter, N. N., Ketko, N. V., Simonov, A. B., & Makarevich, I. V. Digital twins as a tool for systemic integration of innovative digital technologies in agriculture. *IOP Conference Series. Earth and Environmental Science*, 1069(1), 012042(2022). https://doi.org/10.1088/1755-1315/1069/1/012042

Shamia, D., Suganyadevi, S., Satheeswaran, V., & Balasamy, K. Digital twins in precision agriculture monitoring using artificial intelligence. In *Elsevier eBooks*, 243–265(2023). https://doi.org/10.1016/b978-0-323-99205-3.00004-3

Van Der Burg, Simone van der, Sanneke Kloppenburg, Esther J. Kok, and Mariska van der Voort. "Digital Twins in Agri-Food: Societal and Ethical Themes and Questions for Further Research." NJAS: Impact in Agricultural and Life Sciences 93, no. 1, 98–125(2021). https://doi.org/10.1080/27685241.2021.1989269.

Verdouw, C., Sundmaeker, H., Tekinerdogan, B., Conzon, D., & Montanaro, T. Architecture framework of IoT-based food and farm systems: A multiple case study. *Computers and Electronics in Agriculture*, 165, 104939(2019). https://doi.org/10.1016/j.compag.2019.104939

Verdouw, C., Tekinerdogan, B., Beulens, A., & Wolfert, S. Digital twins in smart farming. *Agricultural Systems*, 189, 103046(2021). https://doi.org/10.1016/j.agsy.2020.103046

9 Blockchain-Enabled Security Framework for Smart Agriculture

Harvesting Smart Agriculture Supply Chains Transformed by Blockchain Efficiency

Maheswari R, Pattabiraman V, and Basim Alhadidi

9.1 INTRODUCTION

Blockchain technology has become increasingly popular in recent years due to its ability to provide secure, transparent, and decentralized record-keeping capabilities. This technology has many potential applications in various industries, including supply chain management. A supply chain is a complex network of interconnected parties, including manufacturers, suppliers, distributors, retailers, and customers. The process involves the exchange of goods, services, and information between these parties. The traditional supply chain system is often plagued by inefficiencies, including lack of transparency, fraud, and errors. Blockchain technology can potentially address these inefficiencies by creating a secure and transparent platform for supply chain management (Al-Jaroodi and Mohamed 2019). Distributed ledger technology allows all parties in the supply chain to have access to the same data in real time, thereby reducing the risk of fraud and errors. Additionally, the immutability of blockchain records makes it possible to trace the movement of goods and services from their origin to their final destination. Incorporating blockchain technology in supply chain management can also help with inventory management, order tracking, and payment processing, among other benefits. With the potential to improve supply chain efficiency, reduce costs, and increase transparency, it is no wonder that more and more companies are exploring the use of blockchain technology in their supply chain management systems (Astarita et al. 2020).

9.1.1 Agricultural Supply Chain

In the complex web of agricultural supply chains, the presence of multiple intermediaries poses significant challenges. These intermediaries, while essential for facilitating transactions, often contribute to inefficiencies, opacity, and complexities within the system. At the heart of these challenges lies the difficulty in monitoring and administering the activities of all participants, ensuring seamless information connectivity, and accurately identifying the origin of products. One of the primary hurdles encountered in agricultural supply chains is the lack of transparency and accountability resulting from the involvement of various intermediaries (Astill et al. 2019). As products traverse through multiple hands, from farmers to distributors to retailers, the opacity of transactions often obscures the true origin, quality, and handling practices of agricultural goods. This lack of transparency not only undermines consumer trust but also hampers efforts to ensure food safety and traceability (Saranya et al. 2021).

DOI: 10.1201/9781003507390-9

Furthermore, the monitoring and administration of participants' activities present a daunting task for supply chain managers (Luu et al. 2016). Coordinating the numerous activities of farmers, suppliers, transporters, and distributors while ensuring compliance with regulations and standards is a complex endeavor. Traditional methods of record-keeping and monitoring often prove inadequate, leading to inefficiencies, errors, and discrepancies in inventory management, logistics, and quality control (Maxim 2015).

Information connectivity stands out as another critical challenge plaguing agricultural supply chains. The fragmented nature of the industry, coupled with disparate data systems and communication channels among stakeholders, impedes the seamless flow of information. As a result, critical data related to crop yields, market demand, pricing, and inventory levels often remain siloed, inhibiting timely decision-making and coordination across the supply chain (Saranya et al. 2023).

In the research work, Saranya and Maheswari (2023) suggested the Proof of Transaction (PoTx) framework to improve traceability and transparency in the agricultural supply chain, tackling inefficiencies, and delays. Developed the PoTx framework to improve traceability and transparency in the agricultural supply chain, utilizing blockchain integration to alleviate inefficiencies and delays.

Moreover, accurately identifying the origin of agricultural products is paramount for ensuring quality, authenticity, and compliance with regulatory requirements. However, the lack of robust traceability systems makes it challenging to trace products back to their source, particularly in cases of contamination outbreaks, food fraud, or quality issues (Azzi et al. 2019). This inability to pinpoint the origin of products not only undermines consumer confidence but also complicates efforts to implement targeted recalls or quality assurance measures.

In addressing these persistent challenges, blockchain technology emerges as a transformative solution offering unprecedented transparency, security, and efficiency to agricultural supply chains. By leveraging blockchain's decentralized ledger system, all transactions and data exchanges within the supply chain can be recorded immutably and transparently, providing a tamper-proof audit trail of product movement, ownership, and attributes (Shrier et al. 2016).

Through blockchain-enabled smart contracts, stakeholders can automate and enforce predefined rules and agreements, streamlining transactions, and eliminating the need for intermediaries. Smart contracts facilitate seamless and instant payments between parties, reducing transaction costs and delays while enhancing trust and transparency in financial transactions.

Furthermore, blockchain enables real-time monitoring and traceability of products throughout the supply chain, from farm to fork. By embedding unique identifiers or digital tags into products, stakeholders can track the journey of agricultural goods at every stage, ensuring authenticity, quality, and compliance with regulatory standards.

Additionally, blockchain facilitates seamless information connectivity by providing a shared platform for data exchange and collaboration among supply chain participants. By standardizing data formats and protocols, blockchain fosters interoperability and integration across disparate systems, enabling better visibility, coordination, and decision-making.

Thus the adoption of blockchain technology holds immense promise for revolutionizing agricultural supply chains and overcoming the challenges posed by intermediaries, information silos, and traceability issues. By embracing blockchain's transformative capabilities, stakeholders can unlock new opportunities for transparency, efficiency, and trust in agricultural supply chains, ultimately ensuring the delivery of safe, high-quality, and sustainable food products to consumers worldwide.

9.1.2 Smart Contract

A smart contract is a self-executing contract with the terms of the agreement between buyer and seller being directly written into code. The code and the agreements contained therein exist across a distributed, decentralized blockchain network. The code controls the execution, and transactions are trackable and irreversible.

Smart contracts are created by developers using programming languages specifically designed for blockchain platforms, such as Solidity for Ethereum.

Once the code is written, the smart contract is deployed to the blockchain network. This typically involves submitting the code along with an associated transaction to be included in a block.

Transaction Trigger: Smart contracts are triggered by predefined conditions or actions. These triggers can be events on the blockchain, such as receiving cryptocurrency or reaching a specific date. When a smart contract is triggered, the code automatically executes the predefined actions without the need for intermediaries. For example, in a simple payment smart contract, funds are automatically transferred from one party to another when a specified condition is met.

The execution of the smart contract is verified and validated by nodes on the blockchain network through a consensus mechanism. Once verified, the results of the contract execution are recorded on the blockchain and cannot be altered. Smart contracts operate on a transparent and decentralized blockchain network, meaning that the code, as well as the outcomes of contract execution, are visible to all participants on the network. Smart contracts are secured by cryptographic algorithms and consensus mechanisms inherent to the blockchain network. Once deployed, they are tamper proof and resistant to censorship or manipulation (Maheswari Raja et al. 2024).

By automating contract execution and removing the need for intermediaries, smart contracts can significantly reduce transaction costs and eliminate the risk of fraud or human error. Thus smart contracts enable trustless and decentralized execution of agreements, opening up a wide range of applications across industries such as finance, supply chain management, real estate, and more.

9.1.3 Supply Chain in Agriculture

The supply chain in agriculture encompasses the process of producing, processing, distributing, and selling agricultural products from farm to consumer. It involves various stakeholders, including farmers, suppliers, manufacturers, distributors, retailers, and consumers, who collaborate to ensure the efficient flow of goods from production to consumption.

1. Production: The supply chain begins with agricultural production, where farmers cultivate crops, raise livestock, or grow other agricultural products. This stage involves activities such as planting, irrigation, fertilization, pest control, and harvesting.
2. Processing: Once agricultural products are harvested or collected, they may undergo processing to transform them into value-added goods. Processing activities can include cleaning, sorting, grading, packaging, and preservation to enhance the quality and shelf life of the products.
3. Distribution: After processing, agricultural products are distributed to various markets or destinations. This stage involves transportation, warehousing, and logistics to ensure timely delivery of products to retailers, wholesalers, or other intermediaries.
4. Retailing: Agricultural products reach consumers through retail outlets such as supermarkets, grocery stores, farmers' markets, or online platforms. Retailers play a crucial role in merchandising, marketing, and selling agricultural products to end consumers.
5. Consumption: Finally, consumers purchase and consume agricultural products for food, feed, fiber, or other purposes. Consumers may also provide feedback on product quality, preferences, and purchasing behaviors, which can influence future production and distribution decisions.

Throughout the agricultural supply chain, various factors such as weather conditions, market demand, regulatory requirements, and technological advancements can impact the efficiency and effectiveness of operations (Agrawal et al. 2021).

Additionally, challenges such as price volatility, supply chain disruptions, food safety concerns, and sustainability issues may arise, requiring collaboration and innovation among stakeholders to address.

To overcome these challenges and optimize the agricultural supply chain, stakeholders can leverage technologies such as blockchain, Internet of Things (IoT), artificial intelligence (AI), and data analytics to improve transparency, traceability, efficiency, and sustainability. By embracing digitalization and adopting innovative solutions, the agricultural supply chain can better meet the demands of a growing population, ensure food security, and promote economic development in rural communities (Ma et al. 2020).

9.2 LITERATURE REVIEW

This research work (Mukherjee et al. 2021) provides a comprehensive overview of the potential applications of blockchain technology across various industrial domains. By exploring the opportunities, benefits, and challenges associated with incorporating blockchain, you shed light on the diverse possibilities for leveraging this technology to enhance efficiency, transparency, and security in different sectors. Additionally, identifying the specific requirements for implementing blockchain in each industrial application helps provide clarity on the necessary steps and considerations for successful deployment (Y. Wang et al. 2022).

Despite the numerous opportunities presented by blockchain, it's crucial to acknowledge and address the challenges to fully realize its potential in optimizing operations and creating value across industries. This balanced assessment sets the stage for further exploration and development of blockchain solutions tailored to the unique needs and contexts of different industrial sectors (Y. Wang et al. 2022).

This research paper provides a review of literature on the use of blockchain-based systems in transportation. The main objective was to identify current research trends, gaps in the literature, and future challenges using a multi-step methodology (Prashar et al. 2020).

The study began with a bibliometric analysis to obtain a broad overview of the topic, followed by an in-depth analysis of the most influential contributions in two areas: supply chain and logistics, and road traffic management and smart cities. The results showed that while blockchain technology is still in its early stages, it has promising applications in various fields, including food tracking, regulatory compliance, smart vehicle security, and supply-demand matching. However, much effort is still needed to reach maturity, as many models have been theorized but few have been implemented in real contexts (X. Yang et al. 2021).

The paper also explores the relationship between blockchain and sustainability, highlighting the potential for the technology to limit food waste, reduce exhaust gas emissions, support proper urban development, and improve overall quality of life (X. Yang et al. 2021). A supply chain involves a network of individuals, organizations, and resources responsible for moving goods or services from the supplier to the customer, while maintaining quality throughout the process. However, centralized supply chain management systems can pose risks such as corruption, fraud, and tampering (Sohrabi and Tari 2020). The emergence of blockchain technology offers a new approach to address the challenges of visibility and transparency in supply chain management (S. Kim et al. 2018).

This paper examines how blockchain can be integrated into supply chain architecture to create a reliable, transparent, authentic, and secure system (S. Wang et al. 2019). The study explores the benefits of blockchain in supply chain management and the challenges of implementing it. Both theoretical and real-world applications are examined to develop a comprehensive theory of the requirements for an efficient blockchain-based supply chain (Gao et al. 2020). The manufacturing of goods has become increasingly complex due to the involvement of more intermediaries between producers and consumers. With globalization and market expansion, companies have had to expand their product portfolios and life cycles to meet the demands of new markets (Chen et al. 2017). As a result, there is limited knowledge about the origins, processing, and shipping journeys of products. The supply chain faces challenges in both quantity and quality, with traceability and data management being the primary concerns. In most sectors, including healthcare, finance, food, and education, information systems are managed centrally (Bocek et al. 2017). However, centralized management systems can pose a threat to data integrity, availability, and resiliency, and the system may be vulnerable to corruption, fraud, and tampering. A trustworthy ecosystem between suppliers and customers can be created by implementing policies that focus on supply chain transparency, requiring accurate data collection and secure data storage to ensure product traceability (Korpela et al. 2017).

9.3 OBJECTIVE

The objective of the work is to decentralize the entire supply chain network through the utilization of blockchain technology, offering a range of benefits.

1. Transparency: With blockchain, all stakeholders gain the ability to track products at every stage of the supply chain, fostering transparency and trust among participants.
2. Increased Security: By leveraging blockchain's tamper-proof nature, the system enhances security, mitigating the risk of data manipulation or unauthorized access.
3. Risk Management: Blockchain facilitates real-time monitoring and management of risks within the supply chain, enabling proactive decision-making and risk mitigation strategies.
4. High-Quality Products: Through the implementation of blockchain technology, the supply chain can ensure the delivery of high-quality products by maintaining transparent and traceable records throughout the production and distribution process.
5. Instant Payment: Smart contracts enable automated and instant payment settlements between stakeholders upon completion of buying and selling transactions, eliminating delays and improving cash flow within the supply chain network.

By decentralizing the supply chain network and harnessing the capabilities of blockchain technology, the project aims to optimize operational efficiency, enhance trust and collaboration among stakeholders, and ultimately deliver greater value to all participants involved.

9.4 SCOPE OF THE WORK

This work focuses on leveraging blockchain technology to enhance supply chain visibility and maintain tamper-proof product records. By eliminating the inefficiencies of paper-based trails, such as those leading to counterfeit goods or delayed shipments, blockchain can offer significant cost savings for firms. Furthermore, a blockchain-based system can foster greater transparency and trust between supply chain partners, enabling faster issue resolution and improved collaboration. While challenges like scalability and integration exist, solutions like consortium blockchains can pave the way for a more efficient and secure supply chain ecosystem (Ruta et al. 2017). By ensuring transparency and tamper-proof records, blockchain not only mitigates issues like counterfeit products and delays but also fosters trust and collaboration among supply chain partners. Despite challenges like scalability, solutions such as consortium blockchains offer promising pathways to a more efficient and secure ecosystem (Kim and Laskowski 2018).

9.5 CASE STUDY

First, the study analyzed a generic supply chain management system using an appropriate dataset of a product-based company. The company offered four modes of shipping the product: standard, same, and first and second class. However, every mode of delivery shows some level of inefficiency (Figorilli et al. 2018). Therefore, there is a delivery data variance associated with all the modes of shipping. The study also analyzed the best model for prediction for the given dataset; the model was found to be XGBoost, with an accuracy of 99.65%. Using this model, forecasting was done between predicted and actual sales. To analyze the supply chain performance by different shipping modes, we plotted a bar chart (Nguyen and Kim 2018). Figure 9.1 shows a bar chart illustrating the percentage delay in delivery modes.

Each bar represents the percentage delay in a particular delivery mode, with labels indicating the mode of transport. This visualization enables a quick comparison of delay percentages across different modes, helping to identify trends or areas for improvement within the supply chain.

As a stacked bar chart in which the x axis is taken as different modes of shipping and the y axis as the percentage of deliveries, it shows that the performance of all shipping modes is not so good. In general, 40% of the orders are late delivery. Almost 95% of first-class shipping mode is late delivery. As a result, first-class and second-class shipping modes require improvements in terms of supply chain to deliver orders on time.

In order to compare the shipping days and the variance in date, a line graph has been plotted between the different shipping modes and actual days required for shipping. Figure 9.2 depicts the variance in actual versus estimated delivery, and a line graph is suitable to showcase fluctuations over time.

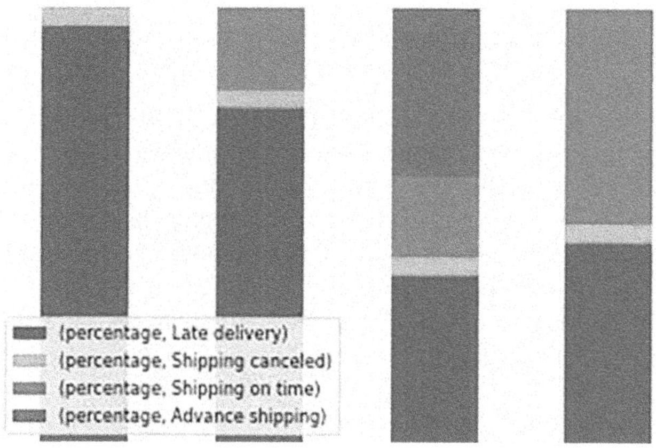

FIGURE 9.1 Bar chart for percentage delay in delivery modes.

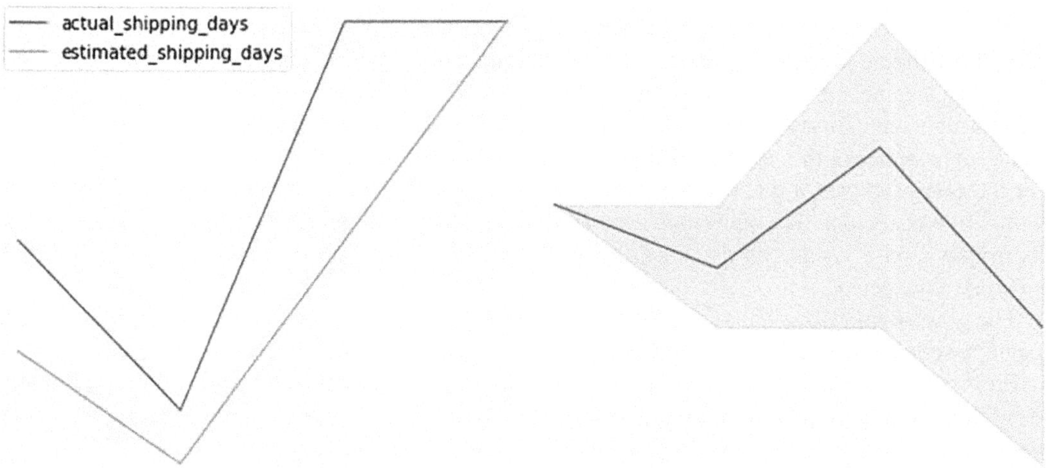

FIGURE 9.2 Graphs showing variance in actual vs estimated delivery.

TABLE 9.1
Analysis of Various Shipping Modes

Mode of Shipping	Analysis
First Class	Takes longer time for shipping than estimated
Second Class	Takes longer time for shipping than estimated
Same Day	Takes longer time for shipping than estimated
Standard	Delivers on time

TABLE 9.2
Accuracy Analysis of Different Algorithms

Algorithm	Accuracy (%)	Precision (%)	Recall (%)
Linear regression	0.7852	78.00	76.50
Random forest	0.9954	99.60	99.40
XGBoost	0.9965	99.70	99.60

Each point on the line graph represents the variance between actual and estimated delivery times for a specific time period. This visualization allows for a clear understanding of the trend in delivery performance over time and highlights any deviations from the estimated delivery times. From the line graph on the left-hand side, it is clear that the actual shipping days are higher than the estimated ones for first-class, same-day, and second-class shipping modes. Only standard class can complete the delivery as estimated. It shows that first class, second class, and same day take longer to be completed than expected. However, standard class normally delivers on time. From the line graph on the right-hand side, the delivery date variance is clearer. Table 9.1 compares the performance of all the modes.

Table 9.1 shows an analysis of various shipping modes. In this table, each row represents a different shipping mode, and the columns display the average delivery time and the percentage of delay associated with each mode. This analysis provides insights into the performance of various shipping modes within the supply chain. In order to identify which model best fits the dataset, we analyze the accuracies of various algorithms.

Table 9.2 shows the accuracy of different algorithms. In this table, each row corresponds to a different machine learning algorithm, and the columns display the accuracy, precision, and recall percentages achieved by each algorithm. This analysis provides insights into the performance of various algorithms in a predictive modeling context. The table shows that the accuracy of XGBoost is the highest; hence this model can be used for making predictions of actual and predicted sales. For the same analysis, a line chart between actual and predicted sales has been made (Figure 9.3). In this line chart, one line represents actual sales, and another line represents predicted sales. The chart allows for a visual comparison between the actual sales figures and the predictions over the specified time period.

The graph depicts line plots for actual and predicted sales of different products. From the graph, it can be seen that there are few discrepancies in the actual and predicted sales.

From the first bar chart, it is clear that all shipping days take more time for shipping than expected. From the line chart, it is clear that only standard class can complete the delivery as estimated. The shadows present in the line graph indicate the delivery date variance,

To conclude we can say that:

$$\text{Delivery Date Variance} = \text{Actual Shipping Days} - \text{Estimated Shipping Days}.$$

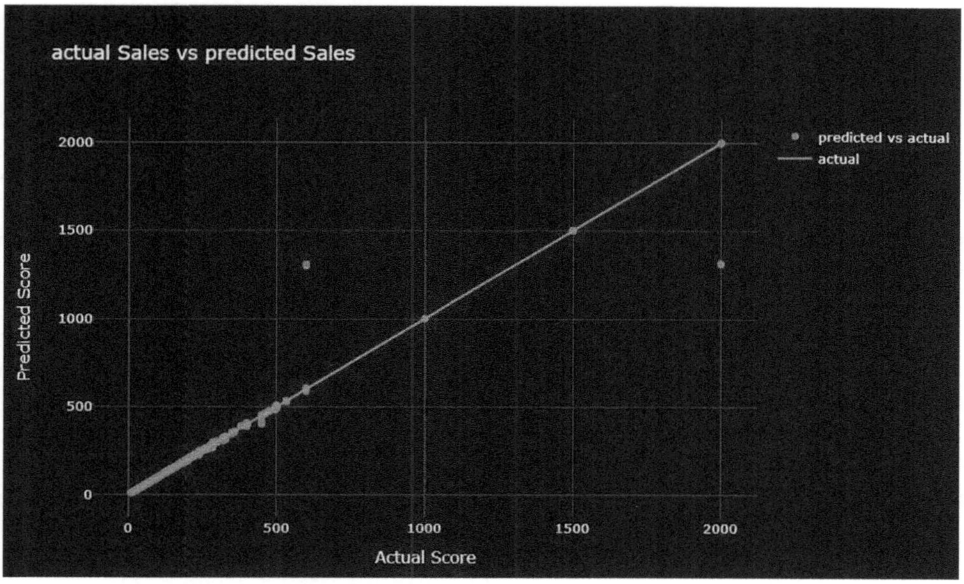

FIGURE 9.3 Line chart for actual vs predicted sales.

It indicates that first class is very stable, always 1 day behind the schedule. The performance of same-day shipping mode is not bad, as it has an average of 0.5 days longer than expected. Second class has the poorest performance, as it usually takes 1.5 days longer to complete the delivery. It also has the highest maximum delivery date variance, which indicates that it has the longest delay. Standard class has the best performance, as it usually delivers on time. Also, it sometimes delivers earlier than expected. Next, in order to predict the sales of various products, we need to first find which model is best suited for the dataset. First, appropriate parameters will be selected for the model based on their *p* and *f* values. After applying various ML algorithms, the study found out that XGBoost has the maximum accuracy. As the accuracy of XGBoost is the highest, the work will use this model for making predictions of the actual and predicted sales. The study found that there are discrepancies in the graph obtained for actual vs predicted sales using the XGBoost model. These are mainly caused by late deliveries and fraudulent orders of products.

Delays on delivery of products can have adverse effects on customers, and delays in shipping can hinder the customer–seller relationship, ultimately leading to the retailer losing their daily customers. If a business fails to deliver products or services on time, it can lead to customer dissatisfaction. Customers may become frustrated and angry if they do not receive their orders when expected, which can harm the business's reputation and lead to negative reviews. A delay in delivery can result in lost revenue, especially if customers decide to cancel their orders or take their business elsewhere. This can be particularly damaging for businesses that rely on repeat business or have a high level of competition.

9.6 IMPLEMENTATION

Blockchain for a given supply chain can be implemented in this way:

1. Identifying the Use Cases: The specific use cases for which blockchain technology can provide value in the supply chain management were determined. The specific pain points or inefficiencies that blockchain can address, such as improving traceability, increasing transparency, reducing fraud, or streamlining payments, were considered. This approach

ensures a more targeted and impactful implementation of blockchain technology, maximizing its benefits for all stakeholders involved.

2. Selecting the Right Blockchain Platform: This means evaluating the available blockchain platforms and choosing the one that best suits the use case. Factors such as scalability, security, interoperability, and community support were considered, and finally an Ethereum-based network was chosen to deploy the smart contract, although the main aim would be to deploy it in a Digital Rupee (INR) network in further research work.
3. Building the Network: Establishing a network of partners for a blockchain-based supply chain is a critical step toward ensuring its success. Clearly defining the roles and responsibilities of each partner is essential to foster collaboration and alignment within the network. This ensures that all participants understand their contributions and commitments, leading to a more effective and harmonious implementation of the blockchain solution. By gaining consensus on participation and responsibilities, the network can operate smoothly and realize the full potential of blockchain technology in optimizing supply chain operations.
4. Building the Smart Contract: Absolutely, defining smart contracts is a crucial aspect of implementing a blockchain-based supply chain solution. These contracts serve as the backbone of automated interactions between various parties within the network, ensuring that transactions are executed seamlessly and in accordance with pre-defined rules and agreements. By encoding business logic into smart contracts, you can streamline processes, reduce manual intervention, and enhance trust among participants. Moreover, the self-executing nature of smart contracts helps enforce transparency and accountability, further bolstering the efficiency and reliability of the supply chain ecosystem.
5. Implementing the Solution: Developing and deploying the blockchain solution involves several key steps to ensure its successful integration with existing supply chain management systems. Acquire and configure the necessary hardware components such as servers, nodes, and storage devices. Install and configure blockchain software platforms like Ethereum, Hyperledger Fabric, or Corda, depending on the specific requirements of the solution. Develop and deploy smart contracts that define the business logic and rules governing interactions within the supply chain network. These contracts should be tailored to address specific use cases and ensure seamless execution of transactions. Integrate the blockchain solution with existing supply chain management systems, such as enterprise resource planning (ERP) or customer relationship management (CRM) software. This may involve developing application programming interfaces (APIs) or middleware to facilitate data exchange and interoperability between the blockchain and legacy systems. Conduct rigorous testing to ensure the stability, security, and functionality of the blockchain solution. This includes unit testing of smart contracts, integration testing with existing systems, and performance testing to assess scalability and responsiveness. Deploy the blockchain solution in a phased approach, starting with pilot projects or limited-scale implementations before expanding to the entire supply chain network. Train personnel on how to use the new system effectively and provide ongoing support to address any issues or challenges that may arise. Establish monitoring tools and procedures to continuously monitor the performance and health of the blockchain solution. Implement regular maintenance and updates to address security vulnerabilities, optimize performance, and incorporate new features or enhancements.
6. Educating and Training the Participants: Educating and training all the participants in the blockchain-based supply chain on how to use the system is one of the most important aspects of the complete model. It ensures that they understand the benefits and limitations of the system and how to use it to improve their business processes.

9.7 WORKING MODEL

In an agricultural environment, farmers produce raw materials like rice and milk. The manufacturer buys the raw materials from these farmers. But this process of buying happens through a smart contract where the transaction is instant and is validated and stored in a block. The manufacturer does the processing and creates a product called kheer from the raw materials. The manufacturer then makes the product in bulk and packages it. In between, there are multiple data points which collect vital information from the manufacturing plant. These are done with the use of IoT devices and sensors, which collect data from the data points. Then these products are bought in bulk by wholesalers/distributors with instant payment with the help of the smart contract. Wholesalers then ship the product to their warehouses, from which retailers order their quantities. The complete shipment process is tracked with data points, and all these are stored in a private/public cloud. The retailer lists their product in the shop, where the customers buy it. So, from the starting until the end point, every movement of the product is tracked and stored in multiple blocks, and the transactions are validated and converted into blocks and added to the blockchain. This blockchain is a distributed ledger system and is available for all the stakeholders. Hence, the complete system becomes transparent, trackable, and much more secure. Figure 9.4 shows the architecture for a supply chain model. Illustrating the architecture for the supply chain model, outlines the flow of data and processes within the supply chain model. It begins with data sources feeding into a data cleaning module for preprocessing. The cleaned data then moves into an analysis module for further processing. Results from the analysis are stored in a data storage module. The processed data is then used by smart contract modules for execution and monitoring of various tasks. These smart contract modules interact with a blockchain network, ensuring transparency, security, and efficiency throughout the supply chain.

1. Smart Contract in Supply Chain
   This smart contract is based upon Ethereum and can be deployed in any Ethereum network for testing purposes. This smart contract can have multiple stakeholders of the supply

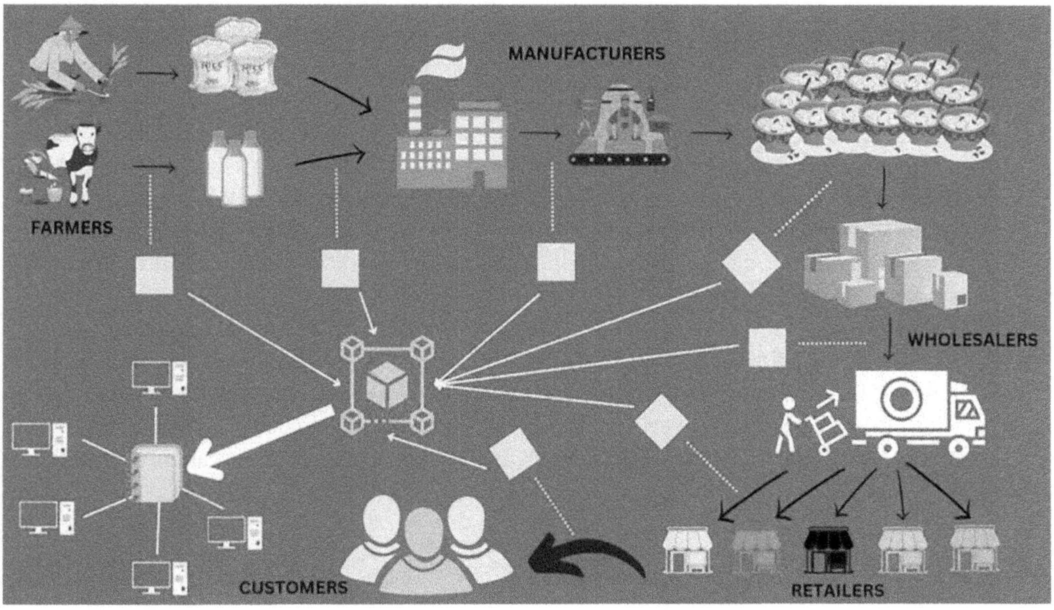

FIGURE 9.4 Architecture for the supply chain model.

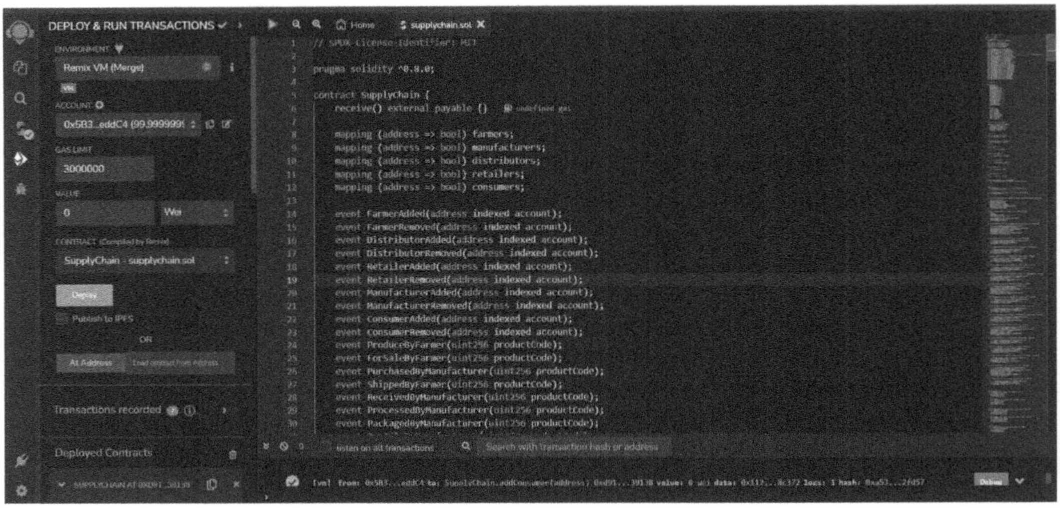

FIGURE 9.5 Working of the smart contract.

chain, including farmer (producer), manufacturer, distributor (wholesaler), retailer, and customer (consumer). The stakeholders can add themselves to the network without the consent of any central authority. Also, each of them is defined with a specific set of roles. The roles of a farmer are: a farmer first produces raw materials and lists them in the contract. A farmer can choose which items to sell and list them for sale with appropriate prices. When a manufacturer buys an amount, the farmer can ship the product to the manufacturer. The roles of a manufacturer are: the manufacturer can buy product from any farmer with an instant payment method. The manufacturer then processes the raw materials. It moves on to packaging. Finally, the manufacturer lists the product for sale with an appropriate price. After a distributor buys an amount, the manufacturer ships the product to the distributor. The roles of a distributor are defined as: the distributor buys the product from the manufacturer with instant payment. The distributor then stores the product in their warehouse. Finally, the distributor lists its product for sale with an appropriate price. After a retailer buys an amount, the distributor ships the product to the retailer. The roles of a retailer are defined as: the retailer buys the product from the distributor with instant payment. The retailers then store product in their individual shops. Finally, the retailer lists the product for sale with an appropriate price. The roles of a customer are as follows: the customer buys the product from the retailer with instant payment. Figure 9.5 illustrates the working of the smart contract.

It outlines the process flow of a smart contract. It begins with the initiation of the contract, where initial parameters and conditions are set. Triggering events such as order placement or delivery then activate the contract. The contract verifies conditions, executes predefined actions if conditions are met, records transaction details on the blockchain, and notifies involved parties about the contract's execution.

9.8 NOVELTY

In India, cryptocurrency is not widely used because the Reserve Bank of India has prohibited banks from dealing with cryptocurrency exchanges, and there is no specific legal framework for cryptocurrencies in the country. So, implementing a blockchain over the generic supply chain without any digital currency is next to impossible. This is where we introduce our novelty. The Digital Rupee is a proposed digital currency in India that will be issued by the Reserve Bank of India. It is also called an E-rupee. This is what would replace Ethereum, and our complete blockchain can be deployed in this Digital Rupee network. Digital Rupee is an innovative solution that leverages blockchain technology to provide a decentralized and secure platform for supply chain management. One of the key features of Digital Rupee is the use of smart contracts to automate and enforce contractual obligations between different stakeholders in the supply chain. These smart contracts are self-executing and programmable, allowing for real-time monitoring and tracking of products as they move through the supply chain.

The novelty of using smart contracts in Digital Rupee lies in their ability to improve supply chain efficiency and transparency. By automating the contract execution process, Digital Rupee eliminates the need for intermediaries and reduces the likelihood of disputes and errors. This leads to faster transaction times and reduced costs for all parties involved. Additionally, the use of smart contracts enables real-time tracking of products and automatic updates to the blockchain ledger, providing stakeholders with complete transparency and accountability throughout the supply chain. Overall, the use of smart contracts in Digital Rupee is a novel and innovative approach to supply chain management that has the potential to revolutionize the complete industry in India.

9.9 RESULTS AND DISCUSSION

Particularly in markets with intense competition, product delivery delays can erode client relationships, impair a company's brand, and result in lost sales. These problems can be resolved by implementing blockchain technology with smart contracts, which would make the supply chain completely transparent and traceable. To stop farmers from being taken advantage of, smart contracts guarantee on-time delivery, verify product damage ownership, and offer immediate payment choices. Predictive analytics, for example, can optimize delivery schedules and minimize delays with algorithms like XGBoost (99.65% accuracy), Random Forest (99.54% accuracy), and Linear Regression (78.52% accuracy). All parties gain from this strategy: farmers are appropriately compensated, retailers retain customers, and customers receive items on schedule. Hence, implementation of a distributed ledger via the use of a smart contract can make all the goods completely trackable and reduce the risks of late deliveries and fraudulent orders. A smart contract can make the complete supply chain trackable with a single source of truth. This ensures that any damage to any product can be validated, and it can be discovered under whose ownership the event happened. Moreover, the smart contract issues an instant payment option which prevents farmers from being exploited. This feature ensures that farmers get their deserved value upon selling raw materials. Also, the customers benefit because the product reaches the retailer on time, and the customer does not need to wait for the product in retail shops. Hence, implementing blockchain with all its parameters makes the supply chain much more transparent and helps to reduce the delay of delivery. Figure 9.6 shows the process flow of a blockchain-implemented supply chain.

It summarizes the sequential flow of processes in a supply chain implemented with blockchain technology. It begins with the capture of relevant data about the product and its journey. The captured data is then validated, encrypted, and recorded as transactions on the blockchain. Through consensus mechanisms, transactions are confirmed and validated by network participants. Smart contracts execute predefined actions based on conditions, ensuring automation and efficiency. The transparency and traceability provided by blockchain technology enable stakeholders to track the product's journey and verify its authenticity throughout the supply chain.

FIGURE 9.6 Process flow of blockchain implemented supply chain.

9.10 CONCLUSION

It is clear that in a generic supply chain system in India, there are a lot of issues, including late deliveries of orders, improper payment schemes, and fraudulent orders. These issues create hindrance in any business in aspects of management as well as analysis. Implementation of blockchain has the potential to transform supply chain management systems by providing an immutable and decentralized ledger that allows for secure and transparent tracking of goods. Blockchain can be used to provide end-to-end visibility of the supply chain, allowing all parties involved to trace the origin of a product or service, track its movement through the supply chain, and verify its authenticity. This can help prevent fraud, counterfeiting, and other types of supply chain risks. Blockchain can be used to automate supply chain processes through the use of smart contracts, which are self-executing contracts with the terms of the agreement between buyer and seller being directly written into lines of code. This can help to eliminate manual processes, reduce administrative costs, and improve efficiency. This smart contract can be implemented in the Digital Rupee network, which would be a revolutionary game changer in terms of the supply chain. The stakeholders can work on their own and within a set of rules and guidelines. Blockchain can also be used to streamline payment processing in the supply chain by enabling secure, instant, and low-cost transactions between buyers and sellers. This can help reduce the risk of fraud, eliminate the need for intermediaries, and improve cash flow for all parties involved. Overall, the use of blockchain technology in supply chain management has the potential to improve efficiency, reduce costs, and increase transparency.

REFERENCES

T.K. Agrawal, V. Kumar, R. Pal, L. Wang, and Y. Chen, "Blockchain based framework for supply chain traceability: A case example of textile and clothing industry," *Comput. Ind. Eng.*, vol. 154, Apr. 2021, Art. no. 107130, doi: 10.1016/j.cie.2021.107130.

J. Al-Jaroodi and N. Mohamed, "Blockchain in industries: A survey," *IEEE Access*, vol. 7, pp. 36500–36515, 2019.

V. Astarita, V.P. Giofrè, G. Mirabelli, and V. Solina, "A review of blockchain-based systems in transportation," *Information*, vol. 11, no. 1, p. 21, 2020.

J. Astill, R.A. Dara, M. Campbell, J.M. Farber, E.D. Fraser, S. Sharif, and R.Y. Yada, "Transparency in food supply chains: A review of enabling technology solutions," *Trends Food Sci. Technol.*, vol. 91, pp. 240–247, 2019.

R. Azzi, R.K. Chamoun, and M. Sokhn. "The power of a blockchain-based supply chain," *Comput. Ind. Eng.*, vol. 135, pp. 582–592, 2019.

T. Bocek, B.B. Rodrigues, T. Strasser, and B. Stiller, "Blockchains everywhere—A use-case of blockchains in the pharma supply-chain," *Proc. IFIP/IEEE Symp. Integr. Netw. Service Manage. (IM)*, pp. 772–777, May 2017.

S. Chen, R. Shi, Z. Ren, J. Yan, Y. Shi, and J. Zhang, "A blockchain-based supply chain quality management framework," *Proc. IEEE 14th Int. Conf. e-Bus. Eng. (ICEBE)*, pp. 172–176, Nov. 2017.

S. Figorilli, F. Antonucci, C. Costa, F. Pallottino, L. Raso, M. Castiglione, E. Pinci, D.D. Vecchio, G. Colle, A. Proto, G. Sperandio, and P. Menesatti, "A blockchain implementation prototype for the electronic open source traceability of wood along the whole supply chain," *Sensors*, vol. 18, no. 9, p. 3133, Sep. 2018, doi: 10.3390/s18093133.

K. Gao, Y. Liu, H. Xu, and T. Han, "Design and implementation of food supply chain traceability system based on hyperledger fabric," *Int. J. Comput. Sci. Eng.*, vol. 23, no. 2, pp. 185–193, 2020, doi: 10.1504/IJCSE.2020.10032813.

H.M. Kim and M. Laskowski, "Toward an ontology-driven blockchain design for supply-chain provenance," *Intell. Syst. Accounting, Finance Manage.*, vol. 25, no. 1, pp. 18–27, Jan. 2018.

S. Kim, Y. Kwon, and S. Cho, "A survey of scalability solutions on blockchain," *Proc. Int. Conf. Inf. Commun. Technol. Converg. (ICTC)*, pp. 1204–1207, Oct. 2018, doi: 10.1109/ICTC.2018.8539529.

K. Korpela, J. Hallikas, and T. Dahlberg, "Digital supply chain transformation toward blockchain integration," *Proc. 50th Hawaii Int. Conf. Syst. Sci.*, vol. 28, pp. 1–5, 2017.

L. Luu, D.H. Chu, H. Olickel, P. Saxena, and A. Hobor, "Making smart contracts smarter," *Proc. ACM SIGSAC Conf. Comput. Commun. Secur.*, pp. 254–269, 2016.

J. Ma, S.-Y. Lin, X. Chen, H.-M. Sun, Y.-C. Chen, and H. Wang, "A blockchain-based application system for product anti counter feiting," *IEEE Access,* vol. 8, pp. 77642–77652, 2020, doi: 10.1109/ACCESS.2020.2972026.

J. Maxim, "Onename launches blockchain identity product passcard," *Bitcoin Magazine*, Nashville, TN, USA, 2015. [Online].

A.A. Mukherjee, R.K. Singh, R. Mishra, and S. Bag, "Application of blockchain technology for sustainability development in agricultural supply chain: Justification framework," *Operations Manage. Res.*, vol. 15, nos. 1–2, pp. 46–61, Jun. 2021, doi: 10.1007/s12063-021-00180-5.

G.-T. Nguyen and K. Kim, "A survey about consensus algorithms used in blockchain," *J. Inf. Process. Syst.*, vol. 14, no. 1, pp. 101–128, Jan. 2018.

D. Prashar, N. Jha, S. Jha, Y. Lee, and G.P. Joshi, "Blockchain based traceability and visibility for agricultural products: A decentralized way of ensuring food safety in India," *Sustainability*, vol. 12, no. 8, p. 3497, Apr. 2020, doi: 10.3390/su12083497.

M. Raja, P. Saranya, S.K. Jagannathan, and G. Bizel, "Demand response analysis in microgrids with blockchain: A comprehensive cyber-physical approach," in *Next-Generation Cyber-Physical Microgrid Systems*, Elsevier, 2024, Pages 1–28, ISBN: 9780443221873, doi: 10.1016/B978-0-443-22187-3.00001-1.

M. Ruta, F. Scioscia, S. Ieva, G. Capurso, and E. Di Sciascio, "Supply chain object discovery with semantic-enhanced blockchain," *Proc. 15th ACM Conf. Embedded Netw. Sensor Syst.*, pp. 1–2, Nov. 2017.

P. Saranya and R. Maheswari, "Proof of transaction (PoTx) based traceability system for an agriculture supply chain," *IEEE Access*, vol. 11, pp. 10623–10638, 2023, doi: 10.1109/ACCESS.2023.3240772.

P. Saranya, R. Maheswari, and T. Kulkarni, "Blockchain and IoT technologies to improve the agricultural food supply chain," in *Recent Trends in Blockchain for Information Systems Security and Privacy*, Boca Raton, FL, USA: CRC Press, 2021.

P. Saranya, R. Maheswari, and R. Ragala, "Blockchain for agri-business: A simulation for assuring product ownership using Hyperledger composer," in *Cyber-Physical System Solutions for Smart Cities*, 2023, doi: 10.4018/978-1-6684-7756-4.ch004.

D. Shrier, W. Wu, and A. Pentland, "Blockchain & infrastructure (identity, data security)," *Massachusetts Inst. Technol.-Connection Sci.*, vol. 1, no. 3, 2016. [Online].

N. Sohrabi and Z. Tari, "ZyConChain: A scalable blockchain for general applications," *IEEE Access*, vol. 8, pp. 158893–158910, 2020, doi: 10.1109/ACCESS.2020.3020319.

S. Wang, D. Li, Y. Zhang, and J. Chen, "Smart contract-based product traceability system in the supply chain scenario," *IEEE Access*, vol. 7, pp. 115122–115133, 2019, doi: 10.1109/ACCESS.2019.2935873.

Y. Wang, H. Peng, Z. Su, T.H. Luan, A. Benslimane, and Y. Wu, "A platform-free proof of federated learning consensus mechanism for sustainable blockchains," *IEEE J. Sel. Areas Commun.*, vol. 40, no. 12, pp. 3305–3324, Dec. 2022, doi: 10.1109/JSAC.2022.3213347.

X. Yang, M. Li, H. Yu, M. Wang, D. Xu, and C. Sun, "A trusted blockchain-based traceability system for fruit and vegetable agricultural products," *IEEE Access*, vol. 9, pp. 36282–36293, 2021, doi: 10.1109/ACCESS.2021.3062845.

10 A Systematic Review of Plant Leaf Disease Identification Using Image Search Engines

D Kavitha, Sanjay M, Pillaram Manoj, Benjamin Roy, and Parth Vadera

10.1 INTRODUCTION

The advent of digital cameras and mobile devices with built-in cameras has led to an explosion in the number of digital images available today. With this vast amount of visual data, efficient and effective methods for image retrieval and analysis are becoming increasingly important. Image search is a fundamental task in computer vision and has various applications, including healthcare, security, and social media.

In recent years, deep learning techniques, specifically Visual Geometry Group (VGG) VGG19, have emerged as a game changer in the field of computer vision. VGG19 has shown amazing performance in applications such as image classification, object identification, and segmentation. The use of trained VGG19 as a feature extractor has opened up new possibilities for image search, allowing for the retrieval of identical pictures without depending on user annotations or text-based queries. This pre-training allows the model to recognize a wide variety of objects and patterns in images, making it well suited for feature extraction in image search.

This work provides a comprehensive analysis of an image search engine created particularly to identify plant leaf diseases in response to the growing interest in home gardening and organic farming. The suggested method uses a trained VGG19 for feature extraction and the K-nearest neighbor (KNN) algorithm for similarity comparison. By estimating cosine distances and selecting the top N photos with the most similar characteristics, the method enables the early diagnosis and subsequent control of plant health problems. In recent years, there has been a lot of interest in developing image search engines specifically for detecting plant leaf diseases. These customized engines use deep learning techniques and trained VGG19 to extract key characteristics from plant leaf pictures, enabling disease detection with great accuracy and speed. These engines help farmers and home gardeners diagnose and manage plant health issues by utilizing machine learning algorithms and large datasets of unhealthy plant photos. This article seeks to conduct a thorough analysis and evaluation of such image search engines, with the ultimate objective of encouraging sustainable agricultural practices and food security. The results demonstrate its effectiveness in retrieving relevant images from a dataset of plant leaf images, even in cases where the query image is only loosely related to the target images. The proposed method has the potential to be applied in various domains, such as healthcare, where image retrieval and analysis are becoming increasingly important.

10.2 BACKGROUND STUDY

10.2.1 Digitizing Agricultural Practices in Africa

Ngongoma et al. (2023) and their research group at Durban University of Technology conducted a detailed study of plant disease detection systems designed for agricultural applications. Their work is

 DOI: 10.1201/9781003507390-10

relevant to the Fourth Industrial Revolution and current advances in precision agriculture, with a special emphasis on disease, insect, and weed identification utilizing artificial intelligence approaches. The process entails taking high-resolution photos of agricultural specimens and using machine learning algorithms to appropriately diagnose observed faults. The study describes the numerous image processing procedures, machine learning classification algorithms, and feature extraction approaches required for effective plant disease monitoring systems. Noteworthy is the employment of varied classification methods such as support vector machines and artificial neural networks in vision-based plant disease monitoring systems. The study represents substantial advancement [1].

Oliva Debnath and Himadri Nath Saha (2022) have described that the crop diseases cause huge losses to the agricultural sector, as the profit decreases by 15–20% every year. Deep learning technology has been used to identify diseases in crops, with a suggested model attaining an amazing accuracy of 97.701%. To fight crop damage, smart farming solutions that combine machine learning and IoT are being implemented. A important part of the study is the early identification of brown spot disease in rice paddy using convolutional neural networks (CNNs), which is a unique method. Real-time data-driven picture identification and pre-processing techniques are exploited, supported by a proprietary image-processing tool and implementation using TensorFlow [2].

10.2.2 Advancements in Apple Leaf Disease Identification

Prakhar Bansal et al. (2021), from the Atal Bihari Vajpayee Indian Institute of Information Technology and Management, investigated disease detection in apple leaves using deep convolutional neural networks Their study focused on autonomously identifying illnesses in plants using deep learning algorithms. The suggested model identified illnesses in apple leaves with an astounding 96.25% accuracy, based on a dataset of 3642 pictures divided into four groups. DenseNet121 and EfficientNetB7 pre-trained models were used, which outperformed previous models and effectively identified several plant diseases. The study intends to widen its coverage by including more types of foliar diseases [3].

Bin Liu et al. (2017) from Northwest A&F University presented a ground-breaking solution for detecting apple leaf disease using deep convolutional neural networks, expanding on AlexNet's foundation. Their model, which was designed to reduce parameters, accelerate convergence, and maintain accuracy, surpassed previous models with an amazing 97.62% accuracy on the testing dataset. They improved the model's generalization capabilities by increasing the dataset size. Nonetheless, seasonal restrictions and biological development patterns continue to provide obstacles for data collecting. Liu et al. propose using advanced deep neural network designs such as faster RCNN, YOLO, and SSD to meet real-time detection demands. They emphasize the need to collect a wide set of apple leaf disease samples as well as high-quality photos in order to enable accurate and prompt identification [4].

Yong Zhong and Ming Zhao (2019) did a study, "Research on Deep Learning in Apple Leaf Disease Recognition," which emphasized the major impact of apple leaf diseases on output and economic losses. To improve illness detection accuracy, the researchers used the DenseNet-121 deep convolutional network with regression, multi-label classification, and focus loss function approaches. The suggested methods outperformed established approaches, reaching accuracies of 93.51%, 93.31%, and 93.71% on the test set [5]

Asif Iqbal Khan et al. (2022) emphasized in their work, "Deep Diagnosis," the considerable impact of apple diseases and pests on agricultural output and economic losses within the sector. To solve this difficulty, they created a collection of 9000 high-quality photos particularly designed for disease identification in apples. Their technique consists of a two-stage system: a classification model and a detection step. Notable outcomes include 88% classification accuracy and a mean average precision (mAP) of 42%. This approach is successful at diagnosing numerous apple illnesses and shows potential as a useful tool for growers. Future endeavors may include expanding this system to include other fruits and vegetables [6].

Anupam Bonkra et al. (2023) led a research project at the Czech University of Life Sciences Prague that focused on apple leaf disease identification utilizing collaborative machine learning/deep learning and artificial intelligence methodologies. The agriculture industry, which is crucial to the global economy, faces issues such as hailstorms and plant diseases that reduce crop productivity. The paper emphasizes the need to identify apple leaf diseases using sophisticated ML/DL algorithms from 2011 to 2023. Bibliometric analysis, aided by tools such as the Bibliometrix R-package and VOSviewer, reveals publishing trends and research contributions in this field, shedding light on the changing landscape of apple disease detection strategies [7].

10.2.3 Advancements in Grape Leaf Disease Detection

Bin Liu et al. (2020) observed that China's grape sector produced 13.083 million tons in 2017. They suggested a deep learning approach based on convolutional neural networks for identifying grape leaf illnesses, with an outstanding 97.22% accuracy rate. The study used data augmentation approaches to train the model on a large dataset. Their suggested deep inception CNN (DICNN) model outperformed other established models like ResNet and GoogLeNet. This study proposes a promising technique for classifying agricultural diseases using sophisticated deep learning algorithms [8].

Xiaoyue Xie et al. (2020) presented the faster DR-IACNN model for real-time grape leaf disease detection. By merging the Inception module with the ResNet structure, the model outperformed the faster R-convolutional neural network in terms of mean average precision (81.1%). It accurately detects four prevalent grape leaf diseases at 15.01 FPS. Data augmentation was added to 4,449 original photos, resulting in a 6.8% gain in accuracy. Notably, the model can recognize many sick regions and disease types on a single leaf. The approach used thresholding, anisotropic diffusion, and K-means clustering to achieve excellent training accuracies. This breakthrough offers a viable approach for real-time grape leaf disease diagnosis. Similarly, Mohanty et al. attained an accuracy of 99.35% [9].

Seyed Mohamad Javidan et al. (2022), in "Diagnosis of Grape Leaf Diseases Using Automatic K-Means Clustering and Machine Learning," investigated several algorithms for efficiently identifying plant diseases, with a special emphasis on approaches appropriate for field situations. They discovered that using gray-level co-occurrence matrix features and principal component analysis improved illness diagnostic accuracy. When compared to deep learning approaches like CNN and GoogleNet, their proposed method had faster processing times. The study emphasizes the importance of automated machine vision systems in helping farmers efficiently identify and control crop diseases [10].

Moh. Arie Hasan et al. (2020) addressed the issue of diagnosing grape leaf diseases, which can drastically reduce grape yield and result in large losses for farmers. The study emphasized the need of correct diagnosis for successful illness treatment. Using convolutional neural network technology, the study offered a three-stage approach: data input, feature learning, and classification. Using Keras libraries in Python, the CNN model attained a noteworthy accuracy rate of 91.37% during the training procedure [11].

Pachade et al. (2023) developed a VGG16-based deep learning system for grape leaf diseases with mean accuracies of 77.57% and 78.28%, offering a cost-effective alternative for grape producers [12].

MS Bala Murugan et al. (2021) from Vellore Institute of Technology University highlighted in their study, "IoT-Based Smart Agriculture and Plant Disease Prediction," the pressing need for technological interventions in agriculture, given the sector's declining contribution to India's GDP over the years. The study team wanted to improve crop monitoring and field management techniques via the use of IoT and machine learning. Their new approach, which combined IoT and deep learning techniques, achieved a stunning accuracy rate of 99.35% in forecasting plant illnesses. This breakthrough offers a potential alternative for agriculturists by providing effective instruments for disease surveillance and crop management in modern agriculture [13].

Sharada P. Mohanty et al. (2016) investigated image-based plant disease identification, a critical topic for ensuring food security in the face of problems such as climate change and dwindling pollinator populations. Traditionally, illness categorization was based on manually created characteristics, but the advent of deep learning, particularly convolutional neural networks, provided a potential alternative. With CNNs, reliable agricultural disease categorization became possible, providing optimism, particularly for smallholder farmers who are sensitive to disease-related output losses. The researchers used a large dataset of 54,306 photos of plant leaves with 38 discrete disease labels to illustrate the usefulness of deep learning algorithms in predicting crop-disease couples with high accuracy. This represented a considerable breakthrough above prior techniques based on predetermined characteristics [14].

G. Balram and K. Kiran Kumar (2022) investigated "Crop Field Monitoring and Disease Detection of Plants in Smart Agriculture using Internet of Things," with an emphasis on the Indian agricultural environment and the use of IoT sensor nodes for real-time data collecting in plant disease diagnosis. The study emphasizes the need for early illness detection, with sensors playing a key role in monitoring environmental conditions. Furthermore, image processing techniques are used for illness detection. The study investigates adaptive clustering and machine learning methods for disease categorization, emphasizing their potential to improve agricultural production and assist farmers. MELM segmentation and manual segmentation results were compared using measures such as the Dice similarity coefficient, mean square error, and structural similarity index. The results revealed that these criteria were acceptable for detecting leaf disease [15].

10.2.4 Literature Review: Leaf Disease Classification in Crops

One study introduces EfficientNet-B7, which can classify healthy and sick tomato leaves with 99.95% accuracy. The model beats previous deep learning methods on the Plant Village dataset. Modified U-net performs best for leaf segmentation, whereas EfficientNet-B7 excels in feature extraction. The models can help detect plant diseases early on by utilizing technology such as cellphones and drones. The paper underlines the relevance of using AI in agriculture for disease control and suggests a feedback mechanism to improve crop management [16].

Prajakta Prashant Bhangale's et al. (2017) paper, "Improved Crop Yield Prediction Using Neural Networks," investigates several data mining approaches and clustering algorithms in agriculture for crop yield prediction. The study emphasizes the need for data normalization, feature selection, and visualization in analyzing agricultural data to help farmers choose the best crops depending on meteorological and soil conditions. Key crop prediction technologies include neural networks and association rule mining. The study investigates the impact of factors including soil type, rainfall, and temperature on crop development. Finally, the project aims to provide a framework that would enable farmers to make educated crop selection decisions [17].

Agriculture is critical in emerging nations such as India, contributing considerably to GDP. Traditional agricultural practices are imprecise in terms of soil qualities, water requirements, and crop profitability. AgriRec, a machine learning system, has been created to propose crops and fertilizers with high accuracy and speed of execution. Crop deterioration can result from issues with crop choices, fertilizers, and pesticides. Researchers are striving to select crops based on soil conditions, water needs, and profitability. The AgriRec algorithm recommends top crops for each season based on predicted profitability and fertilizer combinations. Machine learning has the potential to transform agriculture by raising agricultural yields and incomes. Future work will incorporate pest control prediction [18].

Delwar Hossain et al. (2023) revealed in "IoT Based Smart Soil Fertiliser Monitoring And ML Based Crop Recommendation System" that agricultural productivity is dependent on soil fertility, with nitrogen, phosphorus, potassium, pH, temperature, and moisture all playing important roles. A suggested IoT-based system employs sensors to monitor soil nutrients and machine learning to select crops depending on soil and weather conditions. This strategy lowers the cost of fertilizer usage, boosts productivity, and helps national growth [19].

Thendral and Vinothini (2023) utilized deep learning models for precise crop and fertilizer recommendations, aiming to enhance agricultural sustainability and meet global food demand [20].

10.3 PROPOSED SYSTEM

Figure 10.1 shows the proposed architecture of plant leaf detection and the system, which begins with the gathering of leaf images to serve as training data for a VGG19 model. During this training process, the VGG19 learns to recognize diverse properties such as shapes, colors, and textures that indicate distinct leaf kinds or health conditions. When a user submits an image query of a leaf, the system collects characteristics from the image and feeds them into both the trained VGG19 and a classification model designed specifically for leaf discrimination. With a similarity score that indicates the model's confidence level, this classification model then uses the extracted features to search the database for images that are similar to the query leaf. This allows for automatic testing of the leaf's health state, including probable illnesses. The proposed approach compares factors between query images and those in the dataset using K-nearest neighbors and cosine distances, calculating similarity and returning the top N images with the smallest distances as search results. Using image recognition and machine learning algorithms, this technique allows for efficient plant

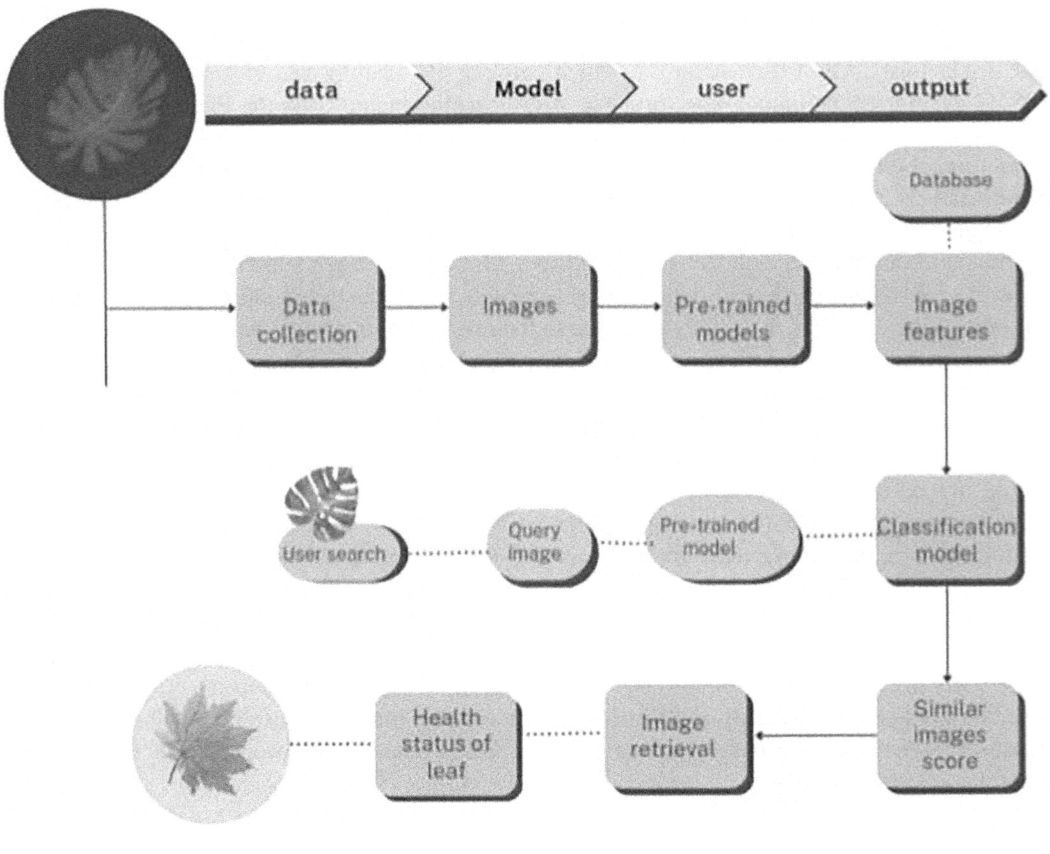

FIGURE 10.1 Proposed architecture of plant leaf disease detection.

leaf health analysis. Furthermore, the system evaluates query images, looks for similarity scores within specified clusters, and alerts users to the plant's health status in four languages, accompanied by emojis. With a threshold of five photos, it facilitates the diagnosis of various plant leaf diseases, improving the user experience. Using machine learning and deep learning techniques provides present efficacy while also allowing for future growth and development of the system's capabilities over time.

10.4 MODULE DECOMPOSITION

The module decomposition of the system involves the following key modules.

10.4.1 VGG19 Model Module

The VGG-19 model is a convolutional neural network. The model is meant to analyze 224 × 224 pictures and categorize them into a variety of categories. The VGG-19 model has been widely applied to a variety of tasks, including image recognition, classification, and feature extraction.

Figure 10.2 shows the variation of VGG model; the VGG-19 model has 19 layers total: 3 fully connected layers, 5 max-pooling levels, 1 SoftMax layer, 16 convolution layers, and 5 max-pooling layers. Its purpose is to categorize photos into different groups. Based on the key components of convolutional neural networks, the model makes use of small convolutional filters. During the model training phase, the images are preprocessed in real time using the ImageDataGenerator from the Keras package. There are three instances of the ImageDataGenerator created: test_datagen, eval_datagen, and train_datagen. These generators preprocess the images by cropping them to 224 pixels in width and height and rescaling the pixel values to a range of 0 to 1. In the class mode, a classification structure with one-hot encoded labels is indicated, and the batch size for processing during training is set to 32.

To specify the trained model, a fine-tuning function is built and the VGG-19 model is used for feature extraction. This function defines new layers on top of the output of the base model and freezes the layers of the base model. These new layers consist of an additional dense layer with a predetermined number of units, a flattening layer, and a fully linked (dense) layer with 64 units and ReLU activation. In order to produce a probability distribution over the possible classes, the function further uses a SoftMax activation function in the last layer. While using the prepare_VGG_model_for_finetuning function, freezing the basic layers is an option. Next, the categorical cross-entropy loss function and Adam optimizer are used to assemble the model. The history variable stores the progress of the model's validation and training. The model is trained using the model.fit method. This module encompasses the loading and use of a trained VGG19 model designed for extracting features from plant leaf images. It instantiates the VGG19 model, freezes the parameters to prevent further training, then extracts features from the dataset. The trained VGG19 is proficient at identifying hidden trends and extracting detailed data from leaves.

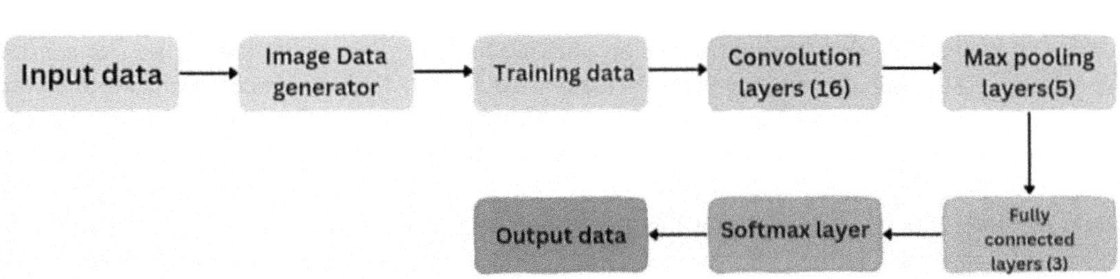

FIGURE 10.2 Architecture of VGG19 model using convolutional layers.

10.4.2 Dataset Loading and Pre-Processing Module

This module is responsible for loading the dataset of plant leaf images and performing necessary pre-processing steps such as resizing, normalization, and conversion to arrays. Proper pre-processing makes the dataset consistent with the model's input requirements, guaranteeing reliable feature extraction.

10.4.3 Feature Extraction Module

This module focuses on extracting features from each plant leaf image in the dataset using the trained model. Shapes, colors, and textures are captured in feature vectors created by the VGG19 by running the pictures through its layers. In order to allow effective comparison and analysis for assessing plant leaf health, these feature vectors function as simplified representations of the leaf images.

10.4.4 Similarity Comparison and Search Module

This module handles the computation of KNN (cosine distances) between the query image's feature vector and the feature vectors of all the leaf images in the dataset. It is responsible for calculating similarity scores between the characteristics of a query images, searches for similarity scores within predetermined clusters, and notifies users about the plant's health state in four languages, accompanied by emojis, with a threshold of five images, which makes it easier to identify different plant leaf diseases. The top N images with the lowest cosine distances are then picked as search results, allowing users to visually evaluate and diagnose the health of plant leaves. By decomposing the system into these modules, it becomes easier to manage and maintain each component separately, leading to a more organized and efficient development process.

10.5 RESULTS AND DISCUSSION

The model was put to the test. Its stronger illness diagnosis abilities appeared to be mostly attributed to its deeper architecture and more complexity. VGG19 continuously outperformed model across a range of illness categories and severities. For VGG-19, the accuracy is 96%.

Figure 10.3 shows the VGG19 model evaluated using leaf photos. The model's accuracy is 0.96, which means that it correctly recognized 96% of the data. The plots show the growth of training and validation loss, as well as accuracy over epochs, providing information about the model's learning efficiency and any overfitting concerns. Monitoring the gap between training and validation metrics is critical for optimizing model performance and avoiding overfitting.

10.5.1 Rice (Farmers)

10.5.1.1 Rice Health

Figure 10.4 represents the distance and index values of KNN model for healthy rice. In a pre-trained k-nearest neighbor model, these attributes are utilized to determine the indices of the five nearest neighbors and to compute similarity scores (distances). The distances between the features of the query image and those of its closest neighbors are represented by the distances array. The indices array represents the positions of these nearest neighbors in the dataset. In this output, the distances are [5.9604645e-08, 0.20226055, 0.2477603, 0.260243, 0.26276714], with corresponding indices [168, 38, 463, 290, 6]. This indicates that the closest neighbor to the query image is at index 168, followed by images at indices 38, 463, 290, and 6, with increasing distances indicating reducing a similarity to the query image. These results provide insights on images that are most comparable to the query, which may be used in further analysis or decision-making processes.

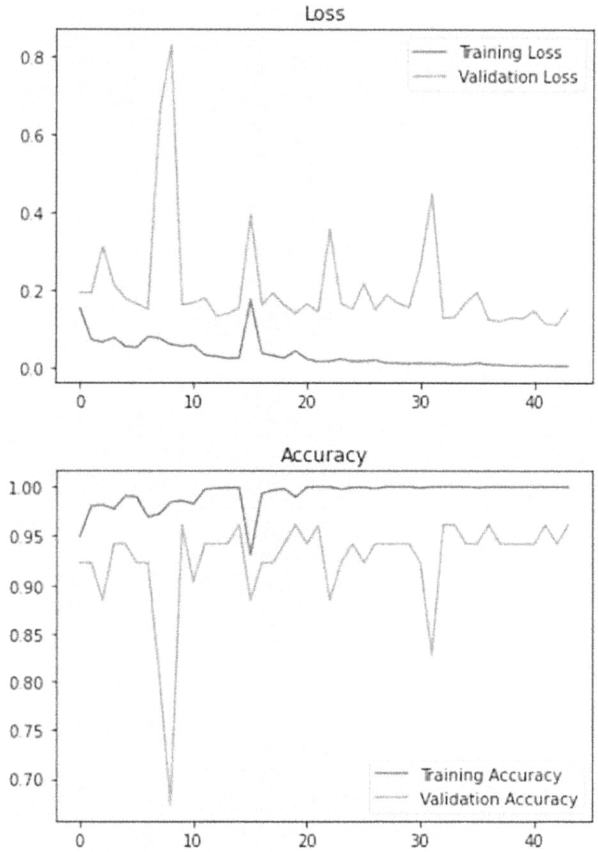

FIGURE 10.3 Model loss and accuracy of VGG19 model.

```
[5.9604645e-08 2.0226055e-01 2.4776030e-01 2.6024300e-01 2.6276714e-01] [168  38 463 290   6]
```

FIGURE 10.4 Distance and index values of KNN model for healthy rice.

Figure 10.5 represents the growth of rice plants through five phases, with the words "Rice healthy" and the numbers 1 through 5 following. A variety of features that reflect the plant's development can be observed at each stage. Plant 2 has longer, greener leaves that indicate future development, whereas Plant 1 has shorter, greener leaves that indicate the early growth period. With a mixture of long, green leaves and a few small, brown leaves toward the base, Plant 3 exhibits a more advanced stage of natural senescence. On the other hand, during its latter phases, Plant 4 has a lot of long, green leaves mixed with a lot of small, brown leaves around the base, which suggests that it is getting close to maturity. When Plant 5 reaches maturity, all of its leaves are brown, signifying that the growth cycle is now complete. It is possible to make observations about leaf health and determine if the leaf is unhealthy or healthy through analyzing these images. By using visual similarities, this method may determine the state of the rice leaf and help with effective diagnosis and possible treatment plans.

Figure 10.6 shows that the image has a message in four languages: Hindi, Tamil, Telugu, and English. It also has a green thumbs-up emoji, which represents a healthy leaf. It means that the leaves in good health are bright green in color. They are free of rips and holes and feel efficient to the touch. This simple representation makes it easier to understand for farmers from a variety of language backgrounds and helps quickly figure out the leaf's condition easily.

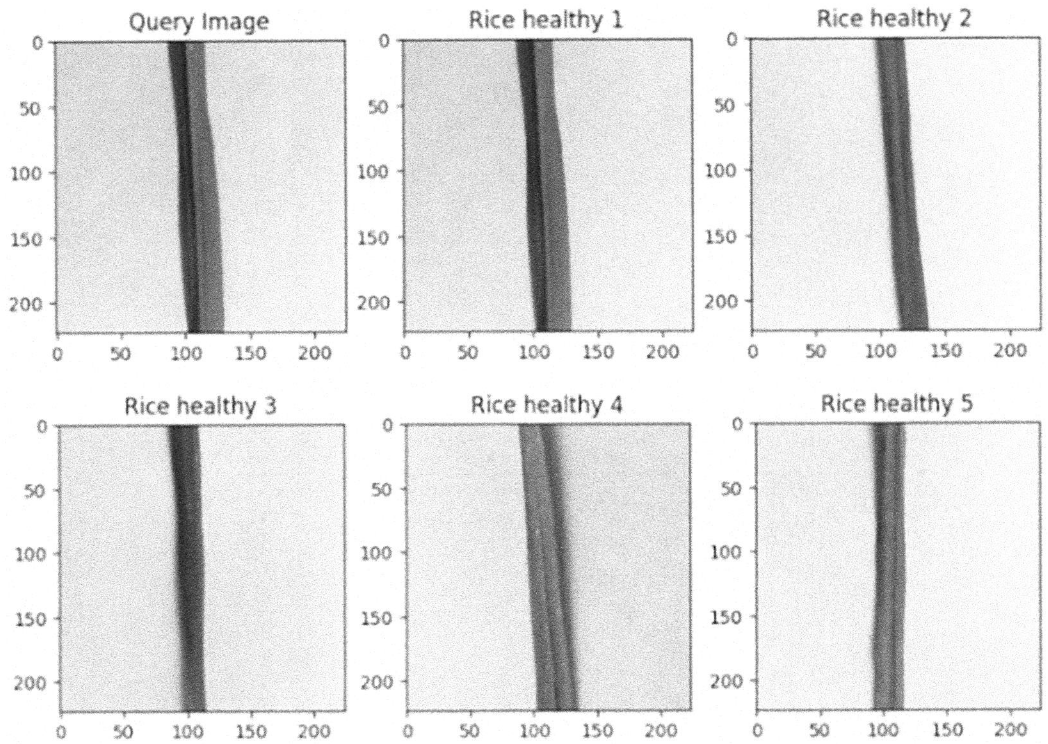

FIGURE 10.5 Healthy rice plants.

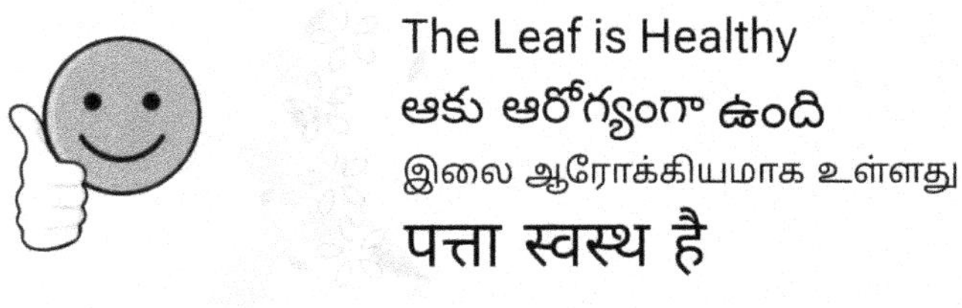

FIGURE 10.6 Status of healthy rice leaf.

```
[1.1920929e-07 2.0841569e-01 2.3500305e-01 2.5354660e-01 2.7593297e-01] [599 560 571 122 414]
```

FIGURE 10.7 Distance and index values of KNN model for diseased rice leaf.

10.5.1.2 Rice with Brown Spots (Diseased)

Figure 10.7 shows the k-nearest neighbors model; comparable images may be found by comparing their characteristics. The indices array shows the precise locations of these closest neighbors in the dataset, while the distances array records the separations between the characteristics of the query picture and those of its closest neighbors. The indices for the distances in this particular output are [599, 560, 571, 122, 414], and the distances are [1.1920929e-07, 0.20841569, 0.23500305, 0.2535466, 0.27593297]. These findings imply that the picture at index 599599 is the closest neighbor to the query image, with photos at indices 560, 571, 122, and 414 following in order of proximity. In general, the distances are getting lower as they get closer to the query image, which helps find similar images for more analysis or classification work.

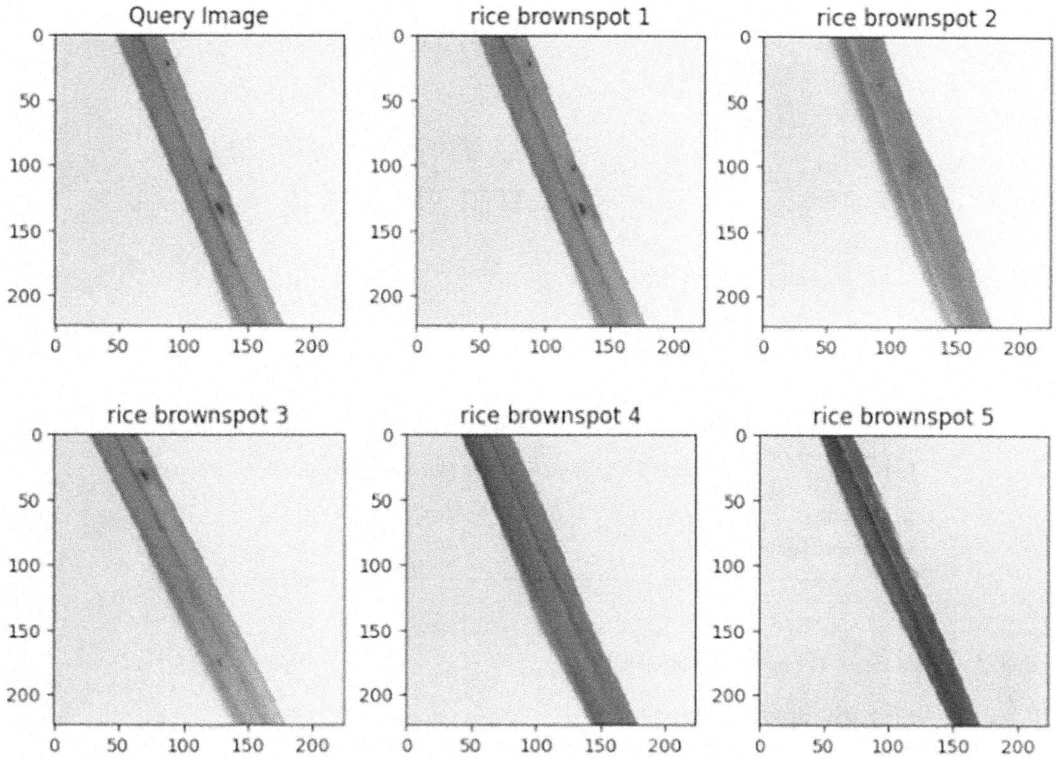

FIGURE 10.8 Rice plants with brown spots (diseased).

Figure 10.8 shows a close-up of a brown-spotted rice stem in the first image, which is the query image. Indicating the existence of rice brown spot disease, the remaining five images, labelled "rice brownspot 1" through "rice brownspot 5," are retrieved depending on how similar they are to the query image. The severity of the illness may vary in these images, from a few little brown spots on the rice stem to bigger, extended brown lesions affecting a significant portion of the stem. To help reduce the disease's negative effects on the health of rice crops, this comparison makes it easier to identify and assess the disease's severity across several cases, which helps with identification and possible treatment options.

Figure 10.9 shows an image that conveys a message in four languages (English, Telugu, Tamil, and Hindi) and includes a red frowning emoji, which globally represents an unhealthy leaf. This symbol system was created to conquer language barriers, allowing people to understand the message regardless of their original language. By giving a clear visual indication of plant health state, it enables individuals to take immediate and appropriate action to treat the leaf's bad condition and ensure their plants' well-being.

Table 10.1 represents the similarity scores of the rice leaf to help farmers, which evaluates the health of the crops. The similarity ratings of an increased risk of brown spot disease infection in rice leaves are generally indicated by a lower score. For example, 168 to 6 is a significant difference when comparing rice leaves that are healthy to those that have brown spots. Better distances, such as 599 and 414, indicate a greater dissimilarity between healthy leaves and brown spot samples. As actionable indicators, these ratings help farmers maintain the health and production of their rice fields by allowing them to quickly detect and contain possible disease outbreaks.

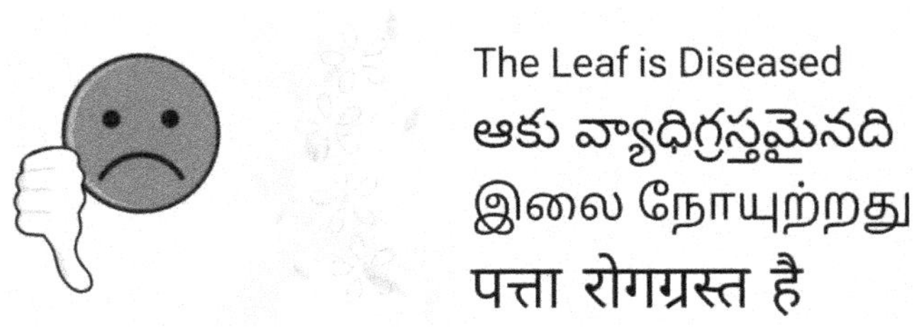

FIGURE 10.9 Status of unhealthy rice leaf.

TABLE 10.1
Similarity Scores for Rice Leaves (Farmers)

Leaf Type	Distance 1	Distance 2	Distance 3	Distance 4	Distance 5
Rice healthy	168	38	463	290	6
Rice brown spots	599	560	571	122	414

10.5.2 Tomatoes (Home Gardening)

10.5.2.1 Leaf Early Blight (Diseased)

Figure 10.10 shows that the VGG19 model extracts features from a query image and then flattens them into a one-dimensional array. In a pre-trained k-nearest neighbor model, these attributes are utilized to calculate similarity scores (distances) and determine the indices of the five nearest neighbors. The distances array represents the distances between the query image's characteristics and those of its nearest neighbors, and the indices array represents the positions of these neighbors within the dataset. In this particular result, the distances range from 0.6758962 to 0.72990847, with indices ranging from 949 to 958. These findings provide insights into images that are most similar to the query, which may be used in future analysis or selection procedures associated to image similarity or classification.

Figure 10.11 starts with the query image and displays a comparison of six images related to home gardening with the labels "tomato early blight 1" to "tomato early blight 5." These images probably show the early stages of tomato plant disease, especially "early blight," which is an increasing issue in home gardening. The portion helps identify and recognize possible disease signs in tomato plants by providing visually comparable images. This makes preventive measures for disease prevention and control easier in home gardening environments. This method enables tomato growers to maintain the health and yield of their plants in a timely manner, which promotes effective growing and harvest.

Figure 10.12 shows a message in four languages: Hindi, Tamil, Telugu, and English. It also features a red frowning emoji, which is a common representation of an unhealthy leaf. The purpose of this symbolic system was to ensure understanding among those who are home gardeners of tomatoes by overcoming language challenges. With the label "Diseased early blight," which gives a clear visual indicator of the plant's health, gardeners are empowered to take immediate action to improve the leaf's bad condition and protect their plants.

10.5.2.2 Leaf Late Blight (Diseased)

Figures 10.13 and 10.14 show the query image's closest neighbors that are located between 0.69165426 and 0.72843087, and their associated indices vary from 953 to 984. Figure 10.12 represents the six images, the first of which is the query image. The remaining five images, labelled

```
[0.6758962  0.702769   0.709795   0.7287446  0.72990847] [949 973 970 955 958]
```

FIGURE 10.10 Distance and index values of KNN model for diseased leaf (early blight).

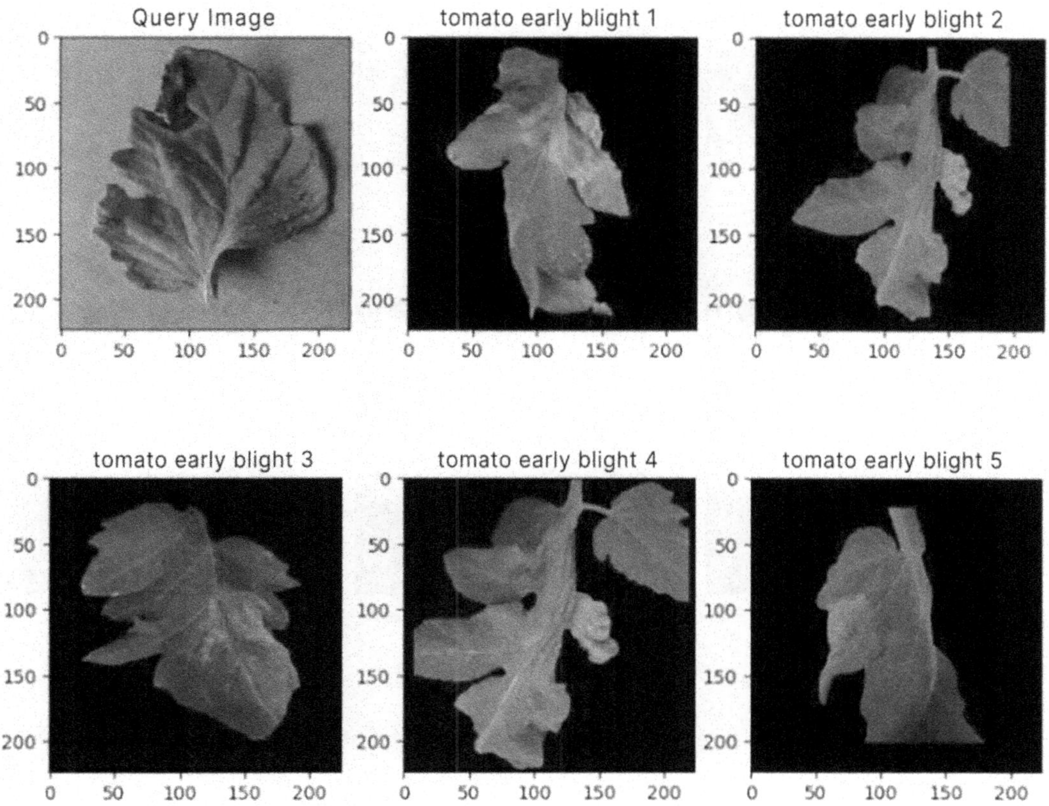

FIGURE 10.11 Tomato leaf with early blight (diseased).

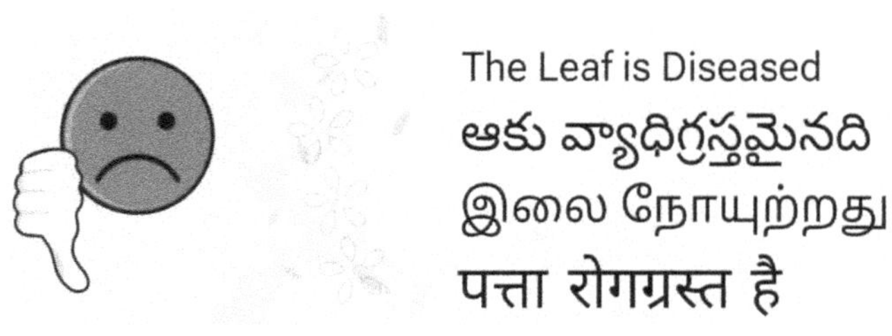

FIGURE 10.12 Status of a unhealthy tomato leaf (early blight).

```
[0.69165426 0.70900655 0.71864855 0.7228104  0.72843087] [963 984 965 972 953]
```

FIGURE 10.13 Distance and index values of KNN model for leaf late blight (diseased).

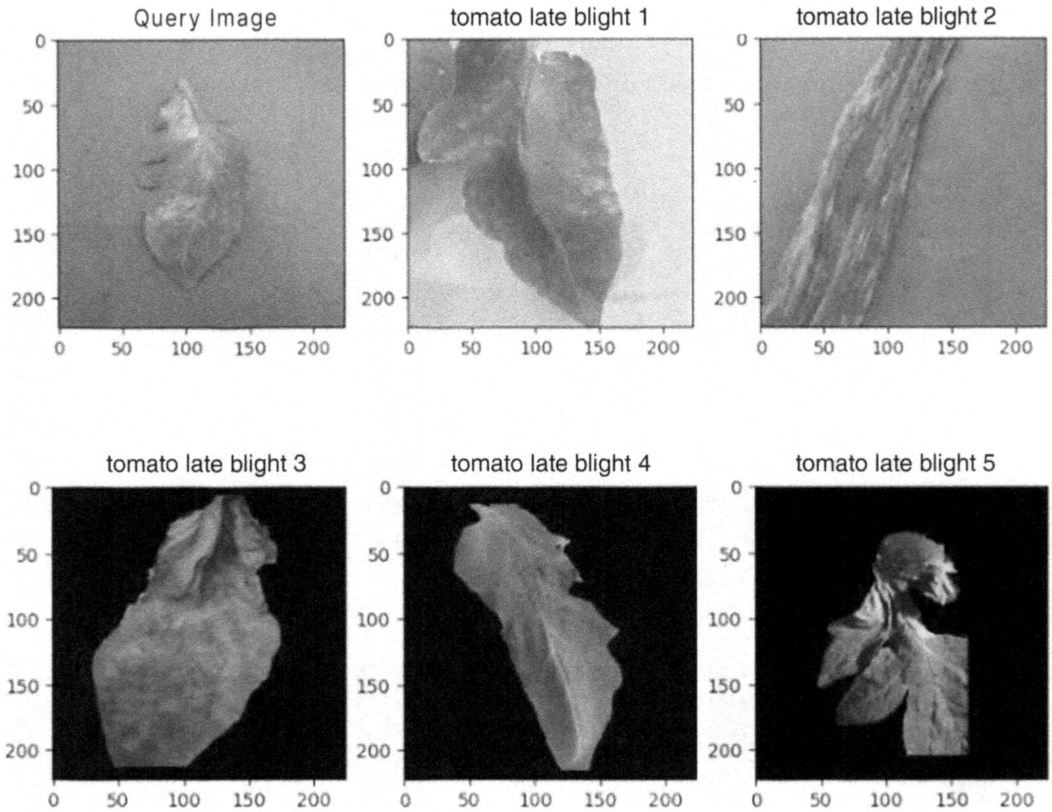

FIGURE 10.14 Tomato leaf with late blight (diseased).

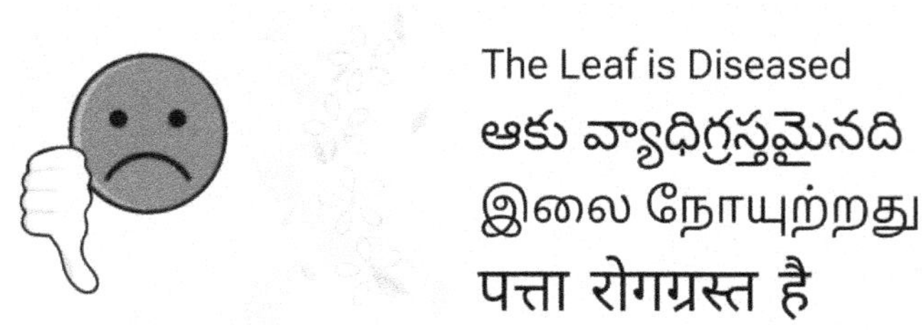

FIGURE 10.15 Status of a unhealthy tomato leaf (late blight).

"tomato late blight 1" to "tomato late blight 5," are obtained on the basis of how similar they are to the query. These photos most likely show advanced stages of "late blight," a tomato plant disease that may seriously harm tomato yields. The detection and analysis of disease development in tomato plants are made easier by this visualization, which enables accurate decision-making and management methods for disease control in home gardening.

Figure 10.15 shows a message in Telugu, Hindi, Tamil, and English. It also features a red frowning emoji, which is well recognized as a representation of an unhealthy leaf. This symbolic approach was developed to help tomato home growers understand one other across language difficulties. With the label "Diseased late blight," which provides a clear visual message of the plant's health, gardeners can act quickly to protect their plants and manage the failing condition of the leaf.

```
[0.74213713 0.7661848  0.77223015 0.77454275 0.77469444] [949 950 973 955 238]
```

FIGURE 10.16 Distance and index values of KNN model for diseased leaf (mold).

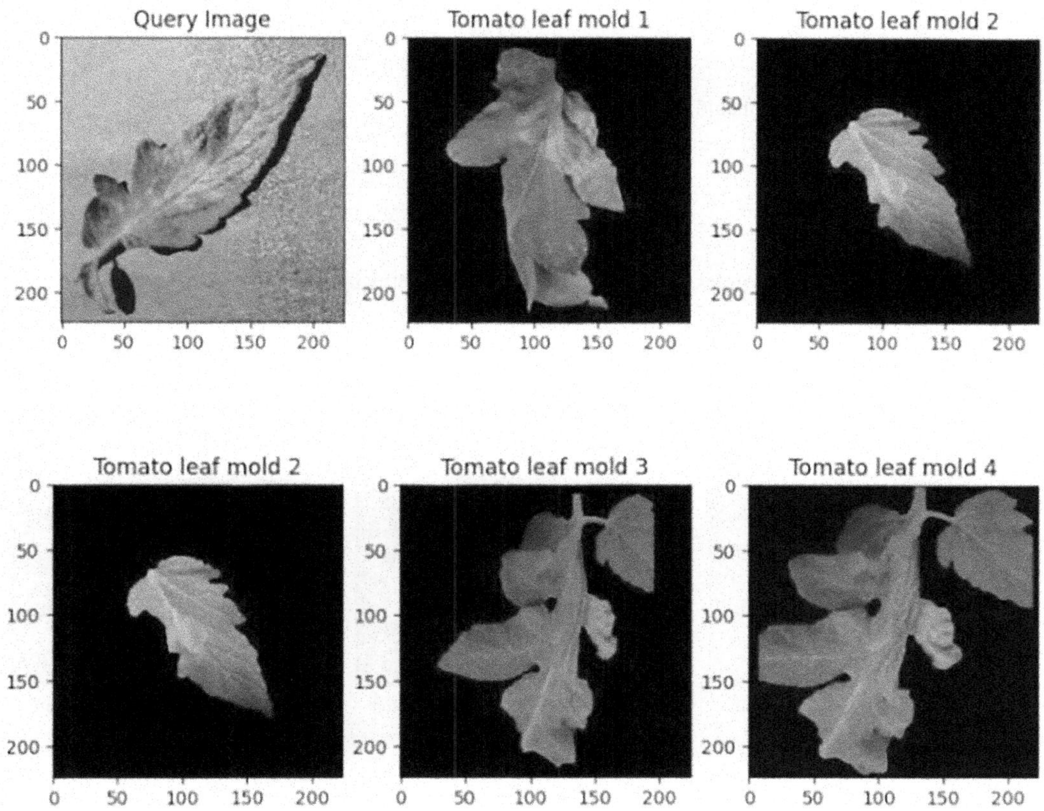

FIGURE 10.17 Tomato leaf with leaf mold (diseased).

10.5.2.3 Leaf Mold (Diseased)

Figures 10.16 and 10.17 represent tomato leaf mold. It is a fungal disease that appears as dark, thick mold on the underside of tomato leaves and yellow dots on their top surface. The image compares six images, the first of which is the query image. The remaining five images, labelled "tomato leaf mold 1" to "tomato leaf mold 5," are obtained on the basis of how similar they are to the query. After using a VGG19-based model to extract features from a query image, it finds visually identical images in a dataset by using a k-nearest neighbor model. The top five nearest neighbors' indices and distances are obtained and displayed. The ideal distance is 0.74213713, and 949 is the matching index. This shows that, based on the set of features extracted by the VGG19 model, the image in the dataset at index 949 is the one that most closely resembles the query image visually. Additionally, the query image is shown next to the dataset's top five identical images by the algorithm, which creates a visual representation.

Figure 10.18 shows the image has content in four different languages: Hindi, Tamil, Telugu, and English. It also has a red worried emoji, which is well recognized as the indication of unhealthy plants. Regardless of language context, this method allows for rapid understanding by addressing language constraints. When there's a visible indicator of a plant's health, people may take immediate

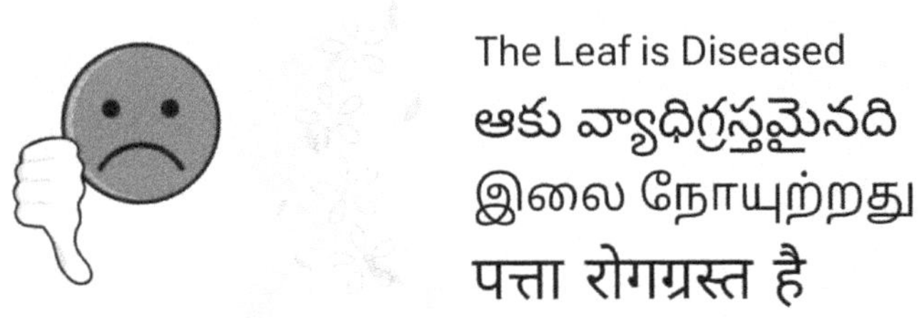

FIGURE 10.18 Status of a unhealthy tomato with leaf mold.

```
[0.6147449  0.6281828  0.6338144  0.65026563 0.6535226 ] [970 973 957 975 958]
```

FIGURE 10.19 Distance and index values of KNN model for healthy tomato leaf.

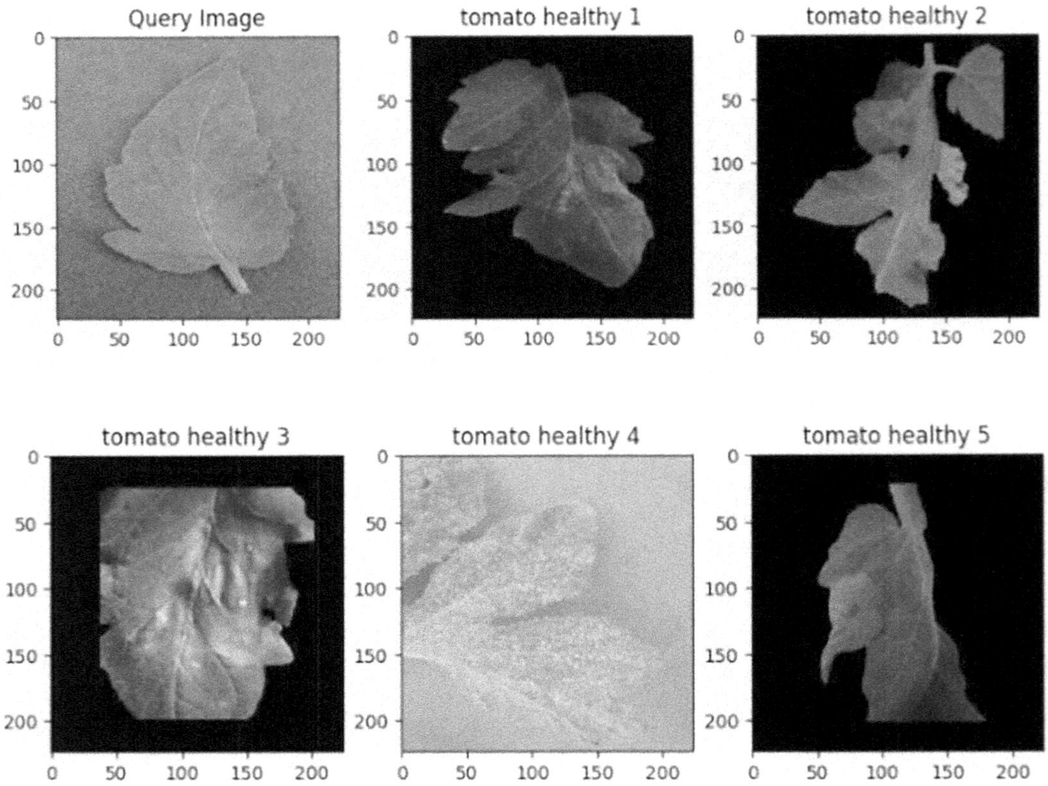

FIGURE 10.20 Healthy tomato leaf.

action, particularly when it comes to unhealthy tomato leaf mold. This approach is especially helpful for home growing with tomatoes since it encourages appropriate gardening strategies.

10.5.2.4 Healthy Leaves

Figures 10.19 and 10.20 show a comparison of six images, beginning with the query image and progressing through five obtained images tagged "tomato healthy 1" to "tomato healthy 5," which

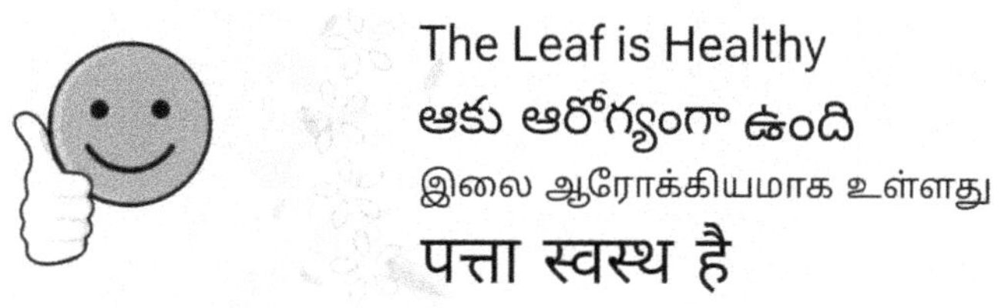

FIGURE 10.21 Status of healthy tomato leaf.

TABLE 10.2
Similarity Scores for Tomato (Home Gardening)

Leaf Type	Distance 1	Distance 2	Distance 3	Distance 4	Distance 5
Leaf early blight	949	973	970	955	958
Leaf late blight	963	984	965	973	953
Leaf mold	949	950	973	955	238
Leaf healthy	970	973	957	975	958

most likely reflect healthy tomato plants. This approach helps to find visually similar pictures and assists a variety of applications, such as image retrieval and classification. A query image's features are extracted using the VGG19 model and then compressed into a one-dimensional array. In a pre-trained k-nearest neighbor model, these attributes are utilized to determine the indices of the five nearest neighbors and to compute similarity scores (distances). The indices array shows the exact positions of these closest neighbors in the dataset, while the distances array shows the separations between the characteristics of the query picture and those of its closest neighbors. The distances in this particular output vary from 0.6147449 to 0.6535226, and the related indices fall between 970 and 958. According to these findings, the picture at index 970 is the closest neighbor to the query image, and it is followed by images at indices 973, 957, 975, and958.

Figure 10.21 shows a message in four languages (Hindi, Tamil, Telugu, and English) and a green thumbs-up emoji, which represents healthy tomato leaves. This widely accepted picture simplifies leaf condition evaluation by highlighting brilliant green color, lack of damage, and firm texture. With a concentration on home gardening, the message helps farmers and beginners quickly recognize healthy tomato plants. The simplest visual indication, labeled "tomato leaves healthy," overcomes language barriers for simple understanding.

Table 10.2 shows the tomato leaf similarity scores for home gardeners, which is useful for keeping an eye on the health of their plants. "Leaf Type" is the header that makes distinctions between different circumstances such as healthy leaves, mold, and late and early blight. The distances show how different leaf types are from one another, indicating similarity. For example, distances between early blight–affected leaves and healthy leaves range from 949 to 970, indicating a moderate degree of dissimilarity. Conversely, the distances between mold-affected leaves and healthy leaves differ greatly, with values as low as 238, suggesting a notable divergence in traits. These scores enable home gardeners to quickly detect and resolve problems like leaf blind or mold, promoting the strength and health of tomato leaves.

10.6 CONCLUSION

The in-depth analysis of plant leaf disease identification using an image search engine, trained VGG19 models, and K-nearest neighbors techniques highlights how technology has the ability to completely change agricultural practices, particularly in organic production and home gardening.

Through multilingual messaging and symbolic representations, this system overcomes language barriers to enable fast and accurate identification of plant health concerns by utilizing machine learning and deep learning approaches. By providing users with timely information and advice, it enables them to take proactive measures to address plant health issues, therefore boosting sustainable agricultural practices and supporting global food security programs. A more sustainable and productive farming future is possible due to the incorporation of modern technologies, which also improve the flexibility and efficiency of agricultural systems and speed the process of identifying diseases.

10.7 FUTURE WORK

Future work will provide specialized pest control tactics for efficient plant health management by including pesticide recommendations in the system in addition to detecting diseases. Preventive treatment in plant health concerns can be enhanced by further improvements in machine learning techniques, which can also improve system accuracy and enable real-time monitoring. The system will be more useful in a variety of climates and crop types if its multiple languages and plant species diversity are increased. This will ensure the system's efficacy and accessibility to a worldwide target audience. These advancements have the potential to improve agricultural sustainability and resilience, therefore supporting farming techniques globally and ensuring their long-term survival.

REFERENCES

1. Ngongoma, M. S. P., Kabeya, M., & Moloi, K. (2023). A review of plant disease detection systems for farming applications. Applied Sciences, 13(10), 5982. https://doi.org/10.3390/app13105982
2. Debnath, O., & Saha, H. N. (2022). An IoT-based intelligent farming using CNN for early disease detection in rice paddy. Microprocessors and Microsystems, 94, 104631. https://doi.org/10.1016/j.micpro.2022.104631
3. Bansal, P., Kumar, R., & Kumar, S. (2021). Disease detection in apple leaves using deep convolutional neural network. Agriculture, 11(7), 617. https://doi.org/10.3390/agriculture11070617
4. Liu, B., Zhang, Y., He, D., & Li, Y. (2017). Identification of apple leaf diseases based on deep convolutional neural networks. Symmetry, 10(1), 11. https://doi.org/10.3390/sym10010011
5. Zhong, Y., & Zhao, M. (2019). Research on deep learning in apple leaf disease recognition. Computers and Electronics in Agriculture, 168, 105146. https://doi.org/10.1016/j.compag.2019.105146
6. Khan, A. I., Quadri, S., Banday, S., & Shah, J. L. (2022). Deep diagnosis: A real-time apple leaf disease detection system based on deep learning. Computers and Electronics in Agriculture, 198, 107093. https://doi.org/10.1016/j.compag.2022.107093
7. Bonkra, A., Bhatt, P. K., Rosak-Szyrocka, J., Muduli, K., Pilař, L., Kaur, A., Chahal, N., & Rana, A. K. (2023). Apple leave disease detection using collaborative ML/DL and artificial intelligence methods: Scientometric analysis. International Journal of Environmental Research and Public Health, 20(4), 3222. https://doi.org/10.3390/ijerph20043222
8. Liu, B., Ding, Z., Tian, L., He, D., Li, S., & Wang, H. (2020). Grape leaf disease identification using improved deep convolutional neural networks. Frontiers in Plant Science, 11. https://doi.org/10.3389/fpls.2020.01082
9. Xie, X., Ma, Y., Liu, B., He, J., Li, S., & Wang, H. (2020). A Deep-Learning-Based Real-Time detector for grape leaf diseases using improved convolutional neural networks. Frontiers in Plant Science, 11. https://doi.org/10.3389/fpls.2020.00751
10. Javidan, S. M., Banakar, A., Vakilian, K. A., & Ampatzidis, Y. (2022). Diagnosis of grape leaf diseases using automatic K-means clustering and machine learning. Smart Agricultural Technology, 3, 100081. https://doi.org/10.1016/j.atech.2022.100081
11. Hasan, M. A., Riana, D., Swasono, S., Priyatna, A., Pudjiarti, E., & Prahartiwi, L. I. (2020). Identification of grape leaf diseases using convolutional neural network. Journal of Physics Conference Series, 1641(1), 012007. https://doi.org/10.1088/1742-6596/1641/1/012007
12. Pachade, R. S., Sharma, D. A., & Patil, D. M. (2023). Grape leaf disease detection using deep learning based VGG16 model. European Chemical Bulletin, 12(5).

13. Murugan, B., Rajagopal, M. K., & Roy, D. (2021). IoT based smart agriculture and plant disease prediction. Journal of Physics Conference Series, 2115(1), 012017. https://doi.org/10.1088/1742-6596/2115/1/012017
14. Mohanty, S. P., Hughes, D. P., & Salathé, M. (2016). Using deep learning for Image-Based plant disease detection. Frontiers in Plant Science, 7. https://doi.org/10.3389/fpls.2016.01419
15. Balram, G., & Kumar, K. K. (2022). Crop field monitoring and disease detection of plants in smart agriculture using Internet of Things. International Journal of Advanced Computer Science and Applications, 13(7). https://doi.org/10.14569/ijacsa.2022.0130795
16. Chowdhury, M. E. H., Rahman, T., Khandakar, A., Ayari, M. A., Khan, A. U., Khan, M. S., Al-Emadi, N., Reaz, M. B. I., Islam, M. T., & Md Ali, S. H. (2021). Automatic and reliable leaf disease detection using deep learning techniques. AgriEngineering, 3(2), 294–312. https://doi.org/10.3390/agriengineering3020020
17. Bhangale, P. P., Patil, Y. S., & Pati, D. D. (2017). Improved crop yield prediction using neural network. IJARIIE, 3(2). https://ijariie.com/FormDetails.aspx?MenuScriptId=3429
18. Patel, K., & Patel, H. B. (2023). Multi-criteria agriculture recommendation system using machine learning for crop and fertilizers prediction. Current Agriculture Research Journal, 11(1). https://www.agriculturejournal.org/volume11number1/multi-criteria-agriculture-recommendation-system-using-machine-learning-for-crop-and-fertilizers-prediction/
19. Hossain, M. D., Kashem, M. A., & Mustary, S. (2023). IoT based smart soil fertilizer monitoring and ML based crop recommendation system. Conference: 2023 International Conference on Electrical, Computer and Communication Engineering (ECCE). https://doi.org/10.1109/ecce57851.2023.10100744
20. Thendral, R., & Vinothini, M. (2023). Crop and fertilizer recommendation to improve crop yield using deep learning□. https://scholar.google.com/citations?view_op=view_citation&hl=en&user=tjyAGtMAAAAJ&citation_for_view=tjyAGtMAAAAJ:ufrVoPGSRksC

11 Precision Agriculture
Navigating the Future of Farming with Artificial Intelligence

Peace Busola Falola, Joseph Bamidele Awotunde, Abidemi Emmanuel Adeniyi, and Folashade Titilope Ogunajo

11.1 INTRODUCTION

Agriculture is the study, practice, and science of plant growth. Agriculture has a long history dating back thousands of years. Depending on the topography and climate, it started separately in different areas of the planet (Raj et al., 2021). Thanks to agriculture, the number of people in Earth's population has grown well beyond what could be sustained by collecting and hunting for food. The characteristics of systems that evolve over time and space encompass the information relevant to the agriculture sector (Raj et al., 2021). It is necessary to take into account hundreds if not thousands, of factors to comprehend how to manage agricultural operations (Linaza et al., 2021). Consequently, one of the hardest industries to quantify statistically is agriculture. Even inside a single field, conditions are always changing from one area to another. There is always a chance of pests and illnesses, the soil's quality fluctuates, and the weather is unpredictable. The knowledge and expertise of farmers have historically governed many of these traits (Linaza et al., 2021). Even if people believe that there is a strong chance of a harvest soon, the results are never guaranteed until harvest day (Linaza et al., 2021).

In today's agricultural world, precision agriculture is crucial. It improves farming efficiency and agricultural production by enabling farmers to make well-informed decisions based on current information. Precision agriculture is a modern farming management approach that uses technology to measure, track, and respond to variability in the field. Precision farming is an agricultural method that uses technology to monitor, assess, and evaluate the requirements of specific areas and crops (Raj et al., 2021). According to Linaza et al. (2021), precision agriculture (PA) is considered a crucial technical advancement that facilitates the optimal utilization of agricultural resources. Precision agriculture aims to increase farmers' profits by reducing input costs and improving harvest and/or quality yields. It also aims to mitigate the negative environmental effects of farming practices, such as the overuse of pesticides and fertilizers and inadequate watering (Linaza et al., 2021). The primary goal of this farming style, in contrast to traditional farming, is to maximize crop yields and profitability by precisely using inputs (Raj et al., 2021). This strategy is essential for creating more resilient food systems since it lowers farming operations' waste and carbon footprint while promoting sustainable agriculture practices. Precision agriculture offers a means of producing more effectively and sustainably, which is crucial for future food security as issues like climate change and population development put increasing pressure on agricultural production.

Utilizing contemporary information and communication technology, precision farming, or precision farming, seeks to increase agricultural yields and revenue (Beloev et al., 2021). Normalized difference vegetation index (NDVI) and photosystems, satellite systems, online weather forecasts, sensor networks and systems, and weed detection systems are some of the technologies used in precision agriculture (Beloev et al., 2021). Precision agriculture can thus give farmers information about the resources required, as well as guidance on how much and when to apply them

 DOI: 10.1201/9781003507390-11

(Beloev et al., 2021). However, the data these solutions provide must first be analyzed to generate agronomic recommendations or instructions. This focused strategy reduces the impact on the environment and conserves resources by precisely applying herbicides, fertilizers, and water. Precision farming also results in a general reduction in the investments and resources required to cultivate crops, including arable land, water, fertilizers, insecticides, and herbicides. Precision farming has improved productivity in modern agriculture since the 20th century (Raj et al., 2021). Precision farming is an agricultural method that uses technology to monitor, assess, and evaluate the requirements of specific areas and crops (Raj et al., 2021).

Artificial intelligence (AI) is quickly emerging as a key component of precision agriculture, completely changing the way we oversee crop production and farm operations. Farmers may now more effectively than ever examine a large amount of information, from crop health and resource utilization to weather patterns and soil conditions, by utilizing AI. This makes it possible to develop prediction models that help guide choices and improve agricultural techniques for increased yields at reduced costs. Precision agriculture uses cutting-edge technologies like artificial intelligence and the Internet of Things (IoT) to increase the efficacy, productivity, and viability of agricultural practices (Krishnababu et al., 2024). Through predictive analytics and machine learning models, AI improves decision-making by facilitating accurate yield prediction, disease diagnosis, and crop and soil monitoring (Krishnababu et al., 2024). Artificial intelligence technology, such as machine learning algorithms, may analyze data from field sensors, satellites, and drones to identify plant illnesses, forecast pest incursions, and evaluate crop health at a finer scale. This feature helps farmers to use far less water, fertilizer, and pesticide by allowing them to administer tailored treatments exactly where they are required. Farmers may now use a data-driven strategy to improve agricultural productivity and lower risks from diseases, weeds, and pests by gathering and evaluating a ton of data about the present condition of their farms. The advent of new technology developments, such as artificial intelligence, makes this conceivable (Linaza et al., 2021). Some variables, including environmental conditions, soil properties, water availability, harvesting techniques, plant diseases, weeds, and other pests, influence agricultural systems, where the development of intelligent sensors, instrumentation, and machinery is starting to play a critical role (Linaza et al., 2021). Cutting-edge tools for precision agriculture will soon be available because of the integration of artificial intelligence algorithms, automated data collection and analysis, and decision support systems. In addition, robotic systems—both terrestrial and aerial—are advancing precision agriculture and its digitalization. AI-driven robotics, such as self-driving tractors and harvesters, are transforming field operations by executing a range of activities with exceptional precision and independence, including weeding, planting, pest management, harvesting, and data collecting (Linaza et al., 2021). These developments not only increase productivity but also lessen the effects of the manpower shortage in the agricultural industry.

All things considered, AI is augmenting the potential of precision agriculture by offering practical insights, mechanizing labor-intensive procedures, and strengthening the sustainability of farming methods, thereby facilitating a transition towards a more efficient and sustainable agricultural sector.

11.1.1 Key Contributions of This Chapter

The primary contributions of this work are:

1. This chapter describes how improving operational efficiency and customizing agricultural methods are two ways that combining AI and precision agriculture may revolutionize farming operations.
2. Additionally, this chapter makes recommendations for future study directions to deepen AI's integration with precision agriculture.

11.1.2 Chapter Organization

Section 11.2 presents the evolution of precision agriculture. Section 11.3 discusses the fundamentals of AI in agriculture. Section 11.4 discusses applications of AI in precision agriculture. Section 11.5 discusses the challenges and solutions. Section 11.6 discusses the future of AL in precision agriculture, and Section 11.7 presents the conclusion for the chapter.

11.2 THE EVOLUTION OF PRECISION AGRICULTURE

By utilizing data and modern technology, precision agriculture is a management approach that takes into account temporal and geographical variability in agricultural fields (Karunathilake et al., 2023). The significance of precision agriculture in modern agricultural research is growing, given that by 2050, there will be between 9 and 10 billion people on the planet (Karunathilake et al., 2023). The amount of food produced globally must increase by at least 70% by 2050 to fulfill this rising demand. But this is a difficult undertaking since it further taxed the environment and already limited resources (Karunathilake et al., 2023). Precision farming is, therefore, necessary to ensure sustainability, minimize negative environmental effects, and optimize yield while utilizing fewer inputs more effectively (Karunathilake et al., 2023). With the advent of yield monitors, geographic information systems (GISs), global positioning systems (GPSs), and other data producers in all three critical stages of agricultural activities, precision farming emerged in the 1990s (Karunathilake et al., 2023). At first, motorized machinery was only utilized for agricultural tasks, and people had to approve the problem-identifying and decision-making procedures (Karunathilake et al., 2023). The Third Industrial Revolution, or Industry 3.0, moved precision agriculture closer to digitalization by combining information technology with improved automated capabilities. Therefore, "farm practices" employing hand tools evolved into "agriculture" from animal traction to motorized mechanization to digital technology (Karunathilake et al., 2023).

According to McFadden et al. (2023), the primary tools used in precision agriculture up to this point include yield monitors, guided farming systems, electronic maps, and variable rate technologies (VRTs). Germany and Denmark were the first countries to demonstrate variable-rate uses of GPSs, which were made available to the civil services in 1988. In 1983, GPS services were widely available for usage on American farms (McFadden et al., 2023). During the following ten years, GPS technology gave farmers the ability to accurately map and identify their fields, giving them the flexibility to manage their farms based on site-specific circumstances and field variability. With the invention of yield monitors at the start of the new century, farmers could track agricultural production in real time using best matching (Hundal et al., 2023). Farmers may now collect high-resolution data on their fields and make educated decisions about crop management because of the advent of remote sensing technologies, which include workers, drones, satellites, and ground-based sensors (Karunathilake et al., 2023). Precision agriculture focuses on crop farming as well as other agricultural production systems including agronomics, animal husbandry, aquaculture, and agroforestry (Karunathilake et al., 2023).

Many problems plague precision agriculture today, including inefficient use of resources, prolonged monoculture, intensive animal husbandry, environmental degradation, unequal digitization distribution, problems with food safety, an unsatisfactory agri-food supply chain, ignorance, and reluctance to try new things (Singh, 2024). These issues make inadvertent impacts on ecosystems worse and make it more difficult to achieve agricultural production goals of productivity, sustainability, and efficiency (Singh, 2024). The introduction of big data, robotics, artificial intelligence, blockchain technology, and the Internet of Things in 2011 marked the beginning of the fourth industrial revolution, or Industry 4.0 (Singh, 2024). To address the aforementioned problems, these cutting-edge technologies were incorporated into agriculture in 2017, changing precision farming into Agriculture 4.0, or smart farming (Karunathilake et al., 2023). As a result of this shift, farmers are using precision agricultural technology to lessen their farming's negative environmental effects and to advance long-term sustainability (Karunathilake et al., 2023). Because of this, agricultural supply chains and industrial processes have developed into more intelligent and autonomous systems. This includes the automation of several activities including soil sampling, planting, seeding, and harvesting. This lowers labor expenses and increases farming efficiency.

11.2.1 Role of AI in Agriculture

According to Abioye et al. (2022) and Liu et al. (2020), the primary characteristic of the agricultural sector is a fluctuation in the geographical and temporal aspects in system attributes. Many factors, numbering in the hundreds or even thousands, must be understood to handle agricultural operations effectively. Put another way, one of the industries where numerical statistic analysis is most difficult is agriculture. Particular areas of the same land continue to have disparate conditions.

The characteristics of systems that change over time and space include the information of importance in the agriculture sector (Karunathilake et al., 2023). Comprehending agricultural process management requires taking into account several hundred, if not thousands, of factors. Consequently, one of the hardest industries to quantify statistically is agriculture. Even within one field, conditions are continually changing from one area to another. There is always a chance of pests and illnesses, the soil's quality fluctuates, and the weather is unpredictable (Rozenstein et al., 2024). Historically, the knowledge and expertise of the farmers have governed many of these traits. Even if they believe that there is a strong chance of a harvest shortly, the result is never assured until harvest day.

These days, precision agriculture is regarded as a crucial technical advancement that makes it possible to use agricultural resources more effectively. The primary goals of PA are to increase farmers' revenue by raising harvest and/or quality yields while decreasing input costs, as well as to lessen the detrimental impacts of farming on the environment, such as those brought on by the overuse of fertilizers and pesticides and inefficient irrigation. Intelligent machinery, sensors, and instrumentation are beginning to play a bigger role in agricultural systems. Numerous elements, including the climate, soil characteristics, water availability, harvesting methods, plant diseases, weeds, and other pests, influence these systems.

Precision agriculture will have access to cutting-edge instruments thanks to the combination of AI algorithms, automated data collecting and analysis, and decision support systems. Furthermore, the integration of robotic systems both terrestrial and aerial will play a significant part in advancing digitalization and precision agriculture in the field, including tasks like data collecting, pest management, and harvesting. To improve decision-making processes through data-driven insights and monitor farm products, artificial intelligence is being integrated into agriculture through automating data collection, analyzing large data using machine learning techniques, and establishing human-computer interfaces. Figure 11.1 shows the process of AI adoption in agriculture. Farmers can now take a data-driven approach to improve farm yield and lower risks from weeds, pests, and diseases. They can collect and analyze large amounts of data about the current condition of their fields, apply artificial intelligence to monitor their produce, receive signals, take appropriate action, and predict a good harvest. The advent of new technology developments, such as artificial intelligence, makes this conceivable.

AI has been used in agriculture to improve overall production, optimize resource consumption, and strengthen decision-making processes (Sood et al., 2022). This shift, also referred to as precision agriculture or data-driven farming, gives farmers the ability to make well-informed decisions by using knowledge gleaned from enormous volumes of data gathered on the farm (Karunathilake et al., 2023). Through precise resource management and environmental impact reduction, artificial intelligence technologies are important in tackling the complexity of agricultural data, enhancing decision support systems, and allowing sustainable agriculture practices.

Because AI algorithms are made to evaluate complicated agricultural data, uncover hidden patterns, and forecast results, they play a critical role in data-driven agriculture. This allows for proactive decision-making based on actionable insight (Supriya et al., 2024). By utilizing data-driven

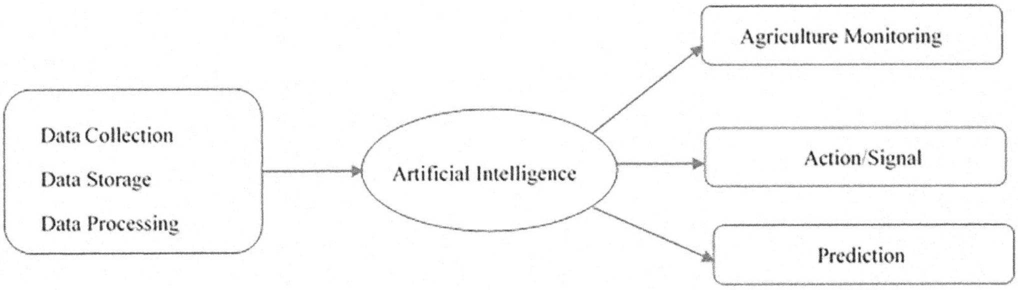

FIGURE 11.1 Process of AI adoption in agriculture.

insights to increase sustainability and production, artificial intelligence in agriculture helps farmers optimize practices, enhance efficiency, and address difficulties (Patil et al., 2024). By providing farmers with up-to-date information on crop conditions, weather patterns, soil health, and machinery performance, data-driven farming—powered by artificial intelligence and data analytics—is revolutionizing the agricultural industry and enabling farmers to make data-driven decisions for sustainable practices and optimized yields (Velamuri, 2023). With data insights enabling more informed decision-making, data-driven farming highlights the role AI technologies play in maximizing resource use, boosting production, and advancing sustainability in agriculture. AI-driven agricultural solutions reduce environmental impact, promote sustainable farming methods for future generations, and allow farmers to make proactive management decisions. Examples of these solutions include real-time monitoring, predictive analysis, and precision farming. Smarter, more effective, and sustainable farming techniques are being made possible by the industry's transformation brought about by the use of AI technologies in agriculture, such as machine learning, IoT devices, and precision farming (Karunathilake et al., 2023).

11.3 FUNDAMENTALS OF AI IN AGRICULTURE

The application of artificial intelligence in agriculture is a revolutionary development that aims to improve farming operations' productivity, efficiency, and sustainability. To improve farm management techniques, artificial intelligence in agriculture applies machine learning methods, computer vision, robots, and data analytics. Improving decision-making procedures, automating tedious work, and improving resource management are the main objectives. Artificial intelligence systems examine enormous volumes of data to generate actionable insights on everything from crop health and market trends to weather and soil issues.

11.3.1 AI Technologies

The use of artificial intelligence technology in robotics, computer vision, and machine learning has become essential to precision agriculture, or "smart farming." This creative method has become a vital instrument for tackling the problems associated with sustainable agriculture. This technology is primarily driven by robotic systems and AI approaches, including machine learning (ML). The application of machine learning, which enables computers to learn without the need for explicit programming, has greatly enhanced agricultural operations (Sharma et al., 2020; Saleem et al., 2021).

The Internet of Things and machine learning are essential for improving agricultural methods. Applications of machine learning have demonstrated potential. in forecasting agricultural yields as well as soil properties like moisture content and organic carbon. Additionally, they support the identification of diseases, weeds, and species in crops (Sharma et al., 2020). Moreover, machine learning and computer vision are combined to categorize various crop picture sets, which aids in yield assessment and crop quality monitoring (Sharma et al., 2020). This method is used for the production of livestock, where ML models are used to study bovine behavior, diagnose eating problems, and forecast reproductive patterns based on information from collar sensors (Sharma et al., 2020).

Human work is greatly reduced by technologies such as intelligent harvesting techniques and irrigation, including drip irrigation (Sharma et al., 2020). Predictive analytics in precision agriculture also heavily relies on machine learning. It forecasts future conditions including weather effects, insect infestations, and agricultural yields by using past data and algorithms. To evaluate crop status and soil health, algorithms also evaluate data from sensors and aerial photos. This allows for prompt actions that maximize growth circumstances (Sharma et al., 2020).

The demand on agricultural systems grows with global population growth and urbanization reduces the amount of land that can be farmed. The need for safe and efficient agricultural food production techniques is growing (Tian et al., 2020). Innovative sensing technologies and enhanced information and communication technology must be used in conjunction with traditional agriculture management. According to Tian et al. (2020), these improvements are essential for accurately boosting agricultural output and fostering the expansion of premium, productive farming.

Computer vision inspection systems have grown indispensable in agricultural operations in the last few decades. In agricultural production management, the use of intelligent and expert systems built on computer vision algorithms is becoming more widespread. Computer vision technology has become much more powerful thanks to technological advancements like deep belief networks (DBNs) and graphics processing units (GPUs). This development ensures the effectiveness of agricultural production by raising productivity and efficiency while also offering insightful information for decision support (Tian et al., 2020). As a result, computer vision technology will keep being essential to agriculture's automation and will gradually advance the shift to intelligent Agriculture 4.0 (Tian et al., 2020).

11.3.2 Data Sources

Different kinds of data are essential for optimizing farming operations in AI-driven agricultural practices. According to Yang (2020), certain data sets, such as satellite images, offer important insights into crop health, growth patterns, and possible problems like insect infestations or illnesses. Artificial intelligence systems are capable of analyzing satellite photos to track crop conditions, spot stressed regions, and allocate resources optimally for higher yields (Ilyas et al., 2023). Real-time data on temperature, humidity, soil moisture content, and other environmental parameters may be obtained via sensor data gathered from Internet of Things devices placed in fields (Lloret et al., 2021). AI systems may increase crop health and resource efficiency by using this data to make intelligent decisions regarding pest management, fertilizer use, and irrigation scheduling (Javaid et al., 2023). According to Hatfield et al. (2020), weather data is crucial for anticipating and controlling how climatic conditions would affect agricultural activity. AI models improve planting schedules, irrigation plans, and harvesting operations by utilizing real-time forecasts and historical meteorological data (Ashraf & Akanbi, 2023; Khan et al., 2023). Farmers may reduce the dangers connected with extreme weather occurrences and adopt preemptive action to safeguard their crops by incorporating meteorological data into AI-driven systems.

Farmers may enhance crop management practices, expedite decision-making processes, and eventually increase agricultural output and sustainability by using artificial intelligence to satellite photos, sensor data, and meteorological data.

11.3.3 AI Models and Algorithms in Precision Agriculture

A range of AI models and algorithms are used in precision farming to process data, forecast outcomes, and automate processes. These AI solutions support resource optimization, environmental impact reduction, and crop production maximization. By evaluating enormous volumes of data gathered from sensors, drones, and satellites, machine learning algorithms are essential to precision agriculture (Shaikh et al., 2022). Based on predictive analytics and historical data, these algorithms let farmers make well-informed decisions about fertilization, watering, and harvesting (Engen et al., 2023). With the least amount of human effort, smart farming prediction models increase crop yield and quality by automating agricultural techniques (Akkem et al., 2023). These models use machine learning approaches to improve decision-making and make agricultural operations more sustainable and efficient.

Software platforms that automate procedures, provide remote monitoring and control, and offer actionable insights are used to incorporate these AI models and algorithms into different farm management systems. The particular application, the kind and quality of data at hand, and the available computer resources all influence the method selection. Precision farming has benefited from more individualized, data-driven management of agricultural operations due to the success of these models, which has increased sustainability and production (Linaza et al., 2021). The precision and accuracy of these models keep getting better as AI technology advances, providing bright prospects for the future of agriculture. These innovations have the potential to change the agriculture sector and guarantee productive and sustainable farming methods going forward.

To categorize the crops into four seasons, several writers, as Katarya et al. (2020), examined the effectiveness of four widely used models: random forest, logistic regression, naive Bayes, and decision table. Using the datasets for the Indian states of Karnataka and Uttar Pradesh, the authors computed the models' performance in terms of accuracy, root relative square error (RRSE), mean absolute error (MAE), relative absolute error (RAE), and root mean square error (RMSE). A thorough examination of machine learning applications in the agriculture industry was presented by Sharma et al. (2020). They focused on forecasting agricultural yield, identifying weeds and crop diseases, identifying species, and forecasting soil parameters such as moisture content and organic carbon. Machine learning and computer vision techniques were used to classify a fresh set of crop pictures to monitor crop quality and yield assessment. This technique may be implemented for higher livestock output by combining data from collar sensors to anticipate reproductive patterns, identify eating disorders, and assess cow behavior using machine learning algorithms. A significant amount of human work is eliminated with intelligent irrigation, which includes drip irrigation and intelligent harvesting systems. Some of the common AI models and algorithms used in precision agriculture are discussed in the following.

11.3.3.1 Regression Models

The basis of agricultural predictive modeling is regression algorithms. Based on different input features, they are used to forecast continuous outcomes including crop yields, soil nutrient levels, and growth rates. Because it is so easy to use, linear regression is frequently employed to predict outcomes based on linear connections between variables. For example, it may forecast agricultural production according to variables like average temperature and rainfall (Dharmaraja et al., 2020; Kang et al., 2020). When there is a nonlinear relationship between the input and output variables, such as when forecasting crop yields that fluctuate significantly with small changes in the pH levels of the soil, polynomial regression is helpful (Ahmed et al., 2022).

11.3.3.2 Decision Trees and Random Forests

These models work especially well for applications involving regression and classification in intricate agricultural datasets with plenty of variables. By organizing scenarios into a model that resembles a tree, decision trees assist in the decision-making process for farm management techniques (Suganya, 2020). They can use past yield data to determine crop rotation or categorize different types of soil according to their properties. A collective method known as random forests uses many decision trees to lower overfitting and boost prediction accuracy (Feng et al., 2020). It may be used to forecast disease outbreaks or pest infestations and is resistant to noise in agricultural data (Domingues et al., 2022).

11.3.3.3 Support Vector Machines

Support vector machines (SVMs) work well for tasks involving classification including determining plant species from photos or differentiating between healthy and unhealthy plants (Hatuwal et al., 2020). They are appropriate for image-based monitoring in precision farming as they perform well with high-dimensional data (Sugumar & Suganya, 2023).

11.3.3.4 Neural Networks and Deep Learning

Precision agriculture uses deep learning models, especially convolutional neural networks (CNNs), for picture analysis. Convolutional neural networks, or CNNs, are perfect for handling the pixel data from satellite and farm photographs. They may be used to track crop and soil conditions, identify weeds, and assess crop health (Hashemi-Beni et al., 2022). When dealing with sequential data, such as weather patterns across time, recurrent neural networks (RNNs) are helpful because they can forecast how the weather will affect agricultural output in the future (Engen et al., 2023).

11.3.3.5 Clustering Algorithms

In precision agriculture, clustering is utilized to identify patterns and groups in data without first categorizing the individual data points. To apply variable rate inputs, agricultural fields can be divided into zones depending on crop performance or soil characteristics using K-means clustering (Filintas et al., 2023). A hierarchy of field zones may be formed using hierarchical clustering, which facilitates in-depth analysis and distinct management strategies.

11.3.3.6 Time Series Forecasting

For the analysis of temporal data, such as weather information or trends of crop development across time, time series models are essential. When predicting future values in a time series, such as rainfall or temperature patterns that influence agricultural choices, the autoregressive integrated moving average (ARIMA) is utilized (Praveen & Sharma, 2020).

11.4 APPLICATIONS OF AI IN PRECISION AGRICULTURE

AI in agriculture can lead to several technical developments. This covers, among other things, data analytics, the Internet of Things, cameras and other sensors, and consultancy services (Javaid et al., 2023). AI in agriculture will advance to the point where it can analyze a variety of data sources, including temperature, crop performance, weather, and soil, to generate more accurate forecast predictions. This study investigated the possible risks to agriculture, such as interoperability, safety and security, data dependability, and unanticipated socio-ecological implications, to maximize yields using machine learning models (Javaid et al., 2023). Through the timely detection of plant diseases and the efficient use of agrochemicals, artificial intelligence has the potential to improve crop management and production in agriculture. Machine learning can be helpful for yield prediction, soil composition analysis, rapid plant phenotyping, agricultural monitoring, and weather prediction, according to Javaid et al. (2023). An increasing number of farmers are employing AI, IoT, and other technological innovations to increase the productivity of their land (Javaid et al., 2023). Figure 11.2 illustrates a few of the major uses of AI in agriculture. Emerging AI-powered technologies are going to drastically alter the agriculture sector. Global farms may benefit from AI and cognitive technology by being able to make better judgments, process data more efficiently, automate labor-intensive tasks, and boost output. It is now feasible to use enormous amounts of data to increase productivity and sustainability in agriculture because of important advancements in computers and technology as well as societies' faith in machine learning. For a variety of reasons, the agriculture sector has opted to employ AI technology to enhance decision-making (Javaid et al., 2023). The data's increased accessibility and the observable increase in its available quantity are the most notable developments. This is made possible by developments in the sector, such as increased usage of sensors, speedier access to satellite images, more inexpensive data loggers, more use of drones, and simpler access to public data

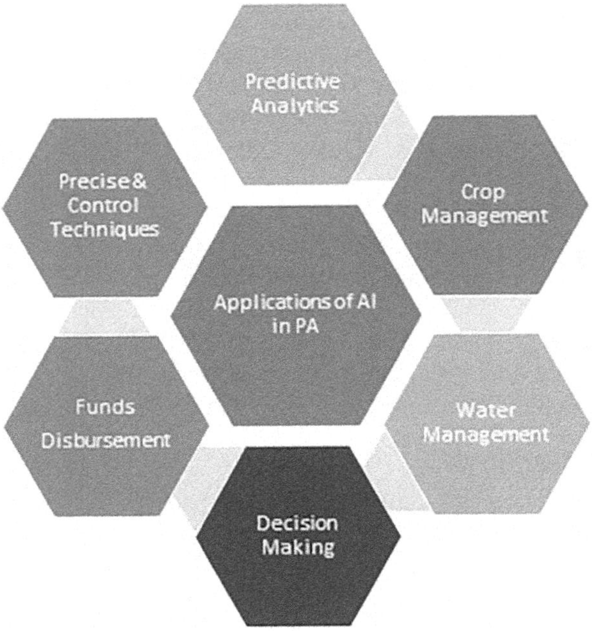

FIGURE 11.2 Applications of artificial intelligence in precision agriculture.

repositories. Banks can evaluate farmers' credit and agricultural experience using AI in agriculture, especially in finance, before granting loans. Because smart farming technology promotes openness during the pre-disbursement phase, it benefits both farmers and financial institutions. Using information about the farms' past performance and present production potential for the given period, banks are able to calculate the projected yield from each plot and supply mortgages with a reduced risk (Javaid et al., 2023). AI is still in its infancy as it relates to agriculture. Before the data it gathers can be utilized to improve decisions about planting, irrigation, input application, and harvesting, some time will need to pass. Both the amount and caliber of training data are inadequate. To identify pests or plant diseases in user-generated photos, artificial intelligence software based on computer vision has to be trained on a large and diverse collection of pest photographs. These images typically feature varied camera angles, lighting setups, and backdrops. It may be feasible to create scalable and useful AI applications for agriculture by gaining access to current farming knowledge (Javaid et al., 2023). An extensive collection of relevant images and data may be used to train the AI model with the assistance of a network of farmers and specialists. The resultant data may then be utilized to improve the knowledge and experience of farmers, helping them make better judgments about which pesticide formula to use and how much to spray at what stage of an infestation. This provides high spatial and temporal resolution techniques and analyses of multisource data for agricultural production management. Agriculture's water management system needs sufficient, timely, and trustworthy data—all necessary for the implementation of successful AI solutions. Richer contextual data might be gathered with the use of participatory and crowdsourced data collection methods. Farmers can use low-tech instruments and training to measure soil moisture and gather data on groundwater through participatory mapping. Hyper-local information on rainfall and other meteorological indicators may be obtained using a variety of remote sensing methods (Javaid et al., 2023).

11.5 CHALLENGES AND SOLUTIONS

To fully realize the promise of AI in precision agriculture, several non-technical and technical obstacles must be overcome. One of the primary technical challenges is ensuring the data required for AI model training is trustworthy and of a high caliber. Model performance can be greatly impacted by variables such as sensor accuracy, data labeling, and treatment of noisy or missing data (Son et al., 2024). Furthermore, AI models need to be resilient enough to manage the intricacy and unpredictability of agricultural systems, taking into consideration elements like crop kinds, soil types, and weather patterns. Another issue is algorithmic bias, which arises from the possibility that AI systems would reinforce preexisting biases in the training data (Ntoutsi et al., 2020).

Additionally, impeding the widespread use of artificial intelligence in precision agriculture are adoption hurdles (Karunathilake et al., 2023). Many farmers, particularly smallholders and those in poor nations, may find the high cost of AI technology, such as sensors, drones, and data processing infrastructure, unaffordable (Javaid et al., 2023). The use of AI-powered products may potentially be slowed down by farmers and agricultural workers' lack of technological expertise. Investments in accessible and inexpensive AI solutions are necessary to overcome these obstacles, as are educational and training initiatives to increase digital literacy and AI proficiency in the agriculture industry (Aggarwal et al., 2024).

When using AI in farming, ethical and environmental factors must also be taken into account (Dara et al., 2022). As artificial intelligence systems often require collecting and evaluating enormous amounts of data from several sources, data security and privacy pose significant difficulties. In the absence of sufficient safeguards and clear policies about data ownership and usage, farmers would be hesitant to reveal confidential data, including financial records or yield figures (Jouanjean et al., 2020). AI technologies' effects on the environment, such as data centers' energy usage and the possibility of higher chemical use as a result of more focused targeting, must also be carefully studied and minimized.

A complex strategy incorporating stakeholder participation, legislative initiatives, and technological advances is required to overcome these difficulties. To ensure that AI is developed and used ethically in agriculture, policymakers may create rules and laws that address concerns like algorithmic bias, data privacy, and environmental sustainability (Dara et al., 2022). Farmers and agricultural workers can be better equipped to make informed decisions by participating in education and training programs that help them comprehend the advantages and constraints of AI technology (Lakshmi & Corbett, 2020). Technological advancements including open-source software, inexpensive sensors, and edge computing might increase AI's accessibility and adaptability to many agricultural situations (Chang et al., 2021). The agriculture industry may use AI to achieve sustainable intensification and guarantee food security for an expanding global population by cooperating to solve these obstacles (Ahmad et al., 2024).

11.6 THE FUTURE OF AI IN PRECISION AGRICULTURE

Changes in the application of AI in precision agriculture is about to occur, fueled by new ideas, scalable solutions, and a vision that supports sustainable practices and global food security. Innovations like hyper-personalized farming, where sophisticated AI algorithms enable personalized farming practices that take into account microclimatic conditions, individual plant health, and particular soil properties to tailor farming strategies accordingly, are leading the way in reshaping precision agriculture through AI. By evaluating genetic data to forecast plant features and improve breeding for desirable qualities like drought tolerance or enhanced yield, the integration of AI with genomics is changing agricultural breeding (Xu et al., 2022). Simultaneously, the advancement of increasingly complex AI-enabled autonomous robots is improving farming operations through autonomous execution of activities like planting, weeding, and harvesting, leading to a rise in production and operational efficiency (Mahmud et al., 2020).

AI at the edge enables live data processing and choice-making without requiring constant cloud connectivity by being implemented locally on devices like agricultural equipment and sensors (Alonso et al., 2020). Improving the operating capacities of remote and rural farm settings is especially important. Global food security may be significantly impacted by AI's ability to increase agricultural production and sustainability in a range of sizes. Javaid et al. (2023) claim that machine learning models for precision planting, pest control, and harvesting, along with drones for aerial imaging, make it possible to manage large regions more effectively using artificial intelligence. With the use of mobile applications, AI technologies may be scaled down and made more inexpensive for smallholder farmers. Providing them with access to essential information like weather forecasts, market prices, and instruments for identifying crop diseases, will boost output and the economy's resilience.

Additionally, AI improves the global food supply chain through AI-driven logistics that lower transportation waste and enhance the distribution of food supplies around the world, as well as predicting crop yields that guide global commodities markets (Pandey & Mishra, 2024). Future farming might be drastically changed by AI, with an emphasis on resilience, efficiency, and sustainability. By using precise application techniques, AI-driven technologies are anticipated to maximize resource utilization and decrease the effect of farming operations on the environment (Akter, 2024). AI also simplifies farm management by optimizing choices based on predictive analytics, such as crop rotation and hybrid selection, and automating jobs. Farms may become more robust to climate variability and extreme weather events by utilizing AI for predictive analytics and live monitoring. This can also equip farms with proactive mitigation methods (Singh & Goyal, 2023).

In conclusion, it should be noted that artificial intelligence technologies have enormous potential to enhance food production and tackle some of the most significant problems that modern global agriculture is confronting, such as food protection, sustainability, and climate change adaptation. These technologies will only grow and become more integrated into the agricultural sector. Future development policies, more accessibility to AI technologies, and ongoing innovation are all necessary for this ambition.

11.7 CONCLUSION

Using AI to solve challenges and boost production and sustainability, this chapter concludes with a revolutionary proposal for agriculture. Artificial intelligence incorporated into precision agricultural methods signifies a fundamental shift away from technology and toward data-driven agriculture, with the promise of enhanced decision-making, efficiency, and resource management. Based on developments that use state-of-the-art technologies like genomics and autonomous robots to tailor agricultural techniques to particular crop needs and weather circumstances, the future of AI in agriculture is depicted as a critical component in tackling global food security challenges. The remarkable ramifications for all stakeholders farmers, technologists, and policymakers in particular indicate the necessity of working together to eliminate barriers and fully realize artificial intelligence's potential in farming. This will encompass the development of new technologies as well as the construction of legislative frameworks that support data privacy and long-term economic viability. Future research directions are advised to focus on refining AI models and algorithms for increased precision and exploring new areas where AI can further benefit agriculture, like integrating more IoT devices and enhancing predictive analytics, to ensure that AI-driven agriculture continues to grow as a cornerstone of sustainable farming practices.

REFERENCES

Abioye, E. A., Hensel, O., Esau, T. J., Elijah, O., Abidin, M. S. Z., Ayobami, A. S., . . . & Nasirahmadi, A. (2022). Precision irrigation management using machine learning and digital farming solutions. *AgriEngineering*, *4*(1), 70–103.

Aggarwal, S., Bansal, S., & Goel, R. (2024). AI in agriculture: A looming challenge, a gleaming opportunity. *International Journal of Engineering Science and Humanities*, *14*(Special Issue 1), 43–52.

Ahmad, A., Liew, A. X., Venturini, F., Kalogeras, A., Candiani, A., Di Benedetto, G., . . . & Martos, V. (2024). AI can empower agriculture for global food security: Challenges and prospects in developing nations. *Frontiers in Artificial Intelligence*, *7*, 1328530.

Ahmed, S. E., Aydin, D., & Yilmaz, E. (2022). Semiparametric time-series model using local polynomial: An application on the effects of financial risk factors on crop yield. *Journal of Risk and Financial Management, 15*(3), 141. https://doi.org/10.3390/jrfm15030141

Akkem, Y., Biswas, S. K., & Varanasi, A. (2023). Smart farming using artificial intelligence: A review. *Engineering Applications of Artificial Intelligence, 120*, 105899.

Akter, M. S. (2024). Harnessing technology for environmental sustainability: Utilizing AI to tackle global ecological challenge. *Journal of Artificial Intelligence General Science (JAIGS), 2*(1), 61–70, ISSN: 3006-4023.

Alonso, R. S., Sittón-Candanedo, I., García, Ó., Prieto, J., & Rodríguez-González, S. (2020). An intelligent edge-IoT platform for monitoring livestock and crops in a dairy farming scenario. *Ad Hoc Networks, 98*, 102047.

Ashraf, H., & Akanbi, M. T. (2023). Sustainable agriculture in the digital age: Crop management and yield forecasting with IoT, Cloud, and AI. *Tensorgate Journal of Sustainable Technology and Infrastructure for Developing Countries, 6*(1), 64–71.

Beloev, I., Kinaneva, D., Georgiev, G., Hristov, G., & Zahariev, P. (2021). Artificial intelligence-driven autonomous robot for precision agriculture. *Acta Technologica Agriculturae, 24*(1), 48–54.

Chang, Z., Liu, S., Xiong, X., Cai, Z., & Tu, G. (2021). A survey of recent advances in edge-computing-powered artificial intelligence of things. *IEEE Internet of Things Journal, 8*(18), 13849–13875.

Dara, R., Hazrati Fard, S. M., & Kaur, J. (2022). Recommendations for ethical and responsible use of artificial intelligence in digital agriculture. *Frontiers in Artificial Intelligence, 5*, 884192.

Dharmaraja, S., Jain, V., Anjoy, P., & Chandra, H. (2020). Empirical analysis for crop yield forecasting in India. *Agricultural Research, 9*(1), 132–138.

Domingues, T., Brandão, T., & Ferreira, J. C. (2022). Machine learning for detection and prediction of crop diseases and pests: A comprehensive survey. *Agriculture, 12*(9), 1350.

Engen, M., Sandø, E., Sjølander, B. L. O., Arenberg, S., Gupta, R., & Goodwin, M. (2023). Farm-scale crop yield prediction from multi-temporal data using deep hybrid neural networks. *Agronomy, 11*(12), 2576.

Feng, W., Ma, C., Zhao, G., & Zhang, R. (2020, August). FSRF: An improved random forest for classification. In *2020 IEEE International Conference on Advances in Electrical Engineering and Computer Applications (AEECA)* (pp. 173–178). IEEE.

Filintas, A., Gougoulias, N., Kourgialas, N., & Hatzichristou, E. (2023). Management soil zones, irrigation, and fertigation effects on yield and oil content of Coriandrum sativum L. using precision agriculture with fuzzy K-means clustering. *Sustainability, 15*(18), 13524.

Hashemi-Beni, L., Gebrehiwot, A., Karimoddini, A., Shahbazi, A., & Dorbu, F. (2022). Deep convolutional neural networks for weeds and crops discrimination from UAS imagery. *Frontiers in Remote Sensing, 3*, 755939.

Hatfield, J. L., Antle, J., Garrett, K. A., Izaurralde, R. C., Mader, T., Marshall, E., . . . & Ziska, L. (2020). Indicators of climate change in agricultural systems. *Climatic Change, 163*, 1719–1732.

Hatuwal, B. K., Shakya, A., & Joshi, B. (2020). Plant leaf disease recognition using random forest, KNN, SVM and CNN. *Polibits, 62*, 13–19.

Hundal, G. S., Laux, C. M., Buckmaster, D., Sutton, M. J., & Langemeier, M. (2023). Exploring barriers to the adoption of Internet of Things-based precision agriculture practices. *Agriculture, 13*, 163.

Ilyas, Q. M., Ahmad, M., & Mehmood, A. (2023). Automated estimation of crop yield using artificial intelligence and remote sensing technologies. *Bioengineering, 10*(2), 125.

Javaid, M., Haleem, A., Khan, I. H., & Suman, R. (2023). Understanding the potential applications of artificial intelligence in agriculture sector. *Advanced Agrochem, 2*(1), 15–30.

Jouanjean, M. A., Casalini, F., Wiseman, L., & Gray, E. (2020). *Issues Around Data Governance in the Digital Transformation of Agriculture: The Farmers' Perspective*. OECD.

Kang, Y., Ozdogan, M., Zhu, X., Ye, Z., Hain, C., & Anderson, M. (2020). Comparative assessment of environmental variables and machine learning algorithms for maize yield prediction in the US Midwest. *Environmental Research Letters, 15*(6), 064005.

Karunathilake, E. M. B. M., Le, A. T., Heo, S., Chung, Y. S., & Mansoor, S. (2023). The path to smart farming: Innovations and opportunities in precision agriculture. *Agriculture, 13*(8), 1593. https://doi.org/10.3390/agriculture13081593

Katarya, R., Raturi, A., Mehndiratta, A., & Thapper, A. (2020). Impact of machine learning techniques in precision agriculture. In *2020 3rd International Conference on Emerging Technologies in Computer Engineering: Machine Learning and Internet of Things (ICETCE)* (pp. 1–6). IEEE.

Khan, A., Hassan, M., & Shahriyar, A. K. (2023). Optimizing onion crop management: A smart agriculture framework with IoT sensors and cloud technology. *Applied Research in Artificial Intelligence and Cloud Computing, 6*(1), 49–67.

Krishnababu, M. E., Devi, B. R., Soni, A., Panigrahi, C. K., Sudeepthi, B., Rathi, A., & Shukla, A. (2024). A review on precision agriculture navigating the future of farming with AI and IoT. *Asian Journal of Soil Science and Plant Nutrition, 10*(2), 336–349. https://doi.org/10.9734/ajsspn/2024/v10i2291

Lakshmi, V., & Corbett, J. (2020). How artificial intelligence improves agricultural productivity and sustainability: A global thematic analysis. *Proceedings of the 53rd Hawaii International Conference on System Sciences*, 5202–5211.

Linaza, M. T., Posada, J., Bund, J., Eisert, P., Quartulli, M., Döllner, J., . . . & Lucat, L. (2021). Data-driven artificial intelligence applications for sustainable precision agriculture. *Agronomy, 11*, 1227. https://doi.org/10.3390/agronomy11061227

Liu, Y., Zou, L., & Wang, Y. (2020). Spatial-temporal characteristics and influencing factors of agricultural eco-efficiency in China in recent 40 years. *Land Use Policy, 97*, 104794.

Lloret, J., Sendra, S., Garcia, L., & Jimenez, J. M. (2021). A wireless sensor network deployment for soil moisture monitoring in precision agriculture. *Sensors, 21*(21), 7243.

Mahmud, M. S. A., Abidin, M. S. Z., Emmanuel, A. A., & Hasan, H. S. (2020). Robotics and automation in agriculture: Present and future applications. *Applications of Modelling and Simulation, 4*, 130–140.

McFadden, J., Njuki, E., & Griffin, T. (2023). *Precision Agriculture in the Digital Era: Recent Adoption on U.S. Farms*. US Department of Agriculture, Economic Research Service 248. Available online: https://www.ers.usda.gov (accessed on 16th May, 2024).

Ntoutsi, E., Fafalios, P., Gadiraju, U., Iosifidis, V., Nejdl, W., Vidal, M. E., . . . & Staab, S. (2020). Bias in data-driven artificial intelligence systems—an introductory survey. *Wiley Interdisciplinary Reviews: Data Mining and Knowledge Discovery, 10*(3), e1356.

Pandey, D. K., & Mishra, R. (2024). Towards sustainable agriculture: Harnessing AI for global food security. *Artificial Intelligence in Agriculture, 12*, 72–84. https://doi.org/10.1016/j.aiia.2024.04.003

Patil, R., Sable, N. P., Mahalle, P. N., Shinde, G. R., Dhotre, P., & Chandre, P. (2024). Role of Data-centric artificial intelligence in agriculture. In *Data-Centric Artificial Intelligence for Multidisciplinary Applications* (pp. 159–176). Chapman and Hall/CRC.

Praveen, B., & Sharma, P. (2020). Climate variability and its impacts on agriculture production and future prediction using autoregressive integrated moving average method (ARIMA). *Journal of Public Affairs, 20*(2), e2016.

Raj, E. F. I., Appadurai, M., & Athiappan, K. (2021). Precision farming in modern agriculture. In Choudhury, A., Biswas, A., Singh, T. P., & Ghosh, S. K. (eds.), *Smart Agriculture Automation Using Advanced Technologies: Transactions on Computer Systems and Networks*. Springer. https://doi.org/10.1007/978-981-16-6124-2_4

Rozenstein, O., Cohen, Y., Alchanatis, V., Behrendt, K., Bonfil, D. J., Eshel, G., . . . & Lowenberg-DeBoer J. (2024). Data-driven agriculture and sustainable farming: Friends or foes? *Precision Agriculture, 25*, 520–531. https://doi.org/10.1007/s11119-023-10061-5

Saleem, M. H., Potgieter, J., & Arif, K. M. (2021). Automation in agriculture by machine and deep learning techniques: A review of recent developments. *Precision Agriculture, 22*(6), 2053–2091.

Shaikh, T. A., Rasool, T., & Lone, F. R. (2022). Towards leveraging the role of machine learning and artificial intelligence in precision agriculture and smart farming. *Computers and Electronics in Agriculture, 198*, 107119.

Sharma, A., Jain, A., Gupta, P., & Chowdary, V. (2020). Machine learning applications for precision agriculture: A comprehensive review. *IEEE Access, 9*, 4843–4873.

Singh, S., & Goyal, M. K. (2023). Enhancing climate resilience in businesses: The role of artificial intelligence. *Journal of Cleaner Production, 418*, 138228.

Singh, V. (2024). Advances in precision agriculture technologies for sustainable crop production. *Journal of Scientific Research and Reports, 30*(2), 61–71. https://doi.org/10.9734/jsrr/2024/v30i21844

Son, N., Chen, C. R., & Syu, C. H. (2024). Towards artificial intelligence applications in precision and sustainable agriculture. *Agronomy, 14*(2), 239. https://doi.org/10.3390/agronomy14020239

Sood, A., Sharma, R. K., & Bhardwaj, A. K. (2022). Artificial intelligence research in agriculture: A review. *Online Information Review, 46*(6), 1054–1075.

Suganya, M. (2020). Crop yield prediction using supervised learning techniques. *International Journal of Computer Engineering and Technology, 11*(2).

Sugumar, R., & Suganya, D. (2023). A multi-spectral image-based high-level classification based on a modified SVM with enhanced PCA and hybrid metaheuristic algorithm. *Remote Sensing Applications: Society and Environment, 31*, 100984.

Supriya, M., Tyagi, A. K., & Tiwari, S. (2024). Sensor-based intelligent recommender systems for agricultural activities. In *AI Applications for Business, Medical, and Agricultural Sustainability* (pp. 197–235). IGI Global.

Tian, H., Wang, T., Liu, Y., Qiao, X., & Li, Y. (2020). Computer vision technology in agricultural automation—a review. *Information Processing in Agriculture*, *7*(1), 1–19.
Velamuri, A. (2023). Data-driven management in agriculture. In *Encyclopedia of Digital Agricultural Technologies* (pp. 267–276). Springer International Publishing.
Xu, Y., Zhang, X., Li, H., Zheng, H., Zhang, J., Olsen, M. S., . . . & Qian, Q. (2022). Smart breeding driven by big data, artificial intelligence, and integrated genomic-enviromic prediction. *Molecular Plant*, *15*(11), 1664–1695.
Yang, C. (2020). Remote sensing and precision agriculture technologies for crop disease detection and management with a practical application example. *Engineering*, *6*(5), 528–532.

12 Smarter Yields, Sustainable Practices

Predictive Modelling for Informed Decision Making in Smart Farming

Ruchi Jakhmola Mani, Saba Masood, Piyush Bhalerao, and Deepshikha Pande Katare

12.1 INTRODUCTION TO PREDICTIVE MODELLING IN AGRICULTURE

The agriculture industry has an essential role in feeding the world's growing population and is considered a major source for increasing the economy and prosperity of a country Elbasi *et al.*, 2023). According to a recent economic survey of 2020–21, agriculture in India was responsible for providing 20.2% of the country's gross domestic product (GDP) (Sunder, 2018). Agriculture is a complex system which depends on a wide range of factors such as temperature; moisture; rainfall; soil fertility; pesticides; and financial considerations like seed prices, equipment costs, and infrastructure maintenance (Sabu & Kumar, 2020). Thus, the agricultural sector must continuously adapt to new systematic and efficient technologies to meet the criteria of high quality, sufficient agricultural commodities. By using data as a valuable resource, the agricultural industry can be made more predictable and modernized, giving rise to smart farms (Sharma *et al.*, 2022).

Implementing predictive modelling in smart farming helps in controlling the agricultural operations based on data and increasing productivity by providing valuable insights on fertilizer consumption, irrigation, soil requirements, weather patterns, crop growth stages, electricity usage, and transportation costs (Wolfert *et al.*, 2017). It utilizes data smartly, provides guidance to farmers throughout the farming cycle according to the individual requirements of their farming lands, and also understands the market trend of the agricultural products (Alfred *et al.*, 2021). This assists farmers, agribusinesses, and governmental organizations to make more informed and profitable decisions related to planting, harvesting, storage, and marketing of crops, thereby promoting sustainable agricultural practices.

12.1.1 Evolution of Predictive Modelling in Farming

Agriculture and cultivation have been a part of human civilization for centuries. Earlier, farmers were not familiar with the concept of integrating science-based technologies with agriculture, but they relied on traditional knowledge and experience to predict crop outcomes. Constellations of stars, the position of the sun in the sky, the blooming of specific flowers and trees, the appearance of migratory birds, rainfall patterns, the direction of the wind, and the local climate were some of the natural indicators which were used to make better farming decisions in ancient times (Zuma-Netshiukhwi *et al.*, 2013). For example, toads, white ants, and birds were used by the local farmers of Kenya to predict the onset of summer season, rainfall, and change in temperature. Also, in some

 DOI: 10.1201/9781003507390-12

areas of South Africa, the blossoming of peach and aloe trees was seen as a natural indicator for forecasting the monsoon in those areas (Zuma-Netshiukhwi, 2013).

In the late 19th and early 20th century, statistical methods offered more potential in analyzing agricultural data with precision and practical results. These techniques were utilized to make basic predictions with the help of weather records, soil conditions, and crop yield data (Parolini, 2015). With the help of crop models, dynamic interactions occurring at point-scale level between plants, weather conditions, soil properties, and various management practices were represented mathematically. This helped the farmers simulate crop growth under varying environmental conditions and management practices. The earliest crop model incorporated light interception and photosynthesis as key parameters to determine crop productivity. It relied on empirical relationships that considered fundamental biochemical and biophysical mechanisms involved in converting solar energy into plant biomass (Ginaldi *et al.*, 2019).

In the late 20th century, geographic information systems (GISs) along with remote sensing technology emerged as a powerful tool for spatial data analysis in farming. It helped to create a comprehensive image of agricultural conditions in a particular area by considering geospatial information (soil maps, weather stations, field boundaries) and data from remote sensing technologies (crop health, growth stage, soil moisture) (Stubert *et al.*, 2020). Crops can be monitored for growth and damage by pests, land use/land cover change can be detected, crop yield and area can be predicted, and soil characteristics can be mapped by incorporating technologies such as GIS and remote sensing into agriculture (Vibhute & Gawali, 2013). Figure 12.1 shows the five different phases involved in the evolution of agriculture.

A revolution is coming into the agriculture industry by introducing cutting-edge technologies like predictive modelling, machine learning, and artificial intelligence. Abundant data on weather, soil health, satellite imagery, and crop yield is helpful in providing real-time updates on crop performance, disease outbreak, and resource requirements. For example, AI sensors developed by Microsoft are employed by Indian farmers of Telangana, Madhya Pradesh, and Maharashtra, which helps them in monitoring their fields for pest attack based on weather data and crop stage (Sabu & Kumar, 2020). It also helps in understanding market trends and consumer preferences, which helps agribusiness and farmers expand their business and gain maximum profits.

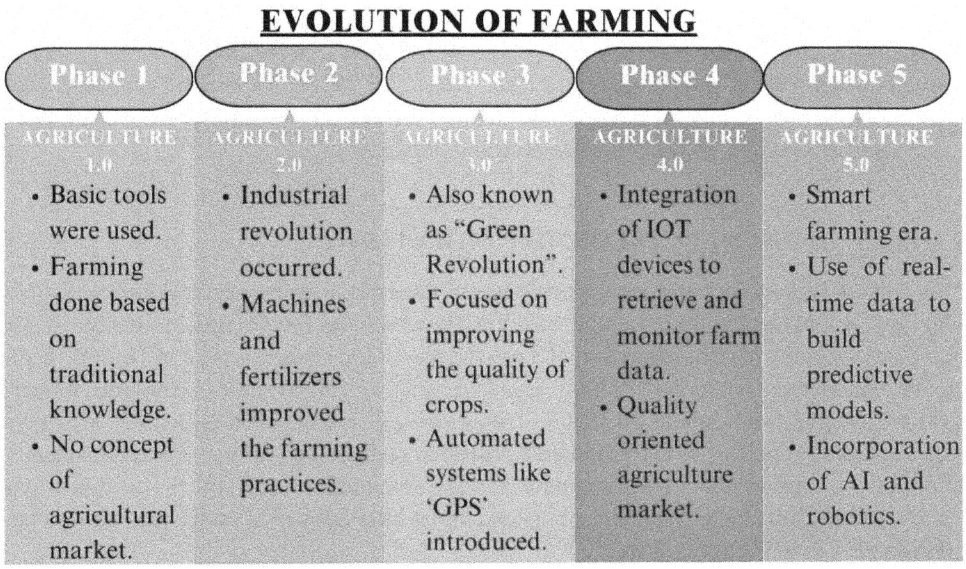

FIGURE 12.1 Different phases involved in the evolution of agriculture (Arrubla *et al.*, 2022).

12.1.2 Importance of and Benefits in Smart Farming

Predictive modelling in smart farming confers a variety of advantages, beginning with predicting crop yields, preventing risks to understanding market trends, and managing logistics. Collecting data related to all the parameters involved directly or indirectly in the growth of a crop such as soil conditions, weather and rainfall patterns, nutrient requirements, irrigation management, fertilizer consumption, and pest attacks can be helpful in analyzing the yield of the crop in the future. These predictions can be utilized by farmers to modify their farming practices such as planting and harvesting schedules and irrigation systems to increase their yield and minimize crop damage. Any hazard to the land due to a sudden change in weather patterns, disease outbreaks, and attack by insects can be minimized or even prevented with the help of predictive models (Mandapuram *et al.*, 2019).

The real-time data retrieved from GIS and remote sensing techniques is useful in monitoring crop health, soil conditions, and temperature. This helps farmers make quick decisions in response to changing conditions. The insights provided by predictive models can be utilized for efficient management of logistics. More systematic planning can be done related to transportation facilities, supply chain management, maintenance of storage areas, and timely delivery to suppliers. Market demands can be understood in depth with the help of predictive analytics (Mandapuram *et al.*, 2019). Thus, farmers can make profitable decisions by increasing the yield of crops required by consumers, increasing the speed of their agriculture process to meet the quantity of the market, and even developing new hybrids of plant varieties.

It is very important to effectively manage parameters and analyze the relationships between these parameters involved in farming. Neo4j, a graph database, has emerged as a novel tool for exploring the intricate relationship between data and creating knowledge graphs (Furqon & Bukhori, 2023). It has found its use in agriculture by linking agricultural data such as crop varieties, crop yield, crop growth, crop damage, weather conditions, soil fertility, irrigation cycles, infrastructure maintenance, and consumer demands.

12.1.3 Overview of Neo4j in Predictive Modelling

Neo4j is based on graph theory, which represents parameters as nodes, and the complex interactions between these nodes are represented as relationships. It has a hierarchical structure, and the layout of the graph facilitates user navigation and data querying (Shi *et al.*, 2021). Neo4j uses Cypher as its query language, which makes it a powerful platform for deep analysis and provides reliable accurate results for making thoughtful decisions (Shi *et al.*, 2021; What Is a Graph Database?—Getting Started, n.d.). It is well suited for handling large amounts of data, including sensor data, historical data, and satellite imagery, making it a scalable and flexible tool.

12.2 ROLE OF NEO4J IN PREDICTIVE MODELLING

With the advent of new technologies, the amount of data has also increased significantly. NoSQL has emerged as a new member in the database family which has the potential to handle, store, and recover large volumes of data effectively. Graph databases are a special type of NoSQL databases which offer an effective solution for data storage and data querying to explore numerous hidden connections within the data (Guia *et al.*, 2017). They store data using graph structures which contain nodes, edges, and properties. It is considered a practical database and offers several advantages such as better optimization of data collection, data storage on the order of petabytes (10^{15}), new types of data, and multiple dimensions to manage big data, and it has high performance when querying deep data (Fernandes & Bernardino, 2018).

Neo4j is the most widely used graph database worldwide. It is open-source software which stores data in the form of graphs instead of tables. It helps in identifying and interpreting interconnections within the data with the help of visually interactive knowledge graphs. This software was released in 2007 and has three versions: Community, Enterprise, and Government. Community is available for any user, while Enterprise is the commercial version of the software. The Government version is an upgrade of the Enterprise version and is mainly designed for government services (Fernandes & Bernardino, 2018).

The interface of the Neo4j database consists of a central image menu where graphs with multiple nodes, edges, and properties can be seen. The node (circles) represents the object or the attribute within the data for which additional descriptive information can be stored as properties, and the relationships or edges (lines) represent the connection between the nodes. The names of the nodes and edges and the version of the program can be found on the left side of the interface. The graph can be easily zoomed in on to analyze more connections within the data. On the top of the central image menu, a textbox is provided to execute queries in Cypher language. With the help of this query language, users can create, insert, and delete nodes, relations, and properties (Guia *et al.*, 2017).

12.2.1 Features Supporting Predictive Modelling

Unlike other relational databases, Neo4j has been found to be very dynamic in handling queries to determine relationships within complex heterogeneous data. It creates index files and uses system memory to read and write those index files. This makes it a superior graph database which can handle large amounts of data. Relational databases store structured data with pre-defined columns. It makes use of 'join' to infer connections between different types of data. With accumulation of different types of data, it becomes difficult to deal with data of complex networks and relationships (Yoon *et al.*, 2017).

The integrity of Neo4j follows ACID (atomicity, consistency, isolation, and durability). It has a very interactive and user-friendly interface. Neo4j provides a rich library of graph algorithms that can be applied to various predictive modeling tasks, including centrality analysis, community detection, and pathfinding (Guia *et al.*, 2017). Neo4j uses Cypher as its query language, which is easy to interpret and helps to explore intrinsic linkages in the data. Table 12.1 describes the different commands in Cypher used for executing queries in Neo4j. It can also provide information on future trends and outcomes. Graph algorithms in Neo4j helps in identifying clusters of nodes with similar characteristics, the most influential nodes and relations, and the shortest or the most efficient paths between the nodes.

The predictive models obtained from Neo4j can be integrated with machine learning. The accuracy of the predictive models can be enhanced by utilizing features from the graph data using

TABLE 12.1
Different Commands in Cypher Used for Executing Queries in Neo4j (NEO4j Basic Commands, n.d.)

Command	Function
MATCH	It helps in identifying the pattern that corresponds to a particular pattern provided by the user.
CREATE	It creates new nodes and relationships in the graph along with their specific properties.
MERGE	It is used for finding a specific pattern in the graph database and, if absent, creates that pattern.
SET	It updates or modifies the existing nodes and relationships with new properties.
ORDER BY	It arranges the query into different orders based on the parameters provided by the user.
CREATE INDEX	It creates an index for the properties present in the nodes for rapid identification of data based on the indexed properties.
RETURN	It specifies the data that must be returned after the execution of the query.

Cypher and graph algorithms, which can be easily combined with machine learning libraries like scikit-learn or TensorFlow (How Continuous-Learning AI Sharpens Predictions, n.d.) (Innovative Ways to Enhance ML Models With Feature Engineering, n.d.). The visually informative graphs in Neo4j help in understanding the reasons behind the linkages seamlessly. This aids in creating more transparent and trustworthy predictive models. The new graph processing engine of Neo4j also helps in real-time analysis of data, which can be helpful in the development of more resilient and adaptive predictive models.

Scalability is another essential feature of Neo4j which facilitates predictive modelling. Neo4j is designed to handle large and heterogeneous data. It provides horizontal scalability (Enterprise version) and vertical scalability (Community version), which help in adding more nodes to the system (Guia *et al.*, 2017). Neo4j's scalability ensures that predictive models remain robust and performant, even as the complexity and volume of data continue to grow. Flexibility is also another critical parameter of Neo4j, where the models adapt to the changing requirements and data structures (Giannakis, n.d.). It facilitates the incorporation of new variables and also modifies existing links, making Neo4j a dynamic model for building more flexible predictive models across various domains.

12.2.2 Role of Neo4j for Predictive Modeling in Agriculture

Effective management and representation of data is one of the common hurdles faced by farmers and agribusiness in the agricultural domain. Graph databases like Neo4j are finding their way in this field as they delve into finding interconnections between agricultural entities with the help of graph algorithms and libraries. Neo4j creates multifaceted knowledge graphs which encompass a variety of agricultural entities such as crops, pests, weather conditions, and soil requirements and provides informative observations for enhancing crop yields, resource management, and understanding the agricultural landscape. Since all the information is available at a single location, farmers and agricultural researchers can take more accurate decisions and enhance their existing farming practices.

Rice is a prominent staple food for billions of people around the globe. It is critical to manage the quality of rice crops and prevent their infestation by pests and disease outbreak. Building knowledge graphs using Neo4j assists farmers in recognizing rice pests and diseases. By consolidating the data on rice crops and pests, knowledge graphs can help in creating predictive models for preventing rice diseases, enhance the quality of the rice crop, and develop new varieties of rice (Furqon & Bukhori, 2023). Figure 12.2 depicts a visual representation. As research and development in agricultural data science advance, Neo4j has the potential to play a significant role in ensuring a more sustainable and disease-resilient future for rice production.

Fruits are considered a very healthy source of nourishment for people, as they are rich in vitamins and trace elements. However, the yield of fruits is significantly disturbed by pest attacks and the occurrence of diseases. Neo4j can be a very helpful tool in an effective pest detection systems based on knowledge graphs (Zhu *et al.*, 2023). The hidden factors responsible for fruit tree disease and pest infestations can be determined with the help of this graph-based software. Thus, high-quality fruits with attractive colors and a fruity smell can be cultivated depending on consumer needs. These dynamic predictive models have the potential to transform the agricultural industry and make a significant contribution to the development of high-yield, efficient, and high-quality agriculture.

12.3 INTEGRATION OF AGRICULTURAL DATA INTO NEO4J

Developing effective predictive models in smart farming requires data from diverse sources, which can be integrated using Neo4j to look into complex relationships within different parameters of farming. Various farm management applications are used by farmers to keep a record of different parameters of the crop such as stages of growth, irrigation schedules, yield of the crop, requirements of fertilizers and pesticides, pests, and disease outbreaks (T & T, 2023). This software are

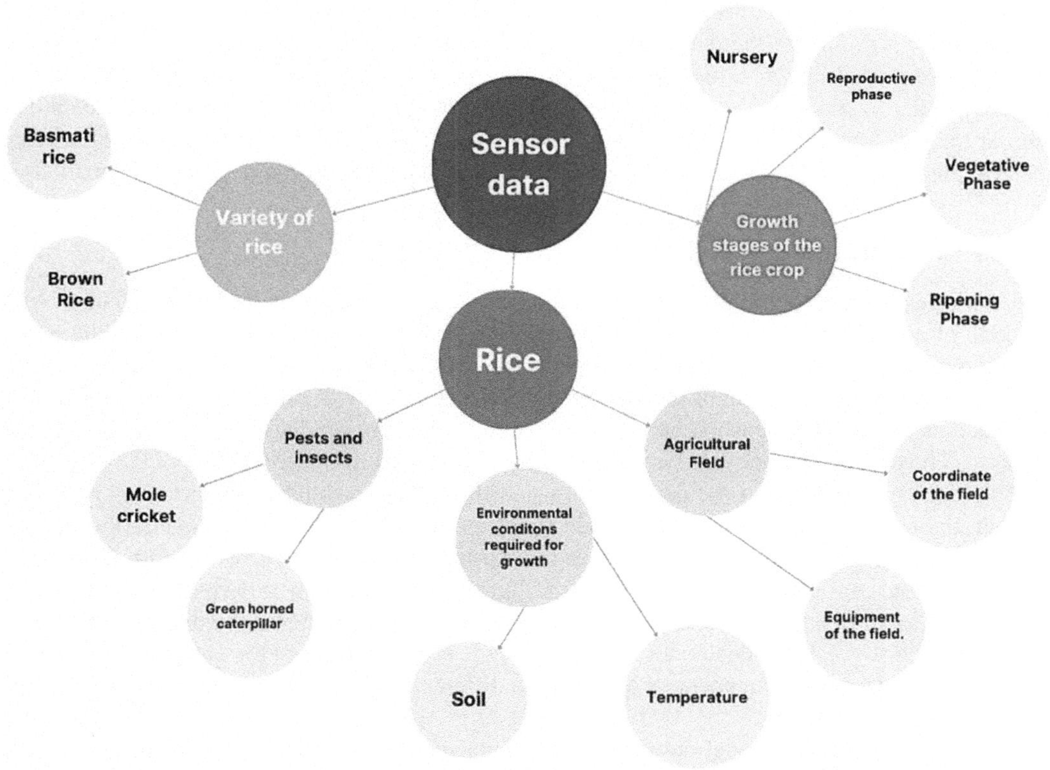

FIGURE 12.2 A visual representation of a knowledge graph for a rice field (Alfred *et al.*, 2021).

also stores data related to the maintenance of the farms, finances, waste management, and storage facility expenditures.

12.3.1 Sources of Agricultural Data

With the advancement of remote sensing and GIS and GPS technology, real-time spatial data is obtained related to crop growth stages, climatic conditions of the agricultural land, type of soil, weeds, hotspots for pests and insects attack, and disease incidences. For instance, it is very challenging to practice smart farming in the hilly areas of Himachal Pradesh due to less land being available, unpredictable climatic conditions, and limited resources. Spatial data, along with references from satellite images, has been advantageous in mapping fields and enhancing farming practices such as resource optimization, yield and farm mapping, and soil sampling (Sood *et al.*, 2015).

Meteorological variables such as temperature, precipitation, velocity of the wind, and frost warnings are of immense importance to farmers for fulfilling the requirements of crop growth, making informed decisions related to farm management, and minimizing damage (Lazarević & Popović, 2020). Also, government agencies like the Department of Agriculture, Extension & Farmers Welfare (DA&FW), and Indian Council of Agricultural Research (ICAR) keep a variety of databases related to crop varieties, pest management practices, and irrigation schedules (Navarro *et al.*, 2020).

12.3.2 Data Quality and Pre-Processing Techniques

Agricultural data obtained from different sources has to be pre-processed to improve its quality. Initially, the data is cleaned by identifying the missing values, inconsistencies, and outliers and replacing them with inferred values or removing them. The data is then reduced both row-wise and column-wise to minimize the data variables. Further, data scaling is done to ensure the validity of predictive modelling when the input variables are of different scales. Max-min normalization, z-score standardization, and logarithmic and sigmoidal methods can be used to scale the data into similar ranges (Ahmad & Aziz, 2019). Last, the data is transformed by creating new features from the available data, and then it is partitioned with the help of clustering and decision tree methods.

12.3.3 Data Integration in Smart Farming Using Neo4j

The working of Neo4j for building predictive models in smart farming can be best understood with the help of an example of monitoring the internal environmental conditions of a greenhouse and reducing the consumption of energy. The information related to crop growth, physiological features of the crops, and microclimate of the greenhouse was collected with the help of different IoT sensors such as temperature sensors, humidity sensors, and sensors for detecting the intensity of the sunlight. The data is then delivered in a Neo4j-compliant format like CSV, JSON, or GeoJSON.

The data is uploaded into the graph database with the help of built-in import tools in Neo4j. Cypher, the query language of Neo4j, is used for cleaning and modifying the data during import. With the help of cleaning, missing values or inconsistencies are rectified, while with the help of standardization, consistency is created across the data. New features can be derived by using feature engineering for in-depth analysis of the data. The agricultural data entities are represented as circles in the knowledge graphs, called nodes. It also stores the characteristics of the entities as properties. The interdependencies or connections between the different nodes are represented as relationships in the knowledge graphs.

The overall predictive model developed for monitoring the conditions of an intelligent greenhouse comprises four subsystems. The greenhouse management information system is responsible for controlling the conditions of the greenhouse. The temperature control and monitoring subsystem regulates the temperature inside the greenhouse, and the seed and production subsystem monitors the growth of the plants inside the greenhouse. The data processing and analysis subsystem collects data from the other three subsystems and provides useful insights to the farmers for effective decision making (Subahi & Bouazza, 2020).

The nodes in the knowledge graph are represented with different colors such as the blue nodes depict the physical components of the greenhouse such as sensors which collect real-time data such as plant growth stages, etc. Similarly, temperature readings are stored in the red nodes. The light blue nodes store structural information related to the greenhouse like the layout and organizations involved, and the pink nodes store historical data related to the crop and the greenhouse (Subahi & Bouazza, 2020).

The relationships between different nodes in the knowledge graphs can be determined by using the commands from the Cypher query language. Commands like CREATE, MATCH, and RETURN can be used to create new nodes, determine relationships between nodes, and retrieve the required data. For example, the relation between the type of seed sown in the greenhouse and their production speed can be extracted with the help of the MATCH command. The influence of the type of seed sown in the greenhouse on the energy consumed can also be determined with the help of the Cypher language (Subahi & Bouazza, 2020).

12.4 BUILDING PREDICTIVE MODELS USING GRAPH ALGORITHMS WITH NEO4J

Neo4j works on graph algorithms to further go into more specific and detailed information from the interconnected data obtained from farming systems. These graph algorithms examine those communities and enable one to see a broader view of relationships among different data points. They help in creating multifaceted models based on complex relationships and help in finding optimal

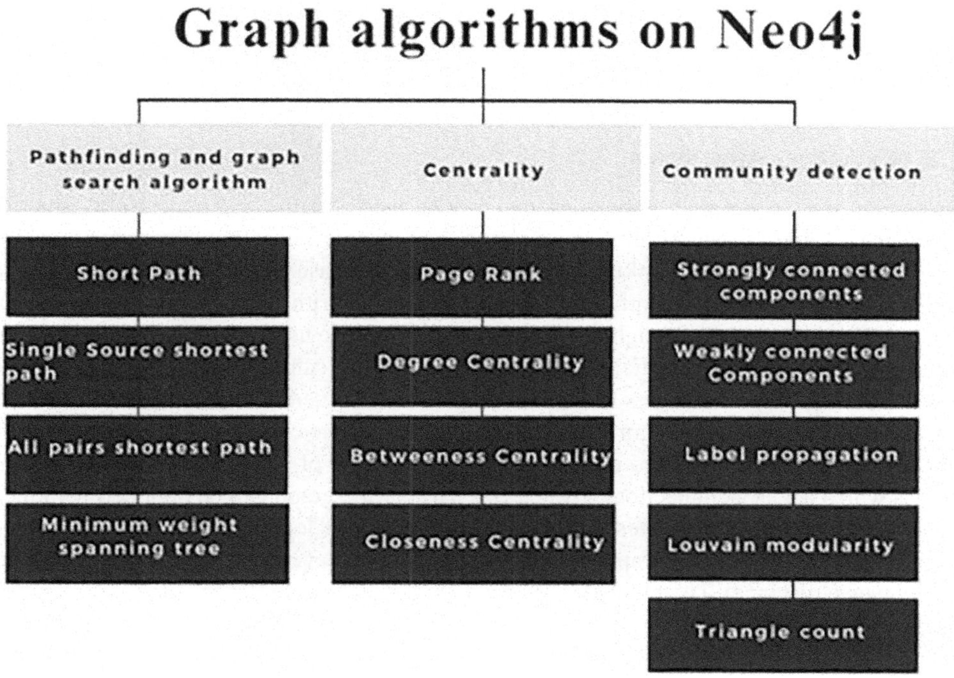

FIGURE 12.3 A flowchart depicting the different categories of graph algorithms used in Neo4j (Needham & Hodler, 2018).

solutions. Graph algorithms used in Neo4j have three main types: the pathfinding and graph search algorithm, centrality algorithm, and community detection algorithm (Needham & Hodler, 2018). Figure 12.3 represents a systematic flowchart of the different graph algorithms used for extracting information from the knowledge graphs in Neo4j.

12.4.1 Path Finding and Graph Search Algorithms

Path finding and graph search algorithms start with a single node and extend the relationships until the target is reached. They include four types of algorithms: shortest path (SST), single source shortest path (SSSP), all pairs shortest path (APSP), and minimum weight spanning tree (Needham & Hodler, 2018). The most popular, shortest path algorithm, namely Dijkstra's algorithm, is a real-time graph algorithm which calculates the shortest distance between two nodes. It is included in the standard user flow in web or mobile applications. SSSP calculates the shortest weighted paths from a node to all paired nodes in the graph. Neo4j uses a delta stepping algorithm, a variation of SSSP which is very efficient in parallel and sequential graphs. The shortest weighted paths amongst all the pairs of a node can be calculated with the help of APSP. This algorithm is advantageous, as it is quicker than the SST algorithm for each pair in the graph. Starting from a node, minimum weight spanning tree finds all the reachable nodes in the tree and links these nodes with the help of minimum weight. Prim's algorithm is known as one of the simplest and most popular minimum weight spanning tree algorithms.

Nowadays, smart farms use autonomous robots for various purposes such as harvesting, spraying, and transportation. These mobile robots minimize the need for labor and reduce labor costs, thereby improving the quality of life for farmers. However, movement of mobile robots in the corridors and gates of smart farms has to be regulated to avoid collisions and damage to the field and the robot. This is where pathfinding algorithms are being utilized. To set a designated path for each robot, this algorithm tracks the path chosen by the robots in different cases and finds the shortest route from the starting point to all other points in the field (Pak *et al.*, 2022). This will help in planning the route more efficiently and save time by reducing the amount of time spent on the field.

12.4.2 Centrality Algorithms

The most significant nodes within the graph can be identified with the help of centrality algorithms. This includes a variety of algorithms such as page rank, degree centrality, betweenness centrality, and closeness centrality. Other ranking algorithms focus on measuring the direct impact of a node, while page rank is one of the transitive, directional ranking algorithms which takes into account the neighbors and neighbors of the neighbors. This algorithm can be used in supply chain management of a farm where the more influential suppliers, agribusinesses, can be determined. The number of incoming and outgoing connections from a node can be determined with the help of degree centrality. It is the standard centrality algorithm and helps in finding the common nodes in a graph.

Betweenness centrality is based on measuring the influence of node to flow of data in a graph. The nodes which act as a connection between two parts of the graph can be obtained with the help of betweenness centrality. Closeness centrality is a method of locating the nodes that distribute information efficiently across a graph. A node's closeness is its average distance from all its neighbors (the inverse distance). The closest nodes have the lowest inverse distances. For each node in a graph, the closeness algorithm calculates the distance between the node and all of its neighbors. The algorithm calculates the distance from each node to all of its neighbors by summing the distances to each node. The resulting distance is inverted to get the node's closeness score.

Managing smart farm operations in difficult-to-reach locations such as hilly areas is very challenging. Also, large farms where natural barriers such as small mountains, steep slopes, trees, etc. are also a challenging task for farmers. The deployment of IoT sensors in such complex agricultural landscapes has made it very easy to gather and record data related to farming operations (Qayyum *et al.*, 2021). With the help of centrality algorithms, the important sensors tracking data related to all the parameters of the farm can be determined. Also, information related to the centrally located sensors in the field can also be obtained.

12.4.3 Community Detection Algorithms

Community detection algorithms help in clustering or segmenting groups. Strongly connected components, weakly connected components, label propagation, Louvain modularity, triangle count, and average clustering coefficient are some of the algorithms falling under the graph algorithm type. A strongly connected component (SCC) is defined as a set of nodes in a directed graph, where each node can be accessed bidirectionally by any node in the same set. SCCs are generally used in the initial phase of graph analysis and help to interpret the structure of the graph. The weakly connected components algorithm finds a set of nodes in an undirected graph. All nodes in the set are reachable from all other nodes in the set. The main difference between weakly connected and strongly connected component algorithms is that weakly connected components require paths in only one direction between pairs of nodes, whereas strongly connected components algorithms require paths in both directions.

The label propagation algorithm (LPA) is a fast algorithm that uses the network structure as a parameter to find communities in a graph. No predefined objective function or prior community information is required. An interesting aspect of LPA is that preliminary labels can be assigned to narrow the range of solutions generated. This means that the initial community can be used as a semi-transparent method to find communities for manual selection. The Louvain community detection algorithm is a method of identifying communities within a network. Each community achieves a modularity score of 100. Modularity evaluates the quality of the assignment of nodes to communities by comparing the node density in the community to the node density in a random network. Triangle count is an algorithm that counts the number of triangles that pass through each node in a community detection graph. A triangle is a collection of three nodes, each of which has a relationship with the other.

The quality of agricultural commodities is given more emphasis than quantity in the agricultural market. Farmers require smart monitoring systems for detecting ripeness, color quality, and infected fruits and vegetables. The data related to the color of the outer skin of vegetables like tomatoes was collected with the help of IoT sensors, and a knowledge graph was constructed. Data related to the ripeness, quantity, and color of the tomatoes was stored in different nodes. Community detection algorithms help in segregating the ripened and unripened tomatoes in the field. It can be helpful in identifying infected tomatoes and creating a separate node for them in the knowledge graph (Siddiquee *et al.*, 2022). This helps farmers in maintaining the quality of their products and increases customer demand.

12.5 PREDICTIVE APPLICATIONS IN SMART FARMING

12.5.1 Crop Yield Prediction

A whole new world in smart farming can be explored by making use of predictive models developed by Neo4j. Sustainable management of crops is one of the most important applications of predictive modelling. It can predict optimal planting times and crop health, as well as future crop yields by considering both the current and historical data on weather, soil, crops, and past yield (Lokhande, 2021). It is estimated that the demand for food is projected to grow from 59% today to 98% in 2050 (Kumar & Menakadevi, 2017). Farmers around the world are growing more crops by increasing the area available for growing crops or by improving the productivity of existing farmland through the introduction of new techniques such as fertilization, irrigation, and precision agriculture. However, the environmental and social costs of clearing more land for agriculture often outweigh the benefits. Crop yields, defined as the number of crops grown per hectare of land, are increasing too slowly to keep up with projected demand (Siddiquee *et al.*, 2022). Other factors, such as climate change, urbanization, and insufficient investment, will make it difficult to produce sufficient food (Krisnawijaya *et al.*, 2022).

The new emerging era of big data and predictive analytics is taking agriculture to the next level. Collecting, storing, and analyzing massive data sets to uncover hidden trends and optimize crop production is no longer just about mining historical crop records, precipitation maps, or diagnostic reports (Krisnawijaya *et al.*, 2022). Neo4j, a powerful graph database platform, is advantageous in developing predictive models by identifying the interconnections in the agricultural data. It not only incorporates historical data from the farms but also integrates real-time data of different sensors, weather predictions, and so on to create comprehensive knowledge graphs. This intertwined structure helps in regular monitoring and comes up with predictive models for evaluating crop yield and resource consumption. It provides a more data-driven approach for decision making as the agricultural industry progresses towards sustainability and productivity.

Electronic farm records hold massive amounts of data related to soil, weather patterns, sensor readings, and historical information. This data is extremely important for building predictive models which are reliable (Yadav *et al.*, 2015). The most unrelated data points can be connected with the help of a knowledge graph created in the environment of Neo4j. For instance, at any given time, the earlier crop yield data can be linked to sensor readings measuring the soil parameters or weather conditions. Likewise, social media insights on the latest agricultural trends can be linked to conventional farming practices and weather parameters such as precipitation. Neo4j promotes real-time analysis of agricultural data due to the interlinked structure of the knowledge graph created by it. Big data scientists can use Neo4j's knowledge graphs to go beyond historical analysis and develop predictive models for increasing crop yields.

12.5.2 Disease Outbreak Forecasting

Traditional knowledge for crop disease or insect pest management is valuable, but it is often based on static expertise (Sishodia *et al.*, 2020). This restricts the ability to respond to changing conditions and integrate large volumes of data produced by modern agricultural processes. Semantic interoperability is the key to unlocking this data's true potential. It enables the full and efficient exchange and integration of information on pests and disease (Yadav *et al.*, 2015). Knowledge graph technology provides a large, open-source approach to obtaining and managing knowledge about pests and disease.

One of the biggest challenges to building predictive models is the complexity of crop disease data and pest data. Crop disease data comes in a variety of formats (textual, sensor, historical, etc.) and in large volumes, making it difficult to integrate and analyze (Xiaoxue *et al.*, 2019). In addition, data is often fragmented across different research institutions, resulting in inconsistent data and limited availability. Finally, data is not stored in standardized ways, formats, or data standards, making it difficult for data to be exchanged and knowledge to be shared. Table 12.2 represents a disease prediction table for determining disease progression in a crop with the help of sensor data.

A graph database platform like Neo4j provides a solution by creating a knowledge graph specifically designed for disease outbreak forecasting. Neo4j excels in representing complex relationships between entities. From sensor readings to historical outbreak data, Neo4j can convert the data into a structured format that allows for effective analysis and integration. In addition, Neo4j combines data from different sources, resolves discrepancies, and establishes connections between entities. This enriched knowledge graph helps in recognizing patterns, which is essential for predicting future disease events based on historical patterns and current conditions. By taking advantage of Neo4j's capabilities, a comprehensive knowledge graph can be created that lays the foundation for precise disease outbreak prediction models within smart farming (Qin & Yao, 2021).

12.5.3 Livestock Monitoring

In the past, large herds were monitored through physical examinations, which was a very difficult task. With increase in the application of graph databases such as Neo4j in farming and livestock management, it helps in developing predictive models for proactive monitoring livestock. Initially, the data associated with an animal's physical parameters is collected, and real-time information of

TABLE 12.2
Disease Prediction Table for Evaluating the Onset of Plant Disease with the Help of Data Collected from Different Sensors (Common Plant Diseases & Disease Control for Organic Gardens, 2021)

Sensor Type	Sensor Data Collected	Expected Outcomes	Observed Outcomes	Prediction Of Disease
Humidity sensor	Humidity level of the surrounding air	50 to 60%	71%	Downy mildew
Soil moisture sensor	Moisture level of the soil	Below 12%	6%	Powdery mildew
Temperature sensor	Temperature of air and soil	14–18 °C	27°C	Fusarium head blight
Light intensity sensor	Intensity of light	Moderate light intensity	High light intensity	Botrytis blight
UV index sensor	Levels of Ultraviolet radiations	Moderate	High	Wheat stripe rust

their vitals, movements and location is monitored with the help of IoT sensors. Neo4j combines this real-time sensor data with the old health records and breed information of the animal and correlates this in the form of a knowledge graph. Machine learning algorithms inside Neo4j can analyze interconnected data to identify patterns and predict health issues based on current readings, historical trends and environmental factors. The system would proactively inform stakeholders (farmers and veterinarians) when health anomalies are identified, allowing early intervention and improved herd health and productivity.

Monitoring large herds through physical examinations can be challenging. With predictive modeling, Neo4j offers digital, proactive monitoring. The first step would be to register each animal's physical parameters and deploy IoT devices to gather real-time information on vitals, movements, and location. Sensor data would be continually fed into Neo4j, where it would be combined with historical health records and breed information, as well as environmental data (Shabani *et al.*, 2022). Because of Neo4j's ability to connect all this data, a knowledge graph can be created. Machine learning algorithms inside Neo4j can analyze interconnected data to identify patterns and predict health issues based on current readings, historical trends and environmental factors. The system would proactively inform stakeholders (farmers and veterinarians) when health anomalies are identified, allowing early intervention and improved herd health and productivity. Figure 12.4 depicts livestock monitoring with the help of IoT devices and remote sensors. This new technology in the field of livestock monitoring is very beneficial, as it helps in the early diagnosis of a disease and helps to take preventive measures such as targeted medication or feed and environmental adjustments. This proactive approach reduces the risk of outbreak and ensures optimal animal health. Second, Neo4j's scalability enables the system to scale with the size of the herd, making it a scalable solution for large livestock operations (Iglesias *et al.*, 2022).

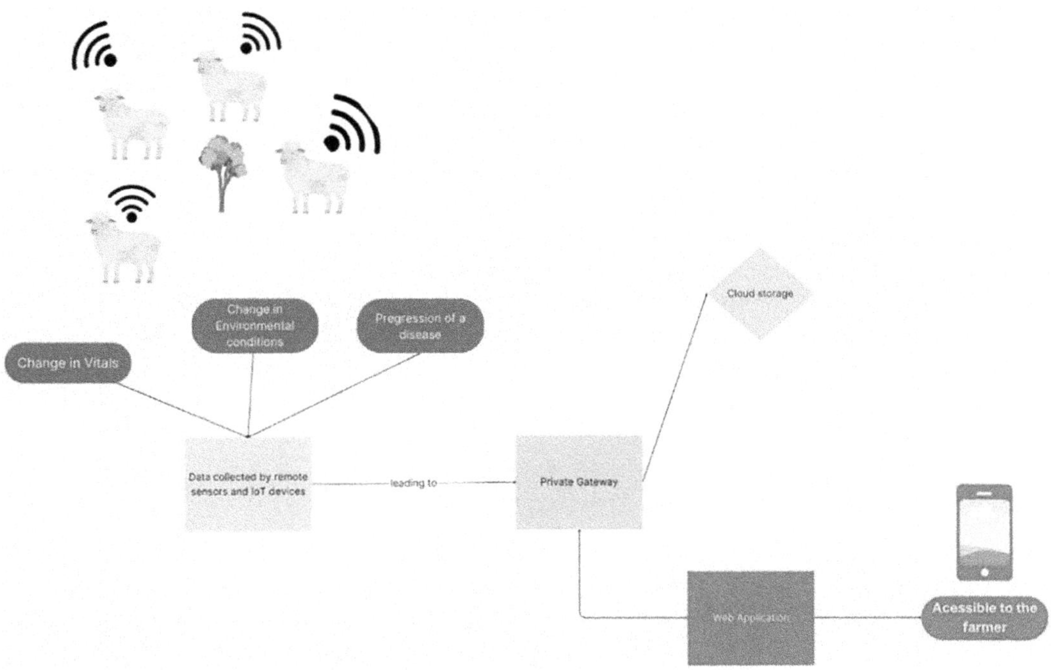

FIGURE 12.4 Livestock monitoring with the help of sensors and IoT devices (Abdullahi *et al.*, 2019).

12.6 REAL-TIME DECISION SUPPORT SYSTEMS

12.6.1 Monitoring Agricultural Systems

The need for smart farming has increased significantly in developing countries such as India. Research on usage of wireless sensor networks for agriculture is reaching new dimensions, including monitoring environmental conditions such as weather, soil moisture, temperature, plant growth monitoring, and automatic irrigation systems. The traditional manual method for collecting data to obtain the desired factors is irregular and led to deviations due to incorrect measurements. Utilizing diverse wireless sensor nodes for agriculture can reduce the time and effort spent on environmental monitoring. Logging data can reduce data loss and inconsistency (Xiang *et al.*, 2016). They can also be installed in critical locations without putting personnel in dangerous situations. Monitoring systems reduce response time to adverse factors and conditions. This technology allows remote measurement of environmental factors such as temperature, humidity, soil moisture, and water level.

This can be best explained with the help of an example of a real-time monitoring system for agricultural machinery operations. This smart operating system, based on a global navigation satellite system, consists of three major units: a data receiving module, a data processing module, and a user interface. This system was designed in a manner such that it could retrieve real-time data related to coordinates of the position, working path, boundary of operating area, total area of the field, operating speed, skip rate, and other essential operating data. This system helped in enhancing the operations of agricultural machinery and also provided insights on minimizing resource consumption (Xiang *et al.*, 2016).

In another example, a smart irrigation system was developed by monitoring the real-time data of soil moisture, level of the water tank, and level of the water well. The data was collected with the help of sensors and was sent to a web service for analysis through a mesh network called ZigBee and a smart gateway. The results provided by the web service were available to the farmers through an online application where farmers could check the values of the soil moisture and water levels in the tanks and wells. It also provided an extra advantage to the farmers to control the opening and closing of the valves in the water tank through their application. This sustainable irrigation system ensured water conservation and less intervention of the farmers and also enhanced quality of the crop (Khan *et al.*, 2021).

Introduction of new electronic equipment is upscaling the field of smart farming by providing new updated technologies to improve their operational efficiencies and implement new farming techniques. Smart farming is based on different technologies and infrastructure: data collection and management systems (DMSs), GIS, GPS, microelectronics, wireless sensor networks (WSNs), and radio frequency identification (RFID) (Ko *et al.*, 2014).

One of the most important applications of precision agriculture is monitoring and traceability systems for the yield and distribution of agricultural products. The outcomes of these systems are used to forecast the yield of the following year. Farmers can also use these systems to provide distribution channels of agricultural products from farmers to customers. Most of the current systems use both RFID and WSN. Generally, RFID systems are used to track and measure the yield of agricultural products, while WSN systems are used to monitor the environment while products are being distributed and stored.

12.6.2 Detecting Anomalies and Adapting Strategies

Smart farming systems help improve product quality and farmer safety by collecting, analyzing, and processing large amounts of field data. These data can be used to guide farmers in making decisions to improve product quality and quantity and can also help eliminate waste, reduce water consumption, and maximize profits. These systems are increasingly used to support and improve the operations, competitiveness, and profitability of the agricultural sector. However, large-scale use of technology also comes with risks and vulnerabilities. Being able to detect anomalies in smart agricultural systems is very important.

For instance, a hybrid anomaly detector called cloud-edge indicator of farming anomalies (CEIFA) was introduced as a solution to these problems. The purpose of CEIFA is to improve data reliability by detecting sensor failures, failures, errors, and some malicious activities and to enable smart farming by analyzing the data. CEIFA filters recognition data to detect anomalies, errors, failures, and other anomalies that affect data quality. It is a hybrid system that operates both at the edge and in the cloud. The edge performs initial analysis. The data can be categorized individually and validated by requesting validation from the cloud. In the cloud, CEIFA can filter all datasets. It can also process only requests sent from the edge (de Araujo Zanella et al., 2022). According to the evaluation results, for the selected anomaly dataset, the detector can detect about 95% of all the anomaly data, the edge can detect about 80% of all the target anomalies, and the cloud can detect about 98% of all the anomaly data.

12.6.3 Utilizing IoT Devices and Sensor Networks

The Internet of Things, a network of different devices, makes everything intelligent and is a cutting-edge technology ushering in a new era of communications and computing. It enables better crop management, better resource management, and lower costs of farming and increases crop quality and quantity. By introducing IoT in agriculture, farmers can remotely manage their field operations more efficiently with the help of data provided by different types of sensors such as soil sensors, temperature sensors, humidity sensors, and UV index sensors. For example, a soil pH sensor helps in detecting the pH of a soil and maintaining the appropriate level for a particular plant variety. The soil pH sensor records data and sends it to a server where the user can view the collected data. The data collected by soil pH sensors can be used to apply chemicals to maintain the appropriate pH for specific crops. Likewise, an air temperature sensor is an IoT device that measures the temperature inside a polyhouse (Doshi *et al.*, 2019). Information from the sensors is sent to the master server, which can perform any necessary operations, such as turning on the air conditioning (cooler, heater, whatever is needed), blowing out the fan, and if necessary spraying water to keep the polyhouse at the ideal temperature.

The agricultural sector is witnessing significant improvements. One of them is the development of automation systems that can use WSNs to monitor plants and control devices. WSNs work by collecting data from a remote location and transmitting the collected data over a wireless network, where it can be monitored by a receiver (Dagar *et al.*, 2018). WSN technology can be mainly used in the agricultural field for distributed data collection of agricultural environments and is important for providing real-time farmland information to farmers.

For example, some researchers are using reflected ultrasound signals to monitor soil moisture content and groundwater levels to predict the likelihood of landslides and slopes. To do this, they first installed an ultrasonic sensor inside a steel pipe and retrieved it from underground. They then calculated the reflection period of the wave signal to measure soil moisture content and water level. Landslides may occur during heavy rain. Further, the measurement and monitoring in agriculture have been widely studied, and technologies for detecting the amount of light intensity, humidity, and temperature through wireless sensor networks are already progressing. These reports and continually developed and processed on computer servers for real-time data access. It is sent to the farmer via a mobile device such as a PDA or cell phone (Kaewmard & Saiyod, 2014).

12.7 SCALABILITY AND PERFORMANCE CONSIDERATIONS

12.7.1 Handling Large-Scale Agricultural Data

Neo4j provides a powerful platform for anomaly detection and predictive modelling in smart farming, but more emphasis should be on scalability and performance when working with large-scale agricultural data. With its ability to process connected data, Neo4j offers a comprehensive solution for managing agricultural data. Data partitioning is one of the methods employed for handling large amounts of agricultural data. It comprises two major types in Neo4j: time partitioning and partitioning by location. Partitioning by time involves segmenting the data into different parts according to time, such as a year, a month, or a week (Tivoli Netcool Performance Manager 1.4.2, n.d.). Historical data related to crop yield, weather forecasts, and sensor readings can be divided with the help of data partitioning. Queries can be executed on specific parts of the data according to the period of interest to improve the performance.

Another approach which uses location for data partitioning divides agricultural data based on the location from which the data is being generated. This method is extremely useful for extracting data, specifically information related to temperature, humidity, and soil from different sensors of an individual field. It also helps in tracking the real-time data and identifying deviations for that specific field. It also promotes scalability by increasing the number of the farms being monitored (Tivoli Netcool Performance Manager 1.4.2, n.d.).

Data partitioning allows faster query execution for frequently accessed data such as historical records. Neo4j scalability allows the addition of more partitions as the data volume grows. This makes it easy to keep up with the growing agricultural operation and data collection. Data management flexibility also helps in managing the data in a specific way (Aggregating Functions—Cypher Manual, n.d.). For example, real-time sensor information can be stored on one partition for fast access and historical data on another partition optimized for longer-term analysis. Partitioning reduces the amount of data on a single server by distributing it across multiple graph instances.

Another method for handling large agricultural data is by data aggregation. Data aggregation is the process of breaking down raw data points into smaller and manageable parts. Neo4j provides various Cypher functions like AVG, SUM, and COUNT to help organize the data within the knowledge graph. AVG aggregates the property's average value over a set of nodes, which can be used to calculate daily or weekly average values for sensor readings such as temperature and humidity. SUM calculates the total value of property across the set of nodes. For example, it can be used to track total precipitation over a certain period or cumulative fertilizer application over a given growing season. COUNT helps in counting the number of nodes in each category. This helps in analyzing the prevalence of certain events, such as how many days have excessively high temperatures or how many irrigation events occurred per week (Scale a Neo4j Deployment—Operations Manual, n.d.).

With the help of data aggregation, the amount of data can be reduced drastically as the raw data is segregated into smaller groups. This creates more storage space and improves the system performance. It improves the query performance, as it processes a fraction of data points for queries, resulting in rapid query execution. It can be helpful for historical analysis which covers long periods of time. Long-run trends and patterns can be better understood with the help of data aggregation. For instance, analyzing average soil moisture levels for a month instead of just looking at individual readings can help detect anomalies and understand seasonal variations.

Large amounts of agricultural data can be archived with the help of Neo4j. Archiving agricultural data within Neo4j can be done in one of two ways: time-based or data lifecycle management. With time-based archiving, raw data is stored for a specific period of time, such as one month or one year, and then exported from the Neo4j system. With data lifecycle management, data is classified according to its importance and frequency of access. Frequently accessed operational data remains in the Neo4j active graph, while less frequently accessed data, such as historical yield data, is archived in a data lake or other secondary storage system.

More storage space can be obtained with the help of data archiving. This helps in storing the critical data for real-time analysis and model training at a fraction of the cost. Queries can be executed more rapidly by reducing the amount of data volume in Neo4j specifically for frequently used operational data. Archiving makes it easier to monitor and detect anomalies in real time. Certain agricultural practices have data retention periods that need to be met. Archiving helps in meeting these regulations while managing the data storage resources efficiently.

12.7.2 Optimizing Performance of Predictive Models Using Neo4j

Neo4j has the ability to manage large amounts of data and complex models, but it is very important to optimize the models for optimal performance. This can be done with the help of different methods such as improving the performance of the Cypher queries, optimizing the storage of the system, training the models, and tuning the performance of the model. With the help of Neo4j's indexing features, indexes based on key properties/relationships can be created. Neo4j will be able to extract relevant data to train and predict models, which will ultimately improve the query performance. They are generally used in the beginning of the query to find the most appropriate position in the knowledge graph for initiating pattern matching. These indexes must be monitored and updated on a regular basis with change in data and queries for optimal performance (Improving Cypher Performance—Neo4j Aura, n.d.).

It has been observed that executing many cypher queries with exact values is responsible for inefficient Cypher processing, as it doesn't involve the use of parameters. By using parameters, a common query can be used for different values which minimizes the time required for query execution and also minimizes the use of resources (The Impact of Indexes on Query Performance—Cypher Manual, n.d.). This creates more space in the memory, as it doesn't have to store the same executed query for each unique value. The amount of data required in the knowledge graphs can be reduced by modifying the structure of the Cypher query. Techniques such as pattern matching and filters can help to focus on the data required for model training or predictions. Making smart use of Cypher queries optimizes the performance of the predictive model by saving resources, memory, and time.

12.7.3 Scalable Deployment Architectures

For handling real-time data and handling complex graph queries in predictive models, Neo4j should have an appropriate amount of RAM. A solid-state drive (SSD) can also be used for storage of data. They provide faster read and write speeds, which improves the performance for accessing data and training models (Improving Cypher Performance—Neo4j Aura, n.d.). The performance of the model can also be optimized with the help of model training. Under this, batches of data are continuously fed to the model. This prevents retraining of the model on the whole dataset when new data is introduced and results in shorter training cycles. Models should be continuously evaluated and tuned to check the performance of the predictive models. On the basis of the evaluation results, the model parameters and the pre-processing steps must be adjusted to improve the accuracy and performance.

With the growth in smart farming business and increase in data collection, it becomes important to ensure that the system has the potential to handle it. For facilitating smooth running of the operations and easy handling of large amounts of data, a scalable architecture needs to be deployed for the system. Neo4j makes use of a cluster-based deployment where a master-slave cluster architecture is used. This architecture comprises two components, the database and the cluster management component. The cluster management component keeps a track of all other clusters (slave) and even elects a cluster as the master automatically. The master controls all the other slaves in the architecture for making any change in the database. The database handles the rest of the system (Montag, 2013). In the event of a single instance failure, the other instances can continue operations, reducing downtime and improving data availability. This approach also allows for scalability, as more instances can be added to the cluster to handle larger data volumes and processing needs.

Accessibility and scalability can be provided by cloud platforms such as AWS and Microsoft Azure for Neo4j deployment. Elasticity, scalability, automatic updates and maintenance, and disaster recovery are some of the advantages of using cloud platforms. For providing elasticity and scalability, Neo4j instances can be scaled up or down depending upon the data processing needs in real time. Cloud providers can automatically update systems and maintain infrastructure to ensure smooth running of the smart farming app. Cloud platforms also have built-in features to ensure data redundancy and minimize downtime in the event of an unexpected disaster.

Microservice architecture is an approach where the applications are broken down into smaller services that can be scaled independently. These microservices can have their own instance of Neo4j or use a shared cluster based on their data needs. An advantage of modular design is that each microservice can scale independently based on its specific needs. If one microservice experiences a problem, it won't affect the performance of others, increasing system resilience. Microservices have been helpful in predicting crop yields based on historical data, weather patterns, and soil properties by making using of decision trees and machine learning algorithms. Deep learning and image processing techniques are employed by microservices to classify different types of plant diseases based on images of the leaves (Aitlmoudden *et al.*, 2023).

12.8 CHALLENGES AND LIMITATIONS

12.8.1 Data Privacy and Security Concerns

While Neo4j provides an impressive solution for creating knowledge graphs and implementing predictive models in smart agriculture, there are a number of challenges that must be addressed in order to ensure widespread adoption and responsible use of this technology. Data privacy and security is one of the major concerns when storing agricultural data on graph databases like Neo4j. Attacks on agricultural data in smart farming have been divided into five categories based on privacy, authentication, confidentiality, availability, and integrity properties (Ferrag *et al.*, 2020).

Real-time data from IoT sensors is collected at regular intervals at different times of the day. This detailed information can provide the exact location of the sensors, insights predicted by these sensors, and optimal strategies for farmers for improving field operations. The privacy of this data can be compromised due to unauthorized access by an external factor. For example, sensor readings related to soil pH can guide farmers regarding the addition of ammonium in the soil. This data can be misused by an adversary by disrupting the pH settings of the soil by using drones or robots.

Masquerade attacks and replay attacks can lead to violations of authenticity for agricultural data and sensors employed for retrieving the data. A masquerade attack involves accessing private farming data by acting as a trusted device of that particular farming system. Replay attacks extract private inaccessible information from the nodes and sensors used in that particular agricultural system. It can be considered a type of man-in-the-middle attack. Maintaining the confidentiality of data is of prime importance when it comes to smart farming. Tracing attacks, brute force attacks, and known-key attacks are some of the attacks disrupting the confidentiality of agricultural data. Tracing attacks extract all the possible information from the sensors and connect the data to specific devices or people. Brute force attacks involve making a list of all the potential passwords for gaining access to sensor information and applying them one by one until the right password is obtained. A known key attack involves creating a new password by figuring out the old one (Ferrag *et al.*, 2020).

Attacks on availability of agricultural data can be categorized into denial of service (DOS) attacks, botnets, and man-in-the-middle (MITM) attacks. Denial of service occurs when a user or device is denied access to a resource such as a network, server, or communication link (Kolias *et al.*, 2017; Antonakakis *et al.*, 2017). For example, radio frequency (RF) interference occurs when a network uses radio frequencies to block the communication services of connected nodes (Ye & Zhang, 2019). A botnet, on the other hand, is a group of internet-connected devices (remote sensors). These can be used for various purposes, such as DDoS, information theft, and spam advertising (Gorli & Yamini, 2017). In a MITM attack, data is sent over a connection, transparently stored, and possibly modified (Al Shorman *et al.*, 2020).

Forgery attacks, trojan horse attacks, and biometric template attacks are categories of attacks that target the integrity of the agricultural data by modifying the inaccessible information of an agriculture system. Attacks on the code include software update attacks, buffer overflow attacks, and indirect attacks. Software update attacks destroy system availability and integrity by disrupting the software update process. Malware injection attacks infect nodes and devices with malicious code. Buffer overflow attacks are software coding errors or vulnerabilities that hackers can exploit to gain access to corporate systems (Gruschka & Jensen, 2010). Indirect attacks (SQL injection) use code injection techniques to trick a database server into inserting malicious SQL code into a database input field. These attacks violate trust (Stoica *et al.*, 2004).

12.8.2 Overcoming Technological and Resource Constraints

Although Neo4j is well equipped to handle complex relationships in the knowledge graph, the underlying algorithms used to create predictions can be complex and opaque. There can be several reasons for this problem, one of them being understanding the logic behind the errors due to incorrect predictions made by a model. It is a tough task to troubleshoot and improve the model when there is lack of interpretability. Second, the practical use of predictive models in smart farming and user acceptance is limited due to lack of explanation about the model's rationale (Ballas *et al.*, 2020). Techniques such as feature importance analysis help determine which data points in the knowledge graph are most influential in the predictions, providing some level of clarity. With the help of a co-development model to machine learning models, farmers can get a better understanding of the logic behind the predictions in the model.

The implementation and maintenance of a Neo4j smart farming system may require a certain level of technical knowledge and resources. There are certain limitations that have to be addressed; for example, the management of Neo4j requires specialized knowledge of graph database administration and the Cypher query language, which can act as a barrier for smaller farms with fewer technical resources (Ballas *et al.*, 2020). A large amount of computational power is required for complex predictive models trained on large datasets. These demands can be fulfilled by smaller farms only if they invest in cloud or physical resources. Also, maintaining the integrity and quality of data can be a challenging task, as agricultural data varies greatly in consistency.

Collaborative efforts will be required to find an apt solution for these challenges. To ensure that data is easily integrated into the system, the implementation of standardization to encourage uniform data collection methods and formats in the agricultural sector has to be encouraged. Comprehensive training and education facilities for using Neo4j can help farmers and agricultural professionals create data-driven knowledge graphs. Advances in explainable AI can help create more open machine learning models that can be trusted and used in agricultural applications more widely.

12.9 FUTURE DIRECTIONS AND EMERGING TRENDS

12.9.1 Digital Twins in Smart Farming

Neo4j is well known for its ability to create knowledge graphs highlighting data linkages and support data-driven decisions in smart agriculture. However, its future in agricultural applications is being shaped by research and development efforts. The advancement of technologies such as artificial intelligence and robotics is bringing a new era in the agriculture industry, sometimes referred to as Agriculture 4.0. These technologies have been used to introduce digital twins in the field of smart farming. A digital twin in smart farming can be understood as a virtual copy of a real field environment wherein the real-time data related to temperature, rain, moisture content of soil, pH, etc. is continuously updated. These digital twins can be helpful in assessing insights in real time and considering all the pros and cons of different parameters. They validate the results obtained with the help of artificial intelligence and help in developing better predictive models (Mallinger & Baeza-Yates, 2024). Figure 12.5 represents the integration of robotics, artificial intelligence, and predictive analytics with the help of a Venn diagram.

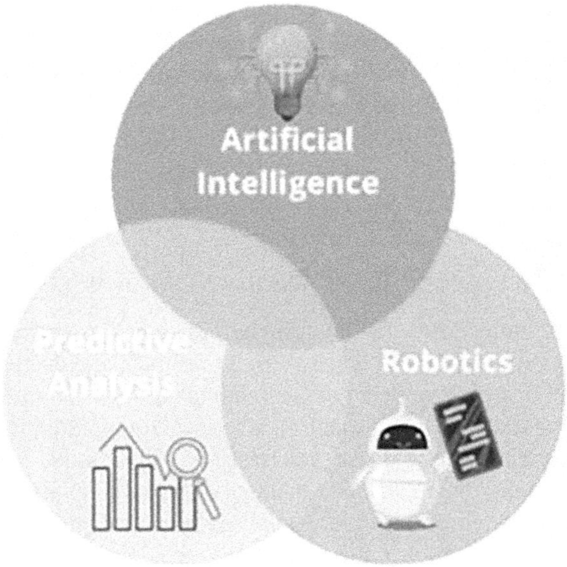

FIGURE 12.5 Integration of AI, robotics and predictive analytics in smart farming.

The application of IoT devices and modern sensors for monitoring and collecting real streamed data such as soil pH levels, crop growth stages, and air quality index is modifying the scenario of the working of agricultural fields. More precise and reliable predictive models can be developed by incorporating this real-time farming data into the knowledge graph of Neo4j (Gorli & Yamini, 2017). Also, this data will help predictive models upgrade their operations and make their modes more refined. Real-time data obtained at periodic intervals will help models give more accurate insight and even identify variations by considering the history of the data collected.

12.9.2 Robotics in Smart Farming

Nowadays, intelligent mobile robots and drones are being deployed in agricultural fields to monitor farming processes and make smart decisions. Real-time data and historical data stored in the knowledge graphs of Neo4j are utilized by robots to identify deviations in the farm and perform tasks to maintain the quality of the crops grown in the field. This heterogenous data is also being used to program the robots to maximize the productivity of the fields and protect them from damage (Yépez-Ponce *et al.*, 2023). For examples, robots have been widely used in disease hotspot areas to determine the severity of an attack by pests and take informed decisions for the usage of pesticides in those areas.

12.9.3 Edge Computing in Smart Farming

The problem of scalability persists when it comes to handling large amount of data and devices used in the management of the smart farms. Edge computing has emerged as an effective solution for processing and examining real time data by being closer to the source. It reduces the amount of data for storage in the knowledge graphs and improves the performance of the models created with the help of Neo4j. Since a large amount of agricultural data is involved, edge computing minimizes the dependency on cloud computing facilities for data storage. It is also beneficial in detecting anomalies or variations in the data and eases the process of decision making for farmers (Zamora-Izquierdo *et al.*, 2019).

In order to expand the agriculture market and provide guidance to the farmers of the rural areas, it is important to share knowledge and useful insights on a common platform which is accessible to everyone. Blockchain technology has shown promising results to gain access to agricultural data essential for improving the yield of the farm. It also excludes the risk of breaching privacy and confidentiality of any agricultural system (Praveen *et al.*, 2021). Blockchain has been helpful to farmers in selecting the appropriate insurance plan in case of damage to the farm. It has been found that farmers do not prefer indemnity-based insurances for their farms. This technology guides them in providing alternative insurance plans with more returns, such as index-based insurance.

12.10 CONCLUSION

The vast amount of agricultural data obtained from agricultural fields can be used smartly using graph databases such as Neo4j. It is very beneficial in building predictive models for farms by considering all the factors involved in the sustainable management and escalating the yield of the crop. It explores the association between different nodes representing the different agricultural entities along with their properties using the Cypher query language. Building predictive models by using powerful tools like Neo4j assists farmers in making smart and sustainable decisions related to optimal use of resources, improving the quality of crops, and systematic management of daily farm activities. Also, Neo4j has the ability to utilize real-time data, which is helpful in monitoring the agricultural field at regular intervals, thereby enhancing the crop yield and efficient usage of resources. Farmers can also work in association with research scientists and agricultural specialists by sharing this predictive data on a common platform.

Implementing Neo4j in smart farming has its own challenges and limitations. First, working on a specialized tool like Neo4j requires prior technical knowledge, which is not possessed by most Indian famers. The insights obtained from predictive models are not easily accepted by farmers and require integration of more technologies for their validation, which makes it a time-consuming process. Although Neo4j has the ability to collect heterogenous data, the accuracy of the model developed by it depends on the quality of the agricultural data provided by the farms. Not all farms have sufficient resources and finances to support the installation of sensors or IoT devices to get high-quality information regarding their farms.

Neo4j has still a long way to go in evolving the field of smart farming. The complex relations identified by using Neo4j can be integrated with artificial intelligence to validate and identify more patterns and trends from the connected data. Also, developing highly precise digital twins which replicate the original conditions of the farm can be helpful in observing the effect of the information provided by the predictive models using Neo4j. This tool enables in-depth analysis of the relations between different parameters involved in farming and opens the way for endless opportunities to maximize the yield and profitability of the agriculture industry.

REFERENCES

Abdullahi, U. S., Nyabam, M., Orisekeh, K., Umar, S., Sani, B., David, E., & Umoru, A. A. (2019). Exploiting IoT and LoRaWAN technologies for effective livestock monitoring in Nigeria. *Arid Zone Journal of Engineering, Technology and Environment*, *15*, 146–159.

Aggregating Functions—Cypher Manual. (n.d.). Neo4j Graph Data Platform. https://neo4j.com/docs/cypher-manual/current/functions/aggregating/

Ahmad, T., & Aziz, M. N. (2019). Data preprocessing and feature selection for machine learning intrusion detection systems. *ICIC Express Letters*, *13*(2), 93–101.

Aitlmoudden, O., Housni, M., Safeh, N., & Namir, A. (2023). A microservices-based framework for scalable data analysis in agriculture with IoT integration. *International Journal of Interactive Mobile Technologies*, *17*(19).

Alfred, R., Obit, J. H., Chin, C. P. Y., Haviluddin, H., & Lim, Y. (2021). Towards paddy rice smart farming: A review on big data, machine learning, and rice production tasks. *IEEE Access*, *9*, 50358–50380.

Al Shorman, A., Faris, H., & Aljarah, I. (2020). Unsupervised intelligent system based on one class support vector machine and Grey Wolf optimization for IoT botnet detection. *Journal of Ambient Intelligence and Humanized Computing*, *11*(7), 2809–2825.

Antonakakis, M., April, T., Bailey, M., Bernhard, M., Bursztein, E., Cochran, J., . . . & Zhou, Y. (2017). Understanding the Mirai botnet. In *26th USENIX Security Symposium (USENIX Security 17)* (pp. 1093–1110). https://www.usenix.org/conference/usenixsecurity17/technical-sessions/presentation/antonakakis

Arrubla-Hoyos, W., Ojeda-Beltrán, A., Solano-Barliza, A., Rambauth-Ibarra, G., Barrios-Ulloa, A., Cama-Pinto, D., . . . & Manzano-Agugliaro, F. (2022, September 26). Precision agriculture and sensor systems applications in Colombia through 5G networks. *Sensors*, *22*(19), 7295. https://doi.org/10.3390/s22197295

Ballas, I., Tsakanikas, V., Pefanis, E., & Tampakas, V. (2020, November). Assessing the computational limits of GraphDBs' engines: A comparison study between Neo4j and Apache Spark. In *Proceedings of the 24th Pan-Hellenic Conference on Informatics* (pp. 428–433). https://www.researchgate.net/publication/349827145_Assessing_the_computational_limits_of_GraphDBs'_engines_-_A_comparison_study_between_Neo4j_and_Apache_Spark

Common Plant Diseases & Disease Control for Organic Gardens. (2021, November 5). Earth's Ally—safe for people, pets & planet. https://earthsally.com/disease-control/common-plant-diseases.html

Dagar, R., Som, S., & Khatri, S. K. (2018, July). Smart farming-IoT in agriculture. In *2018 International Conference on Inventive Research in Computing Applications (ICIRCA)* (pp. 1052–1056). IEEE.

de Araujo Zanella, A. R., da Silva, E., & Albini, L. C. P. (2022). CEIFA: A multi-level anomaly detector for smart farming. *Computers and Electronics in Agriculture*, *202*, 107279.

Doshi, J., Patel, T., & Kumar Bharti, S. (2019). Smart farming using IoT, a solution for optimally monitoring farming conditions. *Procedia Computer Science*, *160*, 746–751.

Elbasi, E., Zaki, C., Topcu, A. E., Abdelbaki, W., Zreikat, A. I., Cina, E., . . . & Saker, L. (2023). Crop prediction model using machine learning algorithms. *Applied Sciences*, *13*(16), 9288.

Fernandes, D., & Bernardino, J. (2018). Graph databases comparison: AllegroGraph, ArangoDB, InfiniteGraph, Neo4J, and OrientDB. *Data*, *10*. https://doi.org/10.5220/0006910203730380

Ferrag, M. A., Shu, L., Yang, X., Derhab, A., & Maglaras, L. (2020). Security and privacy for green IoT-based agriculture: Review, blockchain solutions, and challenges. *IEEE Access*, *8*, 32031–32053.

Furqon, M. A., & Bukhori, S. (2023). Knowledge graph construction for rice pests and diseases. *International Journal of Artificial Intelligence Research*, *7*(1), 88–97.

Giannakis, T. (n.d.). *Neo4j Knowledge Graph | Tom Sawyer Software*. https://blog.tomsawyer.com/neo4j-knowledge-graphs-data-connectivity-and-intelligence

Ginaldi, F., Bajocco, S., Bregaglio, S., & Cappelli, G. (2019). Spatializing crop models for sustainable agriculture. *Innovations in Sustainable Agriculture*, 599–619.

Gorli, R., & Yamini, G. (2017). Future of smart farming with Internet of things. *Journal of Information Technology and Its Applications*, *2*(1).

Gruschka, N., & Jensen, M. (2010, July). Attack surfaces: A taxonomy for attacks on cloud services. In *2010 IEEE 3rd International Conference on Cloud Computing* (pp. 276–279). IEEE.

Guia, J., Soares, V. G., & Bernardino, J. (2017, April). Graph Databases: Neo4j Analysis. In *International Conference on Enterprise Information Systems* (vol. 1, pp. 351–356). ICEIS.

How Continuous-Learning AI Sharpens Predictions. (n.d.). Algolia. https://www.algolia.com/blog/ai/how-continuous-learning-lets-machine-learning-provide-increasingly-accurate-predictions-and-recommendations/

Iglesias, I., Kanankege, K., Valencia, E. J., Liu, C. J., Blanco, C. E., Lim, S., . . . & Perez, A. (2022). DiFLUsion: A new spatiotemporal early warning system for HPAI. *International Journal of Infectious Diseases*, *116*, S101.

The Impact of Indexes on Query Performance—Cypher Manual. (n.d.). Neo4j Graph Data Platform. https://neo4j.com/docs/cypher-manual/current/indexes/search-performance-indexes/using-indexes/

Improving Cypher Performance—Neo4j Aura. (n.d.). Neo4j Graph Data Platform. https://neo4j.com/docs/aura/tutorials/performance-improvements/

Innovative Ways to Enhance ML Models with Feature Engineering. (n.d.). https://www.markovml.com/blog/feature-engineering-best-practices

Kaewmard, N., & Saiyod, S. (2014, October). Sensor data collection and irrigation control on vegetable crop using smart phone and wireless sensor networks for smart farm. In *2014 IEEE Conference on Wireless Sensors (ICWiSE)* (pp. 106–112). IEEE.

Khan, Z., Zahid Khan, M., Ali, S., Abbasi, I. A., Ur Rahman, H., Zeb, U., . . . & Huang, J. (2021). Internet of things-based smart farming monitoring system for bolting reduction in onion farms. *Scientific Programming*, *2021*, 1–15.

Ko, D., Kwak, Y., & Song, S. (2014). Real time traceability and monitoring system for agricultural products based on wireless sensor network. *International Journal of Distributed Sensor Networks, 10*(6), 832510.

Kolias, C., Kambourakis, G., Stavrou, A., & Voas, J. (2017). DDoS in the IoT: Mirai and other botnets. *Computer, 50*(7), 80–84.

Krisnawijaya, N. N. K., Tekinerdogan, B., Catal, C., & van der Tol, R. (2022). Data analytics platforms for agricultural systems: A systematic literature review. *Computers and Electronics in Agriculture, 195*, 106813.

Kumar, H., & Menakadevi, T. (2017). A review on big data analytics in the field of agriculture. *International Journal of Latest Transactions in Engineering and Science, 1*(4), 1–10.

Lazarević, B., & Popović, T. (2020). Meteorological data aggregation system for application in agriculture 2.0. In *7th International Conference on Electrical, Electronic and Computing Engineering IcETRAN 2020 at Belgrade*. IcETRAN.

Lokhande, S. A. (2021, March). Effective use of big data in precision agriculture. In *2021 International Conference on Emerging Smart Computing and Informatics (ESCI)* (pp. 312–316). IEEE.

Mallinger, K., & Baeza-Yates, R. (2024). Responsible AI in farming: A multi-criteria framework for sustainable technology design. *Applied Sciences, 14*(1), 437.

Mandapuram, M., Mahadasa, R., & Surarapu, P. (2019). Evolution of smart farming: Integrating IoT and AI in agricultural engineering. *Global Disclosure of Economics and Business, 8*(2), 165–178.

Montag, D. (2013). *Understanding Neo4j Scalability*. White Paper, Neotechnology. http://info.neotechnology.com/rs/neotechnology/images/UnderstandingNeo4jScalability(2).pdf

Navarro, E., Costa, N., & Pereira, A. (2020). A systematic review of IoT solutions for smart farming. *Sensors, 20*(15), 4231.

Needham, M., & Hodler, A. E. (2018). A comprehensive guide to graph algorithms in Neo4j. *Neo4j.com*. https://go.neo4j.com/rs/710-RRC-335/images/Comprehensive-Guide-to-Graph-Algorithms-in-Neo4j-ebook-EN-US.pdf

NEO4j Basic Commands. (n.d.). DBA genesis support. https://support.dbagenesis.com/post/neo4j-basic-commands

Neo4j Graph Data Platform. (n.d.). What is a graph database?—Getting started. https://neo4j.com/docs/getting-started/get-started-with-neo4j/graph-database/

Pak, J., Kim, J., Park, Y., & Son, H. I. (2022). Field evaluation of path-planning algorithms for autonomous mobile robot in smart farms. *IEEE Access, 10*, 60253–60266.

Parolini, G. (2015). The emergence of modern statistics in agricultural science: analysis of variance, experimental design and the reshaping of research at Rothamsted experimental station, 1919–1933. *Journal of the History of Biology, 48*, 301–335.

Praveen, P., Shaik, M. A., Kumar, T. S., & Choudhury, T. (2021). Smart farming: Securing farmers using block chain technology and IoT. In *Blockchain Applications in IoT Ecosystem* (pp. 225–238). Springer International Publishing.

Qayyum, T., Trabelsi, Z., Malik, A., & Hayawi, K. (2021). Trajectory design for UAV-based data collection using clustering model in smart farming. *Sensors, 22*(1), 37.

Qin, H., & Yao, Y. (2021, February). Agriculture knowledge graph construction and application. In *Journal of Physics: Conference Series* (Vol. 1756, No. 1, p. 012010). IOP Publishing.

Sabu, K. M., & Kumar, T. M. (2020). Predictive analytics in agriculture: Forecasting prices of arecanuts in Kerala. *Procedia Computer Science, 171*, 699–708.

Scale a Neo4j Deployment—Operations Manual. (n.d.). Neo4j Graph Data Platform. https://neo4j.com/docs/operations-manual/current/kubernetes/operations/scaling/

Shabani, I., Biba, T., & Çiço, B. (2022). Design of a cattle-health-monitoring system using microservices and IoT devices. *Computers, 11*(5), 79.

Sharma, K., Sharma, C., Sharma, S., & Asenso, E. (2022). Broadening the research pathways in smart agriculture: Predictive analysis using semiautomatic information modeling. *Journal of Sensors, 2022*, 1–19.

Shi, Y. X., Zhang, B. K., Wang, Y. X., Luo, H. Q., & Li, X. (2021). Constructing crop portraits based on graph databases is essential to agricultural data mining. *Information, 12*(6), 227.

Siddiquee, K. N. E. A., Islam, M. S., Singh, N., Gunjan, V. K., Yong, W. H., Huda, M. N., & Naik, D. B. (2022). Development of algorithms for an IoT-based smart agriculture monitoring system. *Wireless Communications and Mobile Computing, 2022*, 1–16.

Sishodia, R. P., Ray, R. L., & Singh, S. K. (2020). Applications of remote sensing in precision agriculture: A review. *Remote Sensing, 12*(19), 3136.

Sood, K., Singh, S., Rana, R. S., Rana, A., Kalia, V., & Kaushal, A. (2015, October). Application of GIS in precision agriculture. In *Paper Presented as Lead Lecture in National Seminar on Precision Farming Technologies for High Himalayas* (pp. 4–5). https://www.researchgate.net/publication/295858552_Application_of_GIS_in_precision_agriculture

Stoica, I., Adkins, D., Zhuang, S., Shenker, S., & Surana, S. (2004). Internet indirection infrastructure. *IEEE/ACM Transactions on Networking*, *12*(2), 205–218.

Stubert, L., Martín i Oliveras, A., Märker, M., Schernthanner, H., & Vogel, S. (2020). Viticulture in the Laetanian region (Spain) during the Roman period: Predictive modelling and geomatic analysis. *Geosciences*, *10*(6), 206.

Subahi, A. F., & Bouazza, K. E. (2020). An intelligent IoT-based system design for controlling and monitoring greenhouse temperature. *IEEE Access*, *8*, 125488–125500.

Sunder, S. (2018, January 29). India economic survey 2018: Farmers gain as agriculture mechanisation speeds up, but more R&D needed. *Financial Express*. https://www.financialexpress.com/budget/india-economic-survey-2018-for-farmers-agriculture-gdp-msp-1034266/

T. & T. (2023, December 13). *Farm Management Software: Shaping the Future of Agriculture. Blockchain for Food Safety, Traceability and Supplychain Transparency*. https://tracextech.com/farm-management-software-for-sustainable-agriculture/

Tivoli Netcool Performance Manager 1.4.2. (n.d.). https://www.ibm.com/docs/en/tnpm/1.4.2?topic=data-aggregation

Vibhute, A. D., & Gawali, B. W. (2013). Analysis and modeling of agricultural land use using remote sensing and geographic information system: A review. *International Journal of Engineering Research and Applications*, *3*(3), 81–91.

Wolfert, S., Ge, L., Verdouw, C., & Bogaardt, M. J. (2017). Big data in smart farming: A review. *Agricultural Systems*, *153*, 69–80.

Xiang, M., Wei, S., Zhang, M., & Li, M. Z. (2016). Real-time monitoring system of agricultural machinery operation information based on ARM11 and GNSS. *IFAC-PapersOnLine*, *49*(16), 121–126.

Xiaoxue, L., Xuesong, B., Longhe, W., Bingyuan, R., Shuhan, L., & Lin, L. (2019). Review and trend analysis of knowledge graphs for crop pest and diseases. *IEEE Access*, *7*, 62251–62264.

Yadav, R., Rathod, J., & Nair, V. (2015). Big data meets small sensors in precision agriculture. *International Journal of Computer Applications*, *975*(1), 8887–8895.

Ye, D., & Zhang, T. Y. (2019). Summation detector for false data-injection attack in cyber-physical systems. *IEEE Transactions on Cybernetics*, *50*(6), 2338–2345.

Yépez-Ponce, D. F., Salcedo, J. V., Rosero-Montalvo, P. D., & Sanchis, J. (2023). Mobile robotics in smart farming: Current trends and applications. *Frontiers in Artificial Intelligence*, *6*.

Yoon, B. H., Kim, S. K., & Kim, S. Y. (2017). Use of graph database for the integration of heterogeneous biological data. *Genomics & informatics*, *15*(1), 19.

Zamora-Izquierdo, M. A., Santa, J., Martínez, J. A., Martínez, V., & Skarmeta, A. F. (2019). Smart farming IoT platform based on edge and cloud computing. *Biosystems Engineering*, *177*, 4–17.

Zhu, D., Xie, L., Chen, B., Tan, J., Deng, R., Zheng, Y., . . . & Andrew, W. H. (2023). Knowledge graph and deep learning based pest detection and identification system for fruit quality. *Internet of Things*, *21*, 100649.

Zuma-Netshiukhwi, G. N. (2013). *The Use of Operational Weather and Climate Information in Farmer Decision Making Exemplified for the South Western Free State, South Africa* (Doctoral dissertation, University of the Free State).

Zuma-Netshiukhwi, G. N., Stigter, K., & Walker, S. (2013). Use of traditional weather/climate knowledge by farmers in the South-Western Free State of South Africa: Agrometeorological learning by scientists. *Atmosphere*, *4*(4), 383–410.

13 Tech-Driven Harvest
Blockchain's Role in Sustainable Food Systems and Waste-Free Future

Ruchi Jakhmola Mani, Suyashi Singh, Subham Sonu, and Deepshikha Pande Katare

13.1 INTRODUCTION

13.1.1 The Global Food Waste Crisis: Scope and Impact (Environmental, Economic, Social)

A food supply chain (FSC) in food systems is a network of related organizations working together to govern the flow of products and services throughout the value chain to provide customers with high value while keeping costs under control. In contrast, a sustainable food supply chain (SFSC) preserves the integrity of natural ecosystems through the use of ecologically friendly food production and consumption methods. The SFSC is a holistic concept driven by customer preferences. The Food and Agriculture Organization (FAO) claims that this approach protects social, economic, and environmental aspects while guaranteeing food security and nutrition, so guaranteeing the supply of healthy food for future generations (Lopez Barrera & Hertel, 2020).

To achieve sustainability, social justice, environmental protection, and economic prosperity must be balanced as shown in Figure 13.1. This can be done by using robust and effective processing techniques that satisfy present needs without jeopardizing the capacity to satisfy future ones. Businesses that use these strategies can increase their profitability by meeting client expectations, cutting waste, producing high-quality goods, and improving long-term performance. Food waste (FW) is a complex problem that integrates many facets of sustainable production; nevertheless, it is hampered by varying definitions and concepts that lack agreement. A careful balance between environmental protection and quality of life is required because food waste has an impact on the economic, social, and environmental pillars of sustainability. The last phases of the production cycle generate food waste, which is notable for its high costs. These costs include energy consumption, labour costs, and cleaning supplies. A major amount of the waste is accounted for by consumption and food services ("Global Effects of Food Waste," 2020). Understanding the impacts of food waste can inform strategies to mitigate it, bolster local sustainability, and advocate for public policies that promote sustainable practices in the food industry, thus addressing the issue comprehensively. Blockchain technology (BCT) emerged as a transformative tool, introduced by Satoshi Nakamoto, offering a decentralized, transparent, and secure platform for peer-to-peer transactions. BCT enhances the efficiency and traceability of supply chains, addressing challenges such as food recalls and aligning with Sustainable Development Goal 12 (Responsible Consumption and Production) by reducing waste and pollution throughout the FSC (Vishkaei & De Giovanni, 2022).

DOI: 10.1201/9781003507390-13

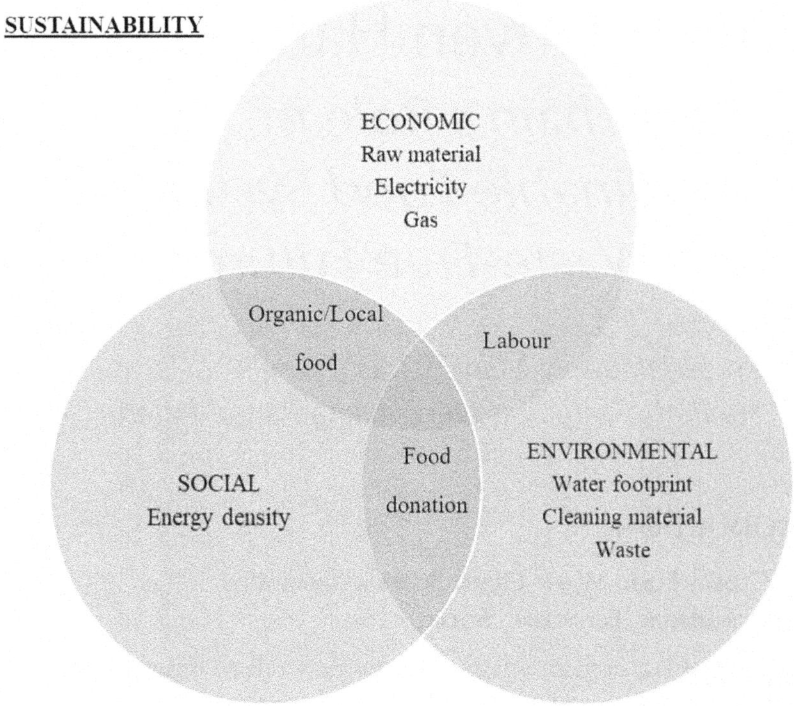

FIGURE 13.1 Impacts of food waste in food service (Lins et al., 2021).

13.1.2 The Need for Innovation: Smart Technology as a Solution

In today's rapidly evolving global community, characterized by dynamic shifts in globalization and technological advancements, organizations are facing new market dynamics propelled by major trends. In this context, innovation has become a necessity for organizations not only to thrive but also to survive. However, the significance of innovation transcends its traditional role of creating value for individuals, organizations, or societies (Farooq et al., 2022). Instead, its ultimate aim should be to contribute to the creation of a smarter future, where people can experience the highest possible quality of life. Throughout history, human survival and progress have relied heavily on continuous innovation efforts. Today, innovation must not only address pressing social challenges and uncertainties about the future but also break down barriers hindering the realization of a smarter future (Lee & Trimi, 2018). A smarter future involves developing intelligent solutions to complex problems, thus ensuring a humane and sustainable environment. One significant challenge in this pursuit is food waste, influenced by various factors such as consumer behaviour, retail strategies, and confusion over product expiration dates. Despite the widespread acknowledgement of this challenge, the scientific community has yet to fully explore effective methods and tools for preventing food waste and fostering a shift in consumer attitudes. However, preventing food waste is pivotal for achieving sustainability goals, emphasizing the importance of proactive measures over waste management (Makanjuola et al., 2020).

It is crucial to inform customers about sustainable food practices for everything from buying to throwing away. The environmental impact of food storage can be minimized by taking easy yet effective procedures, like maximizing refrigerator settings, which can greatly lengthen food storage life and reduce waste (Panghal et al., 2023). Researchers have made significant progress in creating smart refrigerators with sophisticated capabilities that check food inventory, track consumption trends, and give consumers personalized waste-reduction advice. These developments, which range from sophisticated data tracking systems to interactive displays, are meant to enable customers to live more sustainably and to reduce food waste (Farooq et al., 2022).

13.2 UNDERSTANDING FOOD WASTE: A BREAKDOWN BY STAGE

13.2.1 Production and Storage

Food waste is a major worldwide problem, with supply chain inefficiencies accounting for up to one-third of the food that is purposefully created for human use going to waste. This chapter examines the various aspects of food waste, with a particular emphasis on its occurrence in the stages of production and storage (Raak et al., 2017). Agricultural techniques are vulnerable to several factors that lead to food loss, such as variations in the weather, pest infestations, illnesses, and the high aesthetic standards that consumers and retailers have established. The problem is made worse by overproduction, which is motivated by contractual obligations and market uncertainties (Bosona & Gebresenbet, 2023). While efficient storage is essential for reducing waste, problems with supply chain disruptions, low inventory management, and inadequate infrastructure all lead to large losses. Food waste must be addressed with a comprehensive strategy that includes all supply chain participants (Mondol, 2021). Blockchain technology provides traceability and transparency across the production and storage processes, which presents intriguing alternatives. Blockchain technology improves accountability, efficiency, and transparency in production and storage processes, which eventually results in a more robust food system with less waste (Trabold & Nair, 2018).

13.2.2 Processing and Distribution

Because blockchain technology allows for accountability, traceability, and transparency, food waste might be significantly reduced. It does this by generating an unchangeable ledger that documents each food transaction and movement from farm to table, including vital tasks like packing, transportation, processing, and harvesting (Ahmed et al., 2018). Access to comprehensive information on the origin, processing methods, and expiration date of food products is possible for consumers and stakeholders, fostering trust and enabling them to make well-informed decisions (Srinivasan et al., 2017). Blockchain-based smart contracts simplify quality control processes by defining standards for food products at every stage of distribution and processing. Notifications are generated by any discrepancies, guaranteeing that clients only receive high-quality merchandise. The blockchain may also be used to record real-time environmental variable monitoring, such as temperature and humidity, during storage and transit. This would enable stakeholders to promptly address any unfavourable conditions and prevent spoiling and waste. Blockchain speeds up ownership transfers between stakeholders, reduces paperwork, and eliminates human error in supply chain operations. Food supplies may be delivered to their destination more quickly and with less risk of degradation during transit thanks to smart contracts, which also make payment settlements simpler and ensure contractual compliance (Ahmed et al., 2018). Blockchain allows food products to be redistributed to food banks, charities, or other groups via a decentralized marketplace when they are due to expire or are deemed unsellable. By facilitating the transfer of edible food from surplus food producers to those in need, this open network helps to prevent food waste. By identifying trends, patterns, and inefficiencies in food production and distribution, blockchain data analysis can help stakeholders enhance operations, reduce waste, and promote sustainability (Yousefi et al., 2019). Blockchain-enabled systems can offer predictive analytics by integrating artificial intelligence (AI) and machine learning. This would enable stakeholders to foresee fluctuations in demand and adjust production and distribution, accordingly, minimizing food waste (Srinivasan et al., 2017).

13.2.3 Retail and Consumption

Understanding the intricacies of food waste requires investigating its occurrence at several stages, notably during the retail and consumption phases. At the retail level, a variety of variables contribute to food waste. Retailers frequently overstock perishable commodities to suit consumer demand, resulting in surplus produce that may go unsold and thrown. Furthermore, rigorous aesthetic criteria imposed by retailers result in the rejection of imperfect-looking fruits and vegetables, increasing waste (Friha et al., 2021). Furthermore, regardless of their potential for consumption, products are sometimes discarded prematurely due to expiration dates and sell-by labels. In the consumption stage, customers make a major contribution to food waste. Poor meal planning and impulse buying lead to purchasing more food than is necessary, often resulting in things expiring before they are consumed. Misunderstandings concerning expiration dates, such as the difference between "best before" and "use by" labels, lead to the early disposal of edible products. Leftovers from meals are routinely discarded due to poor storage or negligence, aggravating the food waste problem (Sayem et al., 2022). To effectively minimize food waste, a complete solution encompassing both the retail and consumption stages is required. To reduce waste, shops could enhance inventory management, relax cosmetic standards, or implement donation programs. Meanwhile, consumer education initiatives that focus on meal planning, proper storage techniques, and reading expiry labels may help individuals make more informed decisions and prevent waste in their homes (Yousefi et al., 2019). By fully comprehending the complexities of food waste throughout the retail and consumption phases, stakeholders may devise tailored interventions to prevent its negative economic and environmental consequences.

13.3 LEVERAGING SMART TECHNOLOGY FOR FOOD WASTE REDUCTION

13.3.1 Smart Farming and Precision Agriculture (Optimizing Yield and Quality, Minimizing Spoilage in Pre-Harvest Stages)

Climate change exacerbates food loss in agricultural output by adding unpredictable variables. These losses are the result of limited resources, faulty harvesting practices, and inadequate post-harvest management and storage. Traditional agricultural procedures lack the accuracy required for accurate plant development tracking. In the future, agricultural procedures such as sowing, harvesting, irrigation, fertilization, seed selection, crop diversity, and new management approaches will need to be adjusted (Mondol, 2021). Various algorithms evaluate the networked data, providing decision-makers with actionable insights via mobile devices in real time. Industry 4.0 has brought digital technologies that function as data conduits, assisting decision support platforms (Kayikci et al., 2022).

The Internet of Things (IoT) uses networked smart devices with sensors to share data and initiate actions. IoT, when combined with other technologies, allows for real-time data flow and improves process and service management across sectors. Geographical information systems (GISs) handle location-based data, making it easier to acquire, manage, and analyze. GIS allows for easy data integration, resulting in detailed insights through map display (Kayikci et al., 2022). AI simulates human thought processes and decision-making, enhancing reliability through algorithmic learning and comparison data. AI supports complex data processing, accelerating decision-making mechanisms and enhancing overall efficiency. Remote sensing involves collecting and analyzing information about Earth and its objects using instruments on platforms such as satellites, planes, and drones. It enables monitoring of physical properties in various regions, including cities, forests, and agricultural areas, without direct contact. Image processing involves manipulating images using software to extract meaningful information and produce visually, numerically, and graphically represented results after adjustments (Nasir et al., 2018).

13.3.2 Intelligent Inventory Management in Supply Chains (Real-Time Tracking and Monitoring, Demand Forecasting, and Dynamic Logistics)

Supply chain management performance refers to the design of activities aimed at meeting the needs of end users, emphasizing product availability and secure delivery (Annosi et al., 2021). Evaluating supply chain performance involves parameters like inventory turnover rate, shipment timing, and cash-to-cash cycle time. These metrics enable organizations to develop and maintain an efficient supply chain management system that is both cost-effective and capable of tracking all activities. Inventory management poses challenges for organizations due to the difficulty of predicting demand, which varies across industries. Smart inventory systems address these challenges by monitoring inventory levels, ensuring scalability, and providing security and backups. The effectiveness of supply chain management lies in meeting the needs of end users, ensuring product availability, and secure delivery. Key metrics such as inventory turnover, shipment timing, and cash-to-cash cycle time gauge supply chain performance, guiding organizations in building efficient systems (Buisman et al., 2019). However, forecasting demand and managing inventory pose challenges, especially across diverse industries. Smart inventory systems, utilizing advanced technology, enable accurate tracking of inventory levels and scalability while ensuring security and backup. Supply chain success heavily relies on blockchain and smart inventory systems. A smart inventory system, characterized by innovation and advanced technology, aims to manage inventory flow to meet customer demands efficiently. Supported by robust software, these systems enhance efficiency, synchronize inventory, and minimize costs while maximizing profits. Conversely, supply chain performance centres on customer-focused activities, ensuring timely product availability to enhance satisfaction (Mastorakis et al., 2024).

Smart inventory management in supply chains combines real-time tracking, demand forecasting, and dynamic logistics to maximize efficiency and responsiveness. Using technology and data analytics, these systems keep proper inventory levels, reducing stockouts and surplus inventory (Friha et al., 2021). Real-time monitoring uses sensors, RFID tags, or GPS to monitor inventory movement and position, allowing for better decision-making and risk reduction. Demand forecasting uses historical data and predictive analytics to effectively estimate future demand, maximizing inventory levels and reducing expenses. Dynamic logistics adjusts transportation and distribution tactics in reaction to changes in demand or supply, reducing delivery time and cost. Integrating intelligent inventory management principles into supply chain operations increases efficiency, lowers costs, and boosts customer satisfaction (Nasir et al., 2018).

13.3.3 Smart Packaging and Food Storage Solutions (Extending Shelf Life, Monitoring Freshness and Spoilage)

A comprehensive assessment of food quality encompasses various elements such as nutritional content, adherence to consumer standards, stability, and other factors that vary across different regions. Ensuring food quality is pivotal in safeguarding consumer health, prompting the food industry to strive for accurate quantification methods (Gemtou et al., 2024). Presently, expiration dates serve as a primary tool for gauging food quality and determining recall timelines. Even when regulatory criteria are followed, a lack of real-time knowledge regarding food sources and conditions increases the risk of foodborne disease. Conversely, premature food recalls result in significant economic losses annually, underscoring the need for more precise methods beyond predetermined expiration dates. Traditionally, packaging materials have acted as passive barriers, shielding food products from environmental factors like moisture, oxygen, and contaminants. However, advancements in materials and detection methods have broadened the functionality of packaging, giving rise to active and intelligent packaging solutions. These innovations aim to bolster food safety by monitoring and enhancing food quality in real time, potentially reducing reliance on preservatives. Active packaging actively interacts with the product, package, and environment to improve food conditions

TABLE 13.1
Developed Sensors for Monitoring Packaged Food Quality (Yousefi et al., 2019)

Detection Target	Sensing Mechanism	Reporting Method	Substrate/Material
Time-temp	Redox reaction to oxygen diffusion	Colour change	Paper or plastic
CO_2	pH	Colour change	Plastic film, aqueous solution
	Luminescence dye	Fluorescence signal	Plastic film, organically modified silica matrix
O_2	Redox dye, reducing agent, alkaline environment	Colour change	Plastic, organic polymer
Humidity	Change in the dielectric constant	Wireless detector device	Paper
	Change in the dielectric constant	RFID reader	Paper

and extend shelf life. Intelligent packaging, on the other hand, provides consumers with insights into food quality and safety through various signals (Gemtou et al., 2024).

To integrate sensors into food packaging, particular indications must be used to statistically or qualitatively analyze the status of the food product. Physical or chemical changes inside the product are frequently cited as indicators of food deterioration. Key indicators as mentioned in Table 13.1, commonly utilized for assessing food quality and safety include oxygen and carbon dioxide levels, pH changes, humidity, temperature, and specific chemical markers. Oxygen and carbon dioxide levels influence oxidation reactions and microbial growth, while pH changes reflect microbial fermentation. Maintaining optimal humidity levels is crucial for preserving food texture and preventing microbial growth. Temperature fluctuations, particularly in refrigerated products, impact shelf life by creating favourable conditions for bacterial growth. Additionally, animal products are prone to pathogenic microbial growth, leading to the release of specific chemicals during spoilage (Mastorakis et al., 2024).

Methods for Monitoring Food Quality: As previously indicated, various signs can signify a decline in the quality of consumer food items.

Tracing Devices

- Hand-held devices: The utilization of handheld devices, like smartphones, for real-time testing, is gaining traction due to their accessibility and reliability. Integrating software with sensors allows smartphones to process data gathered by the sensor into user-friendly information. While smartphones boast significant technological capabilities, they cannot replace laboratory-grade equipment on their own. Augmenting them with specific attachments is necessary to make them suitable for real-time food monitoring applications (Mastorakis et al., 2024).
- RFID: RFID technology takes three approaches: passive, semi-active, and active. In passive RFID, tags use radio waves from a reader to generate a magnetic field, which powers the tag and transmits data back to the reader. Semi-active RFID employs a battery to store memory and modify reader antenna waves, which improves communication. Active RFID tags have a battery that powers the microchip circuitry for signal transmission, resulting in longer ranges and more communication capabilities (Mastorakis et al., 2024).

13.4 APPLICATIONS OF SMART TECHNOLOGY IN FOOD WASTE REDUCTION

13.4.1 Smart Refrigeration Systems

When discussing the Internet of Things, particularly household technology, the concept of a "smart kitchen" often arises. This is because the kitchen tends to generate a significant amount of waste and consumes a considerable amount of energy within the home. Consequently, manufacturers continually strive to develop smarter kitchen appliances that not only minimize waste and energy usage but also enhance convenience for users. Among these appliances, the refrigerator stands out as a staple in most households. It comprises an insulated compartment and a mechanism (whether mechanical, electronic, or chemical) that transfers heat from inside the fridge to its surroundings, thus cooling the interior to a temperature lower than the ambient room temperature. Refrigeration is vital for food storage in developed countries, as lower temperatures inhibit bacterial growth, thereby reducing the rate of food spoilage. However, even with expensive, branded refrigerators, food freshness can be challenging to maintain over extended periods, potentially leading to wastage (Casino et al., 2020).

Moreover, modern lifestyles often leave users with limited time to regularly inspect the condition of stored food. Consequently, they may forget what items are in the fridge and for how long they've been there, leading to unpleasant discoveries of spoiled food when needed for cooking. To address this issue, smart refrigerator systems have been developed. Leveraging the Internet of Things, these systems utilize sensors and actuators embedded within the appliance to monitor food conditions and alert users accordingly (Kayikci et al., 2022). The Internet of Things enables the seamless communication and interaction of intelligent devices and systems. Through embedded technology, IoT-enabled products can transmit data to users directly or via the internet. In the context of smart refrigeration, these systems inform users of the duration certain foods have been stored in the fridge. When a user-defined time limit is reached, the system issues a warning, prompting users to utilize the food promptly to prevent spoilage and minimize waste (Casino et al., 2020).

13.4.2 Automated Inventory Management Systems

The convergence of the Internet of Things and cloud computing heralds a new era in Internet technology, aimed at seamlessly connecting objects via wired or wireless means, regardless of location or time, and facilitating communication without constraints. IoT facilitates this communication over the internet through protocols like MQTT, enabling smart functionality in connected devices. Data exchange among objects via various platforms supports tracking, control, and management. Implementing an IoT framework within cloud computing can address challenges in supply chain management, while efficient management of agricultural data aids decision-making for farmers and consumers alike (Buisman et al., 2019). In agriculture, an automated system for arranging food distribution, accessible via mobile phones, simplifies operations for distributors and farmers while optimizing demand and supply dynamics. Warehouse capacity planning, which is a vital part of this system, helps to cut harvesting and shipping costs. Internet-enabled Android handsets with NFC readers and GPS antennae may use cellular networks to send data to the cloud, improving vehicle tracking capabilities (Andrieu et al., 2017).

13.4.3 Dynamic Pricing and Discounting Strategies

The Internet of Things (IoT) and other technological advances allow for the collection of large amounts of data to monitor the freshness of bulk products. Using data-driven decision-making and big data methodologies improves corporate performance, especially in the food sector. An IoT-driven pricing model that integrates data from numerous sources such as sensors and mobile apps enables shops to properly monitor and manage perishable products, hence decreasing food wastage. Furthermore, an IoT sensor data-driven pricing model enables edge computing analytics, including real-time data production and automated work order issuing. As a result, retailers may set their product prices based on freshness rankings, without relying on centralized decisions. IoT devices play an important role in giving vital data to stakeholders along the supply chain, allowing for better decision-making and revenue creation. These digitally linked gadgets, which use sensors and communication technologies, expedite operations and enable real-time monitoring, notably in retail, where food loss is predicted and reduced. The following table lists the many types of sensors used in retail to avoid food spoiling and assure product quality and safety. The growing demand for fresh, high-quality items with extended shelf lives has driven shops to install freshness sensors in their supermarket departments (Schedler et al., 2019).

13.4.4 Food Waste Tracking and Analytics Platforms

The degradation of food is caused by a variety of reasons, including microbiological, enzymatic, chemical, physical, and mechanical. As a result, food supply networks require effective logistical systems. The integration of computerization and technology platforms allows for online communication within these supply chains, which aids in the management of agricultural resources. According to research, new technological platforms in supply chains can avert up to 50% of these losses (Ma et al., 2018). Emerging smart agriculture solutions, such as food waste tracking and analytics platforms, have the potential to significantly reduce food loss. These systems use data-collecting and analysis approaches to detect and address waste-related issues, providing actionable insights for prevention (Lin et al., 2018).

Food waste tracking and analytics platforms operate in three main steps:

1. Data Collection: Sensors in farms gather real-time data on parameters like temperature, humidity, and soil moisture. Drones capture imagery to assess crop health and identify potential issues affecting yield.
2. Data Analysis: Analytical models process collected data to detect patterns and trends related to food waste, pinpointing stages in the agricultural process prone to spoilage.
3. Actionable Insights: Based on the analysis, platforms offer farmers recommendations to minimize food waste, including optimizing storage, adjusting harvesting practices, and improving transportation logistics.

The benefits of these platforms include reduced food waste, improved resource management, enhanced profitability, and promotion of sustainable agricultural practices. Examples of such platforms include Trutaste, which predicts shelf life using AI; Too Good To Go, a mobile app connecting surplus food with consumers; and Wasteless, aiding businesses in tracking and reducing food waste (Ahmed et al., 2018; Yousefi et al., 2019).

13.5 BUILDING A SUSTAINABLE FOOD SYSTEM: COLLABORATION AND EDUCATION

13.5.1 Integrating Smart Technologies across the Chain

Digital technology in smart farming uses big data analysis, IoT, AI, remote sensing, robots, and sensors to improve real-time information accuracy, increase production, and enhance farm management tools and decision-making (Srinivasan et al., 2017). Digitalization enables detailed characterization of soil properties and components, revolutionizing processes like precision irrigation. Despite challenges, machine learning holds promise for handling agricultural big data effectively. Integration of IoT and digitalization has improved crop productivity and minimized losses. Mobile apps now offer predictive models for crop yield, productivity, and health, integrating historical data and weather forecasts with government support schemes. Smart digital technologies combine data on soil, weather, irrigation, plant health, and fertilizer use to provide tailored crop management strategies. Integration of smart digital tools with farm machinery enhances precision and efficiency (Kayikci et al., 2022). In precision livestock farming, digitalization aids in disease prediction, welfare analysis, and monitoring of animal behaviour and environmental impacts. However, there's a need for farmers to better understand and utilize data, requiring simplified access to reliable information. Integration of diverse data sources, development of new analysis techniques, and governance structures are crucial for the effective utilization of agricultural data, leading to enhanced economic resilience and environmental sustainability. The proposed concept involves integrating real-time data from various sources, developing an integrated data platform, effectively utilizing AI, and creating decision support systems for robust and reliable farm management across diverse scenarios (Mondol, 2021).

13.5.2 Consumer Awareness and Behaviour Change

To achieve a comprehensive solution, it's essential to address consumer awareness and behaviour (Galli et al., 2019). Here's how focusing on educating consumers and altering their purchasing habits can significantly contribute:

1. Educating Consumers: Smart labelling technologies, such as QR codes, may give customers product origin, storage suggestions, and recipes for near-expiry foods, allowing them to make educated decisions and manage food safely.
2. Promoting Sustainable Consumption: Meal Planning and Portion Control: Educational efforts and apps can assist consumers in planning meals and controlling portions, reducing impulse purchases and food waste.
3. Supporting Businesses with Waste Reduction Initiatives: Consumers can choose to support businesses that implement practices like selling close-to-expiry items at a discount, fostering a culture of waste reduction.
4. The Power of Technology for Consumers: Food waste tracking apps such as mobile apps can help consumers track their purchases and expiry dates, reminding them to use items before they spoil.
5. Recipe Apps with Leftover Utilization: Apps suggesting dishes incorporating leftover ingredients promote creativity and reduce food waste.
6. The Consumer-Farmer Connection: Smart agriculture technologies can facilitate a stronger connection between consumers and farmers. Initiatives like community-supported agriculture (CSA) programs and subscription services delivering fresh produce directly to homes foster appreciation for food and encourage responsible consumption practices.

By integrating smart agricultural technologies with consumer awareness campaigns and behaviour changes, a sustainable food system can be established, minimizing waste from farm to fork (da Costa et al., 2022).

13.5.3 Policy and Regulatory Incentives

Governmental and regulatory measures play crucial roles in promoting the reduction of food waste within the framework of smart agriculture. Here are various strategies authorities and regulatory bodies can employ:

1. Mandatory Reporting and Monitoring: Governments may require food producers, processors, and retailers to report statistics on food waste generation, disposal methods, and diversion efforts. Monitoring systems provide insights into where food waste occurs, allowing for more focused solutions.
2. Setting Waste Reduction Targets: Setting national or regional targets might guide action towards decreasing food waste. Governments can work with industry stakeholders to

establish ambitious but attainable targets, as well as systems for tracking progress and guaranteeing compliance.

3. Tax cuts and subsidies encourage businesses to invest in waste-reduction technology and infrastructure. This can take the form of tax credits for giving leftover food, incentives for purchasing waste-reduction technologies, or lower taxes for meeting waste-reduction targets.
4. Regulation of Expiry Date Labelling: Governments can govern expiry date labelling to avoid customer confusion and prevent safe-to-eat products from being dumped prematurely. Standardizing label language and providing clear packaging information enables people to make informed decisions about food safety.
5. Facilitating voluntary agreements between food supply chain stakeholders promotes collaboration on waste reduction activities. These agreements may include data sharing, the implementation of best practices, and investments in innovation.

Implementing these legislative and regulatory measures could help the government reduce food waste in smart agriculture. This will promote accountability, innovation, and teamwork throughout the whole food supply chain (Aschemann-Witzel et al., 2017).

13.6 CASE STUDIES AND SUCCESS STORIES

13.6.1 Implementations of Smart Technology in Food Waste Reduction

13.6.1.1 A Case Study of Real-Time Monitoring Technologies and Their Potential Use to Reduce Food Loss and Waste: Key Elements of Food Supply Chains and IoT Technologies

Real-time sensors present a viable way to continuously monitor food quality and reduce FLW since they can track quality characteristics including temperature and humidity. There is still a lack of knowledge on the technique, scope, and efficacy of sensor technology when it comes to FLW monitoring, despite the wealth of literature on the subject. This research offers a comprehensive analysis of 59 studies that address sensor technologies intending to decrease food waste at different points in the food supply chain. It examines the many kinds of IoT technologies that are used and the features of the supply chains in which they are implemented, taking into account variables like product category, supply chain phase, and location. It also explores the intricacies of sensor technology, including communication protocols, application layers, monitored parameters, and data storage techniques, providing insights into the many strategies and lessons discovered throughout implementation.

It focuses on high-perishability goods such as fruits, vegetables, meat, and seafood and emphasizes a practical approach to avoiding food waste from several food categories. Applications for real-time food monitoring may be possible with the combination of spectroscopic and imaging methods in Internet of Things networks. Raman, near-infrared (NIR), Fourier transform infrared (FTIR), 3D fluorescence, and laser-induced breakdown spectroscopy (LIBS) have all been widely studied for real-time food monitoring applications. Wireless communication technologies were employed, with cellular technology being the most popular. The applicability of network protocols is determined by real-world applications, and criteria such as network dependability, roaming capabilities, affordability, and installation costs warrant additional investigation. IoT solutions are predicted to have an impact not just on food production, management, transportation, and storage, but also on social, environmental, and economic outcomes, as mentioned in Table 13.2 (da Costa et al., 2022).

TABLE 13.2
IoT Architecture: Representing the Sensing Layer, Communication Layer, Storage Layer, Application and Control Layer (da Costa et al., 2022)

Stages of Food Loss and Waste		Food Supply Chain	Production Transport Manufacturing Storage Retail	With the Help of IoT, Food Quality Can Be Monitored in Real Time
4 stages of IoT solutions architecture	1.	Sensing layer	Sensors Microcontroller GPS Temperature indicator Humidity indicator	It senses physical changes and creates electronic signals or data
	2.	Communication layer	Gateway Wi-Fi Bluetooth LoRa Mobile system RFID	It makes the data ready to transmit to the cloud or server through protocol
	3.	Storage layer	Cloud Data centre Big data USB MicroSD	It stores all the data sent from the sensing layer
	4.	Application and control layer	Computers Laptops Smartphones Visualization Analytics	It observes the status and analytics through applications and software

13.6.2 Real-World Applications in Retail, Hospitality, and Food Service Industries

13.6.2.1 Blockchain-Based Traceability of Food Supply Chains: A Case Study in the Dairy Sector

In industries where customer safety is paramount, such as food and pharmaceuticals, traceability has developed as an essential component of supply chain management. Consumers have high standards for food supply chains, emphasizing elements linked to safety. The dispersion and complexity of today's FSC networks, however, make it difficult to develop accurate tracing systems. The case study examines Pagonis Sisters and Co., a Greek dairy firm that sells products at around 360 points in Greece and has expanded to 12 European and African countries. The company follows traditional dairy production practices to ensure consumer safety and quality. They provide technical assistance to breeders and conduct daily quality checks on raw materials. The company conducts microbiological and physicochemical investigations on its products. A blockchain-based framework for FSC traceability is presented, focusing on distributed, tamper-proof chain-of-custody technology and smart contracts for managing stakeholders and processes (Casino et al., 2020).

13.6.3 Impact Assessment and Lessons Learned

13.6.3.1 Prioritizing Investments for Climate-Smart Agriculture: Lessons Learned from Mali

Mali faces challenges in agricultural output due to erratic rainfall and climate change, with agriculture accounting for half of the GDP and 75% of the population living in rural areas. Climate-smart agriculture (CSA) aims to increase yields, enhance resilience, and promote a low-emission agriculture economy. Two CSA portfolios and action plans were developed for the Malian-Sudan zone, involving collaboration with national and international partners. The participatory CSA-PF strategy in Mali involves 30 decision-makers from various sectors, including government, academia, research institutes, NGOs, and donors, as shown in Figure 13.2. The process has resulted in priority practices in research and development activities and demands for CSA inclusion in mainstream agricultural policies (Andrieu et al., 2017).

The study distils five key lessons from this endeavour:

1. The initiative is effective in most countries with supportive institutions cognizant of climate change and committed to CSA implementation, facilitating alignment with national and regional contexts.
2. Local leadership familiar with ongoing and future agricultural projects is crucial for successful CSA investment prioritization, enabling the identification and capitalization of opportunities throughout the process.
3. Research scientists play a vital role in overseeing the process and ensuring impartial facilitation.
4. While precise outcome calculations are secondary, the value of indicator assessments lies in facilitating discussions, comparing practices, and fostering stakeholder networking.
5. Constructing agricultural development portfolios based on CSA objectives promotes synergy with sectorial goals, and economic assessments help pinpoint practices with greater potential for farmer adoption.

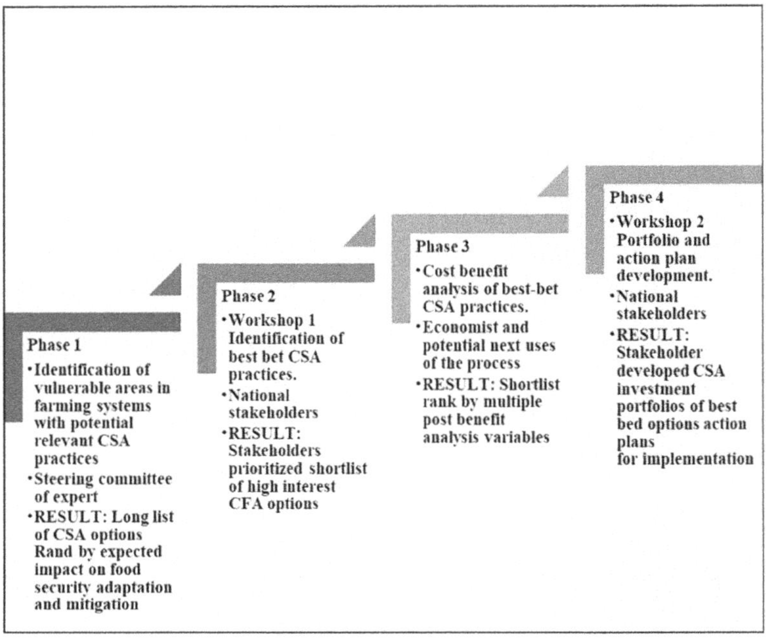

FIGURE 13.2 Overview of the CSA-PF (Andrieu et al., 2017).

13.7 CHALLENGES AND LIMITATIONS OF SMART TECHNOLOGY ADOPTION

13.7.1 Technological Barriers and Infrastructure Requirements

Smart agriculture has the potential to transform food production and minimize food waste, but its road is not without challenges. Smart agriculture technology, including sensors, drones, and automated systems, confronts several technological challenges. Cost and accessibility are major impediments, particularly in underdeveloped countries. Farmers may find these technologies daunting due to their intricacy, which necessitates knowledge and additional expenditures. Data security and privacy problems are especially crucial since these systems collect massive volumes of data. Moreover, not every farmer possesses the necessary computer literacy to effectively utilize these instruments. To close this gap and promote broader use of smart agriculture, training and capacity-building initiatives are essential. Consistent internet connectivity is necessary for smart agriculture solutions, although this might be challenging in rural areas due to limited internet access. Another problem with energy availability is that some systems require a continuous power source. Farmers also need technical help to build, maintain, and troubleshoot systems. Adoption may be hampered by a lack of support services (Mastorakis et al., 2024).

To address the difficulties surrounding smart agriculture, a diversified strategy is required. Government assistance, such as investment in R&D, farmer subsidies, and rural infrastructure development, can all contribute to improving accessibility and affordability. Public-private partnerships between governments, technological companies, and research institutes can help to speed up innovation and create cost-effective solutions. Training programs can help farmers gain digital skills that will enable them to use smart agricultural technologies more effectively. Standardization activities are crucial for enabling smooth system integration and data transmission. Addressing these challenges can lead to a more resilient and accessible smart agricultural ecosystem (Aschemann-Witzel et al., 2017).

13.7.2 Cost Considerations and Return on Investment

The use of smart agriculture technologies may be impeded by financial concerns and an ambiguous return on investment (ROI), despite the promise of reducing food wastage. Particularly for small and medium-sized farmers with limited funding, the initial cost of obtaining these technologies such as sensors, drones, automated irrigation systems, and data management platforms can be high. The costs of upkeep and installation, such as employing specialists to perform installations and handle technological problems, can also add up. Depending on variables like crop yields, market pricing, and farm operation size, the return on investment of smart agricultural systems can be unpredictable and change (Gemtou et al., 2024). It may take time for these technologies' long-term benefits—like better resource management and decreased spoiling—to materialize financially. Farmers must evaluate the long-term benefits of these technologies to ensure sustainable food production. Another difficulty is data reliance, as the performance of smart agriculture solutions is frequently determined by the quality and interpretation of data collected. If farmers lack the experience or resources to successfully analyze and use this data, their ROI may be lower than expected. To overcome these challenges, government incentives, cost-sharing and leasing models, shifting focus from short-term financial gains to long-term benefits, participating in pilot projects and demonstrations, and focusing on cost-effective solutions can help create a more financially accessible smart agriculture ecosystem. This will empower farmers to leverage technology for waste reduction, improve their bottom line in the long run, and contribute to a more sustainable food system (Aschemann-Witzel et al., 2017).

13.7.3 Cultural and Behavioural Shifts

Smart agriculture technologies have the potential to reduce food waste, but they face challenges and limitations due to cultural and behavioural shifts. Traditional practices may be resisted by farmers due to the perceived disruption of established routines. Generational differences may also impact adoption, with younger generations more comfortable with technology. Trust and transparency concerns about data privacy and ownership can be significant hurdles.

Behavioural changes include overcoming a learning curve and modifying decision-making processes to integrate data-driven insights. Another consideration is risk aversion, since farmers may be unwilling to invest in new smart solutions if the potential advantages are unknown. Labour factors are also essential, as smart agriculture can result in higher efficiency and lower labour requirements. Concerns about employment relocation in rural regions may arise, prompting retraining and skill-building activities. Overall, the successful implementation of smart agriculture necessitates a combination of cultural and behavioural adjustments (Ma et al., 2018).

To address cultural and behavioural issues in smart agriculture, options include farmer capacity building and training, community outreach and involvement, culturally sensitive solutions, user-friendly technologies, and emphasis on long-term benefits. Training programs, workshops, and instructional resources can help farmers develop the skills they need to operate and use smart agriculture technologies effectively. Creating solutions that are respectful of local cultural practices and knowledge systems can boost their acceptance among farmers, a concept map for the same is shown in Figure 13.3. Prioritizing user-friendly interfaces, intuitive design, and local language support can help farmers of various digital literacy levels gain access to smart agriculture.

13.8 FUTURE DIRECTIONS AND OPPORTUNITIES

13.8.1 Emerging Trends in Smart Technology for Food Waste Reduction

Food waste reduction is being significantly aided by innovative smart technologies such as machine learning and artificial intelligence, robotics, and automation. AI-powered harvesting robots can harvest fruits and vegetables at optimal ripeness, minimize damage, and extend shelf life. AI-powered sorting and grading systems can reduce food waste due to imperfect appearance. Blockchain technology enhances food traceability, improves safety, and reduces recalls. It tracks

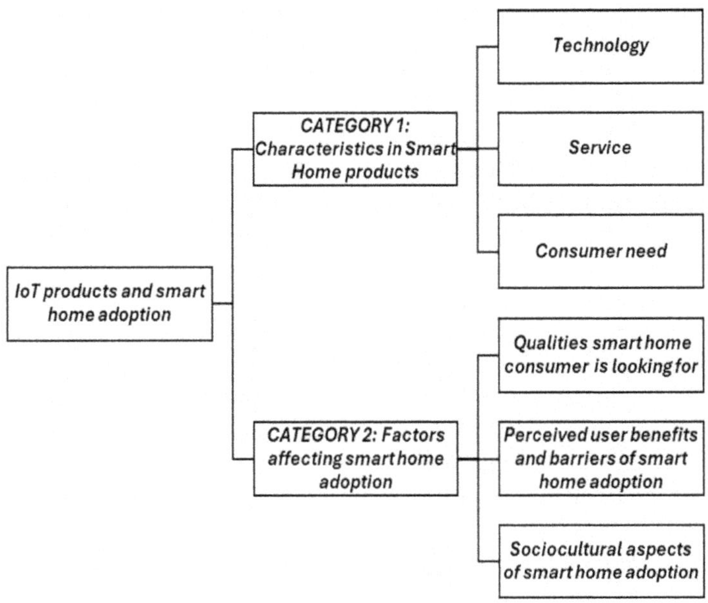

FIGURE 13.3 Concept map for critical analysis (Vaidya & Kalita, 2021).

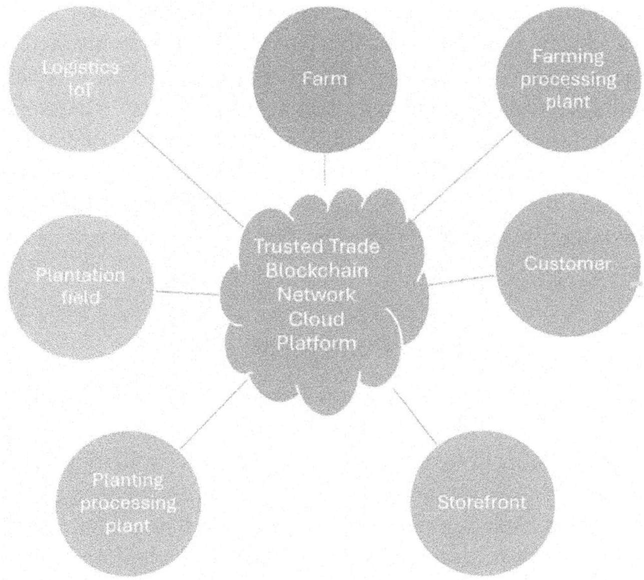

FIGURE 13.4 A smart agriculture ecosystem based on blockchain and IoT Lin et al. (2018).

food contamination outbreaks more effectively, leading to targeted recalls and minimizing food waste. Blockchain empowers consumers by providing information about food origin, storage conditions, and recipes, promoting informed purchasing and reducing food waste. IoT and sensor technology optimize storage and warehouse operations, while smart appliances with waste-reduction features prevent spoilage. Digital platforms and mobile apps connect consumers with discounted food, promoting creativity and reducing household waste. Blockchain technology also improves food safety and consumer confidence, reducing foodborne illnesses and waste. It can promote sustainable practices, creating a more informed consumer base. Challenges for adoption include scalability, infrastructure, standardization, interoperability, and cost (Nasir et al., 2018). Blockchain technology holds great promise for reducing food waste in smart agriculture, as it matures, costs decrease, and interoperability standards improve, a cloud network platform for blockchain and IoT is shown in Figure 13.4. This will create a more transparent, accountable, and sustainable food system benefiting all stakeholders, including farmers, businesses, consumers, and the environment. IBM Food Trust and the World Wildlife Fund are leading blockchain platforms for food traceability and food transparency (Lin et al., 2018).

13.8.2 Innovations in Food Packaging and Preservation

Blockchain technology is being incorporated into smart agriculture, with the UN FAO suggesting that ICT e-agricultural infrastructure components meet both ICT and blockchain technology standards. This is a response to the global issue of food waste, which is being addressed through innovative solutions in food packaging and preservation. Smart packaging uses biodegradable materials like cellulose or algae, which decompose naturally, reducing waste. Active packaging with sensors monitors freshness, allowing for targeted storage and transportation, and minimizing spoilage. Consumer information is also provided through colour changes or indicators on the packaging, promoting informed purchasing and consumption practices (Nasir et al., 2018). Time-temperature indicators on labels indicate unsafe temperatures, reducing spoilage risk. Self-preserving packaging releases natural antioxidants or antimicrobial agents to extend shelf life and reduce food spoilage. Through the incorporation of data on the product's provenance, manufacturing processes, and storage circumstances into a QR code on the packaging, blockchain integration with smart packaging provides more transparency and traceability. This data can be connected to a blockchain platform, giving consumers verified information about the food's journey. Blockchain technology can also be combined with sensor data from smart packaging to display a dynamic "best before" date based on real-time temperature and storage conditions. This helps to combat food theft and reduces waste caused by rotting or safety concerns. Smart packaging and blockchain technology work well together, giving real-time data on product quality and legitimacy. This combination allows for informed decision-making, prevents spoilage, and decreases food waste across the supply chain (Nižetić et al., 2019).

13.8.3 Collaborative Initiatives and Partnerships

Governments need to take the lead in promoting digitalization within public administration, reducing barriers to its adoption. This involves investing in research, innovation, education, and training to showcase the benefits of technology. Policy actions could include nurturing blockchain ecosystems in agricultural chains, integrating the technology to enhance competitiveness and sustainability, and establishing clear regulations. While blockchain offers potential for transparent agricultural supply chains, constraints still exist, limiting its adoption by farmers and food systems (Khedkar & Khedkar, 2020). Future initiatives by governments and the commercial sector will determine how these difficulties are addressed, to establish blockchain as a secure and transparent instrument for food safety. Furthermore, integrating blockchain with other emerging technologies such as big data and IoT has the potential to further automate and improve food supply systems while maintaining transparency and traceability (Nižetić et al., 2019). Food waste must be combated in several ways, including the use of smart agricultural technology and blockchain. Public-private partnerships (PPPs) can help to make this a reality. Governments may provide financial incentives and funds to help develop low-cost, scalable smart agricultural and blockchain technology. Joint initiatives may establish pilot projects on farms to demonstrate the effectiveness of these technologies in reducing food waste. Standardization efforts between public and commercial stakeholders are crucial for defining and implementing standardized data formats and communication protocols to ensure the smooth integration of smart agriculture systems and blockchain platforms (Galli et al., 2019).

Industry collaborations and consortiums can employ blockchain technology to decrease inefficiencies and food waste, resulting in more transparent and traceable supply chains. These collaborations can assist in accelerating the application of blockchain technology and smart agriculture for food waste reduction by allowing knowledge exchange and cooperation across a wide range of food system stakeholders. Furthermore, cooperation can result in the creation of sector-specific solutions to certain problems in many food industries, including fruits, vegetables, cereals, and meat products. Farmer cooperatives and organizations can invest in smart agriculture and blockchain technology to increase their accessibility to small and medium-sized farms. They may offer training programs to assist farmers with the skills and knowledge required to efficiently operate and employ this technology, therefore addressing digital literacy problems. Blockchain-enabled traceability may be used to negotiate higher rates with purchasers based on the verifiable quality and provenance of their goods, potentially decreasing food waste caused by economic causes. Educational institutions and research partnerships can accelerate innovation in smart agriculture and blockchain technologies specifically designed to tackle food waste. By incorporating blockchain technology and smart agriculture into their curricula, educational institutions can equip the next generation of farmers and food industry experts with the knowledge and abilities they need to reduce food waste. Collaborative projects include the World Wildlife Fund's Blockchain for Food Transparency, the IBM Food Trust, and the Regen Network, a blockchain network centred around regenerative agriculture techniques (Khedkar & Khedkar, 2020).

13.9 ETHICAL AND ENVIRONMENTAL CONSIDERATIONS

13.9.1 Social Equity and Food Access

Smart agriculture and blockchain technology can significantly reduce food waste, improve food security for vulnerable populations, and empower small-scale farmers. However, challenges such as accessibility, the digital divide, implementation costs, data ownership, privacy concerns, and inclusive design need to be addressed. Policymakers can provide financial assistance, tax breaks, and training programs to help small-scale farmers adopt these technologies. Open-source solutions can reduce costs and increase accessibility. Community-based initiatives and farmer cooperatives can facilitate collective investment and utilization of these technologies. Information sharing and education programs are essential for informed decision-making and promoting equitable adoption. Collaboration between policymakers, technology companies, researchers, and community organizations is crucial for developing solutions that benefit all stakeholders (Hairie Hanis & Fernando, 2024).

13.9.2 Environmental Impacts of Food Waste

Food waste is a widespread and serious global issue and raises ethical and environmental concerns, resulting in hidden costs and severe consequences for the biosphere. Several UN Sustainable Development Goals are hampered by food waste, which results from emissions and the use of both renewable and non-renewable resources. The UN 2030 Agenda for Sustainable Development emphasizes the importance of addressing this issue, with Goal 2 seeking to eliminate hunger, ensure food security, and enhance nutrition while encouraging sustainable agriculture. Ensuring sustainable patterns of consumption and production is the goal of Goal 12, which includes objectives like cutting the amount of food waste produced globally per person in half, ethically handling chemicals and wastes, and minimizing waste creation through prevention, use, recycling, and reuse. Food waste has a substantial environmental impact, influencing both resource consumption and climate change. Key environmental challenges include rising greenhouse gas emissions, depletion of water supplies, land usage and deforestation, soil degradation, and biodiversity loss. Water pollution and virtual water waste are also serious challenges. Land usage and deforestation are also major challenges, with agricultural land clearing causing deforestation and animal habitat loss. Intensive agricultural operations increase soil erosion and deterioration. Food waste reduction has many advantages beyond the bottom line when it comes to environmental sustainability. Reducing food waste can help to prevent climate change, preserve biodiversity for future generations, and conserve resources. Reducing food waste and promoting environmental sustainability can be achieved by following sustainable patterns of production and consumption (Hairie Hanis & Fernando, 2024).

13.9.3 Ethical Dilemmas in Redistribution and Donation Programs

The redistribution and donation programs are imperative in addressing food waste and ensuring food security. However, they present several ethical dilemmas such as food spoilage and the risk of illness, perception of second-rate food, logistical challenges, and the issue of food banking. Since food transportation and refrigeration increase greenhouse gas emissions, food redistribution has a major environmental impact. Prioritizing fresh, nutrient-dense food, enforcing stringent quality control and safety measures, growing distribution networks and partnerships, encouraging food literacy and waste reduction techniques, and pushing for systemic change to address the underlying causes of food insecurity are all necessary to ensure ethical food redistribution. By placing a strong emphasis on wholesome foods and fresh produce, implementing robust procedures, and promoting food literacy, organizations may decrease food waste and promote sustainable solutions.

FIGURE 13.5 Food waste hierarchy.

In conclusion, programs aimed at redistribution and donation are critical to preventing food waste and ensuring food security as illustrated by a food waste hierarchy in Figure 13.5 (Galli et al., 2019).

13.10 CONCLUSION AND RECOMMENDATIONS

13.10.1 Summary of Key Findings and Insights

Blockchain technology is leading the charge in the battle against food waste, transforming the food chain via the advancement of accountability, traceability, and transparency. It records a product's path from farm to fork irreversibly, enabling us to spot inefficiencies that cause spoiling and to carry out targeted recalls in the event of contamination outbreaks. We can decrease foodborne diseases and waste by improving food safety and consumer confidence through improved food traceability. By giving consumers access to data on shelf life and storage conditions, blockchain technology empowers customers to make educated decisions about what to buy and how to plan their meals. By drawing attention to how food production affects the environment, it also promotes sustainable decisions (Trabold & Nair, 2018). We can reduce food waste by putting waste reduction initiatives into practice, such as effective food donation and redistribution and verifying biodegradable packaging for appropriate disposal. However, there are challenges that we need to address, such as scalability and infrastructure, standardization and interoperability, and cost and integration. As technology matures, costs decrease, and interoperability improves, we expect wider adoption of blockchain technology, creating a more transparent, accountable, and sustainable food system. We can leverage blockchain technology by pairing it with smart packaging, which monitors freshness and integrates with the blockchain for targeted storage, transportation, dynamic shelf-life tracking, and combating food fraud. Collaboration initiatives and partnerships, such as public-private partnerships, industry alliances, farmer cooperatives, educational institutions, and research partnerships, can promote this technology (Hairie Hanis & Fernando, 2024). We need to address challenges such as the digital divide, accessibility in rural areas, cost considerations for small-scale farmers, data ownership and privacy concerns, and strategies for equity such as government support, open-source solutions, community-based initiatives, and information sharing for informed decision-making. With the right approach, we can create a more sustainable food system that benefits everyone.

13.10.2 Implications for Future Research and Practice

Through waste reduction and creating a waste-free future, blockchain technology can completely transform food systems. Future studies should focus on standards and interoperability, consumer behaviour and the uptake of blockchain-enabled food traceability, the environmental impact of blockchain technology, and the application of blockchain to avoid food loss. The development of standardized protocols and data formats is necessary to enable smooth data exchange and communication between various blockchain systems. Understanding consumer attitudes and behaviour towards blockchain-enabled food traceability can guide the development of user-friendly applications and communication strategies. A comprehensive life cycle assessment (LCA) of blockchain solutions can assess the environmental impact of blockchain technologies, including energy consumption and potential benefits from reduced food waste. Blockchain-enabled traceability and artificial intelligence–powered demand forecasting can optimize production and distribution, avoiding waste from oversupply, and they can further improve efficiency and reduce waste by integrating with the Internet of Things and blockchain (Hairie Hanis & Fernando, 2024). Blockchain-based solutions are essential for sustainable food systems, but they require large-scale pilot projects across diverse regions and sectors to demonstrate their practical value. Capacity building and training programs are needed to equip farmers, food businesses, and consumers with the necessary knowledge and skills to effectively utilize blockchain technologies. Strategies to make blockchain solutions more affordable and accessible for small-scale farmers, who often face resource constraints, are also needed. Collaboration between governments, research institutions, technology companies, industry associations, and NGOs is essential for innovation and addressing social equity concerns. Consumer engagement through interactive platforms and educational campaigns can empower them to make informed choices that contribute to waste reduction. By prioritizing these research areas and implementing effective practices, we can unlock the full potential of blockchain technology, leading to a future where food systems are more sustainable, transparent, and efficient, leading to a significant reduction in food waste and a more responsible relationship with food resources (Trabold & Nair, 2018).

Waste reduction and sustainable food systems could be greatly enhanced by blockchain technology. It can improve food safety, cut down on food waste, encourage transparency in the supply chain, and foster customer confidence. Blockchain can also be used to create efficient smart contracts that automate tasks like compliance, quality assurance, and payment. This can lower the chance of fraud or mistakes as well as operating expenses. Data analytics and decision support systems can be developed by leveraging blockchain data for advanced analysis and decision-making. Proactive steps to reduce waste can be taken when blockchain is integrated with IoT devices and sensors to enable real-time monitoring and data collection. Collaborative governance models enabled by blockchain can foster cooperation among stakeholders in sustainable food systems, incentivizing transparency, responsible practices, and equitable distribution of benefits. Blockchain-based solutions can also be used to engage consumers and promote responsible consumption habits. Understanding how transparent access to information influences consumer behaviour and purchasing decisions can inform strategies to promote sustainability and waste reduction at the consumer level. Regulatory and policy implications of blockchain adoption in the food industry are crucial. Research can explore legal frameworks, standards, and certifications needed to ensure compliance, data privacy, and security while fostering innovation and industry-wide adoption. Continued research on successful implementations of blockchain in sustainable food systems can provide valuable insights and best practices for practitioners. Case studies across different regions, scales, and supply chain segments can highlight key challenges, lessons learned, and strategies for overcoming barriers to adoption (Nižetić et al., 2019).

13.10.3 Recommendations for Stakeholders and Policymakers

Blockchain technology has the potential to create an environmentally sustainable food system by reducing waste. To encourage adoption, tax incentives or subsidies can be used. Standards and laws about data gathering, storage, and security should be established to maintain confidence and promote broader adoption. Infrastructure modifications should be made to enable the scalability and efficiency of blockchain networks in the food sector. Government agencies, academic institutions, technology companies, and industry players should collaborate through public-private partnerships to improve traceability and waste reduction (Lin et al., 2018). Technology companies should create affordable and easy-to-use blockchain systems that cater to the needs of small and medium-sized farmers and businesses involved in the food supply chain. Open-source and interoperable solutions should be created to encourage broader use and smooth data sharing throughout the food system. Farmers and other industry participants should participate in trials and exhibits to demonstrate the benefits of blockchain technology in reducing food waste. Investments in capacity building and training should be made to provide farmers and other stakeholders with the skills and knowledge needed to run and efficiently use blockchain-based solutions. Forming cooperatives or industry consortiums can also be established to share best practices and leverage collective bargaining power to negotiate better deals with technology providers. Customers should be aware of the benefits of blockchain technology in reducing food waste and ensuring food safety. Legislators should create clear legislative frameworks for blockchain applications in the food sector. Establishing industry standards and certification programs for blockchain-based solutions can protect consumer interests while fostering interoperability, data integrity, and compliance with sustainability principles (Trabold & Nair, 2018).

REFERENCES

Ahmed, Nurzaman, Debashis De, and Iftekhar Hussain. 2018. "Internet of Things (IoT) for Smart Precision Agriculture and Farming in Rural Areas." *IEEE Internet of Things Journal* 5 (6): 4890–9. https://doi.org/10.1109/jiot.2018.2879579.

Andrieu, N., B. Sogoba, R. Zougmore, F. Howland, O. Samake, O. Bonilla-Findji, M. Lizarazo, A. Nowak, C. Dembele, and C. Corner-Dolloff. 2017. "Prioritizing Investments for Climate-Smart Agriculture: Lessons Learned from Mali." *Agricultural Systems* 154 (June): 13–24. https://doi.org/10.1016/j.agsy.2017.02.008.

Annosi, Maria Carmela, Federica Brunetta, Francesco Bimbo, and Marianthi Kostoula. 2021. "Digitalization within Food Supply Chains to Prevent Food Waste. Drivers, Barriers and Collaboration Practices." *Industrial Marketing Management* 93 (93): 208–20. https://doi.org/10.1016/j.indmarman.2021.01.005.

Aschemann-Witzel, Jessica, Ilona E. de Hooge, Harald Rohm, Anne Normann, Marilia Bonzanini Bossle, Alice Grønhøj, and Marije Oostindjer. 2017. "Key Characteristics and Success Factors of Supply Chain Initiatives Tackling Consumer-Related Food Waste—A Multiple Case Study." *Journal of Cleaner Production* 155 (2): 33–45. https://doi.org/10.1016/j.jclepro.2016.11.173.

Bosona, Techane, and Girma Gebresenbet. 2023. "The Role of Blockchain Technology in Promoting Traceability Systems in Agri-Food Production and Supply Chains." *Sensors* 23 (11): 5342. https://doi.org/10.3390/s23115342.

Buisman, M.E., R. Haijema, and J.M. Bloemhof-Ruwaard. 2019. "Discounting and Dynamic Shelf Life to Reduce Fresh Food Waste at Retailers." *International Journal of Production Economics* 209 (March): 274–84. https://doi.org/10.1016/j.ijpe.2017.07.016.

Casino, Fran, Venetis Kanakaris, Thomas K. Dasaklis, Socrates Moschuris, Spiros Stachtiaris, Maria Pagoni, and Nikolaos P. Rachaniotis. 2020. "Blockchain-Based Food Supply Chain Traceability: A Case Study in the Dairy Sector." *International Journal of Production Research* 59 (19): 1–13. https://doi.org/10.1080/00207543.2020.1789238.

Costa, Tamíris Pacheco da, James Gillespie, Xavier Cama-Moncunill, Shane Ward, Joan Condell, Ramakrishnan Ramanathan, and Fionnuala Murphy. 2022. "A Systematic Review of Real-Time Monitoring Technologies and Its Potential Application to Reduce Food Loss and Waste: Key Elements of Food Supply Chains and IoT Technologies." *Sustainability* 15 (1): 614. https://doi.org/10.3390/su15010614.

Farooq, Muhammad, Jie Cheng, Noor Ullah Khan, Roselina Ahmad Saufi, Nagina Kanwal, and Hanieh Alipour Bazkiaei. 2022. "Sustainable Waste Management Companies with Innovative Smart Solutions: A Systematic Review and Conceptual Model." *Sustainability* 14 (20): 13146. https://doi.org/10.3390/su142013146.

Friha, Othmane, Mohamed Amine Ferrag, Lei Shu, Leandros Maglaras, and Xiaochan Wang. 2021. "Internet of Things for the Future of Smart Agriculture: A Comprehensive Survey of Emerging Technologies." *IEEE/CAA Journal of Automatica Sinica* 8 (4): 718–52. https://doi.org/10.1109/jas.2021.1003925.

Galli, Francesca, Alessio Cavicchi, and Gianluca Brunori. 2019. "Food Waste Reduction and Food Poverty Alleviation: A System Dynamics Conceptual Model." *Agriculture and Human Values* 36 (2): 289–300. https://doi.org/10.1007/s10460-019-09919-0.

Gemtou, Marilena, Konstantina Kakkavou, Evangelos Anastasiou, Spyros Fountas, Soren Marcus Pedersen, Gohar Isakhanyan, Kassa Tarekegn Erekalo, and Serafin Pazos-Vidal. 2024. "Farmers' Transition to Climate-Smart Agriculture: A Systematic Review of the Decision-Making Factors Affecting Adoption." *Sustainability* 16 (7): 2828. https://doi.org/10.3390/su16072828.

Hairie Hanis, Muhammad, and Yudi Fernando. 2024. "Smart Logistics Solutions for Reducing Food Waste: A Case of D Nipah Catering." *International Journal of Industrial Management* 18 (1): 11–21. https://doi.org/10.15282/ijim.18.1.2024.10404.

Kayikci, Yasanur, Sercan Demir, Sachin K. Mangla, Nachiappan Subramanian, and Basar Koc. 2022. "Data-Driven Optimal Dynamic Pricing Strategy for Reducing Perishable Food Waste at Retailers." *Journal of Cleaner Production* 344 (April): 131068. https://doi.org/10.1016/j.jclepro.2022.131068.

Khedkar, Deepak, and Renu Khedkar. 2020. "New Innovations in Food Packaging in Food Industry." *Emerging Technologies in Food Science* 165–85. https://doi.org/10.1007/978-981-15-2556-8_15.

Lee, Sang M., and Silvana Trimi. 2018. "Innovation for Creating a Smart Future." *Journal of Innovation & Knowledge* 3 (1): 1–8. https://doi.org/10.1016/j.jik.2016.11.001.

Lin, Jun, Zhiqi Shen, Anting Zhang, and Yueting Chai. 2018. "Blockchain and IoT Based Food Traceability for Smart Agriculture." *Proceedings of the 3rd International Conference on Crowd Science and Engineering—ICCSE'18*. https://doi.org/10.1145/3265689.3265692.

Lins, Maísa, Renata Puppin Zandonadi, António Raposo, and Veronica Cortez Ginani. 2021. "Food Waste on Foodservice: An Overview Through the Perspective of Sustainable Dimensions." *Foods* 10 (6): 1175. https://doi.org/10.3390/foods10061175.

Lopez Barrera, Emiliano, and Thomas Hertel. 2020. "Global Food Waste Across the Income Spectrum: Implications for Food Prices, Production and Resource Use." *Food Policy* 98 (March): 101874. https://doi.org/10.1016/j.foodpol.2020.101874.

Ma, Zheng, Alla Asmussen, and Bo Jørgensen. 2018. "Industrial Consumers' Smart Grid Adoption: Influential Factors and Participation Phases." *Energies* 11 (1): 182. https://doi.org/10.3390/en11010182.

Makanjuola, Oyenike, Temiloluwa Arowosola, and Chenyu Du. 2020. "The Utilization of Food Waste: Challenges and Opportunities." *Journal of Food Chemistry & Nanotechnology* 6 (4): 182–8. https://doi.org/10.17756/jfcn.2020-0100.

Mastorakis, George, Ioannis Kopanakis, John Makridis, Christina Chroni, Katerina Synani, Katia Lasaridi, Konstadinos Abeliotis, Ioannis Louloudakis, Ioannis N. Daliakopoulos, and Thrassyvoulos Manios. 2024. "Managing Household Food Waste with the FoodSaveShare Mobile Application." *Sustainability* 16 (7): 2800. https://doi.org/10.3390/su16072800.

Mondol, Edward Probir. 2021. "The Impact of Block Chain and Smart Inventory System on Supply Chain Performance at Retail Industry." *International Journal of Computations, Information and Manufacturing (IJCIM)* 1 (1). https://journals.gaftim.com/index.php/ijcim/article/view/30.

Nasir, Haidawati, Wan Basyar Wan Aziz, F. Ali, K. Kadir, and S. Khan. 2018. "The Implementation of IoT Based Smart Refrigerator System." *2018 2nd International Conference on Smart Sensors and Application (ICSSA)*. https://doi.org/10.1109/ICSSA.2018.8535867.

Nižetić, Sandro, Nedjib Djilali, Agis Papadopoulos, and Joel J.P.C. Rodrigues. 2019. "Smart Technologies for Promotion of Energy Efficiency, Utilization of Sustainable Resources and Waste Management." *Journal of Cleaner Production* 231 (September): 565–91. https://doi.org/10.1016/j.jclepro.2019.04.397.

Panghal, Anupama, Sharmistha Pan, Priyanka Vern, Rahul S. Mor, and Sandeep Jagtap. 2023. "Blockchain Technology for Enhancing Sustainable Food Systems: A Consumer Perspective." *International Journal of Food Science & Technology* (December). https://doi.org/10.1111/ijfs.16835.

Raak, Norbert, Claudia Symmank, Susann Zahn, Jessica Aschemann-Witzel, and Harald Rohm. 2017. "Processing- and Product-Related Causes for Food Waste and Implications for the Food Supply Chain." *Waste Management* 61 (61): 461–72. https://doi.org/10.1016/j.wasman.2016.12.027.

Sayem, Ahmed, Pronob Kumar Biswas, Mohammad Muhshin Aziz Khan, Luca Romoli, and Michela Dalle Mura. 2022. "Critical Barriers to Industry 4.0 Adoption in Manufacturing Organizations and Their Mitigation Strategies." *Journal of Manufacturing and Materials Processing* 6 (6): 136. https://doi.org/10.3390/jmmp6060136.

Schedler, Kuno, Ali Asker Guenduez, and Ruth Frischknecht. 2019. "How Smart Can Government Be? Exploring Barriers to the Adoption of Smart Government." *Information Polity* 24 (1): 3–20. https://doi.org/10.3233/ip-180095.

Srinivasan, S.P., D. Sorna Shanthi, and Aashish V. Anand. 2017. "Inventory Transparency for Agricultural Produce Through IoT." *IOP Conference Series: Materials Science and Engineering* 211 (June): 012009. https://doi.org/10.1088/1757-899x/211/1/012009.

Trabold, Thomas A., and Vineet Nair. 2018. "Conventional Food Waste Management Methods." *Elsevier EBooks* (January): 29–45. https://doi.org/10.1016/b978-0-12-811157-4.00003-6.

Vaidya, Gaurav, and Pratul Ch Kalita. 2021. "Understanding Emotions and Their Role in the Design of Products: An Integrative Review." *Archives of Design Research* 34 (3): 5–21. https://doi.org/10.15187/adr.2021.08.34.3.5.

Vishkaei, Behzad Maleki, and Pietro De Giovanni. 2022. "Blockchain-Based Smart Contracts for Sustainable Food Waste Management and Collaborations." *5th European International Conference on Industrial Engineering and Operations Management* (July). https://doi.org/10.46254/eu05.20220115.

Yousefi, Hanie, Hsuan-Ming Su, Sara M. Imani, Kais Alkhaldi, Carlos D.M. Filipe, and Tohid F. Didar. 2019. "Intelligent Food Packaging: A Review of Smart Sensing Technologies for Monitoring Food Quality." *ACS Sensors* 4 (4): 808–21. https://doi.org/10.1021/acssensors.9b00440.

14 Digital Twins for the Agriculture Sector

Savita Sheoran and Ritika

14.1 INTRODUCTION

Agriculture is undergoing a technological revolution with the adoption of smart farming methods. Smart farming essentially optimizes agricultural operations via cutting-edge technologies like artificial intelligence (AI), machine learning (ML), and the Internet of Things (IoT) (Javaid et al., 2023; Karunathilake et al., 2023). In real-time synchronization with their physical equivalents, virtual copies of actual assets, procedures, or systems are called digital twins (DTs) (Ashtari Talkhestani et al., 2019; Chaux et al., 2021; Karunathilake et al., 2023). Enhancing agricultural production, sustainability, and efficiency through integrating DT has become a potential strategy (Purcell et al., 2023).

With DTs, farmers can continuously monitor, analyze, and simulate agricultural processes, improving overall performance and empowering them to make data-driven decisions. DTs are essential to smart farming because they give farmers detailed information about crop growth, soil quality, weather patterns, and machinery performance (Cesco et al., 2023). DTs generate dynamic models of agricultural settings using data from sensors, drones, satellites, and other sources. This enables farmers to see and comprehend intricate interconnections inside their farming systems.

One of DTs' main benefits is that they make precision farming techniques easier. Farmers may customize fertilization, irrigation, and pest management tactics by modeling individual plants or field sections to meet unique needs, avoid resource waste, and optimize production.

DTs also make predictive analytics possible, allowing farmers to foresee and reduce possible hazards like crop illnesses, harsh weather, or changes in the market.

In the agricultural industry, DTs benefit crop management, animal husbandry, supply chain optimization, and infrastructure upkeep. Farmers may maximize animal welfare, energy use, and logistical procedures by modeling animals, buildings, and equipment digitally. This will ultimately increase efficiency and sustainability across the food production chain.

However, there are obstacles to the broad use of DTs in smart farming. To guarantee fair access and ethical technology usage in agriculture, concerns around data privacy, interoperability, and infrastructure accessibility need to be addressed. Furthermore, stakeholders such as farmers, academics, policymakers, and technology providers must constantly innovate and collaborate due to the complexity of combining multiple data sources and analytical tools.

In summary, smart farming with DTs has great potential to transform agriculture in the 21st century. Farmers may improve decision-making, productivity, and environmental stewardship by using cutting-edge technologies to build virtual representations of their farming operations. The integration of DTs is set to play a pivotal role in determining the future of digital transformation in the agriculture sector as it continues to embrace them.

14.1.1 Digital Twins

A digital twin is a virtual replica or model of a real-world system, process, or entity that is created to monitor and analyze data from sensors, instruments, and other sources of the corresponding

DOI: 10.1201/9781003507390-14

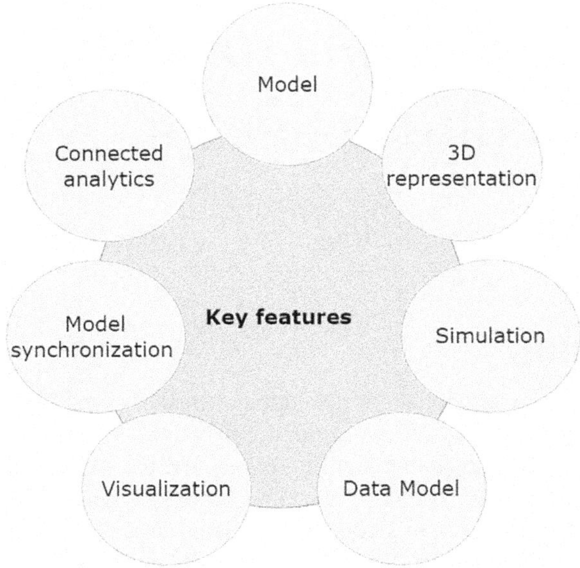

FIGURE 14.1 DT key features.

real-world system (Purcell et al., 2023). After that, a digital model that mimics the real-time motion of the actual item or system is created using this data. DTs that operate close to real time are employed for several tasks, such as tracking, modeling, analyzing, forecasting, and optimizing the physical objects they depict. DTs make it possible to manage, comprehend, and make more effective decisions on the real-world equivalent. Key features of DTs are shown in Figure 14.1.

- **Model**
  A digital clone of a physical device accurately imitates its features and attributes. One may make three-dimensional simulations of real-world things with DTs and observe their behavior.
- **3D representation**
  The term 3D representation refers to a visual representation of a real device. Creating a virtual experience, like real life, allows one to understand the data in analytical reports more quickly.
- **Simulation**
  The DT simulation software places a virtual representation of a physical object within a virtual environment. Additionally, it modifies the environment's parameters to monitor the device's functioning under various circumstances.
- **Data model**
  A real-time data model is something that people can access through a DT. It shows how the components of a physical system or device are connected. It provides a transparent picture of activities and processes and allows you to label data.
- **Visualization**
  It allows customers to see models on a mobile device or standalone desktop. Access to information about a system's or device's operational data is provided.
- **Model synchronization**
  This exceptional characteristic makes the digital model match the parameters of the real environment. It can be done in real time using modern modeling tools. Instantaneously, you may see a DT structure representing every physical upgrade.

- **Connected analytics**
  Analytics provides information based on a physical device's measurable attributes. DT platforms use machine learning calculations and algorithms. Hence, it provides insights into the behavior and operation of your system or device.

DT is a remarkable technology that has revolutionized how farmers manage their agricultural operations. They function as virtual time machines that provide insights into crucial resources and activities at specific times in a farm's history, present, and future. This technology is rapidly changing how agribusinesses run their daily operations, enabling them to manage their facilities, systems, and procedures more efficiently and accurately. As a result, farmers can make informed decisions and take proactive measures to ensure their crops' optimal health and growth, ultimately leading to higher yields and profits (Ukwuoma et al., 2024).

14.1.2 Background

DTs appeared when NASA created a virtual version of its spacecraft in 2010 that altered in tandem with its actual twin (Katsoulakis et al., 2024). One of the newest and most popular developing technologies is DTs. It transforms whole sectors by fusing the virtual and real worlds to produce insights and solutions that generate enormous economic value.

The worldwide DT industry is predicted to reach over $73.5 billion by 2027, driven by a strong increase in the industrial sector's need for DT technologies (Moreno López et al., 2024). A DT is a virtual replica of an actual thing in a virtual environment. It is utilized to collect and evaluate data, simulate actual situations and their results, make data-driven decisions, and, if required, offer remedial recommendations. It can be used to track the real object's performance around the clock remotely.

DT is a cutting-edge technology that harnesses the latest advancements in AI, cloud computing, and IoT technologies to create a virtual replica of a physical object. Using environmental sensors, a DT gathers vast amounts of data that are then analyzed using a cloud-based processing methodology to derive insights about the object. The DT interface gives users a powerful tool to explore an object's past, present, and anticipated future (Cesco et al., 2023). It allows users to examine every aspect of the object, from its physical properties to its performance metrics and maintenance history.

A DT enables bidirectional communication between the physical object and its digital replica. Any changes made to the physical object are immediately reflected in the DT, and vice versa (Ashtari Talkhestani et al., 2019; Ullah et al., 2024). This allows organizations to test different scenarios and evaluate the impact of potential modifications before implementing them in the real world.

Data exchange between a real object's digital doppelganger and its digital replica is shown in Figure 14.2.

14.2 ARCHITECTURE OF DIGITAL TWINS

Due to environmental concerns and the challenge of producing enough food to feed a growing global population, producers are pressured to increase food's nutritional value and yield efficiency, lower maintenance and logistical costs, maintain farm sustainability, and lower unpredictable risks like crop diseases and extreme weather. DT use in agriculture has great potential to help with these issues.

Enough data enables DTs to replicate a physical thing accurately. Many real-time data sources, including satellites, unmanned aircraft, integrated sensors, Radio Frequency Identification (RFID) tags, and past information on soil management, projected crops, and weather patterns, are used

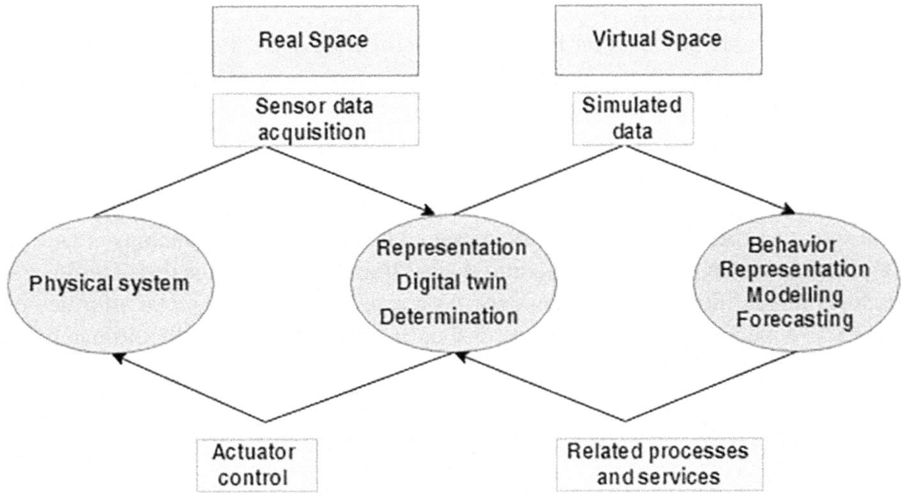

FIGURE 14.2 Data interaction between the physical object and its digital representation.

extensively in creating DTs (Peladarinos et al., 2023). A DT's architecture is composed of three primary parts (Rolle et al., 2019):

- **Physical Object:** This represents the system the DT imitates. It might be a procedure, a physical good, or even infrastructure. Actuators and sensors frequently gather data from the actual object in real time.
- **Digital Model:** This is the actual object's digital representation. It comprises the physical entity's spherical, structural, and behavioral traits. The virtual model may be used for modeling, analysis, and prediction. It is generated using data gathered from the physical object.
- **Connectivity:** This part allows the real object and its virtual duplicate to communicate in both directions. The virtual model receives data from the physical entity's sensors for analysis and display. The physical entity can then use the insights and suggestions gained through the simulation to make decisions and exercise control.

To summarize, a DT's architecture involves building a virtual model replicating a real entity; establishing a link between the two via communication and data exchange channels; and utilizing this connection to track, evaluate, and improve the physical entity's performance in real time.

14.3 COMPONENTS OF DIGITAL TWINS

A DT is a digital representation of a real system or thing. It comprises several parts that come together to represent the actual thing visually. The essential elements of a DT are (VanDerHorn & Mahadevan, 2021) are shown in Figure 14.3.

- **Sensors:** These tools gather information from the system or actual item. Sensors can record different kinds of data, including vibration, pressure, and temperature.
- **Data Acquisition System:** The data acquired from the sensors is gathered, processed, and stored by the data acquisition system component, which guarantees its accuracy and dependability.

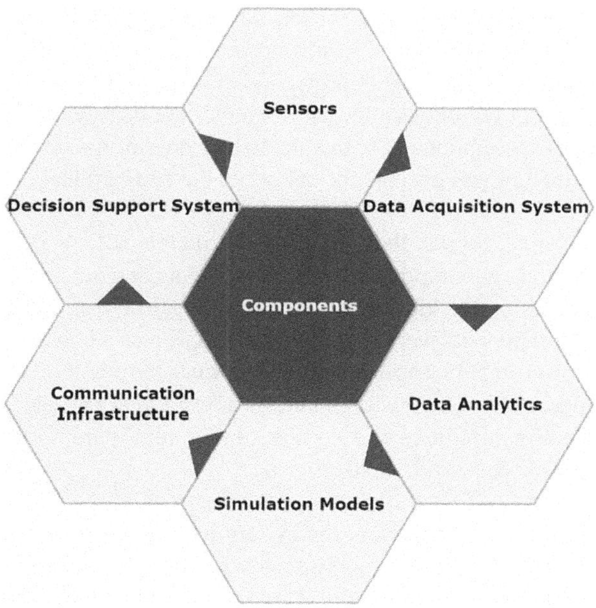

FIGURE 14.3 Components of a DT.

- **Data Analytics:** The data gathered by the DT is examined using data analytics technologies. Designs, patterns, oddities, and other insightful information may be found using these technologies.
- **Simulation Models:** In a virtual environment, simulation models mimic the behavior of an actual item or system. These models aid in forecasting how the real-world entity will perform in the future.
- **Communication Infrastructure:** This part allows the actual device and its digital duplicate to exchange data. It guarantees continuous system monitoring and control.
- **Decision Support System:** This component uses the insights produced by the DT to aid in making well-informed judgments and maximize the physical system's performance.

14.4 TYPES OF DIGITAL TWINS

A technique that makes it possible to create a virtual duplicate of any entity is known as a DT. Still, a DT is far more than just an exact duplicate of a product, piece of machinery, or workplace. The technology's fundamental idea is that the DT and the actual object exchange data in real time. This guarantees that the information gathered via the use of technology is pertinent and aligns with the dynamic real-world circumstances and outside factors (M. Liu et al., 2021).

DTs are applied at different phases of the manufacturing process and used in diverse ways by manufacturers. However, the fundamental benefit of these solutions is still the same: regardless of their variations, all DTs aim to optimize processes, which sets off a series of improvements and modifications that include predictive analytics and intelligent automation.

Specifically, there are several kinds of DTs, yet the primary distinction between them is the extent of coverage. Some solutions could mimic the entire marketplace or even the plant, while others might depict the tiniest unit of the manufacturing line. The four primary categories of DT are:

- **Component Twins**
  The most basic form of DT technology is the component twin, or components twin on occasion. It is equivalent to the tiniest components of the system, such as a particular piece

of machinery or a product, such as a sensor, switch, or valve. In the virtual representation, such smart parts may have their performance monitored and real-time circumstances simulated to assess their stability, durability, and efficiency.

Although this kind of DT only somewhat resembles the factory architecture, it allows for better monitoring of the equipment components and prompt repair, which guarantees the stability of the production process and the caliber of the final product (X. Liu et al., 2023).

- **Asset Twins**
  A DT's product or asset type is the next level in this technological hierarchy. Typically, they are made up of many component twins or model a complex asset, such as an engine, pump, or structure, using the information produced by part twins. An asset twin examines how well various components work together and function as a cohesive whole.

  By using asset/product DTs, engineers may find areas for possible development and gain insights into the operation of their equipment. As a result, productivity increases, KPIs like the average time between failures and average time to repair are optimized, and resource usage decreases (Dietz & Pernul, 2020).
- **System Twins**
  System twins are asset duplications at the system level that illustrate how various assets come together to form functional units. System twins provide a comprehensive perspective of the factory or plant, enabling the testing of various system configurations to attain maximum efficiency or identifying new business prospects to bring in supplementary revenue streams.

  A group of assets used in a certain activity, such as the manufacturing of an essential good within the plant or the delivery of energy, are included in the coverage of the system twin. On this level, continuous tracking and simulation go beyond identifying errors and malfunctions made possible by less complex kinds of DTs. It is a way to acquire information useful for making strategic decisions and achieving complete process visibility (Fan et al., 2021).
- **Process Twins**
  Process twins, the ultimate DT, combine system twins into a single entity to investigate system synchronization and cooperation. This system provides the most comprehensive picture of the workflows and processes inside the factory or plant, enabling a far more in-depth and flexible output analysis. Process twinning provides output data and allows for the adjustment of inputs, such as raw material feeding rate and production temperature, without interfering with the manufacturing process or lowering production quality. Consequently, instead of depending on gut feeling or conjecture, executives can easily and risk-free test various business hypotheses, effectively monitor key company KPIs, and facilitate data-driven decision-making (Xiong & Wang, 2022).

14.5 APPLICATIONS OF DIGITAL TWINS

DT technologies are revolutionizing agriculture and transforming traditional farming practices across various domains. Integrating DT technology in agriculture has numerous benefits, such as increased productivity, improved crop yields, and reduced environmental impact. By creating digital replicas of physical assets, such as farms and crops, DT technology enables farmers to simulate and optimize various scenarios, such as plant growth and soil quality. This technology is also helping farmers to monitor their crops and livestock in real time, enabling them to make informed decisions that can enhance crop quality and yield. The agricultural sector is rapidly embracing DT technology, and it is expected to play a critical role in meeting the growing demand for food in a sustainable manner (Böttjer et al., 2023; Xiong & Wang, 2022). A few applications of DTs illustrating agricultural potential are displayed in Table 14.1.

TABLE 14.1
Applications of DTs Illustrating Agricultural Potential

Reference	Agriculture Application	Year	Results
(Tsolakis et al., 2019)	Precision farming	2019	Developed AgROS, Agricultural Robotics' Robot Operating System-Based Emulation Tool.
(S.-K. Jo et al., 2018)	Smart livestock farms	2018	Conducted a digital twin-based feasibility study for smart livestock farms.
(Moghadam et al., 2019)	Orchard production	2019	Built a model for a future version of orchard production utilizing a digital twin.
(Verdouw & Kruize, 2017)	Farm management	2017	Made a digital twin to manage farms.
(R. G. Alves et al., 2019)	Smart farming	2019	Constructed a digital twin for smart farming.
(Barnard, 2019)	Indoor gardening	2019	Developed a digital twin for indoor gardening.

- **Field Monitoring**
  DT technology transforms agricultural field management by giving farmers and other agricultural professionals access to intricate virtual models of their agriculture, crops, and machinery. In real time, farmers can monitor insect infestations, crop development, soil moisture levels, and nutritional content with these DTs. Precision agricultural techniques are made possible by DTs, which offers comprehensive insights into crop conditions and variability. Farmers may develop specialized planting, fertilizing, and irrigation plans based on the unique requirements of various fields by analyzing the geographical data obtained by DTs.

  A DT is an effective tool for producers, agricultural scientists, and farm advisers to aid in decision-making. DTs make the effective management of agricultural resources, including water, electricity, and fertilizers, possible. They facilitate the accurate allocation of resources and conservation measures by combining data on crop requirements, weather patterns, and soil characteristics. Farmers can remotely monitor and manage their fields through DTs with web connectivity. Farmers may access real-time data, get notifications about important occurrences, and modify operational settings using web-based or mobile applications. A DT combines information from various sources and formats, such as sensor readings, historical documents, satellite pictures, and expert knowledge, and data is examined using sophisticated analytics methods like machine learning and statistical modeling (Wu et al., 2023).

  To summarize, the application of DTs is revolutionizing managing fields in agriculture by giving farmers access to precise insights, prediction tools, and decision-support resources to maximize crop yield, sustainable resource use, and sustainability. By utilizing DTs, farmers may increase agricultural operations' production, resilience, and efficiency to new heights (Wu et al., 2023).
- **Livestock Monitoring**
  Across various sectors and industries, DT technology is now increasing operating efficiency and lowering costs. A promising new frontier in the cattle farming sector is the usage of DTs. Large-scale precision farming techniques may be enhanced by this technology, which can also be used to maximize the utilization of machinery and equipment and enhance the health and welfare of various farm animals. Advanced AI technology is

already being used by several trailblazers in this sector to continuously monitor the actions of animals and the surrounding environment. This helps the farmer make wiser financial decisions and improves knowledge of animal health, illness prevention, and control. With the use of recognition technology, one may keep an eye on an animal's emotional and mental states by analyzing facial traits like eye white areas and ear positions. DTs can assist farmers in constructing more sustainable housing structures, predicting cycles of heat for breeding, discouraging unfavorable livestock habits, and much more by utilizing simulation, modeling, and virtual reality technology. Individual farms must conduct extensive cost and benefit analyses to use DT technology, as is the case with other novel technical developments. Although there are obstacles and constraints to overcome, DTs have enormous promise to revolutionize the cattle agricultural sector (Zhang et al., 2023).

- **Land Cultivation**

  DT technology transforms agricultural field management by giving farmers and other agricultural professionals access to intricate virtual models of their agriculture, crops, and machinery. In real time, farmers can monitor insect infestations, crop development, soil moisture levels, and nutritional content with these DTs. Precision agricultural techniques are made possible by DTs, which offer comprehensive insights into crop conditions and variability. Farmers may develop specialized planting, fertilizing, and irrigation plans based on the unique requirements of various fields by analyzing the geographical data obtained by DTs.

  Farmers can use DTs to monitor their agricultural machinery and equipment. DTs are virtual representations of these machines that can help with various tasks. For instance, they can assist in predicting agricultural yields, identifying potential issues such as insect infestations or disease outbreaks, and optimizing planting and harvesting schedules. This is achieved by analyzing historical data and real-time information gathered from sensors around the field. DTs can also optimize resource allocation by suggesting the appropriate quantity of inputs based on the specific requirements of different geographical regions. This can save farmers money and reduce their impact on the environment. Additionally, DTs provide a testing and scenario analysis framework that enables farmers to try out different farming techniques virtually before implementing them in real life. This reduces the likelihood of errors and helps farmers make informed decisions.

  In conclusion, there are several advantages to utilizing DT technology in land cultivation, including increased efficiency, sustainability, and production. Farmers may improve results and yields by using DTs to obtain important information regarding their land and crops (Monteiro et al., 2018, 2023).

- **Crop Improvement**

  DT technology may assist farmers in more efficient crop management by simulating and predicting the outcomes of various crop management approaches. Farmers must precisely scan and analyze their crop areas to produce a DT. They can do this by compiling comprehensive data on variables, including geography, soil properties, microclimates, and other pertinent information. Drones, weather stations, soil sensors, and satellite images can all be used to do this.

  Farmers may increase production and optimize resource use by knowing which crops to grow and where to sow them using DTs. They suggest the best timing, spacing, and seeding rates using actual-time data and predictive modeling. They suggest the best timing, spacing, and seeding rates based on actual time information and forecasting.

  To ensure consistent crop emergence and establishment, farmers might modify planting dates and practices to consider variations in soil moisture, temperature, and other external conditions.

  DTs also assist farmers in optimizing their fertilization techniques by considering crop nutritional requirements, soil nutrient levels, and nutrient absorption patterns. They recommend the best fertilization strategies—including timing, costs, and distribution techniques—to optimize crop output while reducing the loss of nutrients and environmental impact. They do this by evaluating historical data and predicting models (Kim & Heo, 2024).

- **Supply Chain and Logistics Optimization**
  The agriculture business is quickly adopting DT technology because it provides a practical way to improve supply chain and logistical procedures. A DT provides the smooth and optimal flow of agricultural goods by accurately mapping each supply chain step, from production to distribution. With the help of this cutting-edge technology, items can be continuously managed and monitored in real time, guaranteeing that they are delivered effectively and in ideal condition.

  When a DT is used in supply chain and logistics, logistics expenses are significantly reduced because it optimizes routes and efficiently handles any delays that may occur along the way (Abideen et al., 2021).

  Moreover, it offers improved visibility across the supply chain, enabling stakeholders to track product movement and condition in real time. This enhanced efficiency and transparency ultimately result in a shorter time to market, guaranteeing that agricultural goods are supplied promptly and in good condition, satisfying the exacting requirements of retailers and customers alike.
- **Weather Prediction and Risk Management**
  DT is a cutting-edge technology that is transforming the agriculture sector. To increase yields and improve agricultural operations, they employ sophisticated weather modeling. DTs help farmers predict weather patterns and modify operations by fusing real-time meteorological data with agricultural models. Farmers may reduce their chance of suffering losses from unfavorable weather conditions by using this prediction skill to handle weather-related risks proactively.

  A primary advantage of DTs is enhanced risk management. Accurate weather forecasts enable farmers to plan ahead and take preventative action. Because irrigation can be scheduled around expected rainfall, water resources may be conserved, allowing optimal water usage. Improved weather-related planning guarantees that agricultural operations are planned for the best weather, increasing yields and decreasing losses.

  DTs are revolutionary for farmers because they enable data-driven decision-making and operational optimization. They also provide proactive risk management, enhanced water efficiency, and optimal scheduling of agricultural operations through the utilization of sophisticated weather models (Henriksen et al., 2023).

14.6 DT-DEVELOPED TECHNOLOGIES FOR AGRICULTURE

Using a DT in farming is a significant step toward sustainable and more effective methods. With this strategy, farmers can improve operations from planting to harvesting, as it provides a comprehensive perspective of the farm. In addition, the DT offers elements for forecasting future events and a thorough comprehension of agricultural operations (Henriksen et al., 2023). Resources like fertilizer, water, and insecticides may be used by farmers more effectively. This lowers waste and operating expenses while meeting the unique requirements of crops at various development stages at the appropriate moment. Each of these aids in optimizing yields (Kim & Heo, 2024; Peladarinos et al., 2023).

Integrating and evaluating data from many sources is one of the DT's key components. Actual information from connected devices placed across the land, which keep an eye on important variables like moisture in the soil and nutrient levels, is integrated into this system. Weather patterns play a significant role in agricultural decision-making (Henriksen et al., 2023) Data gathered from nearby weather stations is an important source of information about potential changes to weather patterns. High-resolution satellite photography is another tool the DT uses to track more significant alterations in crop conditions and environmental parameters. By gaining access to agricultural databases and gathering information on crop types, pest/disease trends, and historical data, the DT increases the accuracy of its predictions. This can help farmers make better-informed decisions in the context of contemporary agriculture.

In addition to simulating and forecasting the results of various farming techniques in various environmental circumstances, the DT also updates its model in real time to account for changes on the farm. By ensuring that the virtual land environment is constantly in sync with the actual farm, this dynamic model enables farmers to see the effects of their decisions before making them. Farmers can effectively manage risks, including bugs or epidemics of diseases, and test various scenarios, including the best and worst-case scenarios, using the decision support tool. Ultimately, this strategy encourages ecologically friendly and sustainable farming methods, giving farmers the resources they need to succeed in the rapidly changing agricultural landscape. Angin et al. (2020) created AgriLoRa, a DT framework, for smart agriculture using UNet and MobileNet models to provide high-precision and affordable solutions. Chaux et al. (2021) created DT architecture with a controlled environment approach to offer tactics and design to boost output. Alves et al. (2023) used an Internet of Things platform built on FIWARE to create an intelligent irrigation control system. A DT framework for wheat was developed by P. Skobelev et al. (2020) using ontologies, multi-agent technology, linked states, and transition rules for wheat. For precision farming, DT models were developed by P. O. Skobelev et al. (2020) using a multi-agent module with a knowledge base. Melesse et al. (2022) developed an ML-based DT to monitor changes in fruit quality utilizing a thermal camera, SAP, and ML to enhance food quality. Lan et al. (2023) created DT framework for intelligent fish farm management with the use of IoT, ML, AI, and computer vision. Monteiro et al. (2018) created a DT technology prototype and an IoT-based concept for vertical farming. A few use cases illustrating DTs' agricultural potential are displayed in Table 14.2.

TABLE 14.2
DTs' Agricultural Potential

Reference	Objective and Scenario of Application	Methodology	Results
(Angin et al., 2020)	Creating a smart agriculture DT framework.	UNet and MobileNet models.	Created AgriLoRa to provide a high-precision, low-cost smart agriculture solution.
(Chaux et al., 2021)	Developing DT design to maximize productivity.	Controlled environment method.	Provided techniques and architecture to increase productivity.
(Alves et al., 2023)	Creating a smart irrigation management system.	Internet of Things platform based on FIWARE.	Developed a DT-based system for irrigation management.
(P. Skobelev et al., 2020)	Developing a DT of wheat.	Ontologies, multi-agent technology, linked states, and transition rules.	Created a multi-agent approach to develop a DT for wheat.
(P. O. Skobelev et al., 2020)	DT model generation for precision agricultural management.	A multi-agent module with a knowledge base.	Constructed DT models for precision agriculture.
(Melesse et al., 2022)	ML-based DT to track the evolution of fruit quality.	Thermal camera and SAP.	Developed DT using ML for fruit quality evaluation.
(Lan et al., 2023)	Development of DT architecture for intelligent management of fish farms.	Modified analytic hierarchy process.	Developed DT for smart fish farm management.
(Monteiro et al., 2018)	Creating DT technology for vertical farming.	Use light and misting to sense and regulate the environment.	Developed an IoT-based model and a prototype DT technology.

14.7 CHALLENGES AND OPPORTUNITY

DTs are used for various agricultural applications, such as post-harvest processing, robotics, farm machinery, soil, irrigation, and crop monitoring. With the quick advancement of technology and sensors, DTs in agriculture could support plant production, health, and yield while conserving water and utilizing fewer chemicals by considering the soil's quality and characteristics. Numerous aspects of agricultural land's soil, water supply, and surroundings can be continuously observed and examined using massive amounts of data, ML models, and systems for decision-making integrated into the DT principles management techniques. Crop farming performance may be enhanced by combining DT technologies for irrigation and soil to document, track, and analyze changes in agricultural land. Data-driven updating models and soil structure simulations might use IoT technology to link farmers to their farms and provide detailed images of parameters affecting crop output, irrigation, and soil conditions. Less research, nevertheless, concentrates on creating DT concepts for agricultural soil that are more flexible and take into account a larger range of operations than current simulation models. The dynamic conditions of arable soil, such as temperature, moisture content, and water-holding capacity, might be continuously measured and recorded using soil sensors.

This data can be used to create DT concepts, coupled with information from soil structure and simulation approaches. Real-time soil and water management and control can also benefit from continual input from the digital world. Digital soil mapping, remote sensing, digital farming scenarios, and software platform development have all grown quickly in the last few years. Future research must address the requirement to integrate the current methodologies with IoT, edge computing, AI, ML, data analysis, and simulation methods that may contribute to creating a DT paradigm. Additionally, because DTs are multidisciplinary methodologies requiring viewpoints from systems engineering, researchers must consider the real-world challenges of DT-based systems in agriculture. Real-time agricultural equipment and robotic simulation are made possible by state-of-the-art DT technology. Using this simulation, the best way to design these items, how they interact with the environment, how much energy they use, and how to maintain them may all be greatly improved. The technique can assist with making choices in the cultivation of plants and forecast malfunctions in agricultural machinery. Farm owners have more control and oversight when they can electronically link to the machines in their agricultural farms and track and monitor them remotely using DTs.

Farms, especially those in rural locations with inadequate internet and technology infrastructure, face challenges in accumulating and evaluating the vast amounts of data generated by DT systems and the information they exchange among various assets.

Other options, such as long-range wireless sensor networks and edge computing, could mitigate issues with internet connectivity in remote areas in the DT concept. Creating guidelines and methods for data exchange and communication may enhance the future integration of DT systems with crop farm technologies.

Continuous monitoring of crops in DT systems can be achieved by simulating dynamic farm conditions and taking into account the effects of climate, management, and environmental conditions on plant growth (Angin et al., 2020; Cesco et al., 2023). Data-driven models and sensor fusion techniques can detect abnormal plant conditions and predict growth stages, thus minimizing the risk of environmental and management impacts. DT concepts could be used to replicate the intricate physical system of crop farming in the digital realm. This would integrate modeling, forecasting, simulation, and variable sensor technologies.

DT concepts can be utilized to facilitate food tracking, monitoring, and analysis across the entire supply chain, resulting in minimized labor costs, food waste, and energy utilization through the use of data gathered from multiple sensors and simulation models. Future research should focus on incorporating additional environmental and post-harvest product attributes to generate dependable DTs. Developing DTs for post-harvest processing to reduce quality losses and increase product shelf life is a significant challenge that requires consideration of the agricultural product value chain from farm to fork. To achieve this, it is necessary to employ comprehensive experimental and data-gathering methods and numerical modeling and validation procedures in post-harvest processing. These measures help reduce DT uncertainty and enable users to trust the output of DT concepts.

14.8 CONCLUSION AND FUTURE SCOPE

DT technology is a creative approach with the potential to completely transform the agriculture industry. It provides farmers with up-to-date information and insights that can aid in better decision-making. Farmers may improve their crop management and resource allocation tactics, gain a better understanding of their operations, and make digital replicas of real assets like crops, animals, and equipment.

In addition, DTs use sensors, IoT devices, and sophisticated analytics to monitor farming operations. As a result, farmers can better manage their operations, boosting production, sustainability, and efficiency. Farmers with enhanced decision-making skills can take preventive measures to tackle possible problems, leading to better productivity and efficient resource use.

Another important advantage is DTs' capacity to digitally test solutions and recreate various scenarios. This can assist farmers in reducing risks associated with bad weather, insect infestations, and crop failures. Farmers may predict possible problems before they arise and try multiple approaches by modeling various scenarios to see which works best.

To sum up, using DT technology has great promise for converting conventional farming methods into intelligent and data-driven enterprises. Thanks to the current information, insights, and capacity for simulation it offers, farmers may benefit from enhanced efficiency, production, and sustainability. This might pave the way for a more prosperous future for the agriculture industry.

REFERENCES

Abideen, A. Z., Sundram, V. P., Pyeman, J., Othman, A. K., & Sorooshian, S. (2021). Digital Twin Integrated Reinforced Learning in Supply Chain and Logistics. *Logistics*, *5*(4). https://doi.org/10.3390/logistics5040084

Alves, R. G., Maia, R. F., & Lima, F. (2023). Development of a digital twin for smart farming: Irrigation management system for water saving. *Journal of Cleaner Production*, *388*, 135920. https://doi.org/10.1016/j.jclepro.2023.135920

Alves, R. G., Souza, G., Maia, R. F., Tran, A. L. H., Kamienski, C., Soininen, J.-P., Aquino, P. T., & Lima, F. (2019). A digital twin for smart farming. *2019 IEEE Global Humanitarian Technology Conference (GHTC)*, 1–4. https://doi.org/10.1109/GHTC46095.2019.9033075

Angin, P., Anisi, M. H., Göksel, F., Gürsoy, C., & Büyükgülcü, A. (2020). AgriLoRa: A digital twin framework for smart agriculture. *Journal of Wireless Mobile Networks, Ubiquitous Computing, and Dependable Applications*, *11*(4), 77–96. https://doi.org/10.22667/JOWUA.2020.12.31.077

Ashtari Talkhestani, B., Jung, T., Lindemann, B., Sahlab, N., Jazdi, N., Schloegl, W., & Weyrich, M. (2019). An architecture of an Intelligent Digital Twin in a Cyber-Physical Production System. *Automatisierungstechnik*, *67*(9), 762–782. https://doi.org/10.1515/auto-2019-0039

Barnard, A. (2019). *In the digital indoor garden*. Report. Available Online: https://New.Siemens.Com/Global/En/Company/Stories/Research-Technologies/Digitaltwin/Digital-Indoor-Garden. Html (Accessed on 22 October 2021).

Böttjer, T., Tola, D., Kakavandi, F., Wewer, C. R., Ramanujan, D., Gomes, C., Larsen, P. G., & Iosifidis, A. (2023). A review of unit level digital twin applications in the manufacturing industry. *CIRP Journal of Manufacturing Science and Technology*, *45*, 162–189. https://doi.org/10.1016/j.cirpj.2023.06.011

Cesco, S., Sambo, P., Borin, M., Basso, B., Orzes, G., & Mazzetto, F. (2023). Smart agriculture and digital twins: Applications and challenges in a vision of sustainability. *European Journal of Agronomy*, *146*, 126809. https://doi.org/10.1016/j.eja.2023.126809

Chaux, J. D., Sanchez-Londono, D., & Barbieri, G. (2021). A digital twin architecture to optimize productivity within controlled environment agriculture. *Applied Sciences*, *11*(19), Article 19. https://doi.org/10.3390/app11198875

Dietz, M., & Pernul, G. (2020). Digital twin: Empowering enterprises towards a system-of-systems approach. *Business & Information Systems Engineering*, *62*(2), 179–184. https://doi.org/10.1007/s12599-019-00624-0

Fan, Y., Yang, J., Chen, J., Hu, P., Wang, X., Xu, J., & Zhou, B. (2021). A digital-twin visualized architecture for flexible manufacturing system. *Journal of Manufacturing Systems*, *60*, 176–201. https://doi.org/10.1016/j.jmsy.2021.05.010

Henriksen, H. J., Schneider, R., Koch, J., Ondracek, M., Troldborg, L., Seidenfaden, I. K., Kragh, S. J., Bøgh, E., & Stisen, S. (2023). A new digital twin for climate change adaptation, water management, and disaster risk reduction (HIP digital twin). *Water*, *15*(1). https://doi.org/10.3390/w15010025

Javaid, M., Haleem, A., Khan, I. H., & Suman, R. (2023). Understanding the potential applications of artificial intelligence in agriculture sector. *Advanced Agrochem*, *2*(1), 15–30. https://doi.org/10.1016/j.aac.2022.10.001

Jo, S.-K., Park, D.-H., Park, H., & Kim, S.-H. (2018). Smart livestock farms using digital twin: Feasibility study. *2018 International Conference on Information and Communication Technology Convergence (ICTC)*, 1461–1463. https://doi.org/10.1109/ICTC.2018.8539516

Karunathilake, E. M. B. M., Le, A. T., Heo, S., Chung, Y. S., & Mansoor, S. (2023). The path to smart farming: Innovations and opportunities in precision agriculture. *Agriculture*, *13*(8). https://doi.org/10.3390/agriculture13081593

Katsoulakis, E., Wang, Q., Wu, H., Shahriyari, L., Fletcher, R., Liu, J., Achenie, L., Liu, H., Jackson, P., Xiao, Y., Syeda-Mahmood, T., Tuli, R., & Deng, J. (2024). Digital twins for health: A scoping review. *NPJ Digital Medicine*, *7*(1), 77. https://doi.org/10.1038/s41746-024-01073-0

Kim, S., & Heo, S. (2024). An agricultural digital twin for mandarins demonstrates the potential for individualized agriculture. *Nature Communications*, *15*(1), 1561. https://doi.org/10.1038/s41467-024-45725-x

Lan, H.-Y., Ubina, N. A., Cheng, S.-C., Lin, S.-S., & Huang, C.-T. (2023). Digital twin architecture evaluation for intelligent fish farm management using modified analytic hierarchy process. *Applied Sciences*, *13*(1), Article 1. https://doi.org/10.3390/app13010141

Liu, M., Fang, S., Dong, H., & Xu, C. (2021). Review of digital twin about concepts, technologies, and industrial applications. *Journal of Manufacturing Systems*, *58*, 346–361. https://doi.org/10.1016/j.jmsy.2020.06.017

Liu, X., Jiang, D., Tao, B., Xiang, F., Jiang, G., Sun, Y., Kong, J., & Li, G. (2023). A systematic review of digital twin about physical entities, virtual models, twin data, and applications. *Advanced Engineering Informatics*, *55*, 101876. https://doi.org/10.1016/j.aei.2023.101876

Melesse, T. Y., Bollo, M., Pasquale, V. D., Centro, F., & Riemma, S. (2022). Machine learning-based digital twin for monitoring fruit quality evolution. *Procedia Computer Science*, *200*, 13–20. https://doi.org/10.1016/j.procs.2022.01.200

Moghadam, P., Lowe, T., & Edwards, E. J. (2019). Digital Twin for the future of orchard production systems. *Proceedings*, *36*(1). https://doi.org/10.3390/proceedings2019036092

Monteiro, J., Barata, J., Veloso, M., Veloso, L., & Nunes, J. (2018). Towards sustainable digital twins for vertical farming. *2018 Thirteenth International Conference on Digital Information Management (ICDIM)*, 234–239. https://doi.org/10.1109/ICDIM.2018.8847169

Monteiro, J., Barata, J., Veloso, M., Veloso, L., & Nunes, J. (2023). A scalable digital twin for vertical farming. *Journal of Ambient Intelligence and Humanized Computing*, *14*(10), 13981–13996. https://doi.org/10.1007/s12652-022-04106-2

Moreno López, G. A., Recaman Chaux, H., & Molina Parra, P. A. (2024). Digital twins, metaverse, and learning: Review and proposal of conceptual framework. In M.-L. Bourguet, J. M. Krüger, D. Pedrosa, A. Dengel, A. Peña-Rios, & J. Richter (Eds.), *Immersive learning research network* (pp. 515–532). Springer Nature.

Peladarinos, N., Piromalis, D., Cheimaras, V., Tserepas, E., Munteanu, R. A., & Papageorgas, P. (2023). Enhancing smart agriculture by implementing digital twins: A comprehensive review. *Sensors*, *23*(16). https://doi.org/10.3390/s23167128

Purcell, W., Neubauer, T., & Mallinger, K. (2023). Digital twins in agriculture: Challenges and opportunities for environmental sustainability. *Current Opinion in Environmental Sustainability*, *61*, 101252. https://doi.org/10.1016/j.cosust.2022.101252

Rolle, R., Martucci, V., & Godoy, E. (2019). *Digitalization of manufacturing processes: Proposal and experimental results* (p. 431). IEEE. https://doi.org/10.1109/METROI4.2019.8792838

Skobelev, P., Laryukhin, V., Simonova, E., Goryanin, O., Yalovenko, V., & Yalovenko, O. (2020). Multi-agent approach for developing a digital twin of wheat. *2020 IEEE International Conference on Smart Computing (SMARTCOMP)*, 268–273. https://doi.org/10.1109/SMARTCOMP50058.2020.00062

Skobelev, P. O., Mayorov, I. V., Simonova, E. V., Goryanin, O. I., Zhilyaev, A. A., Tabachinskiy, A. S., & Yalovenko, V. V. (2020). Development of models and methods for creating a digital twin of plant within the cyber-physical system for precision farming management. *Journal of Physics: Conference Series, 1703*(1), 012022. https://doi.org/10.1088/1742-6596/1703/1/012022

Tsolakis, N., Bechtsis, D., & Bochtis, D. (2019). AgROS: A robot operating system based emulation tool for agricultural robotics. *Agronomy*, *9*(7). https://doi.org/10.3390/agronomy9070403

Ukwuoma, H. C., Dusserre, G., Coatrieux, G., & Vincent, J. (2024). Analysis of digital twin and its physical object: Exploring the efficiency and accuracy of datasets for real-world application. *Data Science and Management*. https://doi.org/10.1016/j.dsm.2024.04.002

Ullah, I., Adhikari, D., Su, X., Palmieri, F., Wu, C., & Choi, C. (2024). Integration of data science with the intelligent IoT (IIoT): Current challenges and future perspectives. *Digital Communications and Networks*. https://doi.org/10.1016/j.dcan.2024.02.007

VanDerHorn, E., & Mahadevan, S. (2021). Digital Twin: Generalization, characterization and implementation. *Decision Support Systems*, *145*, 113524. https://doi.org/10.1016/j.dss.2021.113524

Verdouw, C., & Kruize, J. W. (2017). Digital twins in farm management: Illustrations from the FIWARE accelerators SmartAgriFood and fractals. *7th Asian-Australasian Conference on Precision Agriculture*. https://doi.org/10.5281/zenodo.893662

Wu, Y., Sun, J., Yang, G., Cui, L., Wang, Z., & Wang, M. (2023). Research on digital twin based temperature field monitoring system for LNG storage tanks. *Measurement*, *215*, 112864. https://doi.org/10.1016/j.measurement.2023.112864

Xiong, M., & Wang, H. (2022). Digital twin applications in aviation industry: A review. *The International Journal of Advanced Manufacturing Technology*, *121*(9), 5677–5692. https://doi.org/10.1007/s00170-022-09717-9

Zhang, Y., Zhang, Y., Gao, M., Dai, B., Kou, S., Wang, X., Fu, X., & Shen, W. (2023). Digital twin perception and modeling method for feeding behavior of dairy cows. *Computers and Electronics in Agriculture*, *214*, 108181. https://doi.org/10.1016/j.compag.2023.108181

15 Smart Farming with "Blockchain"
Revolutionizing Medicinal Plant Cultivation

Rati Kailash Prasad Tripathi

15.1 INTRODUCTION

Medicinal plants have been revered for their healing properties across diverse cultures for millennia. These plants contain bioactive compounds with therapeutic effects, offering treatments for various ailments ranging from common colds to chronic diseases. As the demand for natural remedies continues to rise, driven by factors such as increasing awareness of holistic health practices, rising healthcare costs, and concerns about the side effects of synthetic drugs, the cultivation and trade of medicinal plants have become significant economic activities worldwide, valued at billions of dollars annually. However, the industry faces numerous challenges related to quality control, authentication, and sustainable sourcing that threaten its integrity and sustainability. These challenges include adulteration and mislabelling, lack of transparency, and unsustainable practices. Due to high market demand and limited regulation, there is a risk of adulterating medicinal plant products with inferior or substitute ingredients. Mislabelling also occurs, where products are falsely labelled or misrepresented, leading to consumer deception and safety concerns. Furthermore, the medicinal plant supply chain is often opaque, with limited visibility into the cultivation, processing, and distribution processes. This lack of transparency makes it difficult to authenticate their quality. In addition, the indiscriminate harvesting of medicinal plants can lead to habitat destruction, loss of biodiversity, and depletion of wild populations. Unsustainable cultivation practices, such as overuse of pesticides and fertilizers, can also degrade soil health and water resources, threatening the long-term viability of medicinal plant cultivation. Smart farming addresses these challenges by employing precise, data-centric techniques to track and regulate the cultivation process (Govindaraghavan, 2008; Aung & Chang, 2014; Simmler et al., 2018).

Smart farming, also referred to as precision agriculture, is a contemporary approach that utilizes technology to improve agricultural practices and boost productivity. This strategy holds particular promise for medicinal plant cultivation, where maintaining consistent quality and maximizing yield are paramount. Smart farming integrates a range of cutting-edge technologies, including the artificial intelligence (AI), drones, Internet of Things (IoT), and big data analytics. These technologies gather and analyze real-time data from farms, with the goal of maximizing crop yield, minimizing resource waste, and improving the overall efficiency in agricultural activities. The essence of smart farming revolves around the idea of making decisions based on data. Sensors and connected devices collect real-time information on conditions of the soil, weather forecasts, health of crops, and equipment status. With this abundance of data, farmers can make well-informed decisions about fertilization, pest management, irrigation, and harvest schedules, resulting in enhanced productivity and sustainability. Thus, smart farming can revolutionize the way medicinal plants are grown, harvested, and processed (Karunathilake et al., 2023; John & Arul Leena Rose, 2024).

DOI: 10.1201/9781003507390-15

On the other hand, blockchain technology, originally devised for secure cryptocurrency transactions, has found a myriad of applications beyond the financial sector, including agriculture. Within the realm of smart farming, blockchain fosters trust and transparency across the entire agricultural supply chain, spanning from farm to fork. By recording every stage of the production process, from planting and harvesting to processing, packaging, including distribution, blockchain ensures the traceability and authenticity of agricultural products. This provenance tracking capability is particularly valuable for premium or organic produce, empowering consumers to confirm the origin and quality of the food they procure. Furthermore, blockchain enables the implementation of smart contracts, which are automated agreements programmed to enforce pre-defined conditions automatically when specific criteria are fulfilled. In agriculture, smart contracts can streamline transactions between farmers, suppliers, distributors, and retailers by automating payment settlements, inventory management, and quality assurance processes. By eliminating intermediaries and reducing administrative overheads, smart contracts help optimize resource allocation and improve profit margins for all stakeholders involved (Bosona & Gebresenbet, 2023; Li et al., 2023; Panwar et al., 2023).

In this era of digital transformation, the convergence of smart farming and blockchain holds immense potential to address pressing challenges facing the agricultural industry, including food security, environmental sustainability, and supply chain inefficiencies. Using data analytics, IoT devices, and decentralized ledgers, farmers can access fresh avenues for invention, resilience, and expansion within the contemporary agricultural landscape (Iqbal & Butt, 2020; Bhat et al., 2022).

This chapter offers a comprehensive understanding of basic fundamentals, and benefits of smart farming and blockchain technology in the cultivation of medicinal plants. Furthermore, the chapter highlights the integrated potential of smart farming and blockchain in revolutionizing the agriculture industry along with the challenges and limitations of these techniques. It also explores the technologies involved, their applications and real-world case studies, the broader implications, and future prospects of this innovative approach, ultimately contributing to the advancement of sustainable and efficient agricultural practices.

15.2 SMART FARMING IN MEDICINAL PLANT CULTIVATION

Smart farming is revolutionizing the agricultural landscape by integrating cutting-edge technologies to enhance productivity, efficiency, and sustainability. This transformation is particularly significant in the cultivation of medicinal plants, which require precise conditions to produce high-quality bioactive compounds. Medicinal plants have been a cornerstone of traditional medicine and are increasingly being utilized in modern pharmaceuticals. However, their cultivation presents unique challenges that can be effectively addressed through smart farming techniques (Unal, 2020; Khanna et al., 2022; Javaid et al., 2022).

15.2.1 Components of Smart Farming

Smart farming encompasses a variety of technological innovations that work in synergy to revolutionize traditional medicinal plant cultivation methods. Figure 15.1 illustrates the principal elements of smart farming (Javaid et al., 2022; Walter et al., 2017).

15.2.1.1 Internet of Things

IoT devices play a crucial role in smart farming by furnishing operational data on diverse environmental framework. Soil sensors measure moisture content, pH, and nutrient levels, while weather stations monitor temperature, humidity, and light intensity. This data is crucial for maintaining the specific conditions required for different medicinal plants, ensuring consistent growth and high-quality yields.

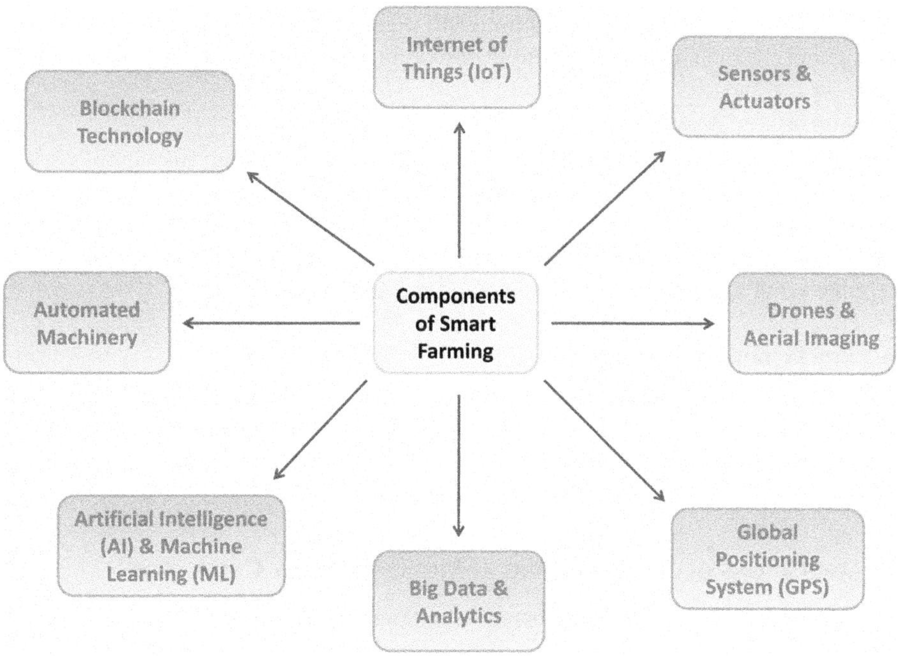

FIGURE 15.1 Critical components of smart farming in medicinal plant cultivation.

15.2.1.2 Sensors and Actuators

Sensors measure various parameters like soil moisture, nutrient levels, and environmental conditions. Actuators, on the other hand, are used to control farm equipment and systems in response to the data acquired from the sensors. For example, they can automate irrigation systems to provide precise amounts of water based on soil moisture data.

15.2.1.3 Drones and Aerial Imaging

Drones, outfitted with powerful, high-definition cameras and multi-spectral sensors, offer intricate aerial images of fields. This technology allows for early detection of plant stress, disease, and pest infestations, enabling targeted interventions. In medicinal plant cultivation, drones help ensure that plants are harvested at the peak of their medicinal potency.

15.2.1.4 Global Positioning Systems

Global positioning system (GPS) technology is used for precision mapping and navigation of farm equipment. It ensures that planting, fertilizing, and harvesting are carried out with high accuracy, minimizing waste and optimizing field operations.

15.2.1.5 Big Data and Analytics

Big data analytics is employed to analyze the extensive volume of data entailed by IoT devices and sensors. Incorporating big data analytics allows for the consolidation and examination of data from multiple origins, offering extensive insights into the entire cultivation process. This helps in predicting trends, improving crop management practices, and making data-driven decisions. Predictive analytics can forecast weather conditions, pest outbreaks, and optimal harvest times.

15.2.1.6 Machine Learning and Artificial Intelligence

Machine learning (ML) and AI algorithms analyze vast amounts of data collected by IoT devices, offering predictive insights and optimization strategies. For instance, AI can forecast weather patterns, predict pest outbreaks, and recommend irrigation schedules. Machine learning models continuously improve their accuracy over time, providing increasingly precise guidance for farmers.

15.2.1.7 Automated Machinery

Automated machinery, such as robotic planters and harvesters, and autonomous tractors, are increasingly being used in smart farming, which lowers the labour costs and enhances operational effectiveness. With minimal human involvement and exceptional precision, these machines can execute tasks like planting in addition to weeding and harvesting, minimizing damage to delicate medicinal plants and ensuring uniformity in planting and harvesting.

15.2.1.8 Blockchain Technology

Blockchain enhances transparency and traceability within the medicinal plant supply chain. It provides an immutable record of every stage of the cultivation process, from planting to harvesting and processing. This traceability is essential for validating the genuineness and quality of medicinal products, guaranteeing adherence to regulatory standards.

15.2.2 Applications of Smart Farming in Medicinal Plant Cultivation

Smart farming has several applications in medicinal plant cultivation, as illustrated in Figure 15.2 (Javaid et al., 2022).

15.2.2.1 Precision Irrigation

Smart irrigation systems use soil moisture sensors and weather predictions to provide the exact amount of water needed by medicinal plants. This not only conserves water but also prevents overwatering, which can harm plant health and reduce the potency of medicinal compounds.

15.2.2.2 Nutrient Management

Soil sensors monitor nutrient levels and provide real-time feedback to automated fertilization systems. This ensures that plants receive the right amount of nutrients at the right time, promoting healthy growth and maximizing the concentration of active compounds. This guarantees that plants acquire the appropriate nutrients at optimal times, fostering robust growth and optimizing the concentration of active compounds.

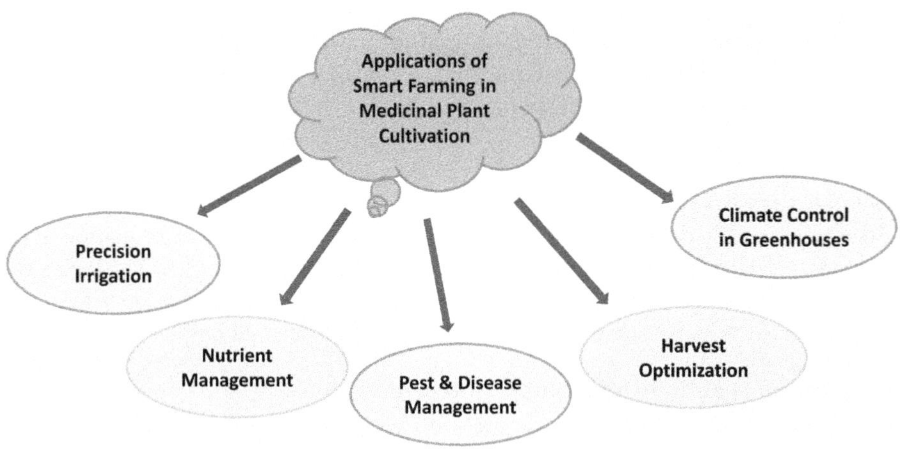

FIGURE 15.2 Applications of smart farming in medicinal plant cultivation.

15.2.2.3 Pest and Pathogen Control

Drones and monitoring systems powered by AI can identify initial indications of pest outbreaks and diseases. This enables the timely and precise administration of biopesticides or alternative treatments, diminishing crop losses and decreasing reliance on hazardous chemicals.

15.2.2.4 Harvest Optimization

Harvest timing is crucial for medicinal plants, as the concentration of active compounds can vary significantly throughout the growth cycle. Smart farming technologies, including drones and AI, help determine the optimal harvest time to ensure maximum potency and quality.

15.2.2.5 Climate Control in Greenhouses

For medicinal plants grown in controlled environments, smart climate control systems regulate temperature, light, humidity, and CO_2 levels. This fosters ideal growth conditions, resulting in increased yields and improved product quality.

15.2.3 Benefits of Smart Farming in Medicinal Plant Cultivation

The integration of these technologies in the cultivation of medicinal plants brings about numerous benefits such as increased productivity, resource efficiency, environmental sustainability, cost reduction, improved crop quality, risk management, and enhanced decision-making (Javaid et al., 2022; Walter et al., 2017; Jones, 2022).

15.2.3.1 Increased Productivity

Smart farming technologies significantly increase crop yields by optimizing growing conditions and mitigating losses caused by pests and pathogens. Higher yields translate to increased profitability for farmers and more affordable medicinal plant products for consumers.

15.2.3.2 Resource Efficiency

Precision agriculture techniques optimize the efficiency of water, fertilizer, and pesticide usage, thereby diminishing waste and environmental repercussions. Efficient resource management is particularly important in medicinal plant cultivation, where the quality of the soil and water directly affects the potency of the active compounds.

15.2.3.3 Environmental Sustainability

By employing fertilizers and pesticides in a precise manner, their runoff into the environment is minimized, thereby curbing pollution and advocating for sustainable farming practices. Besides, it assists in conserving water, and reducing greenhouse gas emissions through efficient resource management. Smart farming also promotes adoption of renewable energy sources and sustainable agricultural techniques.

15.2.3.4 Cost Reduction

Automation and precision agriculture reduce labour costs and the use of inputs. Through resource optimization and waste reduction, smart farming can result in considerable cost savings.

15.2.3.5 Improved Crop Quality

Continuous monitoring and precise management of environmental conditions ensure the consistent production of high-quality medicinal compounds. This is essential for meeting regulatory standards and maintaining consumer trust.

15.2.3.6 Risk Management

Predictive analytics and real-time monitoring enable farmers to anticipate and mitigate risks such as pest infestations, disease outbreaks, and adverse weather conditions. Proactive risk management reduces crop losses and ensures a stable supply of medicinal plants.

15.2.3.7 Enhanced Decision-Making

Insights derived from data empower farmers to make well-informed decisions. Smart farming provides the necessary information to make sound decisions by assisting in determining the best time to plant or harvest, choosing the right crop varieties, or optimizing irrigation schedules.

15.2.3.8 Transparency and Traceability

Blockchain technology furnishes a transparent and unalterable record of the entire cultivation process. This traceability is essential for medicinal plants, where authenticity and quality are paramount. Consumers and regulatory bodies can trace products back to their source, ensuring that they meet safety and quality standards.

15.3 BLOCKCHAIN TECHNOLOGY IN MEDICINAL PLANT CULTIVATION

15.3.1 Fundamentals and Features

Blockchain is a decentralized ledger technology that facilitates the recording and validation of transactions across computer networks. Every transaction is cryptographically connected to the preceding one, establishing an unmodifiable chain of blocks that is resilient to tampering. This architecture eliminates the need for intermediaries, such as regulatory authorities, and facilitates peer-to-peer transactions with increased transparency and security. The key features of blockchain are illustrated in Table 15.1 (Yaga et al., 2018).

15.3.2 Applications of Blockchain in Medicinal Plant Cultivation

There are several applications of blockchain technology in medicinal plant cultivation, as depicted in Figure 15.3 (Kassanuk & Phasinam, 2022; Heinrich et al., 2019).

15.3.2.1 Supply Chain Transparency

One of the primary blockchain applications in medicinal plant cultivation is enhancing supply chain transparency. By recording every stage of the cultivation, harvesting, processing, and distribution on blockchain, the origin of medicinal plants can be traced and their authenticity can be verified. This transparency helps prevent adulteration and ensures that consumers receive genuine products with documented quality and safety standards.

TABLE 15.1
Key Features of Blockchain

Features	Description
Decentralization	Blockchain functions on a peer-to-peer network where multiple nodes verify and record transactions, eliminating the necessity for a central authority.
Transparency	All transactions on blockchain are accessible to network participants, guaranteeing accountability and transparency.
Immutability	Once a transaction is logged on the blockchain, it cannot be modified or erased, securing the integrity and longevity of data.
Security	Blockchain employs cryptographic methods to safeguard transactions and deter unauthorized access, rendering it highly resistant to fraud and hacking.
Smart Contracts	Smart contracts are automated agreements programmed on the blockchain, executing predefined conditions without intermediaries.

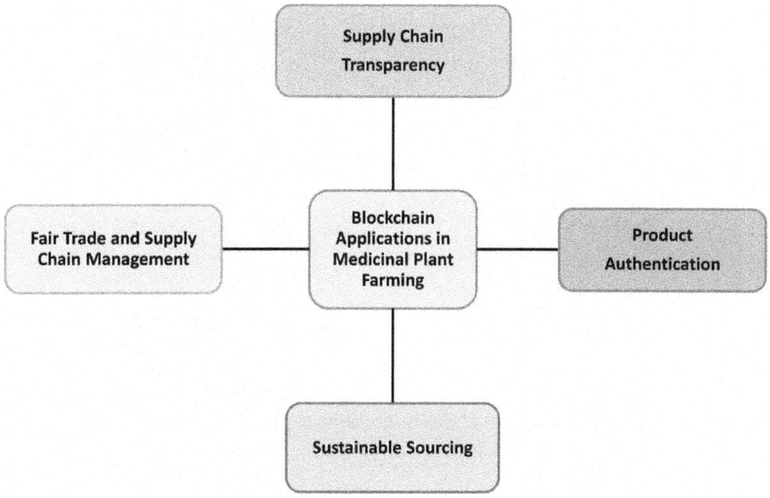

FIGURE 15.3 Blockchain applications in medicinal plant farming.

15.3.2.2 Product Authentication

Blockchain technology can also be used to authenticate medicinal plant products by creating unique digital identities for each batch or individual plant. These digital fingerprints, stored on the blockchain, contain information about the plant's genetic profile, cultivation practices, and quality control measures. Consumers can verify the authenticity of a product by scanning its QR code or accessing its digital certificate, thereby mitigating the risk of counterfeit or substandard products.

15.3.2.3 Sustainable Sourcing

Another application of blockchain in medicinal plant cultivation is promoting sustainable sourcing practices. By recording information about the geographical origin, cultivation methods, and harvesting techniques on the blockchain, stakeholders can ensure compliance with ethical and environmental standards. Moreover, blockchain-based systems can incentivize farmers to adopt sustainable practices by providing transparent incentives and certifications for eco-friendly cultivation methods.

15.3.2.4 Fair Trade and Supply Chain Management

Blockchain technology can facilitate fair trade and supply chain management by automating payments and contracts through smart contracts. Smart contracts can automatically execute payments to farmers based on predefined criteria, such as quality and quantity of produce, thereby ensuring fair compensation and reducing transaction costs. Additionally, blockchain-based supply chain management systems can streamline logistics, inventory management, and regulatory compliance, enhancing efficiency and accountability across the supply chain.

15.3.3 Benefits of Blockchain Integration in Medicinal Plant Cultivation

The integration of blockchain in medicinal plant cultivation offers several benefits, which are listed in Figure 15.4 (Li et al., 2024; M et al., 2021).

15.3.3.1 Enhanced Transparency and Traceability

Blockchain integration in medicinal plant cultivation enhances transparency and traceability by offering an immutable record of transactions together with activities throughout the supply chain. This transparency enables stakeholders to authenticate the genuineness and quality of medicinal plant products, thus fostering trust and confidence among consumers.

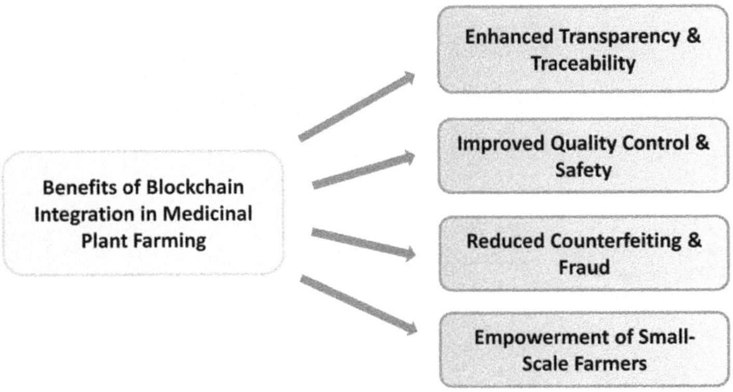

FIGURE 15.4 Benefits of integrating blockchain in medicinal plant cultivation.

15.3.3.2 Improved Quality Control and Safety

By enabling real-time monitoring of cultivation practices, processing techniques, and storage conditions, blockchain technology improves quality control and safety in medicinal plant production. Stakeholders can track the entire journey of a plant, from seed to sale, and identify any potential issues or contaminants that may arise during the process.

15.3.3.3 Reduced Counterfeiting and Fraud

Blockchain-based authentication systems help reduce counterfeiting and fraud in the medicinal plant industry by providing verifiable digital identities for products. Consumers can authenticate a product by accessing its digital certificate or scanning its QR code, thereby mitigating the risk of purchasing counterfeit or substandard products.

15.3.3.4 Empowerment of Small-Scale Farmers

Small-scale farmers are empowered by blockchain technology by granting them access to transparent and equitable markets. By recording their cultivation practices and certifications on the blockchain, farmers can demonstrate compliance with quality and sustainability standards, thereby enhancing their marketability and negotiating power.

15.4 CASE STUDIES AND REAL-WORLD APPLICATIONS OF SMART FARMING INTEGRATED WITH BLOCKCHAIN IN MEDICINAL PLANT CULTIVATION

Integrating blockchain technology with smart farming practices in the cultivation of medicinal plants is revolutionizing the agricultural industry by ensuring traceability, enhancing quality control, resource efficiency, and promoting transparency and regulatory compliance. Table 15.2 illustrates several case studies and real-world applications where these technologies have been successfully implemented (Dutta et al., 2023; Liontakis et al., 2021; Eka Putri et al., 2023; Ane et al., 2023; Vernon et al., 2023; Nowvaratkoolchai et al., 2024; Yoon et al., 2018; Christy & Fu, 2020; Ivanova & Latyshov, 2021; Lama & Rana, 2023; Sistani et al., 2024; Esteki et al., 2019; Mahmoudi et al., 2020; Bannor et al., 2023; Hu et al., 2021; Yang et al., 2021; Yan et al., 2021).

The case studies and real-world applications highlighted here demonstrate the potential of these technologies to revolutionize medicinal plant cultivation, ensuring that consumers receive high-quality, authentic products while promoting sustainable and efficient cultivation techniques. With the ongoing advancement of technology, the fusion of blockchain and smart farming is poised to become more widespread, further transforming the agricultural industry and the cultivation of medicinal plants.

TABLE 15.2
Case Studies and Real-World Applications of Smart Farming Integrated with Blockchain in Medicinal Plant Cultivation

Case Study	Background	Implementation	Technologies Used	Outcomes
Aloe vera cultivation	Aloe vera, known for its medicinal properties, is widely cultivated in several places worldwide. However, the quality of aloe vera gel can vary significantly based on cultivation practices and environmental conditions.	A smart farming project was initiated in Greece and other places, integrating IoT sensors, AI-driven analytics, and blockchain technology.	**IoT Sensors:** Soil moisture sensors and climate monitoring stations were installed to ensure optimal growing conditions. **AI Analytics:** AI algorithms analyzed the data to optimize irrigation schedules and predict pest outbreaks. **Blockchain:** A blockchain-based system was implemented to record every stage of the cultivation process, from planting to harvesting and processing.	**Quality Improvement:** The use of IoT sensors and AI analytics led to a significant improvement in the quality of aloe vera gel, with higher consistency in the active compounds. **Traceability:** Blockchain ensured complete traceability of the aloe vera products, enhancing consumer trust and compliance with regulatory standards. **Resource Efficiency:** Optimized irrigation schedules reduced water usage by 30%, contributing to sustainable farming practices.
Lavender farms	Lavender is widely used in aromatherapy and cosmetics. Ensuring high-quality essential oil content is crucial for its commercial value.	Lavender farms adopted smart farming technologies combined with blockchain to improve the quality and traceability of products.	**Drones and Aerial Imaging:** Drones fitted with multi-spectral cameras monitored plant health and growth stages. **IoT Sensors:** Soil sensors measured moisture and nutrient levels to optimize fertilization. **Blockchain:** A blockchain ledger recorded the cultivation process, including data on environmental conditions, harvesting, and oil extraction.	**Optimized Harvesting:** Drones helped determine the optimal harvesting time, maximizing essential oil yield and quality. **Transparency:** Blockchain provided transparency and traceability, which is essential for organic certification and consumer trust. **Reduced Chemical Use:** Targeted interventions based on drone data reduced the use of pesticides and fertilizers by 25%.
Cannabis cultivation	Cannabis is cultivated for both medicinal and recreational purposes. Ensuring compliance with regulatory standards is critical in the cannabis industry.	Cannabis cultivation was integrated with smart farming and blockchain technology to monitor and document the entire cultivation process.	**IoT Sensors:** Environmental sensors monitored temperature, humidity, CO_2 levels, and light intensity in indoor growing facilities. **AI Analytics:** AI models predicted plant growth patterns and optimized resource use. **Blockchain:** Blockchain was used to record data from seed to sale, ensuring compliance with legal and regulatory requirements.	**Regulatory Compliance:** The blockchain ledger provided a transparent and immutable record, ensuring full compliance with regulations. **Quality Control:** Real-time monitoring and AI optimization led to higher consistency in cannabinoid and terpene profiles. **Supply Chain Efficiency:** Blockchain improved supply chain efficiency by reducing paperwork and streamlining processes.

(*Continued*)

TABLE 15.2 *(Continued)*
Case Studies and Real-World Applications of Smart Farming Integrated with Blockchain in Medicinal Plant Cultivation

Case Study	Background	Implementation	Technologies Used	Outcomes
Ginseng cultivation in South Korea	Ginseng is a valuable medicinal plant with stringent quality requirements. Traditional cultivation methods often result in inconsistent quality.	South Korean ginseng farmers adopted a smart farming approach, integrating IoT, AI, and blockchain technologies.	**IoT Sensors:** Sensors monitored soil health, moisture levels, and temperature to maintain optimal growing conditions. **AI Analytics:** AI analyzed data to provide recommendations on irrigation, fertilization, and pest control. **Blockchain:** Blockchain recorded all cultivation activities, providing traceability and ensuring quality control.	**Improved Quality:** Consistent monitoring and AI-driven optimization improved the quality and potency of ginseng roots. **Traceability:** Blockchain ensured full traceability, which is crucial for exporting high-quality ginseng to international markets. **Sustainability:** The smart farming approach reduced water usage and minimized the environmental impact of ginseng cultivation.
Saffron cultivation	Saffron, one of the most expensive spices and medicinal plants, requires precise cultivation practices to ensure high quality and yield.	Iranian saffron farmers implemented smart farming technologies alongside blockchain to enhance the traceability and quality of their products.	**IoT Sensors:** Soil sensors and climate monitoring stations ensured optimal growing conditions. **Drones:** Drones monitored plant health and identified areas requiring attention. **Blockchain:** A blockchain ledger recorded data from planting to harvesting and processing.	**Enhanced Quality:** The integration of smart technologies improved the consistency and quality of saffron. **Global Market Access:** Blockchain traceability increased consumer confidence and opened up new international markets. **Resource Management:** Optimized irrigation and fertilization reduced resource usage, promoting sustainable farming practices.
Herbal medicine plantation	Traditional Chinese medicine relies heavily on various medicinal plants. Ensuring the authenticity and quality of these plants is essential.	A large-scale herbal medicine plantation in China adopted smart farming technologies and blockchain to enhance quality control and traceability.	**IoT Sensors:** Sensors monitored soil and environmental conditions. **AI Analytics:** AI algorithms optimized cultivation practices based on real-time data. **Blockchain:** Blockchain provided an immutable record of the cultivation process, from planting to harvesting and distribution.	**Quality Assurance:** Improved monitoring and optimization led to higher-quality medicinal plants. **Traceability:** Blockchain ensured full traceability, crucial for traditional Chinese medicine practitioners and consumers. **Efficiency Gains:** Streamlined processes and optimized resource use resulted in cost savings and increased efficiency.

15.5 CHALLENGES OF SMART FARMING AND BLOCKCHAIN IN MEDICINAL PLANT CULTIVATION

15.5.1 Limitations of Smart Farming

Smart farming, while offering numerous benefits, also faces several challenges and limitations, particularly in the context of medicinal plant cultivation. These challenges can be broadly categorized into technological, economic, environmental, social, and regulatory and policy factors (Table 15.3). Addressing these issues is essential for the successful implementation and sustainability of smart farming practices (Walter et al., 2017; Javaid et al., 2022; Sharma et al., 2022).

TABLE 15.3
Challenges Associated with Smart Farming

Specification	Issue	Challenge
		Technological Challenges
Data management and integration	Volume and complexity	The extensive data produced by drones, IoT sensors, and diverse smart devices can be staggering. Managing, integrating, and analyzing this data to derive actionable insights is complex.
	Interoperability	Different devices and platforms often use proprietary protocols, making it difficult to integrate data from various sources seamlessly.
Connectivity issues	Rural areas	Many medicinal plant cultivation sites are in remote or rural areas with limited or unreliable internet connectivity, hindering real-time data transmission and analysis.
	Infrastructure requirements	Setting up the necessary infrastructure for smart farming, including high-speed internet and robust network systems, can be challenging and costly.
Technological adoption	Complexity	The complexity of smart farming technologies can be a barrier for farmers, especially those with limited technical knowledge or experience.
	Training and support	There is a need for continuous training and technical support to ensure farmers can effectively use and maintain smart farming systems.
Cybersecurity risks	Data security	Incorporating digital technologies heightens the vulnerability to cyber-attacks and data breaches. Safeguarding the security and confidentiality of sensitive data is a major concern.
	System vulnerabilities	Smart farming systems can be vulnerable to hacking, which can disrupt operations and compromise the integrity of cultivation data.
Economic Challenges		
High initial investment	Cost of technology	The initial cost of purchasing and installing smart farming equipment, such as sensors, drones, and blockchain infrastructure, can be prohibitively high for many farmers.
	Maintenance costs	Ongoing maintenance and upgrading of smart farming systems add to the financial burden.
Return on investment (ROI)	Uncertainty	The ROI on smart farming techniques can be unpredictable, especially for small-scale farmers. The benefits may not be immediately apparent, and the payback period can be long.
	Market fluctuations	The profitability of medicinal plants can be affected by market fluctuations, which can impact the economic viability of investing in smart farming.
Access to funding	Financial barriers	Many farmers, especially in developing countries, lack access to affordable financing options to invest in smart farming technologies.
	Subsidies and incentives	The availability of government subsidies and incentives varies widely, influencing the adoption of smart farming practices.

(Continued)

TABLE 15.3 ***(Continued)***
Challenges Associated with Smart Farming

Technological Challenges		
Specification	**Issue**	**Challenge**
Environmental Challenges		
Energy consumption	Power requirements	Smart farming technologies, particularly those used in controlled environments like greenhouses, can have high energy demands, contributing to a larger carbon footprint.
	Sustainable energy resources	Incorporating renewable energy resources, like wind or solar power, can mitigate this issue but may require additional investments.
Electronic waste	Disposal of equipment	The rapid advancement of technology can lead to obsolescence of smart farming equipment, generating electronic waste that needs to be managed responsibly.
	Recycling and reuse	Developing efficient recycling and reuse strategies for outdated equipment is essential to minimize environmental impact.
Resource management	Water and soil conservation	While smart farming aims to optimize resource use, improper implementation can still lead to overuse or depletion of natural resources.
	Pesticide and fertilizer use	Ensuring that smart farming practices do not lead to excessive use of chemicals is crucial for maintaining soil health and environmental balance.
Social Challenges		
Digital divide	Access to technology	There is a significant digital divide between urban and rural areas, and between large-scale and small-scale farmers. Ensuring equitable access to smart farming technologies is a major challenge.
	Educational gaps	Farmers with limited education or technical knowledge may struggle to adopt and benefit from smart farming technologies.
Labor dynamics	Job displacement	Automation and the use of advanced technologies can lead to job displacement, particularly for labourers engaged in traditional farming practices.
	Skill requirements	The shift towards smart farming requires a workforce with new skill sets, including data analysis, technology management, and system maintenance.
Community acceptance	Resistance to change	There may be resistance to adopting new technologies due to cultural preferences for traditional farming methods.
	Awareness and perception	Increasing awareness and changing perceptions about the benefits of smart farming are necessary to gain community acceptance.
Regulatory and Policy Challenges		
Regulatory frameworks	Compliance requirements	Ensuring that smart farming practices comply with existing agricultural regulations can be complex. Regulatory frameworks often lag behind technological advancements.
	Standardization	The lack of standardized protocols for data collection, analysis, and blockchain implementation can hinder interoperability and wider adoption.
Intellectual property	Ownership of data	Issues related to data ownership, privacy, and intellectual property rights need to be addressed to protect the interests of farmers and technology providers.
	Licensing and patents	Navigating the legal landscape of patents and licenses for smart farming technologies can be challenging.
Incentives and support	Government policies	Effective government policies and incentives are crucial for promoting the adoption of smart farming technologies. This includes subsidies, tax breaks, and funding for research and development.
	Public-private partnerships	Partnerships between private and public sectors can foster innovation and provide the necessary support for farmers.

While smart farming holds great promise for enhancing the cultivation of medicinal plants, addressing the challenges and limitations outlined previously is critical for its widespread adoption and success. To surmount technological, economic, environmental, and social obstacles, policy-makers, farmers, researchers, and technology providers must work together collaboratively.

15.5.2 Limitations of Blockchain

Blockchain technique stands poised to transform medicinal plant cultivation through its provision of traceability, transparency, and security measures. Nevertheless, there are various challenges and limitations encountered in blockchain adoption and implementation in this particular field. These can be broadly categorized into technological, economic, regulatory, and social factors (Table 15.4) (Li et al., 2023; Torky & Hassanein, 2020).

TABLE 15.4
Challenges Associated with Blockchain

Specification	Issue	Challenge
Technological Challenges		
Scalability and performance	Transaction speed	Blockchain networks, especially public ones, often suffer from low transaction speeds and scalability issues. High latency can hinder real-time data recording and processing, which is critical in smart farming.
	Data volume	The large volume of data generated in medicinal plant cultivation, including environmental conditions, plant growth stages, and supply chain logistics, can overwhelm blockchain networks, leading to performance bottlenecks.
Interoperability	Integration with existing systems	The complexity arises when incorporating blockchain into existing agricultural technologies and databases. Achieving smooth interoperability among diverse platforms and devices is challenging and demands standardized protocols.
	Cross-chain compatibility	Different blockchain networks may not be compatible with each other, complicating data sharing and collaboration across various platforms.
Data confidentiality and security	Privacy	While blockchain maintains data transparency, it can be difficult to balance this with the need for confidentiality in certain aspects of medicinal plant cultivation, such as proprietary cultivation techniques or sensitive business information.
	Cybersecurity threats	Blockchain networks are not immune to cyber-attacks. Potential vulnerabilities in smart contracts or network protocols can be exploited, compromising data integrity and security.
Energy consumption	High energy demand	Blockchain, particularly proof-of-work (PoW) based systems, can consume significant amounts of energy, contributing to a larger carbon footprint. This is at odds with the sustainable practices often associated with medicinal plant cultivation.
Economic Challenges		
High initial costs	Implementation expenses	Setting up a blockchain infrastructure can be expensive. Costs include hardware, software, and the integration of blockchain with existing systems.
	Maintenance and upgrades	Ongoing maintenance, updates, and potential scaling of the blockchain network add to the long-term costs.
Return on investment	Uncertainty	The advantages of blockchain, like improved traceability and transparency, may not immediately translate into financial returns. The ROI can be uncertain and long-term, which may deter investment.
	Cost-benefit analysis	For small-scale farmers, the cost of adopting blockchain technology may outweigh the perceived benefits, especially in regions with tight financial constraints.

(*Continued*)

TABLE 15.4 (*Continued*)
Challenges Associated with Blockchain

Specification	Issue	Challenge
		Technological Challenges
Access to funding	Financial barriers	Securing funding for blockchain projects might be challenging, especially for medium- and small-sized enterprises and individual farmers. Limited access to capital and credit can hinder adoption.
Regulatory Challenges		
Lack of regulatory frameworks	Evolving regulations	Blockchain technology is relatively new, and regulatory frameworks are still evolving. The lack of clear regulations can create uncertainty and deter adoption.
	Compliance issues	Ensuring compliance with local, national, and international regulations regarding data privacy, security, and agricultural practices can be complex and burdensome.
Standardization	Absence of standards	The lack of industry-wide standards for blockchain implementation in agriculture leads to fragmentation and inconsistent practices. Standardization is necessary for interoperability and widespread adoption.
	Certification and quality assurance	Establishing standardized procedures for certifying blockchain-based systems and ensuring quality assurance is challenging.
Legal implications	Liability concerns	Determining liability in the case of errors or failures within a blockchain system can be legally complex. Smart contracts and automated processes add layers of legal considerations.
	Intellectual property	Protecting intellectual property rights while ensuring transparency in blockchain systems requires careful balancing and legal clarity.
Social Challenges		
Awareness and understanding	Knowledge pap	Many farmers and stakeholders in the agricultural sector lack awareness and understanding of blockchain technology. Educational initiatives are needed to bridge this gap.
	Perception issues	Misconceptions about blockchain, such as its association with cryptocurrencies, can create resistance to its adoption in medicinal plant cultivation.
Digital divide	Access to technology	The gap in digital access between rural and urban areas, and between large-scale and small-scale farmers, can hinder the equitable adoption of blockchain technology. Ensuring access to necessary digital infrastructure is crucial.
	Technical skills	Implementing and managing blockchain systems require technical expertise that may be lacking among farmers and agricultural workers.
Cultural resistance	Traditional practices	Farmers accustomed to traditional cultivation practices may resist adopting new technologies, including blockchain. Changing long-standing practices and mindsets can be difficult.
	Trust issues	Building trust in blockchain technology among stakeholders, including farmers, consumers, and regulatory bodies is essential for its successful implementation.
Environmental Challenges		
Sustainability concerns	Energy efficiency	As mentioned earlier, the high energy consumption of some blockchain systems raises environmental sustainability concerns. Transitioning to more energy-efficient blockchain protocols, such as proof-of-stake (PoS), is necessary.
	Environmental impact	The overall environmental impact of blockchain technology needs to be assessed and mitigated to align with the sustainable practices often emphasized in medicinal plant cultivation.
Resource allocation	Focus on technology	Excessive focus on blockchain and other digital technologies may divert resources from other essential areas of medicinal plant cultivation, such as research on plant health and traditional agricultural practices.
	Balanced approach	Ensuring a balanced approach that integrates blockchain with sustainable farming practices and environmental conservation is crucial.

While blockchain technology offers promising solutions for enhancing transparency, traceability, and security in medicinal plant cultivation, addressing the challenges and limitations outlined here is essential for its successful adoption. Overcoming these barriers requires concerted efforts from all stakeholders, including farmers, technology developers, policymakers, and researchers. Surmounting these barriers necessitates unified efforts from all stakeholders, including policymakers, farmers, researchers, and technology developers.

Future efforts should focus on:

1. Technological Advancements: Creating energy-efficient, scalable, and interoperable blockchain solutions customized for the requirements of medicinal plant cultivation.
2. Capacity Building: Providing education and training to farmers and agricultural workers to build their technical skills and knowledge.
3. Policy Development and Regulation: Establishing clear regulatory frameworks, policies and industry standards to support the integration of blockchain in agriculture while ensuring sustainability and fairness.
4. Sustainability Initiatives: Ensuring that blockchain adoption aligns with sustainable farming practices and environmental conservation and integrating renewable energy sources into smart farming systems.
5. Equitable Access: Ensuring that small-scale and marginalized farmers have access to smart farming technologies and the benefits they offer.

By addressing these challenges, smart farming integrated with blockchain technology can significantly enhance the cultivation and supply chain management including the quality, efficiency, and sustainability of medicinal plants, benefiting producers, consumers, and the broader agricultural industry.

15.6 EMERGING TRENDS IN SMART FARMING AND BLOCKCHAIN IN MEDICINAL PLANT CULTIVATION

The cultivation of medicinal plants is increasingly benefiting from advancements in smart farming and blockchain technologies. These innovations are addressing various challenges in cultivation, supply chain management, and quality control. Some of the most notable emerging trends in this specialized field are discussed in the following (Lei et al., 2022; Sharma et al., 2022; Mirabelli & Solina, 2020; Khanna et al., 2022; Choudhary, 2023; Torky & Hassanein, 2020; Javaid et al., 2022).

15.6.1 Precision Agriculture and IoT Integration

1. Smart Sensors

   Environmental Monitoring: IoT sensors used to monitor pH levels, soil moisture content, temperature, light intensity, and humidity. These sensors provide real-time data, helping farmers optimize growing conditions for medicinal plants, which often have specific requirements.

   Automated Irrigation Systems: Integrating IoT with irrigation systems allows for precise water application, minimizing water wastage and safeguarding that medicinal plants gain optimal moisture levels.
2. Drones and Aerial Imaging

   Crop Health Monitoring: Drones outfitted with multispectral and hyperspectral cameras can detect early indications of stress, pest, or disease infestation in medicinal plants, allowing for timely intervention.

   Topographical Mapping: Drones can create detailed maps of cultivation areas, assisting in land management and optimizing planting patterns to enhance yield and quality.

15.6.2 Blockchain for Traceability and Quality Assurance

1. Transparency in Supply Chain
   Provenance Verification: Blockchain technology has the capability to trace the journey of medicinal plants from cultivation to final product, ensuring that consumers and regulatory bodies can verify the origin and authenticity of the plants.
   Immutable Records: Blockchain provides an immutable ledger for recording every transaction and process step, securing transparency and impeding fraud.
2. Smart Contracts
   (a) Automated Compliance: Smart contracts can enforce compliance with agricultural standards and organic certifications by automatically verifying that all conditions and practices meet regulatory requirements.
   (b) Efficient Transactions: These contracts streamline financial transactions and agreements between farmers, suppliers, and buyers, reducing delays and administrative overhead.

15.6.3 Advanced Robotics and Automation

1. Robotic Harvesting
   Precision Harvesting: Robots equipped with AI and computer vision can selectively harvest medicinal plants at their peak potency, improving quality and reducing waste.
   Labor Efficiency: Automation of labour-intensive tasks like planting, in addition to harvesting, and weeding can significantly reduce labour costs and increase operational productivity.
2. Automated Processing
   Post-Harvest Handling: Automated systems for drying, sorting, and packaging medicinal plants ensure consistent quality and reduce the risk of contamination.
   Quality Control: Robotics and AI can be used for quality control in processing plants, ensuring that only high-quality products reach the market.

15.6.4 AI and Big Data Analytics

1. Predictive Analytics
   Yield Forecasting: Advanced AI models will continue to improve, providing more accurate predictions and recommendations for optimizing growth conditions. These models can learn from historical data and adapt to changing environmental conditions, ensuring optimal plant health and productivity.
   Disease Prediction: ML models can anticipate potential disease outbreaks on the basis of environmental data and historical patterns, facilitating proactive measures.
2. Decision Support Systems
   Optimized Cultivation: As data analytics technology advances, farmers will have access to even more detailed and actionable insights, enabling them to refine their cultivation methods further so as to obtain better product quality and efficiency.
   Risk Management: Data analytics help identify and mitigate risks such as adverse weather conditions or market fluctuations, ensuring more stable production cycles.

15.6.5 Sustainable Farming Practices

1. Climate-Smart Agriculture
   Resilient Varieties: Genetic research and biotechnology are developing medicinal plant varieties with increased resilience to pests, pathogens, and climate change.
   Resource Efficiency: Smart farming technologies enhance the utilization of energy, water, and nutrients, fostering sustainable farming methods and diminishing environmental footprint.

2. Organic and Regenerative Agriculture
Soil Health: Practices like crop rotation, organic amendments, and reduced tillage are being integrated with smart farming to enhance soil health and biodiversity.
Carbon Sequestration: Techniques that improve soil carbon sequestration are being employed to mitigate climate change, supported by blockchain for verifying and recording carbon credits.

15.6.6 Vertical Farming and Controlled Environment Agriculture

1. Urban Medicinal Plant Cultivation
Vertical Farms: Vertical farming, combined with smart farming technologies, allows for the cultivation of medicinal plants in controlled environments. This can significantly increase yield per square foot and reduce the impact of adverse weather conditions. Vertical farming systems use LED lighting, hydroponics, and aeroponics to grow medicinal plants in urban settings, ensuring a consistent supply of fresh, high-quality plants year-round.
Controlled Environments: Controlled environment agriculture (CEA) systems precisely control climate conditions, light, and nutrients, creating ideal growing conditions for medicinal plants and improving potency and yield.
2. Blockchain Integration
Traceable Production: Blockchain ensures transparency in the production processes of urban farms, providing consumers with verifiable information about the cultivation conditions and practices.
Resource Optimization: Combining blockchain with IoT in vertical farms helps optimize resource use, track inputs, and enhance operational efficiency.

15.6.7 Collaborative Platforms and Digital Ecosystems

1. Digital Marketplaces
Direct-to-Consumer Sales: Increased collaboration between researchers, farmers, and technology providers will drive innovation in smart farming. Knowledge sharing platforms and global networks will facilitate the dissemination of best practices and new technologies, accelerating the adoption of smart farming techniques. Platforms that connect farmers directly with consumers and businesses reduce the need for intermediaries, ensuring fair prices and fostering trust.
Blockchain-Based Marketplaces: These platforms use blockchain to ensure transparent transactions, secure payments, and verified product information.
2. Innovation Hubs and Research Networks
AgriTech Collaboration: Innovation hubs bring together researchers, startups, and farmers to develop and test new technologies in medicinal plant cultivation.
Public-Private Partnerships: Collaborative efforts between government bodies, research institutions, and private companies drive innovation and facilitate the adoption of smart farming and blockchain technologies.

15.7 FUTURE PROSPECTS

15.7.1 Integration of Biotechnological Advances with Smart Farming of Medicinal Plants

The fusion of biotechnological progress with smart farming offers a ground-breaking agricultural approach, merging genetic understanding with sophisticated farming technologies. This synergistic approach leverages genetic engineering, molecular biology, and advanced agricultural technologies to optimize crop production, improve plant health and efficiency and sustainability of agricultural methods, particularly in medicinal plant cultivation. An in-depth look at how genomics can be integrated with smart farming and its potential impacts is highlighted in Table 15.5 (Janbandhu et al., 2024; Zhang et al., 2023; Badiyal et al., 2024; Clapp & Ruder, 2020; Canter et al., 2005).

TABLE 15.5
Integration of Genomics with Smart Farming and Its Potential Impacts

S. No.	Particulars	Applications	Potential Impact
1	Genomic insights for crop improvement	Genetic selection and breeding	*Precision Breeding:* Utilizing genomic data, breeders can identify plants possessing sought-after traits like high yield, disease resistance, and optimal growth characteristics. This accelerates the development of superior plant varieties. *Marker-Assisted Selection (MAS):* This technology employs molecular markers to pinpoint and choose plants with specific genetic traits, making the breeding process more efficient and accurate.
		Trait mapping	*Identifying Beneficial Traits:* Genomic mapping helps identify genes associated with beneficial traits, enabling targeted breeding and genetic modification to enhance these traits in medicinal plants.
2	Enhanced plant health and disease management	Disease resistance	*Genetic Resistance:* Genomic studies can identify genes responsible for resistance to specific diseases, enabling the development of disease-resistant plant varieties through selective breeding or genetic engineering. *Pathogen Detection:* Integrating genomics with IoT sensors facilitates timely detection and diagnosis of plant diseases by analyzing the genetic material of pathogens presents in the environment.
		Pest management	*Pest-Resistant Varieties:* Genomics can identify genes that confer resistance to pests, facilitating the development of pest-resistant crops and reducing the need for chemical pesticides.
3	Optimizing growth conditions	Precision agriculture	*Customized Growth Plans:* By understanding the genetic makeup of plants, farmers can tailor growth conditions such as nutrient supply, irrigation, and lighting to meet the specific needs of each plant variety. *Environmental Adaptation:* Genomic data can pinpoint plant varieties that are more aptly tailored to specific environmental conditions, allowing for more effective cultivation in diverse climates.
		Soil health management	*Microbiome Enhancement:* Genomics can be used to study the soil microbiome and its interactions with plant roots, leading to the development of strategies that enhance beneficial microbial communities and improve soil health.
4	Quality control and consistency	Phytochemical profiles	*Consistent Quality:* Genomic analysis helps ensure that medicinal plants produce consistent levels of active compounds, which is crucial for their efficacy and safety in medicinal applications. *Metabolic Pathways:* Understanding the genetic basis of metabolic pathways enables the manipulation of these pathways to enhance the production of desired phytochemicals.
		Authentication and traceability	*Genetic Fingerprinting:* Using genetic markers, each batch of medicinal plants can be authenticated and traced back to its origin, ensuring product integrity and combating fraud.
5	Economic and environmental benefits	Cost efficiency	*Reduced Inputs:* Genomics has the potential to generate plant varieties that necessitate reduced inputs like water, pesticides, and fertilizers, thus reducing production costs. *Enhanced Yields:* Improved plant varieties and optimized growth conditions can result in higher yields and better-quality crops, increasing profitability for farmers.

TABLE 15.5 (*Continued*)
Integration of Genomics with Smart Farming and Its Potential Impacts

S. No.	Particulars	Applications	Potential Impact
		Sustainability	*Resource Conservation:* Precision agriculture practices informed by genomic data lead to more efficient use of resources, reducing environmental impact. *Biodiversity Preservation:* By developing plant varieties that are resilient to local environmental conditions, genomics helps preserve biodiversity and promote sustainable farming practices.
6	Sustainable practices	Resource conservation	*Water and Energy Efficiency:* Biotechnological advances help identify plant varieties that are more water efficient or drought resistant, further conserving resources. *Eco-Friendly Inputs:* Biotechnological innovations like biofertilizers and biopesticides diminish the requirement for chemical inputs, promoting environmental sustainability.
		Climate resilience	*Adaptation Strategies:* Genomics can identify genes that confer resilience to climate change-related stresses such as extreme temperatures, drought, and salinity. Integrating this knowledge into smart farming practices helps develop climate-resilient medicinal plants. *Predictive Modelling:* Data analytics and predictive modelling tools use genomic and environmental data to forecast climate impacts on crop performance, empowering farmers to adjust their strategies proactively.

Integrating smart farming with biotechnological advances signifies a fundamental shift in agricultural methods, particularly in the cultivation of medicinal plants. By combining genetic insights with advanced technologies, this approach enhances crop performance, improves plant health, and ensures the consistent quality of medicinal products.

15.7.2 Integration of Renewable Energy with Smart Farming of Medicinal Plants

The integration of renewable energy with smart farming techniques offers a sustainable and economically viable approach to medicinal plant cultivation. By utilizing green energy sources like biomass, solar, geothermal, and wind energy, farmers can enhance the sustainability, efficiency, and productivity of medicinal plant cultivation. This holistic approach reduces carbon footprints, lowers operational costs, and improves the overall quality of medicinal plant production. Table 15.6 depicts the way renewable energy is being integrated into smart farming for medicinal plants (Majeed et al., 2023; Vishnoi & Goel, 2024; Nizetic et al., 2019; Mondejar et al., 2021).

Integrating smart farming practices with renewable energies not only promotes environmental sustainability but also enhances resilience and reliability of farming operations. With ongoing technological advancements and increasing policy backing, the utilization of renewable energy in smart farming is poised to expand, leading to a more sustainable and prosperous future for farming of medicinal plants.

TABLE 15.6
Integration of Renewable Energy with Smart Farming and Its Potential Impacts

S. No.	Renewable Energy	Applications	Potential Impact
1	Solar energy	Solar panels and photovoltaic systems	*Energy Production:* Solar panels installed on farms transform sunlight into electrical energy, furnishing a sustainable and green energy source for a multitude of farm operations. This energy can be used to power irrigation systems, lighting, and another farm machinery. *Solar-Powered Irrigation:* Solar-powered pumps can draw water for irrigation, diminishing dependence on conventional energy sources and guaranteeing a steady water supply, crucial for medicinal plant cultivation.
		Solar greenhouses	*Controlled Environments:* Solar greenhouses utilize photovoltaic cells to generate electricity, which powers climate control systems such as heating, cooling, and artificial lighting. These controlled environments can be optimized for the specific growth needs of medicinal plants, ensuring optimal conditions year-round. *Energy Storage:* Excess energy generated by solar panels can be stored in batteries, providing a consistent energy supply during cloudy days or at night, ensuring uninterrupted operations.
2	Wind energy	Wind turbines	*Electricity Generation:* Small-scale wind turbines can generate electricity for farm operations. This renewable energy source can be used alongside solar energy to power various systems, creating a hybrid energy solution that maximizes energy production. *Wind-Powered Water Pumps:* Wind-powered pumps can be used for irrigation, reducing energy costs and promoting sustainable water management practices essential for medicinal plant cultivation.
3	Biomass energy	Biogas production	*Organic Waste Utilization:* Farms can convert organic waste such as plant residues, manure, and other biodegradable materials into biogas through anaerobic digestion. Biogas can be used for heating, electricity generation, or as fuel for farm machinery. *Nutrient-Rich Fertilizer:* The by-product of anaerobic digestion is a nutrient-rich slurry that can be used as a natural fertilizer, enhancing soil health and reducing the need for chemical fertilizers.
		Bioenergy crops	*Dedicated Energy Crops:* Some farms grow crops specifically for biomass production. These crops can be processed into biofuels or biopower, providing an additional renewable energy source. *Integrated Farming Systems:* Growing bioenergy crops alongside medicinal plants can optimize land use and contribute to the overall sustainability of the farming operation.
4	Geothermal energy	Geothermal greenhouses	*Stable Temperature Control:* Geothermal energy can heat greenhouses, providing a stable and sustainable heat source. This ensures optimal growing conditions for medicinal plants throughout the year. *Efficient Energy Use:* Geothermal systems are highly efficient and have a low environmental impact, making them ideal for sustainable medicinal plant cultivation.
		Geothermal heating and cooling	*Temperature Regulation:* Geothermal heat pumps can regulate temperature of farm buildings and storage facilities, ensuring optimal conditions for both plant growth and storage.

TABLE 15.6 (*Continued*)
Integration of Renewable Energy with Smart Farming and Its Potential Impacts

S. No.	Renewable Energy	Applications	Potential Impact
5	Enhancing medicinal plant cultivation	Sustainable water management	*Efficient Irrigation:* Smart irrigation systems use renewable energy to deliver precise amounts of water to medicinal plants, minimizing water wastage and safeguarding that plants acquire appropriate moisture levels for optimal growth. *Water Recycling:* Renewable energy can power water recycling systems, ensuring that water used in cultivation is purified and reused, promoting sustainability.
6	Economic and environmental benefits	Cost savings	*Reduced Energy Bills:* Renewable energy decreases reliance on grid electricity and fossil fuels, substantially reducing energy bills and operational costs. *Long-Term Investment:* Although the upfront cost of renewable energy systems can be significant, the long-term savings and potential revenue from selling surplus energy make it a financially viable option.
		Environmental sustainability	*Lower Carbon Footprint:* Renewable energy sources generate minimal to no greenhouse gas emissions, greatly reducing the carbon footprint of medicinal plant farming. *Resource Conservation:* Sustainable energy practices help conserve natural resources, such as water and fossil fuels, contributing to overall environmental health.
		Resilience and reliability	*Energy Independence:* Farms generating their own renewable energy are less vulnerable to energy price fluctuations and grid outages, enhancing resilience and reliability. *Climate Adaptation:* Renewable energy systems help farms adapt to climate change by providing stable and sustainable energy sources, regardless of external conditions.
7	Policy and community support	Government incentives	*Subsidies and Grants:* Many governments offer subsidies, grants, and tax incentives to support the adoption of renewable energy in agriculture. These financial incentives can offset initial costs and encourage the integration of renewable energy. *Regulatory Support:* Policies that advocate for renewable energy and sustainable farming methods encourage widespread adoption of these technologies.
		Community and cooperative models	*Energy Cooperatives:* Farmers can form cooperatives to invest in and manage renewable energy projects collectively, reducing individual costs and sharing the benefits among community members. *Knowledge Sharing:* Community-based initiatives and farmer networks facilitate knowledge sharing and best practices, fostering successful integration of renewable energy in medicinal plant cultivation.

15.8 CONCLUSION

Integrating smart farming with blockchain technology marks a major advancement in medicinal plant cultivation, providing numerous benefits, including enhanced transparency and traceability, along with improved supply chain efficiency and sustainability. By exploiting smart farming tools such as IoT sensors, drones, and AI-driven decision support systems in conjunction with blockchain's immutable ledger and smart contract capabilities, stakeholders across the supply chain can ensure the authenticity, quality, and ethical sourcing of medicinal plants. This integration not only empowers farmers, promotes fair trade, and facilitates compliance with regulatory standards but also fosters consumer trust and confidence in the products they consume.

This integrated approach addresses many of the traditional challenges in medicinal plant cultivation, including resource inefficiency, lack of transparency, and difficulties in maintaining consistent product quality. Renewable energy sources, combined with smart farming technologies, reduce reliance on fossil fuels, lower carbon footprints, and ensure more sustainable use of resources. Blockchain's role in securing and transparently recording every stage of the plant life cycle adds a layer of trust and reliability, crucial for both regulatory compliance and consumer confidence.

As the adoption of these technologies increases, supported by government incentives, training programs, and community initiatives, the future of medicinal plant cultivation looks promising. Farmers will benefit from reduced operational costs, increased revenue streams, and the ability to fulfil the increasing demand for superior quality medicinal plants. Ultimately, the integration of smart farming and blockchain technology will create a more reliable, efficient, and adaptable agricultural sector, setting new standards for the cultivation of medicinal plants; driving positive social, environmental, and economic outcomes for all stakeholders involved.

REFERENCES

Ane, T., Nepa, T., Khan, M.R., Smart and intelligent production strategy for the flower market using data mining knowledge-based decision. *Bangladesh Journal of Multidisciplinary Scientific Research*, 7, 35, 2023.

Aung, M.M., Chang, Y.S., Traceability in a food supply chain: Safety and quality perspectives. *Food Control*, 39, 172, 2014.

Badiyal, A., Mahajan, R., Rana, R.S., Sood, R., Walia, A., Rana, T., Manhas, S., Jayswal, D.K., Synergizing biotechnology and natural farming: Pioneering agricultural sustainability through innovative interventions. *Frontiers of Plant Science*, 15, 1280846, 2024.

Bannor, R.K., Arthur, K.K., Oppong, D., Oppong-Kyeremeh, H., A comprehensive systematic review and bibliometric analysis of food fraud from a global perspective. *Journal of Agriculture and Food Research*, 14, 100686, 2023.

Bhat, S.A., Huang, N.-F., Sofi, I.B., Sultan, M., Agriculture-food supply chain management based on blockchain and IoT: A narrative on enterprise blockchain interoperability. *Agriculture*, 12, 40, 2022.

Bosona, T., Gebresenbet, G., The role of blockchain technology in promoting traceability systems in agri-food production and supply chains. *Sensors*, 23, 5342, 2023.

Canter, P.H., Thomas, H., Ernst, E., Bringing medicinal plants into cultivation: Opportunities and challenges for biotechnology. *Trends in Biotechnology*, 23, 180, 2005.

Choudhary, V.K., Applications of emerging smart technologies in farming systems: A review. *Global Journal of Computer Science and Technology*, 23(G1), 49, 2023.

Christy, R.D., Fu, L., *Innovative institutions to accelerate agroindustry development in Asia*. The Asian Productivity Organization, 2020.

Clapp, J., Ruder, S.-L., Precision technologies for agriculture: Digital farming, gene-edited crops, and the politics of sustainability. *Global Environmental Politics*, 20, 49, 2020.

Dutta, M., Metkewar, P.S., Kumar Dhanaraj, R., Gupta, D., Juneja, S., Mapping Smart Vertical *Farming in cultivation of herbaceous medicinal plants using bibliometric analysis*. 2023 International Conference on Advances in Computation, Communication and Information Technology (ICAICCIT), Faridabad, India, 453, 2023.

Eka Putri, R., Darmadi, W., Cherie, D., Puari, A., The design of automatic soil pH control system on aloe vera cultivation with an integration of Internet of Things (IoT). *Jurnal Teknik Pertanian Lampung*, 12, 597, 2023.

Esteki, M., Regueiro, J., Simal-Gándara, J., Tackling fraudsters with global strategies to expose fraud in the food chain. *Comprehensive Reviews in Food Science and Food Safety*, 18, 425, 2019.

Govindaraghavan, S., Quality assurance of herbal raw materials in supply chain: Challenges and opportunities. *Journal of Dietary Supplements*, 5, 176, 2008.

Heinrich, M., Scotti, F., Booker, A., Fitzgerald, M., Kum, K.Y., Lobel, K., Unblocking high-value botanical value chains: Is there a role for blockchain systems? *Frontiers in Pharmacology*, 10, 396, 2019.
Hu, S., Huang, S., Huang, J., Su, J., Blockchain and edge computing technology enabling organic agricultural supply chain: A framework solution to trust crisis. *Computers & Industrial Engineering*, 153, 107079, 2021.
Iqbal, R., Butt, T.A., Safe farming as a service of blockchain-based supply chain management for improved transparency. *Cluster Computing*, 23, 2139, 2020.
Ivanova, S.V., Latyshov, A.V., The Korean experience in implementing the fourth industrial revolution: General and agricultural aspects. In: *The challenge of sustainability in agricultural systems*, Bogoviz, A.V., Ed., Lecture Notes in Networks and Systems, Springer, Cham, 205, 2021.
Janbandhu, M.S., Mehta, A., Beese, S., Pandey, S.K., Singh, B., Patel, A., Singh, B.P., Advances and emerging trends in horticultural production and management. *Journal of Experimental Agriculture International*, 46, 47, 2024.
Javaid, M., Haleem, A., Singh, R.P., Suman, R., Enhancing smart farming through the applications of agriculture 4.0 technologies. *International Journal of Intelligent Networks*, 3, 150, 2022.
John, S., Arul Leena Rose, P.J., Smart farming and precision agriculture and its need in today's world. In: *Intelligent robots and drones for precision agriculture*, Balasubramanian, S., Natarajan, G., Chelliah, P.R., Eds., Signals and Communication Technology, Springer, Cham, 2024.
Jones, A.M., A conceptual model for digitalizing farming of medicinal plant sector in Kerala. *Journal of Management & Entrepreneurship*, 16, 36, 2022.
Karunathilake, E.M.B.M., Le, A.T., Heo, S., Chung, Y.S., Mansoor, S., The path to smart farming: Innovations and opportunities in precision agriculture. *Agriculture*, 13, 1593, 2023.
Kassanuk, T., Phasinam, K., Design of blockchain based smart agriculture framework to ensure safety and security. *Materials Today: Proceedings*, 51, 2313, 2022.
Khanna, A., Jain, S., Maheshwari, P., *Precision agriculture for medicinal plants: A conjunction of technologies*. 2022 International Conference on Electrical and Computing Technologies and Applications (ICECTA), Ras Al Khaimah, United Arab Emirates, 300, 2022.
Lama, A., Rana, N., Emerging technology in advancement of Indian agriculture. In: *Technological advancement in Indian agriculture*, Aggarwal, G., Lama, A., Varma, A., Eds., Sankalp Publication, Chhattisgarh, India, 2023.
Lei, M., Xu, L., Liu, T., Liu, S., Sun, C., Integration of privacy protection and blockchain-based food safety traceability: Potential and challenges. *Foods*, 11, 2262, 2022.
Li, K., Lee, J.Y., Gharehgozli, A., Blockchain in food supply chains: A literature review and synthesis analysis of platforms, benefits and challenges. *International Journal of Production Research*, 61, 3527, 2023.
Li, Q., Wang, J., Zhang, Y., Exploring the public space model of smart recreation for Chinese medicine cultivation based on artificial intelligence and blockchain technology. In *Computer Science and Educational Informatization, CSEI*, Gan, J., Pan, Y., Zhou, J., Liu, D., Song, X., Lu, Z., Eds., Communications in Computer and Information Science, Springer, Singapore, 1900, 2024.
Liontakis, A., Sintori, A., Tzouramani, I., The role of the start-up aid for young farmers in the adoption of innovative agricultural activities: The case of aloe vera. *Agriculture*, 11, 349, 2021.
M, Sandeep Kumar, Maheshwari Venkat, Prabhu Jayagopal, Prasanna Mani, Jothikumar, R., Applying blockchain in agriculture: A study on blockchain technology, benefits, and challenges. In: *Deep Learning and Edge Computing Solutions for High Performance Computing*, Suresh, A., Paiva, S., Eds., EAI/Springer Innovations in Communication and Computing, Springer, Cham, 2021.
Mahmoudi, R., Ghajarbeygi, P., Niaraki, A.S., Kazeminia, M., Survey of fraud in the foods that were used in its production of saffron. *Carpathian Journal of Food Science and Technology*, 12, 145, 2020.
Majeed, Y., Khan, M.U., Waseem, M., Zahid, U., Mahmood, F., Majeed, F., Sultan, M., Raza, A., Renewable energy as an alternative source for energy management in agriculture. *Energy Reports*, 10, 344, 2023.
Mirabelli, G., Solina, V., Blockchain and agricultural supply chains traceability: Research trends and future challenges. *Procedia Manufacturing*, 42, 414, 2020.
Mondejar, M.E., Avtar, R., Diaz, H.L.B., Dubey, R.K., Esteban, J., Gomez-Morales, A., Hallam, B., Mbungu, N.T., Okolo, C.C., Prasad, K.A., She, Q., Garcia-Segura, S., Digitalization to achieve sustainable development goals: Steps towards a smart green planet. *Science of the Total Environment*, 794, 148539, 2021.
Nizetic, S., Djilali, N., Papadopoulos, A., Rodrigues, J.J.P.C., Smart technologies for promotion of energy efficiency, utilization of sustainable resources and waste management. *Journal of Cleaner Production*, 231, 565, 2019.
Nowvaratkoolchai, P., Thawesaengskulthai, N., Viriyasitavat, W., Rangsunvigit, P., Blockchain-based cannabis traceability in supply chain management. *International Journal of Advanced Computer Science & Applications*, 15, 75, 2024.

Panwar, A., Khari, M., Misra, S., Sugandh, U., Blockchain in agriculture to ensure trust, effectiveness, and traceability from farm fields to groceries. *Future Internet*, 15, 404, 2023.

Sharma, V., Tripathi, A.K., Mittal, H., Technological revolutions in smart farming: Current trends, challenges & future directions. *Computers and Electronics in Agriculture*, 201, 107217, 2022.

Simmler, C., Graham, J.G., Chen, S.-N., Pauli, G.F., Integrated analytical assets aid botanical authenticity and adulteration management. *Fitoterapia*, 129, 401, 2018.

Sistani, M.R., Gallo, P., Timoshina, M., Red gold traceability: computer vision and blockchain for saffron quality. *BlockSys '23: Proceedings of the Fifth ACM International Workshop on Blockchain-enabled Networked Sensor Systems, November 2023*, 13, 2024.

Torky, M., Hassanein, A.E., Integrating blockchain and the internet of things in precision agriculture: Analysis, opportunities, and challenges. *Computers and Electronics in Agriculture*, 178, 105476, 2020.

Unal, Z., Smart farming becomes even smarter with deep learning—a bibliographical analysis. *IEEE Access*, 8, 105587, 2020.

Vernon, M., Kouzani, A.Z., Webb, L.D., Adams, S.D., A survey of modern greenhouse technologies and practices for commercial cannabis cultivation. *IEEE Access*, 11, 62077, 2023.

Vishnoi, S., Goel, R.K., Climate smart agriculture for sustainable productivity and healthy landscapes. *Environmental Science & Policy*, 151, 103600, 2024.

Walter, A., Finger, R., Huber, R., Buchmann, N., Smart farming is key to developing sustainable agriculture. *PNAS</APEX-ITALIC>, 114, 6148, 2017.*

Yaga, D., Mell, P., Roby, N., Scarfone, K., Blockchain technology overview. NIST, 2018. https://doi.org/10.6028/NIST.IR.8202

Yan, H.-S., Yang, J.-W., Kim, H.-H., Application of blockchain technology in agricultural products' traceability system. *Asia-Pacific Journal of Convergent Research Interchange*, 7, 55, 2021.

Yang, X., Li, M., Yu, H., Wang, M., Xu, D., Sun, C., A trusted blockchain-based traceability system for fruit and vegetable agricultural products. *IEEE Access*, 9, 36282, 2021.

Yoon, C., Huh, M., Kang, S.-G., Park. J., Lee, C., Implement smart farm with IoT technology. *2018 20th International Conference on Advanced Communication Technology (ICACT), Chuncheon, Korea (South)*, 749, 2018.

Zhang, M., Han, Y., Li, D., Xu, S., Huang, Y., Smart horticulture as an emerging interdisciplinary field combining novel solutions: Past development, current challenges, and future perspectives. *Horticultural Plant Journal*, 2023. https://doi.org/10.1016/j.hpj.2023.03.015

16 Insights into Coral Reefs of the Western Indian Ocean

A Comprehensive Analysis of Habitat Classification, Forecasting Taxa Changes, and Explainable AI-Driven Health Assessment

Tanvi Bahedia, Sparsh Wabhale, S A Sajidha, and Aakif Mairaj

16.1 INTRODUCTION

16.1.1 Coral Reefs

Coral reefs, often referred to as the "rainforests of the sea," represent some of the most diverse and biologically rich ecosystems on our planet. Coral reefs are underwater structures formed by colonies of tiny animals called coral polyps. These polyps secrete calcium carbonate to create hard, rock-like structures that provide the foundation for the reef. Over time, as new generations of coral polyps grow on top of the existing structures, they create the intricate, three-dimensional formations we recognize as coral reefs. Coral reefs offer a wide range of benefits and play crucial roles in marine ecosystems, making them invaluable to both the environment and human society. They are renowned for their excellent biodiversity. They are home to approximately 25% of all known marine species. This richness in species diversity makes coral reefs a vital resource for scientific research and conservation efforts. Coral reefs provide essential habitats for numerous fish species, including those that are commercially important for fisheries. They serve as breeding and nursery grounds for many marine organisms, contributing to the sustainability of fish populations. They also act as a natural barrier to protect the coastlines from damaging effects of nature (storm, waves, erosion). Their rich biodiversity also has helped develop drugs and medicines to treat diseases like HIV and cancer in the medical field.

16.1.2 Environmental Effects on Coral Reefs

Over the past ten years, there has been an exceptional occurrence of coral bleaching and extensive coral die-offs. These events have been triggered by a sequence of years with high temperatures, consecutive El Niño events, and heatwaves. When we account for both local and global influences, coastal reefs are notably vulnerable. This vulnerability arises from the fact that they also face intense and immediate human pressures associated with activities taking place within the reef ecosystem. However, these delicate marine environments face unprecedented threats from the dual forces of climate change and human activities. As rising sea temperatures, ocean acidification, and destructive practices continue to imperil coral reefs worldwide, there is an urgent need for innovative solutions that facilitate their accurate monitoring, prediction, and, most critically, conservation.

DOI: 10.1201/9781003507390-16

16.1.3 Efforts to Conserve Coral Reefs Using Technology

The impact of coastal coral reef ecosystems' topography and underwater features is well documented, and maintaining the complexity of their natural relief is crucial for safeguarding their fundamental functions. Satellite-based remote sensing has been utilized for mapping tropical coral reefs since the 1970s, starting with the launch of the first Landsat satellite. Recent technological advancements in the past decade have made it possible to acquire satellite data with resolutions that can provide detailed mapping of both the biological communities within coral reef areas and the surrounding topographical features. Furthermore, satellite imagery offers a cost-effective means of obtaining important data like bathymetry, terrain complexity, and benthic cover, making it a valuable tool in species distribution models (SDMs) and conservation planning. The integration of machine learning (ML) techniques has enhanced accuracy and the ability to predict, allowing for the identification of previously unmodeled variables.

16.1.4 Recent Trends and Statistics

Technology has evolved since environmental scientists have been involved in global research, from human-based paper calculations to advanced algorithms that automate and provide deep insights and analytics. Recent developments in this field have advanced ahead from normal image and sound analysis. A multi-disciplinary approach is being adopted to understand these marine ecosystems. Conservation and intervention-based approaches, including restoration, or the rebuilding of reef ecosystems harboring complex and interwoven biological, chemical, and physical components, can be facilitated by a system equipped with sensors and technologies to quantitatively characterize the biological, chemical, and physical parameters of the reef, ultimately informing evidence-based management decisions and promoting the long-term sustainability of these critical ecosystems.

1. **Reef Imaging:** Recent developments in underwater camera technology have made it possible to integrate cameras into coral reefs for extended periods, which greatly improves the ability to monitor changes over time. This includes observing events like coral recruitment and post-settlement selection, as well as assessing the diversity, biomass, and abundance of fish and other mobile reef organisms. Additionally, by mounting camera systems on autonomous underwater vehicles (AUVs), we can expand our coverage of reef imaging efforts to include challenging depths that are not easily accessible to divers.
2. **Soundscapes:** Acoustic recorders include sound traps, hydro moths, and others which are typically moored onto a reef. Enhancing the spatial resolution of reef-scale soundscapes is accomplished by mooring multiple recorders on reefs and integrating hydrophones on AUVs.
3. **Reef Water Microbes, Metabolites, and eDNA:** Water surrounding coral reefs contains important information in the form of microbes, metabolites, and environmental DNA (eDNA). These elements provide insights into reef health, environmental conditions, and the presence of different species. Scientists collect water samples, filter them, and analyze them in labs to understand these aspects better. It's a bit complex, but improving software for analyzing this data could help us monitor and protect coral reefs more effectively.
4. **Robots for Enabling Reproducible High-Resolution Observation at Scale:** New robots, powered by artificial intelligence, are being developed to explore coral reefs. These robots can navigate complex underwater environments safely and can collect data at precise locations, similar to what drones do in the air. These robots can help scientists study and protect coral reefs more effectively by focusing on areas that are critical for understanding changes in reef health and biodiversity. They can also operate for longer periods while conserving energy.

This new era of coral reef monitoring will reduce barriers to detecting and predicting the health of coral reefs and bridge gaps in our understanding of overall reef function. These advancements will be critical to the successful mitigation of climate impacts and restoration of reef ecosystems in an uncertain future. The purpose of this research work is to harness the power of machine learning to address key objectives in coral diversity management:

1. Understanding Coral Diversity: To gain insights into coral diversity patterns and composition within reef locations.
2. Early Detection of Threats: To develop predictive models for early identification of stressors affecting coral reefs.
3. Optimizing Conservation Efforts: To prioritize and focus conservation efforts on areas with high biodiversity and vulnerability.
4. Enhancing Resilience: To contribute to the resilience of coral reef ecosystems through data-driven management and restoration recommendations.

Through this research, we intend to provide valuable tools and insights that support the conservation and preservation of coral reefs, ensuring these vital ecosystems thrive in the face of ongoing environmental challenges.

16.2 RELATED WORK

There have been recent developments and improvements on traditional methods to analyze the data from coral reefs and gain insights for multiple purposes and needs. Multiple approaches to understand the data are formulated and applied to deeply discover the potential and possibilities in marine life.

da Silveira et al. [1]: In recent years, the imperative to safeguard coral reefs from mounting threats like climate change, bleaching events, and human activities has grown increasingly clear. These exceptionally biodiverse ecosystems necessitate innovative approaches for mapping, monitoring, and conservation. This work explores the promising fusion of remote sensing and machine learning techniques, drawing insights from a study focused on nurturing and natural factors affecting coral distribution in marine protected areas (MPAs). The study's methodology showcases how remote sensing and machine learning can be employed for coral reef analysis, encompassing steps such as field surveys, ecological and topographical data from satellite imagery, and integration through species distribution modeling. Notably, it takes into account human impacts like fishing and tourism pressure. SVM yielded 15 benthic classes (e.g., seagrass, sand, coral), with an overall accuracy of 79%. The findings underscore the substantial utility of SDMs in coral reef conservation, aiding in the identification of areas conducive to coral growth and offering invaluable guidance for conservation efforts. SDMs, utilizing MaxEnt, were employed to predict suitable habitats for *Millepora alcicornis* in northeast Brazil. Presence data and various environmental variables were integrated, resulting in high AUC scores (>0.97). The habitat suitability map revealed limited areas suitable for large colonies, with 81% of the mapped area deemed highly unsuitable. The study enhances understanding of *M. alcicornis* distribution, providing valuable insights for coral reef restoration and conservation planning in marine protected areas, emphasizing the need for a balance between human activities and ecological sustainability.

This methodology's versatility extends beyond this study, holding potential for broader marine ecosystems. By providing a data-driven view of ecologically crucial regions and emphasizing the importance of no-take zones within MPAs, these findings enhance the comprehension of coral reef dynamics and provide direction for more effective conservation and restoration endeavors. In summary, the amalgamation of remote sensing and machine learning techniques in coral reef mapping represents a significant leap in coral reef conservation and management.

Collin et al. [2]: Continuing this exploration into coral reef diversity prediction and analysis, another study delves into the critical need for detailed coral reef monitoring as coral health continues to decline globally. This study leverages remote sensing techniques, a significant breakthrough by offering high-resolution spectral data vital for assessing coral well-being. Using spectral diversity indices, the study aims to distinguish between various coral assemblages and their health states based on the WorldView-2 satellite data. Focusing on three *Scleractinia* coral species in Moorea's reef-building areas in French Polynesia, including their health states, this work begins with field surveys at chosen sites within Moorea's coral-rich lagoon. The research uses two WorldView-2 satellite datasets, each containing multispectral and panchromatic images, which undergo thorough calibration and enhancement. Coral spectra from two categories, massive boulders and branching coral fields, are analyzed across six distinct health classes using spectral diversity indices. The study reveals promising results, indicating the support vector machine's efficacy in classifying the six coral classes based on five spectral bands, surpassing traditional three-band methods. Integration with spectral diversity indices further enhances classification reaching accuracy up to 82%, demonstrating great potential for assessing coral vitality and ecosystem health, thereby contributing valuable insights for coral management and restoration efforts.

Mahmood et al. [3]: The need to safeguard and monitor coral reefs in the face of mounting threats has led to the development of automated technologies for the analysis and assessment of marine coral species. This paper emphasizes the urgency of such technologies, given the labor-intensive process of manually annotating vast collections of underwater images obtained from remotely operated and autonomous vehicles. The proposed system employs advanced deep learning methods to automatically classify and annotate marine coral species, offering an efficient and accurate solution for marine ecological studies. Its methodology includes using deep convolutional neural networks (CNNs) with pretrained feature representations fine-tuned on domain-specific data, like the Benthoz15 dataset. The Benthoz15 dataset used a three fold cross-validation approach and consistently achieved coral classification accuracy exceeding 90%, demonstrating the robustness of the method across different trials. Through patch-based training and data augmentation, it overcomes challenges related to irregular coral assemblages and multiple species within single images, achieving improved classification results. The study further explored the efficacy of residual features (ResFeats) extracted from deep ResNet-152 layers and demonstrated their superior performance. ResFeats outperforms CNN-based features in terms of classification accuracy, achieving a rate of 91.90% in this experiment, although slightly lower than the first three trials. Notably, the precision and recall values for the coral class are both at 93%. Additionally, the system automates the annotation of coral mosaics, offering high recall rates for coral classes and revealing a decline in coral populations, consistent with previous findings. The framework automatically detected the decreasing trend in the coral population of the region observed from 2011 to 2013, which is consistent with their previous findings. This automated approach not only enhanced efficiency but also contributed significantly to the monitoring and management of marine ecosystems, providing vital insights into coral health and population trends essential for informed ecological management and policy making. The integration of deep learning and automated image analysis in coral reef conservation is poised to revolutionize the field, with potential for future expansion into species-level coral distribution mapping.

González-Rivero et al.: This literature study leveraged diverse methodologies in coral reef conservation. da Silveira et al. [1] focused on remote sensing and machine learning for coral reef analysis, highlighting the utility of species distribution modeling. Collin et al. [2] introduced high-resolution spectral data for coral health assessment using WorldView-2 imagery. Mahmood et al. [3] employed deep learning for automated coral image classification, emphasizing efficiency in annotation and mosaic analysis. González-Rivero et al. [4] explored the use of artificial intelligence, particularly convolutional neural networks, for cost-effective coral reef monitoring. Some applied machine learning for predicting coral reef bleaching, emphasizing the role of support vector machine. This research stands out from previous research by introducing dynamic coral taxa prediction, utilizing time series analysis and deep learning to forecast changes in coral taxa over time, considering environmental variables such as sea surface temperature and human impacts. This innovative approach adds a proactive dimension to coral reef conservation. Unlike prior studies, our work features a comprehensive AI-based framework that incorporates machine learning and explainable AI to classify coral health. The diversity of algorithms explored, including random forest, LSTM, and simplified feed forward neural network, enhances the methodological robustness of our research. By providing a holistic perspective on coral reef conservation that actively predicts and mitigates future threats, our work significantly contributes to the existing body of knowledge in this field.

16.3 METHODOLOGY

The proposed system aims to develop an integrated machine learning solution for monitoring and conserving coral reef biodiversity. It leverages advanced data analytics and predictive modeling techniques to address the following objectives:

16.4 DYNAMIC CORAL TAXA PREDICTION

This component focuses on developing predictive models that can forecast changes in the number of coral taxa (different species) over time. Coral reefs are dynamic ecosystems, and changes in the diversity of coral habitat (crest, lagoon/back reef, slope, flat) are crucial indicators of ecosystem health. These models will help identify trends and fluctuations in coral taxa composition.

From Figure 16.1, it is clear that machine learning techniques such as time series analysis, regression models, and deep learning can be employed to analyze historical data on coral taxa diversity and environmental variables. Classification models including decision tree, random forest, and support vector machine were employed. Time series forecasting models such as ARIMA, SARIMA, and LSTM were utilized for predicting coral taxa dynamics. These models will consider factors like sea surface temperature, nutrient levels, and human impacts to make predictions.

16.5 AI-POWERED DECISION SUPPORT SYSTEM

An AI-powered decision support system assists conservation decision-makers in selecting the most effective strategies. It takes into account data, predictions, and the interpretability of AI models to generate recommendations.

Figure 16.2 clearly depicts the structure of a decision model built using a neural network to determine the health status of a coral reef based on factors like coral cover, sea-surface temperature, dissolved oxygen level, and salinity mean. The system is provided with real-time inputs and readings upon which a neural network is trained to determine whether a coral reef is healthy. The neural network gathers input such as dissolved oxygen, salinity, pH level, surface temperature, and chlorophyll A levels. A data imbalance was observed in the dataset, primarily characterized by a higher representation of healthy coral instances compared to insufficient data for unhealthy corals.

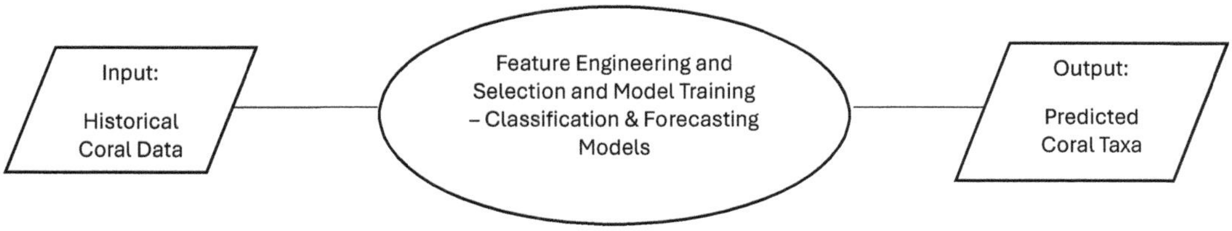

FIGURE 16.1 General diagram for dynamic coral taxa prediction.

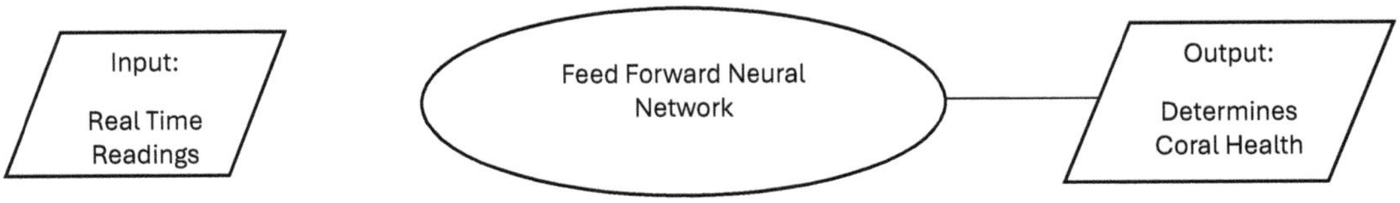

FIGURE 16.2 General diagram for AI-powered decision support system.

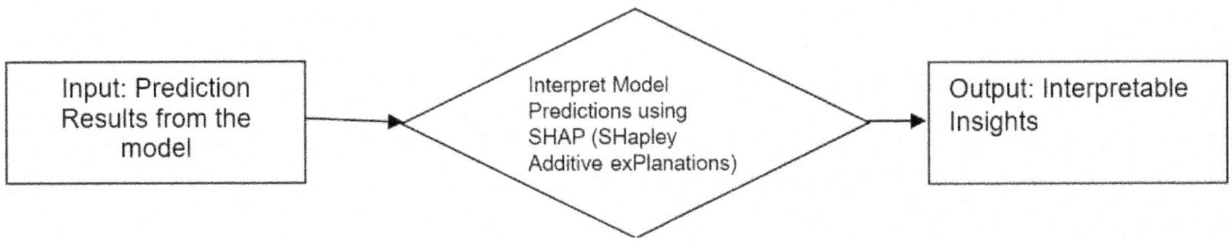

FIGURE 16.3 General diagram for explainable AI for conservation.

In order to address this imbalance and enhance the robustness of the analytical model, the synthetic minority over-sampling technique (SMOTE) was employed. The neural network is a binary classification model for coral health prediction using TensorFlow and Keras. It consists of an input layer with a linear activation, followed by two hidden layers with 64 and 32 neurons, each using a rectified linear unit (ReLU) activation function and dropout for regularization. The output layer employs a sigmoid activation for binary classification. The model is trained for 10 epochs with a batch size of 32, using the Adam optimizer and binary cross entropy loss at output layer to binarize the results. The architecture aims to predict coral health based on specified features, preventing overfitting and introducing non-linearity through ReLU activations.

16.6 EXPLAINABLE AI FOR CONSERVATION

Explainable AI models are designed to make machine learning results more transparent and interpretable. In the context of coral reef conservation, these models aim to provide clear explanations for the recommendations and decisions they make, making it easier for stakeholders to understand and trust the AI's guidance. SHAP (SHapley Additive exPlanations), an explainable AI model, was applied to machine learning models. This technique creates explanations for individual model predictions, making it possible to understand why a model recommends specific conservation actions or predicts changes in coral diversity.

Figure 16.3 shows the utilization of SHAP, an explainable AI model employed to interpret predictions generated by the AI-powered decision model. The SHAP analysis takes as inputs the predictions generated by a neural network classifying the health status of coral reefs using the specified attributes. When the neural network predicts that a coral reef is unhealthy, SHAP is employed to elucidate the contributing factors responsible for this classification.

16.7 PREDICTIVE MAPPING

Predictive risk mapping focuses on creating maps that highlight areas within coral reef ecosystems that are most vulnerable to specific environmental risks or threats. These maps can help conservationists allocate resources where they are most needed. Machine learning algorithms analyze historical data on environmental risks (e.g., high sea surface temperatures, pollution levels, latitude, longitude, coral cover, dissolved oxygen, salinity, etc.) and correlate these with known impacts on

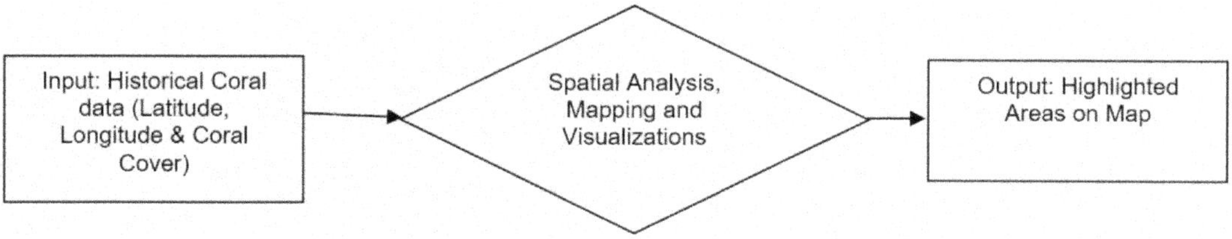

FIGURE 16.4 General diagram for predictive risk mapping.

coral reefs. By mapping these risk factors and overlaying them on reef locations, areas at high risk can be identified using folium (a Python library used for visualizing geospatial data).

Figure 16.4 illustrates the implementation of spatial analysis, mapping, and visualizations derived from historical coral data encompassing latitude, longitude, and coral cover. The resultant map output employs strategic highlighting techniques to accentuate specific features and trends in the geographical distribution of corals.

16.8 DATA SET

These are data on coral cover and number of genera in the Western Indian Ocean [5] collected from a rapid assessment method used to sample coral bleaching. The dataset comprises a comprehensive set of ecological and environmental parameters characterizing coral reefs along Tanzania's East African Coral Coast. It includes geographical coordinates, ecological features, management practices, habitat types, and diverse environmental metrics such as sea surface temperature, dissolved oxygen, and nutrient levels. Socioeconomic aspects, such as travel time to markets and population centers, add a multidimensional layer to the dataset. With temporal information and replicates, the dataset serves as a valuable resource for multidisciplinary research, offering insights into the intricate relationships between environmental conditions, reef health, and human-related factors. Researchers can leverage this dataset to deepen our understanding of coral reef dynamics in the East African Coral Coast region.

16.9 OVERALL ARCHITECTURE

Figure 16.5 provides a deeper insight on the architecture and workflow used for this chapter.

1. Data Collection: The code begins with data collection, loading a dataset (WesternIndian OceanCoralDiversity.csv) containing information about the Western Indian Ocean coral reefs.
2. Data Pre-Processing: The dataset undergoes extensive pre-processing, including handling missing values, standardizing numerical features, and applying techniques like principal component analysis (PCA).
3. Feature Selection: Through principal component analysis, components that contribute significantly to the dataset's variance are extracted. It identifies key features that impact coral health and diversity.
4. Model Selection: Several machine learning models are employed for predictive analytics, including decision trees, random forest, and support vector machines used for classification of taxa. The random forest model is identified as the most promising, considering performance metrics.
5. Training: The chosen models are trained on the selected features using relevant algorithms, such as decision tree, random forest, and SVM.

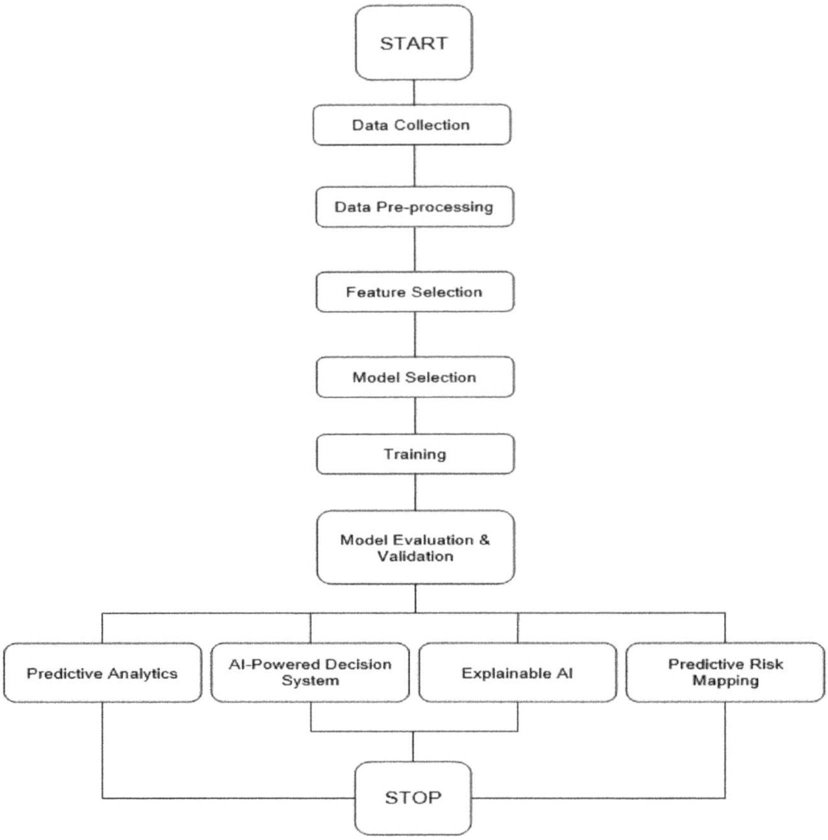

FIGURE 16.5 Architecture diagram for proposed model.

6. Model Evaluation and Validation: Performance metrics, including accuracy, precision, recall, and F1-score, are calculated for each model.
7. Predictive Analytics: The goal is to explore future trends in coral health and diversity through reports and visualizations based on machine learning models.
8. Explainable AI: SHAP generates transparent explanations for model predictions, enhancing interpretability.
9. AI-Powered Recommendations: The system aims to provide actionable suggestions for conservation decisions aligned with conservation goals, leveraging AI insights.
10. Predictive Risk Mapping: Creating maps indicating areas at high risk of environmental threats using machine learning and GIS data. Regular updates to these maps would contribute to ongoing risk assessment and conservation efforts.

16.10 IMPLEMENTATION AND RESULTS

A comprehensive implementation of diverse data analysis and machine learning techniques is demonstrated in the provided code. The initial stages encompass data preprocessing, exploratory data analysis (EDA), and dimensionality reduction using principal component analysis, with a subsequent calculation of the explained variance ratio for each principal component.

From Figure 16.6, it can be seen that, notably, opting for five components captures over 50% of the variance, offering a moderately streamlined representation. On the other hand, choosing ten

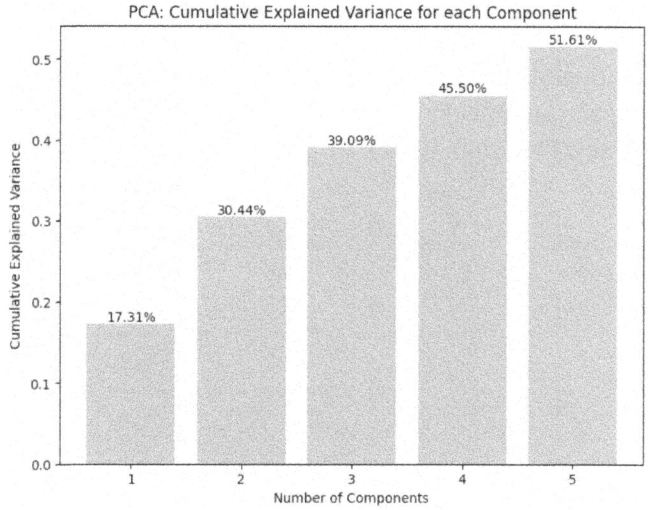

FIGURE 16.6 Applying PCA with various numbers of components to check for variance and coverage from initial attributes.

TABLE 16.1
Top Five Features for Components (*n* = 5) Taken in PCA

Components	Top Five Features
1	['ClimateStressModel', 'PH', 'cumDHW', 'SST_Kurtosis', 'SST_bimodality']
2	['Salinity_mean', 'Coralcover', 'SST_bimodality', 'SST_Skewness', 'TT_market_hrs']
3	['mean.npp', 'ChlorA_median', 'Diff_attenuation', 'SSTRateofRise', 'TT_market_hrs']
4	['max_yr', 'min_yr', 'andrello_sediments', 'Grav_NP', 'Dis_oxygen']
5	['Netflow', 'Depth', 'Outdegree', 'SSTRateofRise', 'min_yr']

components extends the explained variance to around 71.51%, providing a more detailed representation for those, emphasizing a comprehensive outlook.

Table 16.1 shows the top five features of each PCA component; this serves a crucial purpose in feature selection and interpretation. By identifying and highlighting the most influential features within each PCA component, it provides valuable insights into the underlying factors contributing to the variance in the dataset.

Top five unique features across all components:

1. SST_Skewness: Skewness of sea surface temperature.
2. Coral Cover: Percentage of coral cover in a given area.
3. Outdegree: Outbound water flow or interaction in a water network or ecosystem.
4. cumDHW: Cumulative degree heating weeks, a measure of thermal stress on corals related to coral bleaching.
5. ChlorA_median: Median value of chlorophyll-A concentration.

16.10.1 Dynamic Coral Taxa Prediction

Model evaluation involves a meticulous assessment of machine learning models, specifically decision tree, random forest, and support vector machine. The evaluation employs key metrics such

TABLE 16.2
Comparing Machine Learning Models (Decision Tree, Random Forest, and Support Vector Machine) Performance Metrics (Accuracy, Precision, Recall, F1-Score)

Models	Accuracy	Precision	Recall	F1-Score
Decision Tree	0.688	0.688	0.688	0.687
Random Forest	0.732	0.746	0.732	0.735
SVM	0.522	0.28	0.522	0.364

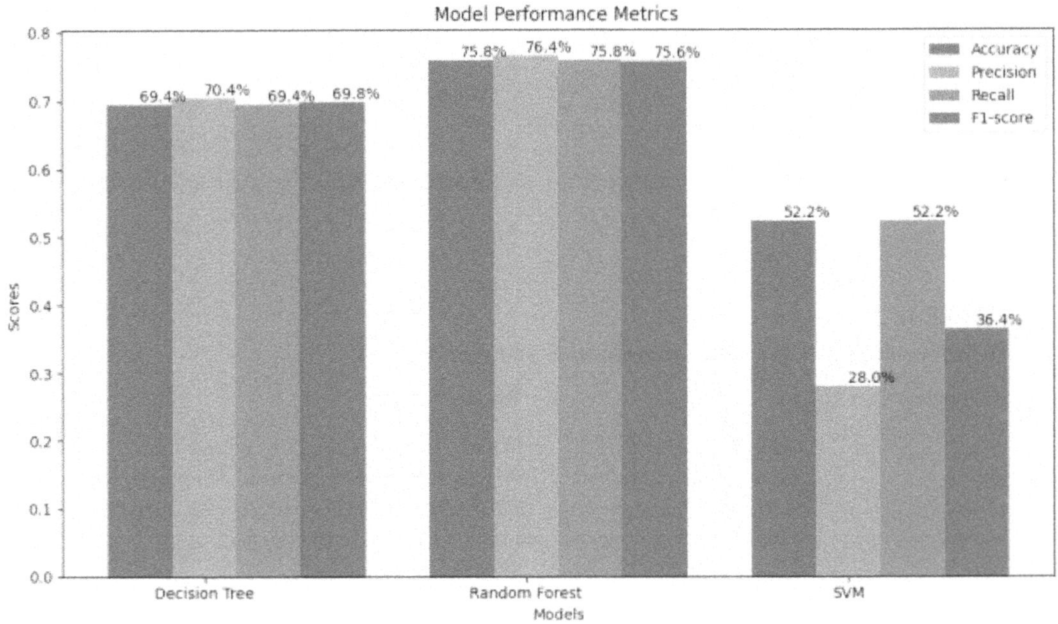

FIGURE 16.7 Visual representation of comparing machine learning model (decision tree, random forest, and support vector machine) performance metrics (accuracy, precision, recall, F1-score).

as accuracy, precision, recall, and F1-score. Results are presented in both tabular and graphical formats.

Upon evaluating the models, from Table 16.2 and Figure 16.7, it is evident that decision tree and random forest exhibit comparable performance, with the latter slightly edging ahead in accuracy, recall, and precision. Meanwhile, SVM lags notably behind, showcasing lower scores across the board. This implies that SVM might not be the optimal fit for our dataset, suggesting the need for further fine-tuning or reconsideration of model choice. From Figure 16.8, random forest, emerging with the highest scores, emerges as the frontrunner among these models, presenting itself as the most promising candidate for effective predictive analytics on this dataset.

Figure 16.8 shows the random forest decision tree obtained with the highest accuracy for classification and prediction.

16.10.2 Time Series Forecasting

Figure 16.9 shows that the time series analysis focuses on the max_yr column, visualizing average coral cover over the years. Time forecasting is executed using both autoregressive integrated

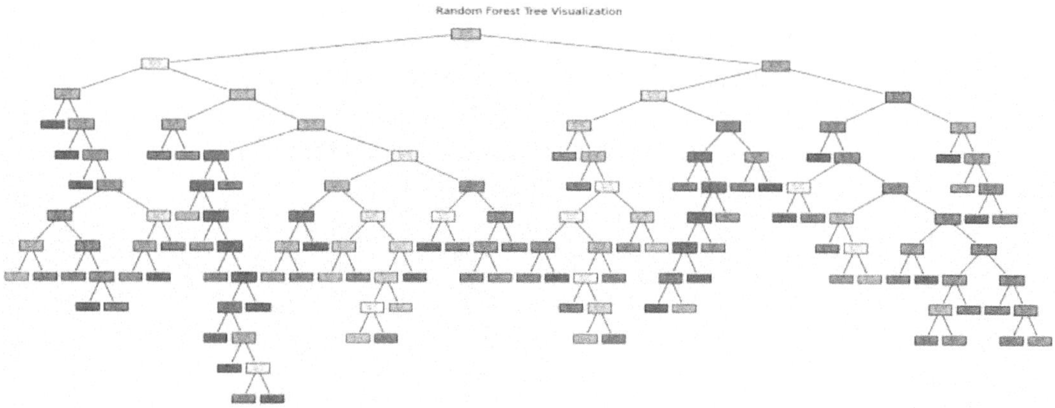

FIGURE 16.8 Random forest application.

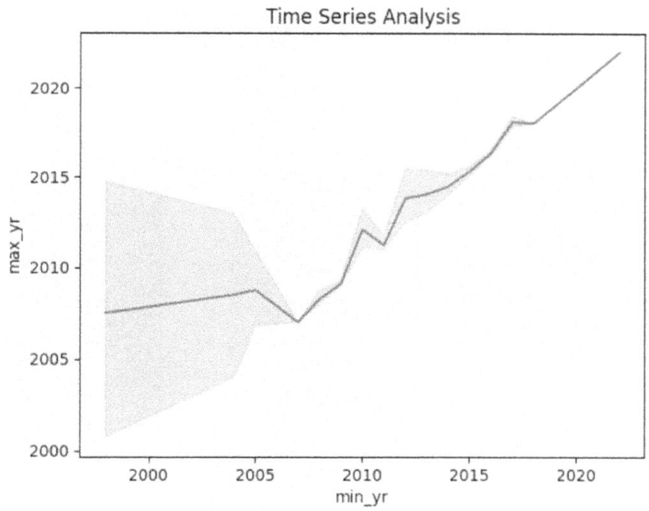

FIGURE 16.9 Time series data for visualizing average coral cover over the years 2000–2020.

moving average (ARIMA), seasonal autoregressive integrated moving average (SARIMA) and long short-term memory models, with model accuracy assessed through mean absolute error (MAE) and root mean squared error (RMSE).

From Table 16.3 and Figures 16.10 and 16.11, the observations clearly indicate that the LSTM model outshines ARIMA and SARIMA for our coral reef dataset, boasting the lowest mean squared error (MSE) and root mean squared error. These metrics signify that the LSTM model's predictions align more closely with the actual coral cover values in the test set. The superior accuracy of the LSTM model positions it as the preferred choice for time series forecasting in this specific context, emphasizing its efficacy in capturing the intricate dynamics of coral cover fluctuations.

16.10.3 AI-Powered Decision Support System

Class imbalance is adeptly addressed using the synthetic minority over-sampling technique to oversample the minority class, ensuring a more balanced representation.

TABLE 16.3
Comparing Time Forecasting Models on Their Average Coral Cover for the Years 2000–2020 Using ARIMA, SARIMA Forecasts and LSTM Predictions, with Historical Data

Model	MAE	RMSE
ARIMA	6.938	7.816
SARIMA	5.215	6.865
LSTM	4.586	5.576

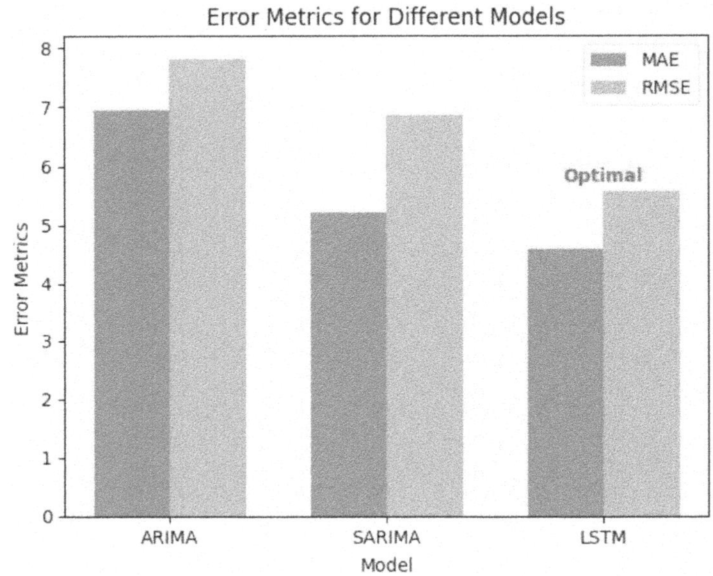

FIGURE 16.10 Bar graph of time-forecasting model metrics.

Table 16.4 and Figure 16.12 show the class distribution of coral cover, categorized as either healthy or not healthy, prior to the application of the synthetic minority over-sampling technique. The purpose of these visualizations is to illustrate the imbalance in the dataset concerning different classes, in this case healthy and not healthy coral cover.

16.10.4 Confusion Matrix

Figure 16.14 shows that the classification report highlights the model's strong performance. With a precision of 1.00 for Healthy and 0.90 for "Not Healthy," the model demonstrates accurate predictions. High recalls of 0.99 and 1.00 for Healthy and Not Healthy indicate effective capturing of true instances. F1-scores are notably high at 0.99 for Healthy and 0.95 for "Not Healthy," balancing precision and recall. The overall accuracy impressively stands at 99%, showcasing the model's proficiency across both classes. The balanced macro average and weighted averages address class imbalances. In summary, the model excels in identifying Healthy instances, displaying robust predictive capabilities and reinforcing its effectiveness in classifying the test data.

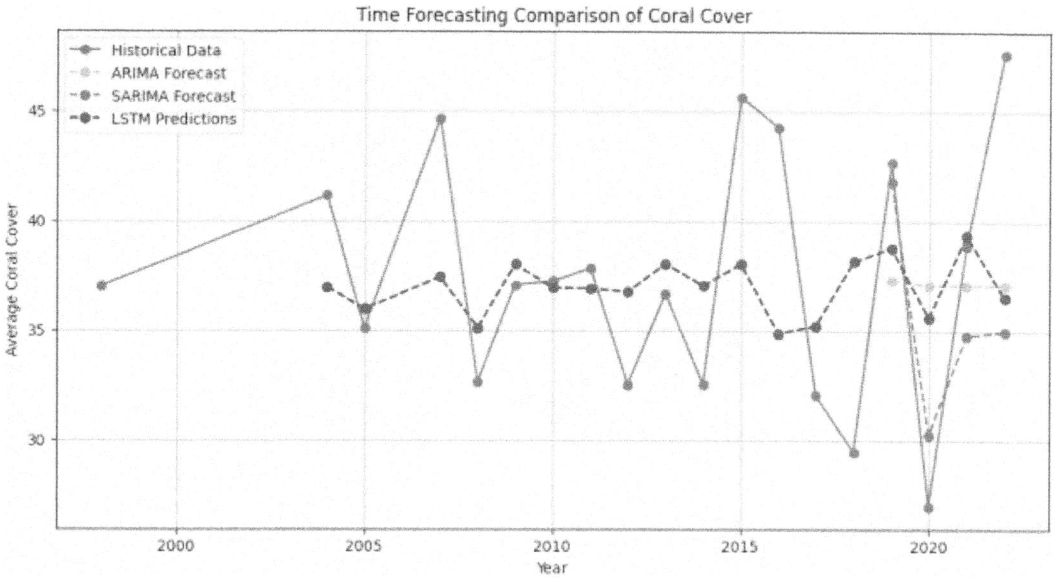

FIGURE 16.11 Visualization of comparing time forecasting models on their average coral cover for the years 2000–2020 using ARIMA and SARIMA forecasts and LSTM predictions, with the historical data.

TABLE 16.4
Class Distribution of Coral Cover: Healthy or Not Healthy, before and after Applying SMOTE

Class Balance	Healthy	Not Healthy
Before	465	57
After	369	369

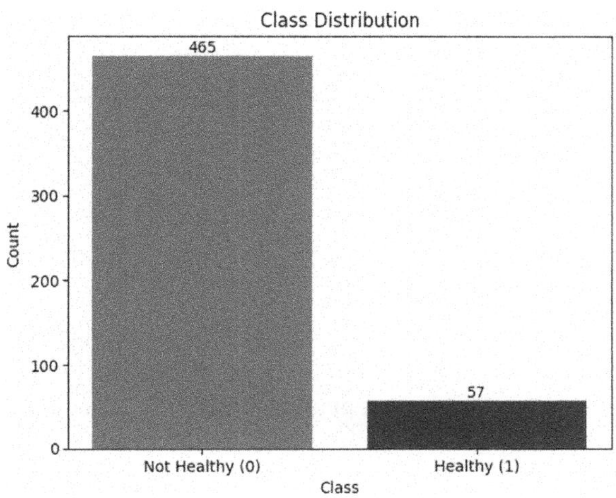

FIGURE 16.12 Class distribution of coral cover: healthy or not healthy, before applying SMOTE.

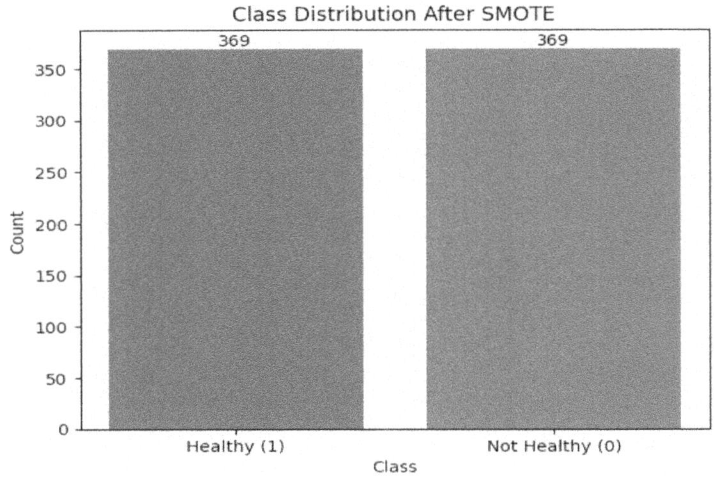

FIGURE 16.13 Class distribution of coral cover: healthy or not healthy, after applying SMOTE

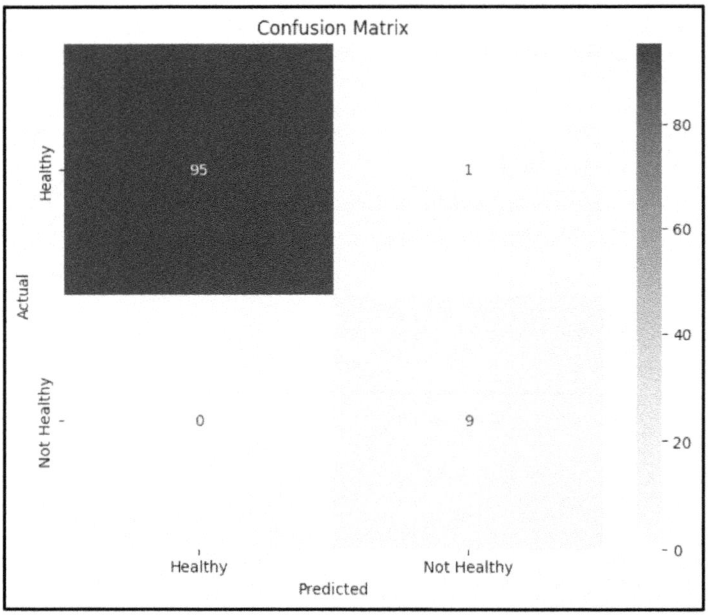

FIGURE 16.14 Confusion matrix of the neural network.

16.10.5 Explainable AI for Conservation

In the context of this study, the threshold value for the decision-making process regarding salinity (as depicted in Table 16.5) was derived from the work of Ou et al [6]. Additional attributes and information were selectively sourced from reputable references, including Wikipedia, to augment the comprehensiveness of the study.

SHAP is a powerful tool for explaining complex model predictions, providing insights into the importance of each feature in making a specific prediction, used as a model-agnostic tool for interpretation. The SHAP values can be interpreted to understand how each feature contributes to the

TABLE 16.5
Health/Optimal Range and Not Healthy Range for a Coral Reef [6]

Attributes Selected	Healthy/Optimal Range	Not Healthy Range
Coral cover	30–100	0–30
cumDHW	< 4	> 4
ChlorA_median	< 1 μg/L	> 1 μg/L
Salinity	32–38 PSU	Others
pH	8–8.5	Others
Dis_oxygen	5–18 mg/L	< 4 mg/L

TABLE 16.6
Test Values for the Mentioned Attributes and Manual Checking if They Are in Range

Attribute	Test Value	In Optimal Range
Salinity_mean	35.0277705	Yes
Dis_oxygen	4.404999971	No
ChlorA_median	0.404617012	Yes
cumDHW	34.6544076	No
pH	8.203000069	Yes

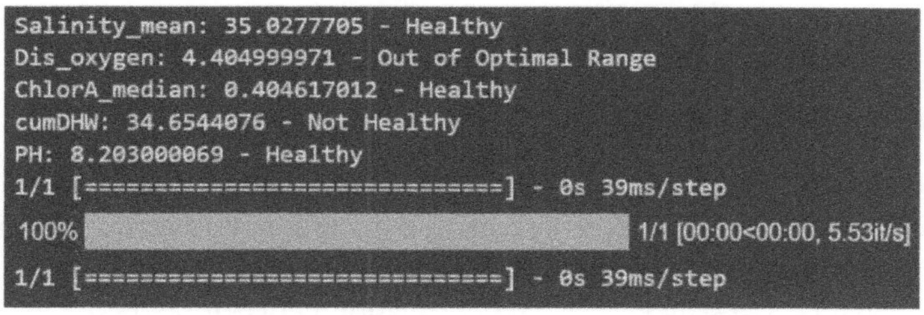

FIGURE 16.15 A screenshot from running simplified feed forward.

model's output for a particular input. Positive SHAP values indicate a positive impact on the prediction, while negative values indicate a negative impact. By analyzing the SHAP values, you can gain insights into which features are driving the model's decision and to what extent.

Table 16.6 presents the test values utilized to assess the accuracy of SHAP interpretations. The manually verified optimal ranges were examined, and the subsequent figure illustrates the results derived from SHAP.

From Figure 16.15, the values indicated show the impact of each feature on the predicted coral health. After applying a feedforward neural network (FFNN) to predict the health status of corals, SHAP can be employed as a post-hoc interpretability tool. Positive SHAP values contribute towards predicting Healthy, while negative values contribute towards predicting Not Healthy. The combination of feature explanations and SHAP values provides a comprehensive understanding of how individual features influence the model's prediction for the given coral health status. This interpretability aids in identifying which features have the most substantial impact and understanding the model's decision-making process.

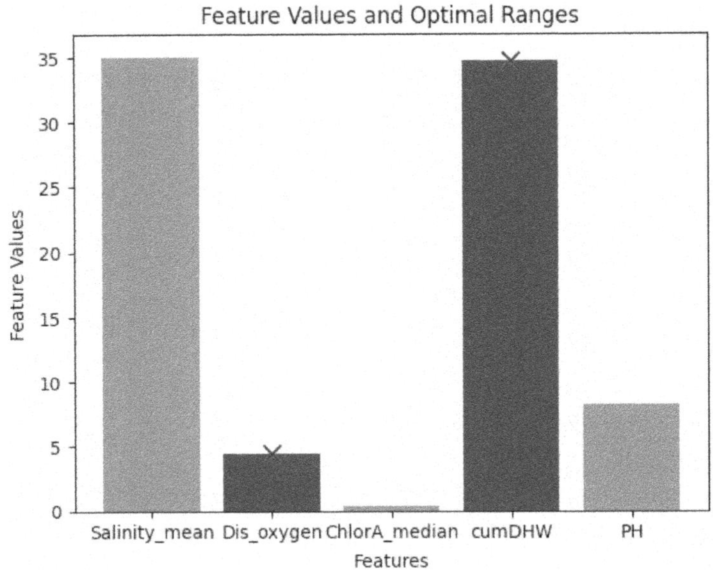

FIGURE 16.16 Featured values (salinity, dissolved oxygen, chlorophyll-A, sea surface temperatures, pH levels) with their optimal ranges.

Figure 16.16 displays a straightforward bar graph visualization that effectively identifies factors contributing to unhealthy corals. This visualization proves valuable in real-time systems, promptly highlighting deficient or out-of-range factors when data is continuously read, facilitating quick and intuitive insights into coral health status.

16.10.6 Predictive Risk Mapping

The exploration of predictive risk mapping for coral covers, situated within specific latitude and longitude coordinates, unveils a spatial representation of coral covers across a distinct region. This mapping not only delineates the geographical distribution of coral reefs but also provides valuable insights into the potential risks they face. The correlation between predictive risk values and the corresponding latitude and longitude allows for a nuanced understanding of the factors influencing coral health in different locations. This observation underscores the significance of geospatial analysis in comprehending and managing the predictive risks associated with coral ecosystems, contributing to the broader understanding of marine conservation and sustainability efforts.

Figure 16.17 illustrates the strategic dispersion of coral cover based on our dataset, visualized and strategically laid out on a map providing insights on the risk posed to coral habitats.

16.11 CONCLUSION

In this comprehensive analysis, we aimed to understand and predict coral health and diversity in the western Indian Ocean. We began by preprocessing and exploring the dataset, using advanced techniques like PCA to identify key features impacting coral ecosystems. The predictive analytics phase employed various machine learning models, revealing that a random forest model outperforms decision trees and SVM in accurately classifying coral health. We further delved into time series analysis, employing ARIMA, SARIMA, and LSTM models to forecast coral cover over time. Our findings suggest that the LSTM model excels in this context, providing the most accurate predictions. To address potential class imbalance, we applied SMOTE, enhancing the model's ability to

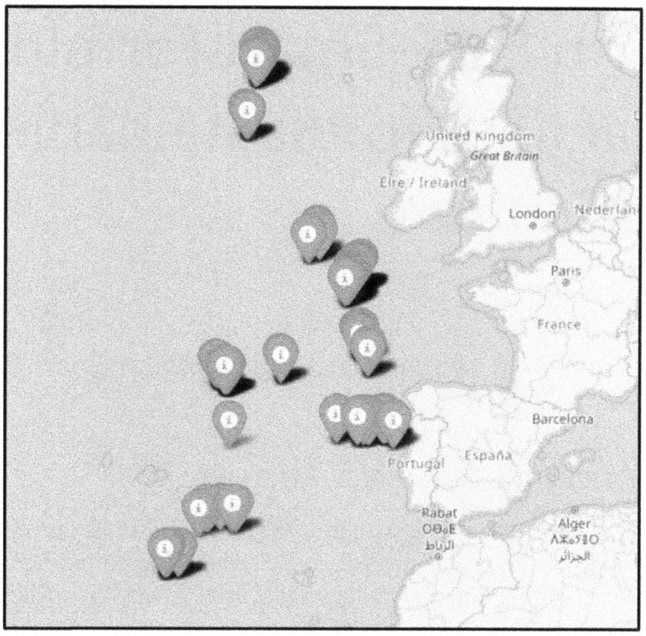

FIGURE 16.17 Coral cover set across the given latitude and longitude present in the dataset.

recognize both healthy and unhealthy corals. Finally, we implemented an explainable AI approach using SHAP, allowing us to interpret and communicate the model's predictions effectively. These insights pave the way for AI-powered recommendations, risk mapping, and informed conservation strategies for preserving coral ecosystems in the western Indian Ocean. The analysis concludes with a detailed model comparison, where the performance of various models is intricately compared and visually presented through bar graphs, map, line graphs, and tables. This multifaceted implementation underscores the robustness of our methodology in addressing diverse aspects of data analysis and predictive modeling.

REFERENCES

1. da Silveira, C.B.L.; Strenzel, G.M.R.; Maida, M.; Gaspar, A.L.B.; Ferreira, B.P. Coral Reef Mapping with Remote Sensing and Machine Learning: A Nurture and Nature Analysis in Marine Protected Areas. *Remote Sensing* 2021, 13, 2907. https://doi.org/10.3390/rs13152907
2. Collin, A.; Planes, S. Enhancing Coral Health Detection Using Spectral Diversity Indices from World View-2 Imagery and Machine Learners. *Remote Sensing* 2012, 4, 3244–3264. https://doi.org/10.3390/rs4103244
3. Mahmood, A.; Bennamoun, M.; An, S.; Sohel, F.A.; Boussaid, F.; Hovey, R.; Kendrick, G.A.; Fisher, R.B. Deep Image Representations for Coral Image Classification. *IEEE Journal of Oceanic Engineering* January 2019, 44(1), 121–131. https://doi.org/10.1109/JOE.2017.2786878
4. González-Rivero, M.; Beijbom, O.; Rodriguez-Ramirez, A.; Bryant, D.E.P.; Ganase, A.; Gonzalez-Marrero, Y.; Herrera-Reveles, A.; Kennedy, E.V.; Kim, C.J.S., Lopez-Marcano, S.; Markey, K. Monitoring of Coral Reefs Using Artificial Intelligence: A Feasible and Cost-Effective Approach. *Remote Sensing* 2020, 12(3), 489. https://doi.org/10.3390/rs12030489
5. McClanahan, T. *Western Indian Ocean Coral Diversity Observations from 1998–2022* [Data set]. Zenodo, 2023. https://doi.org/10.5061/dryad.3xsj3txn1
6. Ou, Z.; Liu, W.; He, M. Temperature and Salinity Adaptability of the Coral Reef Topshell Tectus Pyramis. *Journal of the World Aquaculture Society* 2022, 53(2), 527–541. https://doi.org/10.1111/jwas.1282

17 Farm Easy-Smart Farming Practices for Sustainable Agriculture

N. C. Brintha, J. T. Winowlin Jappes, Arivendan Ajithram, and Yegnanarayanan Venkataraman

17.1 INTRODUCTION TO SMART AGRICULTURE

A new farming method called "smart agriculture," also referred to as "precision agriculture" or "agri-tech," combines technology and data-driven insights to optimize agricultural processes. Smart agriculture utilizes cutting-edge technologies like drones, sensors, artificial intelligence, data analytics, and IoT (Internet of Things) devices to improve farming operations, production, sustainability, and efficiency (Muhie, 2022).

An overview of the essential elements of smart agriculture is shown in Figure 17.1. There are several technologies involved in smart agriculture to improve productivity and efficiency requirements. Some of the elements of smart agriculture are as follows.

- **Sensor Technology:** To keep an eye on a variety of indicators, including crop health, humidity, soil moisture, nutrient levels, and temperature, sensors are placed across fields. Farmers may use the real-time data from these sensors to make well-informed decisions regarding pest control, fertilization, and irrigation.
- **Precision Irrigation:** Utilizing information from weather forecasts and soil moisture sensors, intelligent irrigation systems provide crops with precisely the right amount of water.
- **Remote Monitoring Using Drones:** Drones fitted with cameras and sensors are capable of swiftly and effectively surveying vast tracts of agriculture. They allow farmers to respond promptly by identifying regions of the crop, pest infestations, or disease outbreaks.
- **IoT and Connectivity:** By allowing smooth data interchange and automation, the Internet of Things links farm equipment and gadgets. This allows the farmers to follow the livestock, monitor equipment performance remotely, and run their businesses more effectively.
- **AI and Data Analytics:** These two fields examine enormous volumes of data that are gathered from sensors and other sources. Algorithms for machine learning can forecast yields, spot disease outbreaks, and allocate resources as efficiently as possible. They can also predict and analyze the crop growth trends.
- **Farm Management Software:** By combining data from multiple sources, farm management platforms give farmers a thorough understanding of their activities. Crop planning, inventory control, financial analysis, and compliance monitoring are a few of the functions that these platforms frequently provide.
- **Vertical Farming and Controlled Environments:** Agriculture uses technology to produce crops indoors, usually in stacked layers or hydroponic systems, in metropolitan areas and areas with little arable land. These techniques can produce food closer to consumers, maximize resource utilization, and lower transportation costs.

DOI: 10.1201/9781003507390-17

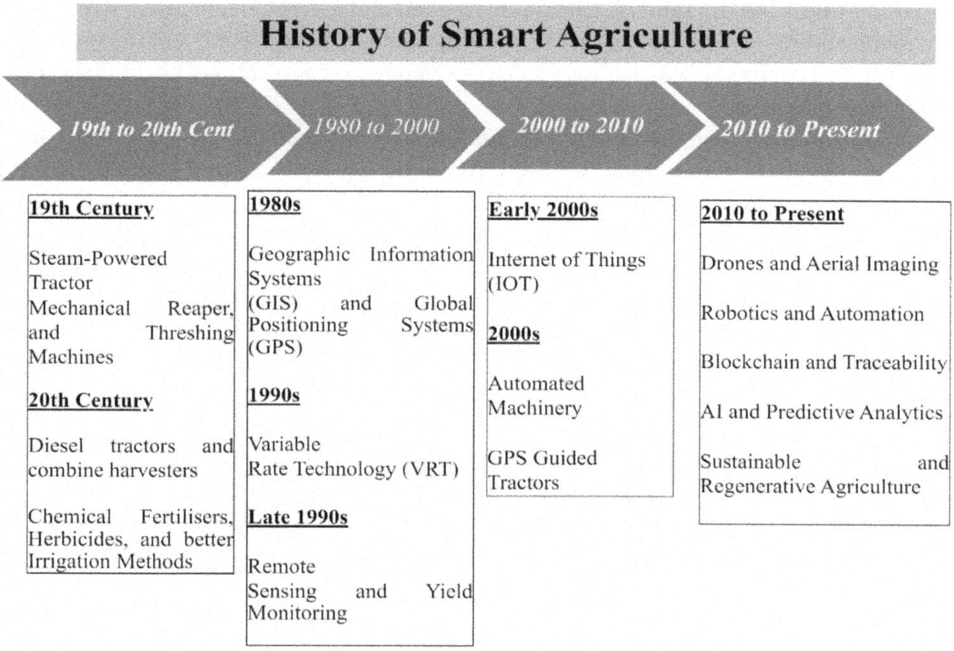

FIGURE 17.1 Essential elements of smart farming with sensors, telecommunication technologies, automation, and other hardware and software.

Smart agriculture has the power to completely transform the way we grow food, making farming more resilient, efficient, and sustainable in the face of rising global food demand and climate change (Trigo, et.al., 2021).

17.1.1 Challenges in Traditional Farming Practices

Although traditional agricultural methods have helped humans survive for generations, they also have several drawbacks, particularly when it comes to population growth, climate change, and other environmental concerns (Mujeyi, 2020). The following are some of the main problems with conventional farming methods:

1. **Depletion of Resources:** Conventional farming practices frequently cause erosion, fertility loss, and soil deterioration. Overuse of chemical pesticides and fertilizers can eventually deteriorate soil health, resulting in lower crop yields and long-term environmental harm.
2. **Water Scarcity and Inefficient Irrigation:** A lot of conventional farming techniques rely on wasteful and inefficient irrigation techniques. This causes runoff to pollute water and contribute to soil salinity, which in turn causes scarcity of water in many areas.
3. **Management of Pests and Diseases:** Chemical pesticides are frequently used extensively in conventional farming methods to combat pests and diseases. Overuse of pesticides, however, can contaminate food, water, and soil, kill beneficial insects, and result in the emergence of pests resistant to the chemicals (Nguyen & Drakou, 2021).
4. **Climate Vulnerability:** The effects of climate change, such as harsh weather, altered precipitation patterns, and temperature swings, are particularly dangerous for traditional agriculture. This may interfere with agricultural cycles, lower yields, and jeopardize food security.

5. **Erosion and Loss of Habitat:** Unsustainable land management techniques, like farming in monocultures and deforestation, can cause habitat destruction, biodiversity loss, and soil erosion. This compromises the long-term productivity of agricultural land in addition to harming ecosystems.
6. **Labour Intensity and Rural Population Decline:** Manual labour is a major component of traditional farming methods, which can be physically taxing and unsustainable economically, especially for small-scale farmers. As younger generations migrate to urban areas in quest of better prospects, this may contribute to the depopulation of rural communities.
7. **Income inequality and market instability:** In earlier days, farmers are frequently exposed to price volatility, market swings, and restricted access to markets and information. This has the potential to worsen rural poverty and maintain income disparity, especially in developing nations.

A move towards smart agriculture, agro ecology, and regenerative farming—agricultural approaches that are more robust and sustainable—is necessary to address these problems. These methods provide creative ways to increase output, preserve resources, and advance food security by fusing technology, ecological concepts, and community involvement.

17.1.2 History of Smart Agriculture

The evolution of technological innovation has propelled the development of smart agriculture to improve farming efficiency, sustainability, and productivity (Harwood, 2020). The overall evolution of different stages of agriculture is depicted in Figure 17.2. The following is a timeline of the major turning points and advancements in the development of smart agriculture (Mohamed Haris, et.al., 2018).

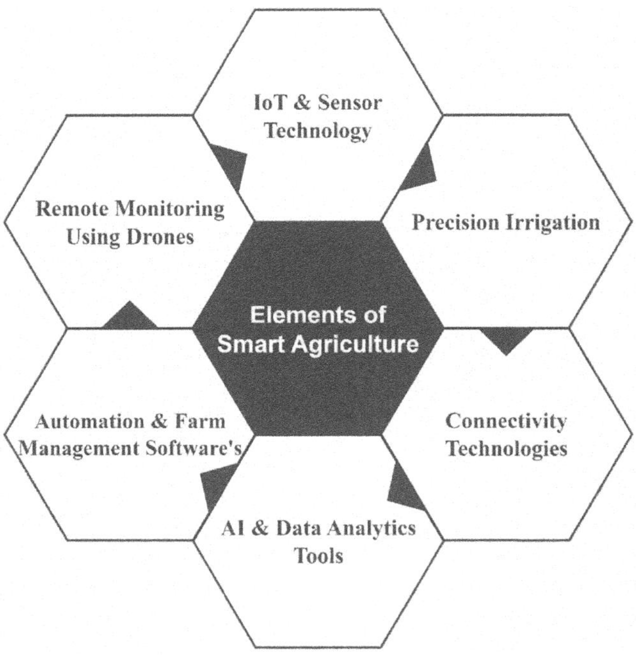

FIGURE 17.2 Transformation of agricultural practices over time.

Early Beginnings: Mechanization and Early Innovations (19th to Mid-20th Century)

Mechanization of Agriculture:

19th Century: The Industrial Revolution introduced mechanized equipment such as the steam-powered tractor, the mechanical reaper, and threshing machines, which significantly altered agriculture. These breakthroughs reduced manual labour requirements and increased output, laying the foundation for contemporary farming methods.

Post-War Technological Advances:

Mid-20th Century: Following World War II, more sophisticated equipment, like diesel tractors and combine harvesters, was introduced. The widespread use of chemical fertilizers, herbicides, and better irrigation methods during the 1960s and 1970s contributed to the Green Revolution.

Digital Revolution and Precision Agriculture (1980s to 2000s)

The Rise of Precision Agriculture:

In the 1980s, the development of geographic information systems (GISs) and global positioning systems (GPSs) allowed precision agriculture to start taking its shape. With the use of this technology, farmers were able to optimize the use of inputs like water, fertilizer, and pesticides by tracking and managing fluctuations in the field.

Variable Rate Technology (VRT):

1990s: Using information from crop investigation and soil monitoring, VRT allowed farmers to apply inputs at different rates throughout a field. This decreased wastage and increased efficiency.

Remote Sensing and Yield Monitoring:

Late 1990s: Detailed data on crop health, soil conditions, and field variability was made possible by the use of remote sensing technology, such as satellite and aerial images. Data on yield could be gathered at real time to monitor the yield at the fields at a faster rate.

Integration of IoT and Advanced Technologies (2000s to 2010s)

Introduction of IoT in Agriculture:

Early 2000s: The Internet of Things started to be incorporated into agriculture in the early 2000s. Sensors were positioned in fields to gather information on crop health, weather, and soil moisture. This helps in achieving real-time data transmission for analysis and decision-making.

Development of Smart Machinery:

2000s: Sophisticated agricultural machinery emerged with the introduction of automated machinery and GPS-guided tractors. These developments improved the precision of farming operations and decreased the demand for physical labour.

Adoption of Data Analytics and Machine Learning:

2010s: More complex analysis of agricultural data was made possible by the growth of big data and improvements in machine learning algorithms. Based on data-driven insights, farmers could now forecast crop yields, improve planting schedules, and allocate resources more effectively.

Modern Era: Advanced Automation and Sustainability (2010s to Present)

Drones and Aerial Imaging:

2010s: With the ability to provide high-resolution airborne imagery for pest detection, crop monitoring, and precise spraying, drones have emerged as a useful tool in agriculture. This provides an affordable means of rapidly collecting comprehensive field data.

Robotics and Automation:

2010s to Present: Developments in robotics have resulted in the creation of autonomous machinery that can carry out difficult jobs like planting, weeding, and harvesting. These robots decreased the need for physical labour while increasing efficiency.

Blockchain and Traceability:

Late 2010s: Agriculture began utilizing blockchain technology to increase the traceability and transparency of the supply chain. With the help of this technology, customers may be guaranteed food's provenance and quality.

AI and Predictive Analytics:

2020s: Predictive analytics and artificial intelligence are currently being utilized to improve agricultural decision-making. Farmers can take preventive measures since AI models can predict weather trends, pest outbreaks, and crop illnesses.

Sustainable and Regenerative Agriculture:

Present: Smart technology-enabled sustainable and regenerative farming methods are becoming more and more important. These methods seek to improve biodiversity, strengthen soil health, and lessen their negative effects on the ecosystem.

17.2 FACTORS DRIVING INNOVATION IN SMART AGRICULTURE

Innovation in smart agriculture is driven by a multitude of factors, encompassing technological advancements, economic needs, environmental concerns, and social dynamics, as shown in Figure 17.3. The key drivers which drive innovation are as follows.

17.2.1 Technological Advancements

1. Internet of Things: IoT devices allow agricultural processes to be monitored and managed at real time, increasing productivity and efficiency.
2. Big Data Analytics: The capacity to examine huge datasets aids in the development of well-informed choices about resource allocation, weather patterns, and crop management.
3. Precision agriculture techniques, automated decision-making, and predictive analytics are made possible.
4. Robotics and Drones: Robotics can automate labour-intensive operations like planting and harvesting, while drones offer aerial observation and data collection.

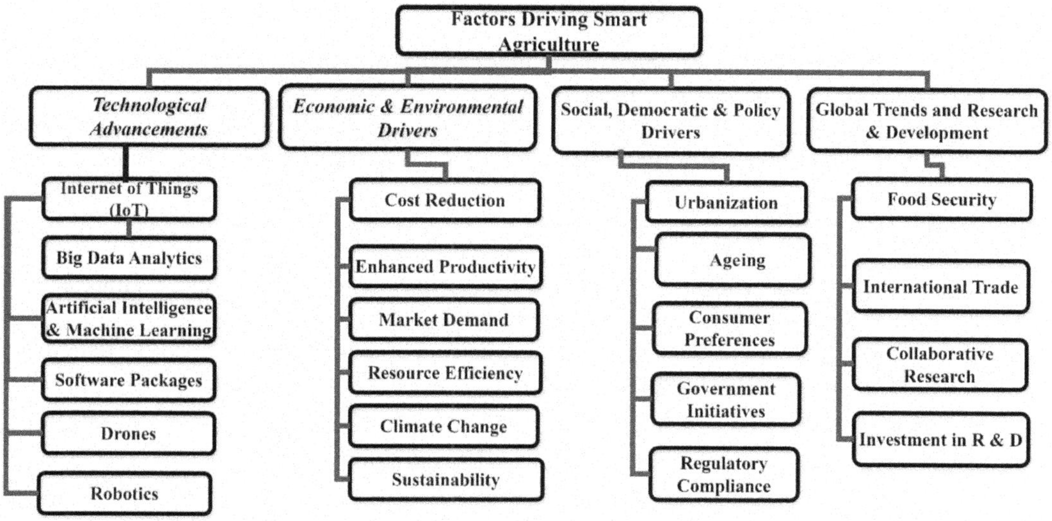

FIGURE 17.3 Different aspects of precision agriculture in terms of technology, R&D, social, democratic, policy, environmental and global trends.

17.2.2 Economic Drivers

1. Cost Reduction: By lowering labour and input costs, smart agriculture technologies results in more economically sound farming techniques.
2. Enhanced Productivity: Technological advancements can raise livestock and crop output, which raises profitability.
3. Market Demand: The adoption of smart agriculture technologies is driven by the growing demand for high-quality, sustainably produced food items.

17.2.3 Environmental Concerns

1. Resource Efficiency: Technologies that maximize the use of pesticides, fertilizers, and water can help save natural resources and lessen their negative effects on the environment.
2. Climate Change: Smart agriculture offers methods for adjusting to shifting weather patterns, guaranteeing stability and food security.
3. Sustainability: Farming methods that minimize environmental deterioration are becoming more and more important.

17.2.4 Social and Demographic Factors

1. Urbanization: As cities become more populated, the demand to produce more food with fewer resources arises, which encourages the use of cutting-edge agricultural techniques.
2. Ageing Farmer Population: One of the main issues brought about by the aging farmer population, is the need for technologies which minimizes physical labour and makes farming jobs simpler for them.
3. Consumer Preferences: Farmers are encouraged to use smart agriculture practices which guarantee quality and traceability as a result of the growing trend towards organic and locally sourced produce.

17.2.5 Policy and Regulatory Drivers

1. Government Initiatives: The adoption of smart agriculture technologies can be greatly accelerated by subsidies, grants, and legislation.
2. Regulatory Compliance: The adoption of cutting-edge technologies is encouraged by regulations about food safety, environmental protection, and sustainable farming methods.

17.2.6 Global Trends

1. Food Security: Improving food production and delivery networks is vital, as the world's population is expected to exceed 9.7 billion people by 2050.
2. International Trade: Smart agriculture innovations can make agricultural products more competitive on the world market.

17.2.7 Research and Development

1. Collaborative Research: New techniques and technologies are developed through collaborations between academic institutions, research centres, and the commercial sector.
2. Investment in R&D: Research and development in agriculture receives more funding, which promotes the development of novel ideas and technology.
   By utilizing these forces, smart agriculture seeks to develop an agricultural sector that is more productive, sustainable, and efficient—capable of satisfying the world's expanding population needs while reducing its negative effects on the environment (Lynch, et.al., 2021).

17.3 BENEFITS OF SUSTAINABLE AGRICULTURE

Precision farming, also referred to as smart agriculture, uses technology to enhance farming methods, boost productivity, and boost efficiency (Ayinde, et.al., 2017). Here are a few advantages it provides:

1. **Higher Crop Yields:** Smart agriculture tracks the health of crops, the state of the soil, and environmental variables in real time by using sensors, drones, and other technology. Accurate data enables farmers to make well-informed decisions about pest management, fertilization, and irrigation, resulting in increased yields.
2. **Resource Efficiency:** Smart agriculture helps farmers use water, fertilizers, and pesticides more efficiently by utilizing methods like drip irrigation, soil moisture monitors, and weather forecasting. This lessens the impact on the environment, saves money, and conserves resources.
3. **Optimized Resource Management:** By accurately matching inputs to the unique requirements of crops and fields, smart agricultural systems help farmers manage resources more efficiently. This entails modifying irrigation schedules, timing the application of fertilizers, employing focused pest control techniques, and decreasing waste.
4. **Data-Driven Decision Making:** Farmers can examine enormous volumes of data gathered from sensors, satellites, and other sources to make data-driven decisions by using data analytics and predictive modelling. This makes it possible to control risks, plan, and spot patterns and trends that over time can enhance agricultural operations.
5. **Better Crop Quality:** Smart agriculture can contribute to the assurance of higher-quality output by closely monitoring crop health and environmental conditions. This results in harvests of superior quality that satisfy market criteria. It also includes elements like consistent ripening, ideal nutritional levels, and decreased exposure to pests and diseases.

6. **Remote Monitoring and Control:** Farmers can use computers or mobile devices to monitor and manage their activities remotely. Because of its adaptability, farmers may make timely interventions even when they are not physically present on the field, which boosts productivity
7. **Sustainability**: Smart agriculture promotes sustainable farming methods by cutting waste, maximizing the use of resources, and minimizing the impact on the environment. This is essential to preserving agricultural systems' long-term sustainability and reducing the impact of climate change on food production (Bruce & Costa, 2019).
8. **Market Competitiveness:** By boosting production, cutting expenses, and satisfying customer demands for food produced sustainably, adopting smart agricultural techniques will help farmers become more competitive in the global market. For agricultural enterprises, this may result in increased profitability and long-term viability (Fisher, et.al., 2018).
9. **The whole impact of the supplier chain:** Each year, there are significant losses in food production due to food waste. The fact that smart agricultural systems may be integrated into the entire supply chain is one of its benefits. Farmers can avoid losses by making sure that specific standards are maintained for production, storage, and transportation from field to fork.
10. **Stop theft and loss, and facilitate recovery:** Due to the high value of farming equipment and their isolated position, farms are often targeted by criminal activity. With the full-farm perspective provided by smart farming technologies, it is possible to monitor the activities constantly and respond quickly to alerts. This covers equipment and vehicle tracking, door and window monitoring, fence line and gate monitoring at several sites, and access control. Assisting in recoverability, averting losses, and capturing all details are all made possible by smart technology, which provides real-time data insights on the condition and whereabouts of the most precious assets (Li, et. al., 2023).
11. **Ideal maintenance plans:** In addition to keeping the machinery operating at peak efficiency, maintenance also avoids operational downtime. It also has an impact on worker safety, pollution, efficiency, and energy consumption. Because of this, creating the ideal maintenance plan is an essential farm chore. Farmers may arrange reactive and preventive maintenance on important gear with the help of smart farm technology, which provides the necessary data (Ataei, et.al., 2021).
12. **Productivity Enhancement and Lower Expenses:** Technology used in farming can increase machine and worker productivity. This keeps farming operations running as efficiently as possible and often lowers costs in the process. This cost savings enables farmers to make essential, strategic, well-informed decisions and free up funds for infrastructure and technology investments that support sustainability.

17.4 AUTOMATION IN AGRICULTURAL PRACTICES

Automation in smart agriculture refers to the use of advanced technologies to perform agricultural tasks with minimal human intervention. This encompasses a wide range of applications aimed at increasing efficiency, productivity, and sustainability in farming operations (Bedeke, et.al., 2019). Here are key areas where automation is making a significant impact in smart agriculture:

17.4.1 Key Areas of Automation in Smart Agriculture

1. **Precision Farming:**
   GPS-Guided Tractors and Equipment: Highly precise planting, ploughing, and harvesting may be accomplished by automated tractors and equipment that are outfitted with GPS technology, cutting down on waste and boosting productivity.

Variable Rate Technology: Enhances yields and maximizes resource usage by enabling the application of inputs (fertilizers, insecticides, and seeds) at different rates throughout a field.

2. **Drones and Aerial Imaging:**
   Crop Monitoring: Real-time airborne imaging is made possible by drones fitted with thermal and multispectral cameras, which are used to monitor crop health, identify diseases, and determine water requirements.
   Field Mapping: Using automated drones, comprehensive maps of fields can be produced, aiding in the more efficient planning and administration of agricultural activities.
3. **Robotics:**
   Planting and Sowing Robots: These machines are capable of precisely planting and sowing, guaranteeing that seeds are positioned and buried at the ideal depth.
   Weeding and Pest Control Robots: These autonomous devices can detect, pull, and apply pesticides precisely where they are needed, saving money on labour and chemicals.
   Harvesting Robots: By minimizing waste and increasing productivity, specialized robots can harvest crops, especially those that need to be handled carefully, such as fruits and vegetables.
4. **IoT and Sensors:**
   Soil Sensors: These automated devices monitor the temperature, moisture content, and nutrient levels of the soil to provide information for accurate fertilization and irrigation.
   Climate Sensors: By keeping an eye on the weather, sensors assist farmers in scheduling planting, irrigation, and harvesting activities.
5. **Automated Irrigation Systems:**
   Smart Irrigation: Systems that maximize crop yields and conserve water resources by ensuring crops receive the appropriate amount of water at the appropriate time through the use of sensors and meteorological data.
6. **Artificial Intelligence (AI) and Machine Learning (ML):**
   Predictive Analytics: AI algorithms analyze data from various sources to predict crop yields, pest outbreaks, and weather patterns, helping farmers make proactive decisions.
   Automated Decision-Making: ML models assist in decision-making processes for planting, fertilization, irrigation, and pest control, optimizing operations based on real-time data.
7. **Automated Livestock Management:**
   Feeding Systems: By analyzing data from several sources, AI algorithms forecast weather trends, pest outbreaks, and agricultural yields, assisting farmers in taking preventative action.
   Health Monitoring: ML models optimize operations based on real-time data and support decision-making processes for planting, fertilization, irrigation, and pest management.
8. **Greenhouse Automation:**
   Climate Control: To provide the best growth conditions for crops, automated systems in greenhouses control temperature, humidity, and CO_2 levels.
   Lighting and Nutrient Delivery: Hydroponic or aeroponic nutrient delivery systems, along with automated lighting systems, guarantee that plants receive the proper light and nutrients for growth.
9. **Sustainable Practices:**
   Regenerative Agriculture: The long-term environmental sustainability of regenerative farming practices, which improve biodiversity, replenish soil health, and sequester carbon, will be aided by smart technologies.
   Climate-Smart Agriculture: Farmers will be better able to manage the effects of climate change and cut greenhouse gas emissions if they implement adaptation and mitigation measures

10. **Biotechnology and Genomics:**
    Genetically Modified Organisms (GMOs): Crops that have undergone genetic engineering to provide characteristics including resistance to pests, tolerance to drought, and enhanced nutritional value.
    CRISPR Technology: This genome editing instrument makes it possible to precisely alter agricultural DNA, leading to the creation of new crop kinds with desired characteristics.

17.4.2 Benefits of Automation in Smart Agriculture

- **Enhanced Efficiency**: Automation increases agricultural processes speed and precision by eliminating the need for manual labour.
- **Greater Productivity:** Optimal resource utilization through automated systems results in increased crop yields and better quality.
- **Cost Savings:** Farmers can save a lot of money because of optimized input use and decreased labour expenses.
- **Sustainability:** Automation lessens waste, saves resources, and lessens farming's negative environmental effects.
- **Data-Driven Decision Making:** By enabling faster and more informed decision-making, real-time data gathering and analysis enhance farm management as a whole.

17.4.3 Challenges and Considerations

- **Initial Investment:** Small-scale farmers may find it difficult to afford the high upfront expenses of automated equipment.
- **Technical Abilities:** To run and maintain automated systems, farmers must develop technical abilities.
- **Data Security:** It's critical to guarantee the confidentiality and security of data gathered by automated systems.
- **Integration:** It can be difficult to integrate different automated systems and technologies seamlessly.

Smart agriculture automation has the potential to revolutionize the agricultural industry by increasing productivity, efficiency, and sustainability (Misniakiewicz, et.al., 2021). Automation in agriculture is predicted to increase the outcomes, resolving many of the issues that contemporary farmers confront.

17.5 TECHNOLOGIES IN SMART FARMING FOR SUSTAINABLE DEVELOPMENT

A company's specialized needs and unique set of requirements may determine the best combination of farming technology equipment (Sharifuddin, et.al., 2020). Typical components of a smart farming system include the following:

- IoT devices to monitor and control temperature, humidity, soil quality, pollution levels, and other variables
- Wi-Fi, cellular, and LoRa connectivity technologies are used to transfer the data collected in fields.
- Software services to integrate, process, and analyze data from Internet of Things devices are either internally produced or offered by outside vendors.
- GPS-based tools are utilized by location-tracking systems to manage and guide vehicles.
- Advances in robotics to control unmanned aerial vehicles (drones).
- Data analytics services help store, structure, and analyze enormous volumes of data, giving decision-makers with analytical insights.

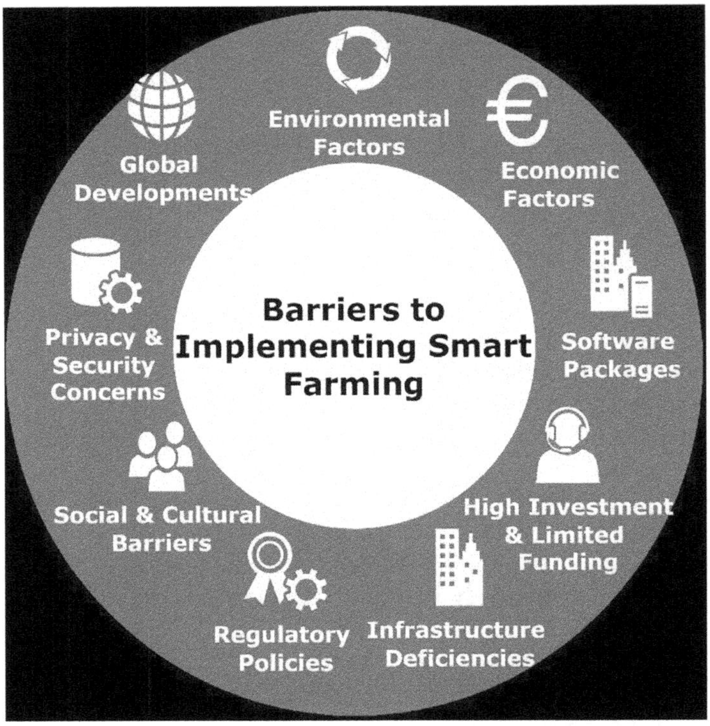

FIGURE 17.4 Limitation and barriers in implementation of best agricultural practices.

Smart farming technologies encompass a range of innovative tools and practices aimed at increasing agricultural efficiency, productivity, and sustainability (Sapbamrer & Thammachai, 2021). With the use of cutting-edge technologies, smart agriculture has the potential to significantly change the food security, environmental sustainability, and economic viability of agriculture.

17.5.1 Barriers to Implementing Smart Farming Technologies

Several obstacles may prevent smart farming technology from being adopted and used effectively (Coulibaly, et.al., 2021). Economic, technological, data-related, social, cultural, and regulatory factors can all be used to group these obstacles, as shown in Figure 17.4.

1. **Economic and Financial Barriers**
   High Initial Investment:
   Capital Costs: Investing in and setting up smart agricultural technology, such as sensors, drones, and automated gear, can come with a hefty upfront cost.
   Return on Investment (ROI): Farmers may find it challenging to defend the expenditures of these investments due to the often-lengthy time lag before realizing a profit (Mgomezulu, et.al., 2023).
   Limited Access to Funding:
   Financial Resources: Many farmers may not have access to the loans or financial resources they need to invest in cutting-edge technologies, especially those who manage small to medium-sized farms.
   Subsidies and Incentives: Many farmers are often unable to receive government incentives and subsidies, which might assist offset costs, or they are handled inconsistently or inadequately.

2. **Technical and Infrastructure Barriers**
   Lack of Technical Skills:
   Training and Expertise: To efficiently operate and maintain smart farming devices, farmers frequently require specialized knowledge and training.
   Complexity: Some technologies might be scary due to their complicated nature, which calls for extensive training and continuing assistance.
   Infrastructure Deficiencies:
   Connectivity: A lot of smart farming technology depends on dependable cell phones and internet access. Nevertheless, a lot of rural places lack sufficient network infrastructure.
   Power Supply: It can be difficult in some areas to maintain a consistent and dependable power source for smart farming equipment.
3. **Data Management and Security Barriers**
   Data Overload and Management:
   Data Volume: Farmers may find it difficult to handle and comprehend the vast amounts of data produced by smart agricultural devices.
   Data Integration: It can be difficult and technically demanding to ensure the interoperability and smooth integration of data from several sources and systems.
   Privacy and Security Concerns:
   Cybersecurity Risks: As farms become more digitally integrated, they are more susceptible to data breaches and cyberattacks.
   Data Ownership and Privacy: Farmers may be discouraged from implementing smart farming technologies due to worries about data ownership, sharing, and usage.
4. **Social and Cultural Barriers**
   Resistance to Change:
   Customized Procedures: A lot of farmers have long-standing customs and could be reluctant to use new technologies.
   Perceived Risks: Investing in and relying on untested or new technology can be discouraged by the perceived risk involved.
   Knowledge Gap:
   Education and Training: The knowledge and abilities needed to use new technologies efficiently and reap their benefits are frequently lacking.
   Extension Services: Insufficient assistance and services for agricultural extension may hinder the successful adoption and diffusion of new technology.
5. **Regulatory and Policy Barriers**
   Regulatory Compliance:
   Complicated Regulations: It can be difficult and time-consuming to navigate the regulatory environment for new agricultural technologies.
   Standardization: The broad acceptance of new technology may be hampered by the absence of established norms and rules.
   Policy Support:
   Inconsistent Policies: Long-term investment and planning may be impacted by inconsistent government policies and support initiatives for smart agriculture.
   Subsidy Allocation: Poorly designed or ineffective subsidies might not get to the farmers who need them the most.
6. **Environmental and Ethical Barriers**
   Environmental Impact:
   E-Waste: Electronic device disposal that is outdated or malfunctioning can contaminate the environment.
   Resource Use: The environmental benefits of some technologies may be countered by the substantial amounts of energy, water, or other resources they need.

Ethical Considerations:
Equitable Access: It's critical to make sure that all farmers, regardless of size or location, can benefit from smart agriculture technologies.
Fair Data Practices: To guarantee that farmers' data is utilized morally and that they gain from the insights produced transparent and fair data practices are crucial.

17.5.2 Addressing the Barriers

A multidimensional strategy involving cooperation between governmental agencies, technological companies, academic institutions, and the farming community is needed to address these obstacles (Aryal, et.al., 2018):

1. **Financial Support:** To assist farmers in affording smart technologies, governments, and financial institutions might offer loans, grants, and subsidies.
2. **Education and Training:** Putting in place extensive training initiatives and extension services to provide farmers with the know-how they need.
3. **Infrastructure Development:** Making investments in rural infrastructure to guarantee dependable power and communication.
4. **Regulatory Frameworks:** To encourage the use of new technologies, clear and encouraging regulatory frameworks should be developed.
5. **Data Security Measures**: To safeguard farmers' data, strong cybersecurity measures and data privacy policies should be established.
6. **Community Engagement:** Getting involved in the farming community to show off the advantages and foster a sense of confidence in emerging technologies.

Through the resolution of these obstacles, the agricultural sector may optimize productivity, sustainability, and resilience through the application of intelligent farming technologies (Dhiman, 2020).

17.6 CURRENT CHALLENGES AND FUTURE OF SMART FARMING PRACTICES

17.6.1 Current Challenges in Smart Farming

1. **Economic and Financial Challenges:**
   High Initial Investment: Small and medium-sized farms may find it difficult to afford the high upfront costs associated with installing and buying smart farming technologies.
   Funding: A large number of farmers do not have access to sufficient funds or reasonably priced credit to invest in cutting-edge technologies. Current incentives and subsidies are sometimes insufficient or not easily available.
2. **Technical and Infrastructure Challenges:**
   Lack of Technical Skills: To efficiently operate and maintain smart agricultural devices, farmers frequently require specialized knowledge and training. Adoption may be hampered by this.
   Infrastructure Deficiencies: A lot of rural locations lack the cellular connectivity and dependable internet that smart agricultural technologies depend on for proper operation.
3. **Data Management and Security Challenges:**
   Data Overload and Management: Farmers may find it difficult to efficiently handle and comprehend the massive amounts of data produced by smart agricultural devices.
   Data Privacy and Security: Farmers may be discouraged from implementing these technologies because of worries about cybersecurity, privacy, and data ownership. One major problem is making sure that data is used securely and responsibly.

4. **Social and Cultural Challenges:**
   Resistance to Change: Many farmers have long-standing traditional practices and may be resistant to adopting new technologies. The perceived risk associated with investing in and relying on unproven or unfamiliar technologies can be a significant deterrent.
   Knowledge Gap: It is frequently the case that people lack the skills and information necessary to utilize new technology to their full potential. Inadequate support and services for agricultural extension can hinder the successful acceptance and spread of innovative technologies.
5. **Regulatory and Policy Challenges:**
   Regulatory Compliance: Due to their ingrained traditions, many farmers may be hesitant to adopt new technologies. The perceived danger associated with investing in and relying on novel or experimental technology is one of the main deterrents.
   Policy Support: Inconsistent government policies and initiatives supporting smart agriculture can have an impact on long-term investment and planning. Subsidies that are inefficient or poorly targeted might not get to the farmers who need them.

17.6.2 Future Expectations in Smart Farming

1. **Technological Advancements:**
   Improved AI and Machine Learning: By integrating increasingly complex AI and machine learning models, deeper insights and predictive capabilities will be made possible, enabling proactive farming methods as opposed to reactive ones.
   Enhanced Connectivity: As 5G networks become more widely used and satellite communication technology advances, IoT devices will become more connected, allowing for faster data transmission and more dependable real-time monitoring (Beaman, et.al., 2018).
2. **Automation and Robotics:**
   Autonomous Machinery: As completely autonomous tractors, drones, and robots for a range of agricultural jobs become more prevalent, the requirement for manual labour will decline and operational efficiency will rise.
   Advanced Robotics: Advances in robotics will result in more intelligent and adaptable robots that can carry out difficult jobs like precise weed control and selective harvesting (Patle, et.al., 2020).
3. **Data Integration and Analytics:**
   Big Data and Advanced Analytics: As data collecting and analysis techniques advance, farmers will have access to complete, real-time information that will enable more accurate and effective farm management.
   Cloud Computing and Edge Computing: By utilizing these technologies, smart agricultural solutions will be more responsive and efficient due to improved data processing and storage.
4. **Sustainability and Environmental Impact:**
   Regenerative and Climate-Smart Agriculture: Regenerative agricultural methods that improve biodiversity, replenish soil health, and store carbon will be aided by smart technologies, ensuring long-term environmental sustainability.
   Resource Efficiency: Upcoming technologies will prioritize cutting waste and lessening its negative effects on the environment while making the best use of resources like electricity, water, and fertilizers (Daberkow, et.al., 2003).
5. **Policy and Regulatory Support:**
   Adopting Smart Farming Technologies: Governments will be essential in developing regulatory frameworks and supportive policies that promote the use of smart farming technologies.

International Collaboration: Governments, academic institutions, and the commercial sector will work together internationally to foster innovation and the spread of smart farming technologies.

6. **Market and Consumer Trends:**
   Traceability and Transparency: Blockchain technology will guarantee food products' traceability from farm to table, improving food safety and enabling customers to confirm the provenance and calibre of their purchases.
   Growth in Consumer Demands: As consumers' desire for food produced sustainably grows, so too will the use of environmentally friendly farming techniques.

17.7 CONCLUSION AND FUTURE SCOPE

This chapter focuses on the necessity of smart farming practices for sustainable agriculture. The recent developments in data analytics, assistive technologies, and other sustainable practices are expected to undergo a significant transformation in the upcoming years. The usage of these cutting-edge technologies has created a sustainable environment in farming with enhanced efficiency, productivity, better decision-making, resource management, improved crop quality, adaptability to climatic conditions, and a lot more social and economic impact. In the upcoming decades, smart farming will become more crucial for tackling issues related to sustainability and global food security due to the advancement in technology.

REFERENCES

Aryal, J.P., Rahut, D.B., Maharjan, S., and Erenstein, O. (2018). "Factors affecting the adoption of multiple climate-smart agricultural practices in the Indo-Gangetic Plains of India", *Natural Resources Forum*, 42 (3), 141–158. DOI:10.1111/1477-8947.12152.

Ataei, P., Gholamrezai, S., Movahedi, R., and Aliabadi, V. (2021). "An analysis of farmers' intention to use green pesticides: The application of the extended theory of planned behavior and health belief model", *Journal of Rural Studies*, 81, 374–384. DOI:10.1016/j.jrurstud.2020.11.003.

Ayinde, E., Abdoulaye, T., Ambali, O.Y., and Akinsola, G.O. (2017). "Determinants of the adoption of quality protein maize among farming households in Niger state, Nigeria", *Annals of Science and Technology*, 2 (1), 8–12.

Beaman, L., and Dillon, A. (2018). "Diffusion of agricultural information within social networks: Evidence on gender inequalities from Mali", *Journal of Development Economics*, 133 (1), 147–161, DOI:10.1016/j.jdeveco.2018.01.009.

Bedeke, S., Vanhove, W., Gezahegn, M., Natarajan, K., and Van Damme, P. (2019). "Adoption of climate change adaptation strategies by maize-dependent smallholders in Ethiopia", *Wageningen Journal of Life Sciences*, 88 (1), 96–104.

Bruce, K., and Costa, H. (2019). "Enabling environment for PPPs in agricultural extension projects: Policy imperatives for impact", *Journal of Rural Studies*, 70 (1), 87–95, DOI:10.1016/j.jrurstud.2019.07.005.

Coulibaly, T.B., Du, J., and Diakite, D. (2021). "Sustainable agricultural practices adoption", *Agriculture*, 67 (4), 166–176.

Daberkow, S.G., and McBride, W.D. (2003). "Farm and operator characteristics affecting the awareness and adoption of precision agriculture technologies in the US", *Precision Agriculture*, 4 (2), 163–177, DOI:10.1023/A:1024557205871.

Dhiman, V. (2020). "Organic farming for a sustainable environment: Review of existing policies and suggestions for improvement", *International Journal of Research and Review*, 7 (2), 22–31.

Fisher, M., Holden, S.T., Thierfelder, C., and Katengeza, S.P. (2018). "Awareness and adoption of conservation agriculture in Malawi: What difference can farmer-to-farmer extension make?", *International Journal of Agricultural Sustainability*, 16 (3), 310–325. DOI:10.1080/14735903.2018.1472411.

Harwood, R.R. (2020). "A history of sustainable agriculture", in Edwards, C.A., Lal, R., Madden, P., Miller, R.H. and House, G. (Eds.), *Sustainable Agricultural Systems*, 3–19. Boca Raton: CRC Press. DOI:10.1201/9781003070474.

Li, X., Yu, G., Wen, L., and Liu, G. (2023). "Research on the effect of agricultural science and technology service supply from the perspective of farmers' differentiation", *Innovation and Green Development*, 2 (3), 100055.

Lynch, J., Cain, M., Frame, D., and Pierrehumbert, R. (2021). "Agriculture's contribution to climate change and role in mitigation is distinct from predominantly fossil CO2-emitting sectors", *Frontiers in Sustainable Food Systems*, 4. DOI:10.3389/fsufs.2020.518039.

Mgomezulu, W.R., Machira, K., Edriss, A.-K., and Pangapanga-Phiri, I. (2023). "Modelling farmers' adoption decisions of sustainable agricultural practices under varying agro-ecological conditions: A new perspective", *Innovation and Green Development*, 2 (1), 100036.

Misniakiewicz, M., Luczak, J., and Maruszewska, N. (2021). "Improvement of organic farm assessment procedures on the example of organic farming in Poland – recommendations for organic farming in Poland", *Agronomy*, 11 (8). DOI:10.3390/agronomy11081560.

Mohamed Haris, N.B., Garrod, G., Gkartzios, M., and Proctor, A. (2018). "The decision to adopt organic practices in Malaysia; a mix-method approach", 92nd Annual Conference, April 16–18, Warwick University, Coventry, UK. DOI:10.22004/ag.econ.273485.

Muhie, S.H. (2022). "Novel approaches and practices to sustainable agriculture", *Journal of Agriculture and Food Research*, 10 (1), 100446.

Mujeyi, M.M. (2020). "Economic analysis of climate-smart agriculture technologies in maize production in smallholder farming systems", *African Handbook of Climate Change Adaptation*, 1–16.

Nguyen, N., and Drakou, E.G. (2021). "Farmers' intention to adopt sustainable agriculture hinges on climate awareness: The case of Vietnamese coffee", *Journal of Cleaner Production*, 303, 126828. DOI:10.1016/j.jclepro.2021.126828.

Patle, G.T., Kharpude, S.N., Dabral, P.P., and Kumar, V. (2020). "Impact of organic farming on sustainable agriculture system and marketing potential: A review", *International Journal of Environment and Climate Change*, 10 (11), 100–120.

Sapbamrer, R., and Thammachai, A. (2021). "A systematic review of factors influencing farmers' adoption of organic farming", *Sustainability*, 13(7), 3842. DOI:10.3390/su13073842.

Sharifuddin, J., Mohammed, Z.A., Ramli, N.N., and Farmata, Y. (2020). "Green revolution's role and impact: Organic farming potential for Indonesian sustainable agriculture forum", *Penelitian Agro Ekonomi*, 37 (2), 115–125.

Trigo, A., Marta-Costa, A., and Fragoso, R. (2021). "Principles of sustainable agriculture: Defining standardized reference points", *Sustainability*, 13 (8), 4086. DOI:10.3390/su13084086.

18 Blockchain and Digital Twin Applications in Precision Agriculture

A Comprehensive Approach

Kanthavel R, S. Krithikaa Venket, Anju A, Adline Freeda R, Dhaya R, Frank Vijay, and Joseph Fisher

18.1 INTRODUCTION TO PRECISION AGRICULTURE

Precision farming, sometimes referred to as satellite farming or site specific crop management, is a revolutionary kind of farming that makes use of cutting-edge technologies to maximize resource utilization and increase crop yields (Drosatos & Efraimidis, 2020). This approach uses a variety of data-driven methodologies along with information technology to monitor and manage agricultural activities with previously unheard-of levels of precision. Precision farming fundamentally depends on gathering data from multiple sources, including weather stations, drones, satellites, and soil sensors.

After that, this data is examined to help with decision-making regarding crop planting, fertilization, irrigation, and harvesting. Ensuring crops receive the appropriate amount of resources at the appropriate time will maximize output and reduce waste. Precision agriculture can benefit greatly from the introduction of new technology such as digital twins and blockchain. A transparent, decentralized, and safe platform for managing agricultural data is offered by blockchain technology. From farm to table, it guarantees the information's legitimacy and traceability across the food chain. This is especially crucial for upholding food safety regulations, confirming organic farming methods, and boosting customer confidence.

Conversely, digital twins build digital copies of real-world agricultural objects and operations. Farmers are able to forecast results and optimize operations without the need for physical trial and error thanks to these computerized models that provide real-time monitoring and simulation. Farmers may monitor crop health, soil conditions, and environmental elements continually by integrating digital twins, which can result in more precise and timely responses. Precision agriculture could undergo a revolution if blockchain and digital twin technologies are combined. Sustainable agricultural methods, increased supply chain efficiency, and improved decision-making can all result from this integrated strategy (Chiaraluce et al., 2024).

The demand for intelligent, effective, and resilient farming practices is greater than ever as the world's population continues to rise and climate change reduces agricultural productivity (Dhaya & Kanthavel, 2020a). This study explores how blockchain and digital twin technologies can be applied comprehensively to precision agriculture, with the goal of transforming farming techniques, resolving implementation issues, and promoting environmental sustainability and global food security. Integration of these technologies can greatly increase farms' economic viability in addition to increasing operational efficiency.

DOI: 10.1201/9781003507390-18

Farmers may cut expenses and boost profits by decreasing waste and making the best use of inputs like water, fertilizer, and insecticides. Additionally, the transparency that blocks chain technology offers may provide new market opportunities, especially in areas where consumers are calling for food that is produced sustainably and with traceability (Camilo et al., 2024). The scalability of these technologies is also discussed in the study. Large-scale farms are frequently the first to implement cutting-edge technologies, but small and medium-sized farms can also benefit from blockchain and digital twin systems due to their modular and flexible design. The widespread distribution of precision agriculture's benefits is made possible by the democratization of technology, which promotes a more equal agricultural environment (Dutta et al., 2020).

Furthermore, implementing precision agriculture technology supports more general environmental objectives. These technologies support the sustainability of agricultural systems by encouraging resource efficiency and minimizing the environmental impact of farming operations. This is significant in light of climate change, as preserving food security and safeguarding natural ecosystems depend on the resilience of agricultural operations. In conclusion, combining digital twin and blockchain technology with precision agriculture has a bright future (Wang et al., 2019). This study shows how these technologies can lead to a better, more efficient and sustainable future for agriculture by providing a thorough overview of their uses, advantages, and obstacles.

18.2 BLOCKCHAIN IN PRECISION AGRICULTURE

Precision agriculture is quickly adopting blockchain technology because it offers a transparent, safe, and decentralized platform for data management. This section explores a number of related subjects that highlight the many uses and advantages of blockchain technology in precision agriculture. Figure 18.1 shows blockchain in precision agriculture.

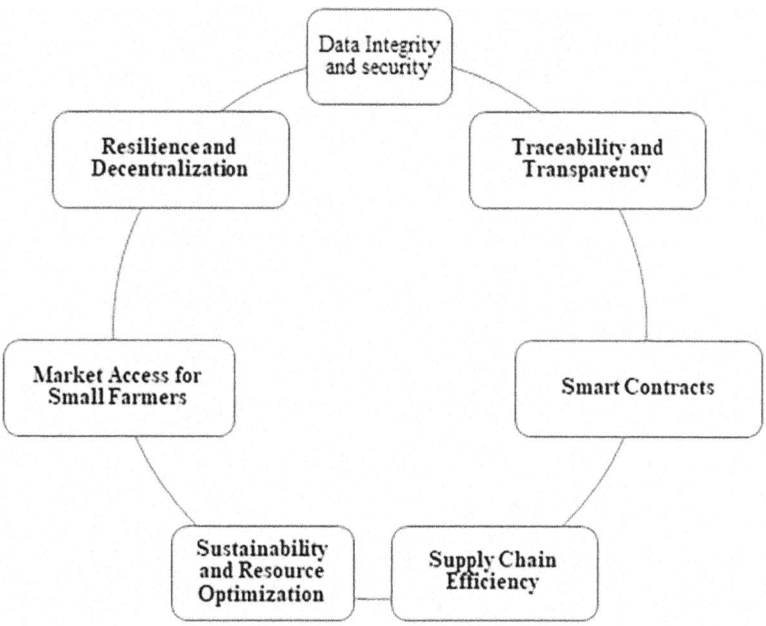

FIGURE 18.1 Blockchain in precision agriculture.

Data Integrity and Security: Large volumes of data are gathered via sensors, drones, and satellites in precision agriculture. Blockchain stores this data in an immutable ledger, guaranteeing its security and integrity. Every data entry is encrypted and time-stamped to guard against illegal changes and guarantee the accuracy and dependability of the information. Making educated decisions regarding crop management and resource allocation requires this.

Traceability and Transparency: Blockchain records each transaction and procedure on a decentralized ledger, improving the traceability of agricultural products. Every stage, from planting and harvesting to distributing, is recorded and verifiable. Ensuring organic practices, upholding food safety regulations, and fostering customer confidence all depend on this traceability. By scanning a QR code, customers can obtain comprehensive details regarding the origin and progression of a product via the supply chain (Li et al., 2020).

Smart Contracts: Various agricultural operations can be streamlined by using smart contracts, which are programmable contracts that automatically execute when predefined circumstances are satisfied. Smart contracts can handle supply chain logistics, track adherence to farming standards, and automate payments in precision agriculture (Dhaya & Kanthavel, 2021). To cut down on intermediaries and expedite transactions, a smart contract might, for instance, pay a farmer only once a crop is delivered and confirmed to fulfil quality criteria.

Supply Chain Efficiency: Supply chain management is made easier by blockchain technology, which offers a single, unchangeable ledger that is available to all parties. The traditional supply chains become less opaque and complex as a result of this transparency, which improves communication between farmers, distributors, retailers, and customers. A more efficient supply chain lowers expenses, cuts down on delays, and improves the quality of the output (Rehman et al., 2023).

Sustainability and Resource Optimization: Blockchain provides transparent and verifiable records of resource use, which helps sustainable farming techniques. With the help of tracking applications of pesticides, fertilizers, and water, farmers can optimize their use and reduce waste and environmental damage. This exact resource management increases farming operations' economic viability and fosters sustainability (Dhaya & Kanthavel, 2020b).

Market Access for Small Farmers: By giving them access to the same degree of technological capability and market transparency as large-scale businesses, blockchain technology can empower small and medium-sized farms (Yuan & Wang, 2018). Because of the democratization of technology, smaller farms are able to compete more successfully, which promotes greater equity in the agriculture industry. Small farmers can benefit from risk mitigation and financial stability through blockchain-enabled micro lending and insurance.

Resilience and Decentralization: The decentralized nature of blockchain technology improves the security and robustness of agricultural data management systems. The distributed ledger of blockchain, in contrast to centralized databases, replicates data across numerous nodes, increasing its resilience against hacking and data loss. Maintaining consistent and dependable access to vital agricultural data requires resilience.

18.3 CASE STUDIES AND REAL-WORLD APPLICATIONS

A plethora of pilot programs and case studies illustrate the usefulness of blockchain technology in precision farming. These practical uses offer insightful information about how blockchain technology can be applied to enhance farming practices' sustainability, efficiency, and transparency (Kumar et al., 2023). The knowledge gained from these initiatives can direct increased industry innovation and wider acceptance. To sum up, precision agriculture may benefit greatly from blockchain technology, which can improve data integrity and traceability, as well as boost efficiency and sustainability. Through the integration of blockchain technology with precision agriculture processes, the agricultural sector may attain unprecedented levels of operational excellence, transparency, and trust. This chapter provides a thorough analysis of these applications and a broad perspective on how blockchain technology might transform precision agriculture (Yang et al., 2024).

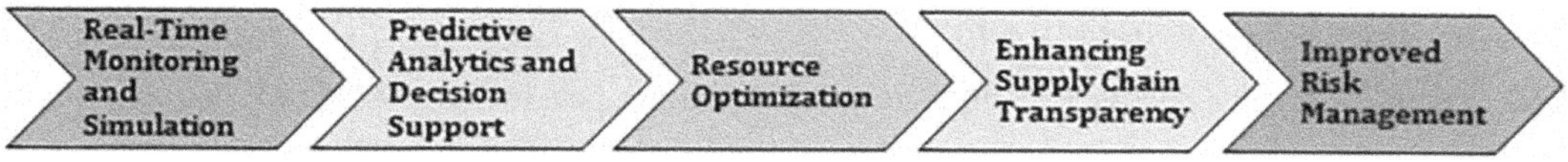

FIGURE 18.2 Digital twin technology in precision agriculture.

18.3.1 Digital Twin Technology in Precision Agriculture

A recent development in precision agriculture is digital twin technology, which provides an effective instrument for improving agricultural process monitoring, management, and optimization. A digital twin is an electronic copy of a physical object or system that can replicate real-world circumstances in real time and offer insightful information for better decision-making and operations in Figure 18.2. The several uses and advantages of digital twin technology in precision agriculture are examined in this section.

18.3.2 Real-Time Monitoring and Simulation

Digital twins make it possible to simulate and monitor agricultural processes in real time. Farmers may continuously monitor weather patterns, crop growth, soil health, and other important elements by building a virtual model of their farm. Data from the physical farm is gathered by sensors and Internet of Things devices and fed into the digital twin. Farmers can quickly detect problems like pest infestations, fertilizer shortages, or water stress thanks to this real-time data, which enables rapid actions to reduce possible losses (Dhaya & Kanthavel, 2022).

18.3.3 Predictive Analytics and Decision Support

The predictive analytics capabilities of digital twin technology are among its main advantages. Digital twins can forecast future events like crop yields, disease outbreaks, and resource requirements by modelling different scenarios and examining historical data. With the use of this predictive power, farmers are better able to plan their irrigation schedules, planting timetables, and pest management tactics (Singh et al., 2023). To ensure maximum production and quality, a digital twin, for example, can estimate the best time to harvest based on weather patterns and crop maturity.

18.3.4 Resource Optimization

In agriculture, digital twins make accurate resource management easier. Farmers may optimize these inputs to avoid waste and lessen their influence on the environment by estimating how much water, fertilizer, and pesticides are used on the farm. In order to assist farmers in selecting the most effective and sustainable practices, the digital twin may simulate various irrigation techniques, fertilizer application rates, and insect management strategies. By lowering input costs, this raises crop productivity while simultaneously improving the farm's economic viability (Kumar et al., 2024).

18.3.5 Enhancing Supply Chain Transparency

Another important function that digital twins can perform is improving supply chain transparency. Digital twins make it possible to track and document product origin and quality more effectively by giving a complete picture of the whole agricultural process—from planting to harvesting to distribution. In order to meet food safety regulations and validate organic or sustainable farming techniques, openness is very crucial (Bermeo-Almeida et al., 2018). By giving consumers and supply chain partners access to comprehensive information about the product's journey, agricultural produce gains credibility and trust.

18.3.6 Improved Risk Management

Risks associated with agriculture include weather variations, pest infestations, and volatile markets. Farmers may better manage these risks by using digital twin technology, which offers a dynamic and interactive platform for risk assessment and mitigation. Farmers can test various risk scenarios, such as pest outbreaks or drought conditions, using digital twins, and then create backup plans based on the simulations. This proactive approach to risk management improves the sustainability and resilience of the farm.

18.4 INTEGRATION OF BLOCKCHAIN AND DIGITAL TWINS IN PRECISION AGRICULTURE

The integration of blockchain and digital twin technologies in precision agriculture offers a holistic solution for optimizing farm operations and supply chain management. Blockchain ensures data security and transparency, while digital twins offer real-time monitoring and predictive analytics. Together, they enhance operational efficiency, improve decision-making, and support sustainable farming practices. The synergy of these technologies presents a promising future for precision agriculture, combining the strengths of secure transactions and advanced data analytics.

Furthermore, increased responsibility and trust are fostered in agricultural operations through the integration of blockchain and digital twin technologies. Every step of the production process, from seed to harvest to distribution, should be publicly documented so that all parties involved in the supply chain can be sure of the legitimacy and caliber of agricultural goods (Saberi et al., 2018). In addition to boosting consumer confidence, this transparency makes it easier to comply with legal requirements and obtain certifications. Farmers can apply precision farming techniques with previously unheard-of accuracy and efficiency thanks to the collaboration of blockchain and digital twin technologies. Farmers can monitor crop health with remarkable precision, optimize irrigation schedules, and customize fertilization tactics thanks to blockchain technology and digital twins' real-time insights and immutable record-keeping. By optimizing agricultural yields while reducing input costs and environmental effect, this precision supports sustainable agriculture (Wang et al., 2019).

18.5 CHALLENGES

Precision agriculture could be revolutionized by the combination of blockchain and digital twin technologies, but there are a number of obstacles that need to be overcome before this potential can be fully realized, as shown in Figure 18.3. Furthermore, examining potential future directions might steer additional developments in this area.

18.5.1 Scalability

As agricultural operations get bigger and more sophisticated, they need to be scalable. It is crucial to make sure that blockchain networks and digital twin models can manage high data and transaction volumes without sacrificing efficiency.

FIGURE 18.3 Challenges of precision agriculture.

18.5.2 Data Security and Privacy

It's critical to guard private agricultural data against illegal access and cyber threats. Although blockchain offers strong security features, there are still difficulties in enforcing data privacy laws and putting encryption methods into practice (Karunathilake et al., 2023).

18.5.3 Cost and Accessibility

Small-scale farmers may not be able to afford the initial expenditure needed to use blockchain and digital twin technologies. It is imperative to make an effort to lower expenses and improve accessibility by means of grants, subsidies, or cooperative projects.

18.5.4 Complexity of Integration

It can be difficult and resource-intensive to integrate blockchain and digital twin technologies into current agricultural systems. For adoption to be effective, user-friendly interface development as well as thorough training and support are essential (Bhadra et al., 2023)

18.6 FUTURE DIRECTIONS

The evolving advancements in blockchain and digital twin technologies hold transformative potential for the future of precision agriculture. Blockchain focuses on enhancing transparency, security, and efficiency in agricultural transactions, while digital twins provide detailed real-time insights and predictive analytics for farm management. The integration of these technologies can lead to a highly resilient and efficient agricultural ecosystem, leveraging the strengths of both approaches to address current and future challenges in precision agriculture.

18.6.1 Integration of AI and Advanced Analytics

Precision agriculture decision making can be improved by integrating artificial intelligence (AI) algorithms and advanced analytics into digital twin models. Large datasets can be analyzed using machine learning techniques to find trends and improve farming methods (Kiruthika et al., 2023).

18.6.2 Edge Computing

By processing data locally at the farm level using edge computing technology, latency can be decreased and real-time monitoring and decision making improved. Digital twin models and edge devices can communicate to deliver quick answers and insights (Alloghani, 2024).

18.6.3 Blockchain Interoperability

Researching interoperability solutions that allow for smooth data exchange and communication between various blockchain networks might encourage wider adoption and cooperation among agriculture ecosystem players.

18.6.4 Internet of Things Integration

By combining blockchain networks, digital twins, and Internet of Things (IoT) devices like sensors and actuators, precision agriculture's capacity for data collecting, monitoring, and automation can be further improved (Praveen et al., 2021). A more thorough and dynamic portrayal of agricultural processes is made possible by this combination.

18.6.5 Decentralized Autonomous Organizations

Investigating decentralized autonomous organizations' (DAOs) potential in agriculture can promote more accountability, transparency, and efficiency. DAOs allow stakeholders to jointly manage and administer farming operations through the use of smart contracts and decentralized decision-making procedures.

18.7 CONCLUSION

In summary, the combination of digital twin and blockchain technology is a novel approach to precision agriculture that provides a thorough resolution to the issues that contemporary farming methods encounter. Farmers are able to ensure the accuracy, confidentiality, and transparency of their data while gaining unparalleled insights into their operations because to the synergy of these advancements. A decentralized and unchangeable platform for tracking agricultural data, improving traceability, and facilitating safe supply chain transactions is offered by blockchain technology. With the use of digital twins, farmers may better manage resources, make better decisions, and predict future situations thanks to their real-time monitoring, predictive analytics, and simulation capabilities. The potential advantages of blockchain and digital twin applications in precision agriculture are enormous, even with the difficulties caused by data interoperability, scalability, privacy, and integration complexity. We can fully realize the potential of these technologies to transform farming methods by tackling these issues and investigating new avenues like decentralized governance models, blockchain interoperability, enhanced analytics, and AI integration. In the end, precision agriculture's all-encompassing strategy of using blockchain and digital twin technologies offers the potential to advance resilience, efficiency, and sustainability in the agricultural industry. These technologies can enable farmers to satisfy the increasing demand for food production while reducing their environmental effect and guaranteeing the long-term viability of agriculture for future generations by encouraging greater transparency, collaboration, and innovation.

REFERENCES

Alloghani, M. A. AI for Sustainable Agriculture: A Systematic Review. In: *Artificial Intelligence and Sustainability*. Signals and Communication Technology. Springer, Cham, 2024. https://doi.org/10.1007/978-3-031-45214-7_3

Bermeo-Almeida, O., Cardenas-Rodriguez, M., Samaniego-Cobo, T., et al. Blockchain in Agriculture: A Systematic Literature Review. In: Valencia-García, R., Alcaraz-Mármol, G., Del Cioppo-Morstadt, J., et al. (eds) *Technologies and Innovation. CITI 2018. Communications in Computer and Information Science*, vol. 883. Springer, Cham, 2018. https://doi.org/10.1007/978-3-030-00940-3_4.

Bhadra, O., Sahoo, S., Halder, R., et al. AgroBLF: Blockchain-Based Framework for Smart Agriculture. *Innovations in Systems and Software Engineering*, 2023. https://doi.org/10.1007/s11334-022-00468-0

Camilo, O. C., Ramírez, G. G., and Juan Carlos, C. Blockchain and Agricultural Sustainability in South America: A Systematic Review. *Frontiers in Sustainable Food Systems*, vol. 8, 2024.

Chiaraluce, G., Bentivoglio, D., Finco, A., et al. Exploring the Role of Blockchain Technology in Modern High-Value Food Supply Chains: Global Trends and Future Research Directions. *Agricultural Economics*, vol. 12, no. 6, 2024. https://doi.org/10.1186/s40100-024-00301-1

Dhaya, R., and Kanthavel, R. Elaborative Investigation of Blockchain Technology in Intelligent Networks. In: *Blockchain Technology and Applications for Digital Marketing*, pp. 93–106. IGI Global, 2020a.

Dhaya, R., and Kanthavel, R. Investigation on Industry Applications of Blockchain Technology. In: *Blockchain Technology and Applications for Digital Marketing*, pp. 60–79. IGI Global, 2020b.

Dhaya, R., and Kanthavel, R. *Developing an Energy-Efficient Ubiquitous Agriculture Mobile Sensor Network-Based Threshold Built-in MAC Routing Protocol (TBMP)*. Springer-Soft-Computing, 2021.

Dhaya, R., and Kanthavel, R. Insights on the Prospects of Multi-Access Edge Computing. In: *Advancing Smarter and More Secure Industrial Applications Using AI, IoT, and Blockchain Technology*, pp. 37–55. IGI Global Publisher, 2022.

Drosatos, G., and Efraimidis, P. *Block Chain Applications in Agriculture: A Scoping Review*. MDPI, 2020.

Dutta, P., Choi, T. M., Somani, S., et al. Blockchain Technology in Supply Chain Operations: Applications, Challenges and Research Opportunities. *Transportation Research Part E: Logistics and Transportation Review*, vol. 142, 2020.

Karunathilake, E. M. B. M., Le, A. T., Heo, S., et al. The Path to Smart Farming: Innovations and Opportunities in Precision Agriculture. *Agriculture*, vol. 13, no. 8, p. 1593, 2023, https://doi.org/10.3390/agriculture13081593

Kiruthika, J. K., Yawanikha, T., Kannika, A., et al. *User Interface of Blockchain-Based Agri-Food Traceability Applications*, pp. 258–261. 2023 4th International Conference on Signal Processing and Communication (ICSPC), Coimbatore, India, 2023. https://doi.org/10.1109/ICSPC57692.2023.10125777.

Kumar, D., Singh, R. K., Mishra, R., et al. Roadmap for Integrating Blockchain with Internet of Things (IoT) for Sustainable and Secured Operations in Logistics and Supply Chains: Decision Making Framework with Case Illustration. *Technological Forecasting and Social Change*, vol. 196, 2023.

Kumar, R., Farooq, M., and Qureshi, M. Advancing Precision Agriculture Through Artificial Intelligence: Exploring the Future of Cultivation. *A Biologists Guide to Artificial Intelligence*, pp. 151–165. Academic Press, 2024.

Li, J., Zhu, S., Zhang, W., et al. Block Chain-Driven Supply Chain Finance Solution for Small and Medium Enterprises. *Front-End Management*, vol. 7, pp. 500–511, 2020. https://doi.org/10.1007/s42524-020-0124-2.

Praveen, P., Shaik, M. A., Kumar, T. S., et al. Smart Farming: Securing Farmers Using Block Chain Technology and IoT. In: Choudhury, T., Khanna, A., Toe, T. T., et al. (eds) *Blockchain Applications in IoT Ecosystem. EAI/Springer Innovations in Communication and Computing*. Springer, Cham, 2021. https://doi.org/10.1007/978-3-030-65691-1_15

Rehman, K. U., Andleeb, S., Ashfaq, M., et al. Block Chain-Enabled Smart Agriculture: Enhancing Data-Driven Decision Making and Ensuring Food Security. *Journal of Cleaner Production*, vol. 427, 2023.

Saberi, S., Kouhizadeh, M., Sarkis, J., et al. Blockchain Technology and Its Relationships to Sustainable Supply Chain Management. *International Journal of Production Research*, vol. 57, no. 7, pp. 2117–2135, 2018. https://doi.org/10.1080/00207543.2018.1533261

Singh, R. K., Mishra, R., Gupta, S., et al. Blockchain Applications for Secured and Resilient Supply Chains: A Systematic Literature Review and Future Research Agenda. *Computers & Industrial Engineering*, vol. 175, 2023.

Wang, Y., Han, J. H., and Beynon-Davies, P. Understanding Blockchain Technology for Future Supply Chains: A Systematic Literature Review and Research Agenda. *Supply Chain Management*, vol. 24, no. 1, pp. 62–84, 2019. https://doi.org/10.1108/SCM-03-2018-0148

Yang, C., Ji, X., Cheng, C., Liao, S., et al. Digital Economy Empowers Sustainable Agriculture: Implications for Farmers' Adoption of Ecological Agricultural Technologies. *Ecological Indicators*, vol. 159, 2024.

Yuan, Y., and Wang, F.-Y. Blockchain and Crypto-Currencies: Model, Techniques, and Applications. *IEEE Transactions on Systems, Man, and Cybernetics: Systems*, vol. 48, no. 9, pp. 1421–1428, 2018. https://doi.org/10.1109/TSMC.2018.2854904

19 Effective Deep Learning-Based Agricultural Models for Classifications of Plant Diseases

Pramit Brata Chanda, Subir Kumar Sarkar, and Surajit Naskar

19.1 INTRODUCTION

Today the agriculture sector has become a crucial area of research throughout the world. The economy of our country is also very dependent on the agricultural sector. Diseases affecting plant leaves are greatly affected in crops, and that can have an impact on the economy of the agricultural sector. These are several types of bacterial diseases that can harm crops that are very important in the agricultural domain, so quicker identification is very much needed to take the necessary steps to reduce the damage to plants and save crops.

Here, image processing, machine learning, and deep learning are widely used to create efficient models that can classify several plant diseases quickly and make earlier decisions to save crops from major damage. Today, several image processing–based algorithms are used to detect plant disease in the agricultural domain. These machine learning–based methods are currently having a huge impact in the agricultural domain of research. Much research has already been performed on plant disease detection using machine learning–based models that are significant in the domain of research. Models like SVM, KNN, regression, and random forest are used to classify several diseases of crops and plants that are crucial for making quicker decisions to save crops from bacterial attacks. Deep learning–based models like ResNet, AlexNet, MobileNet, DenseNet, and VGG16 are also used to classify several important crop diseases more efficiently with higher accuracy rates. Deep learning or transfer learning–based models are based on several layers for processing data, so the execution time significantly increases, but the accuracy of the prediction of disease is higher. Also, parallel processing may be used to give more accurate results for the prediction of disease with higher accuracy, and the chances of loss of crops may be lower if faster identification is done.

19.2 LITERATURE REVIEW

In 2013, Shirsath [1] created a model providing instruction to the user for decision making regarding suitable crops. The system illustrates collected crop samples from several resources with appropriate crops that can be planted.

In [2], Pudumalar gives an ensemble-based technique that combines several models to achieve greater accuracy. Here, methods like random trees, KNN, CHAID, and naïve Bayes are used to ensemble the process for getting better performance and accuracy. The approach has an accuracy score of 88%.

In [3], Khirade et al. use a BPNN-based network for detecting plant disease. Several approaches are illustrated by the authors for plant disease classification. They use several segmentation techniques, such as Otsu's thresholding, to detect infected regions of leaves.

 DOI: 10.1201/9781003507390-19

Madiwalar and Wyawahare use various image-based approaches to plant disease classification [4]. The authors use data from 110 RGB images for more accurate detection of disease and use several algorithms to classify it. The authors use gray level co-occurrence matrix (GLCM) features, a Gabor filter for noise removal, and an SVM classifier for classification. The authors achieve 83.34% accuracy of extracted features.

Moghadam et al. [5] use a k-means approach to segmentation of plant leaves. The authors propose an approach of the removal of grids on hyperspectral images. They reach an accuracy of 83% for the VNIR spectral range and 93% for the full spectrum.

Shrestha et al. create a CNN-based model to identify disease [6]. The authors achieve 88.80% accuracy. Here, the authors use 3000 high-resolution RGB images for research. The network consists of convolution and pooling layer blocks for model construction.

In 2014 [7], plant classification using several image processing and machine learning-based classification approaches are applied on various plant images. Those methods are implied on several leaf images, where leaf diseases classification is performed basis of morphological features. Several classification approaches, such as KNN classifiers, probabilistic neural networks, genetic algorithms, SVM, PCA, and artificial neural networks, are used. The work gives an overview regarding plant classification.

19.3 PROPOSED WORK

Our proposed plant disease detection system will use images of healthy plants and plants with disease. After training the model with this dataset, our proposed system can detect whether the plant is healthy. The research consists of the detection of disease with transfer learning and a deep learning–based network. Thus, it is more effective for farmers.

Here, the dataset that was used was taken from Kaggle and consists of 70,000 values of 38 unique classes. Various models are used for plant disease classification. Primarily, deep learning and transfer learning algorithms are used. Transfer learning is more flexible and allows several pretrained models that separately train and integrate new models. We have applied deep learning such as CNN and transfer learning networks such as InceptionV3, VGG16, MobileNetV2, and ResNet50 algorithms on the dataset. Figure 19.1 illustrates the model architecture.

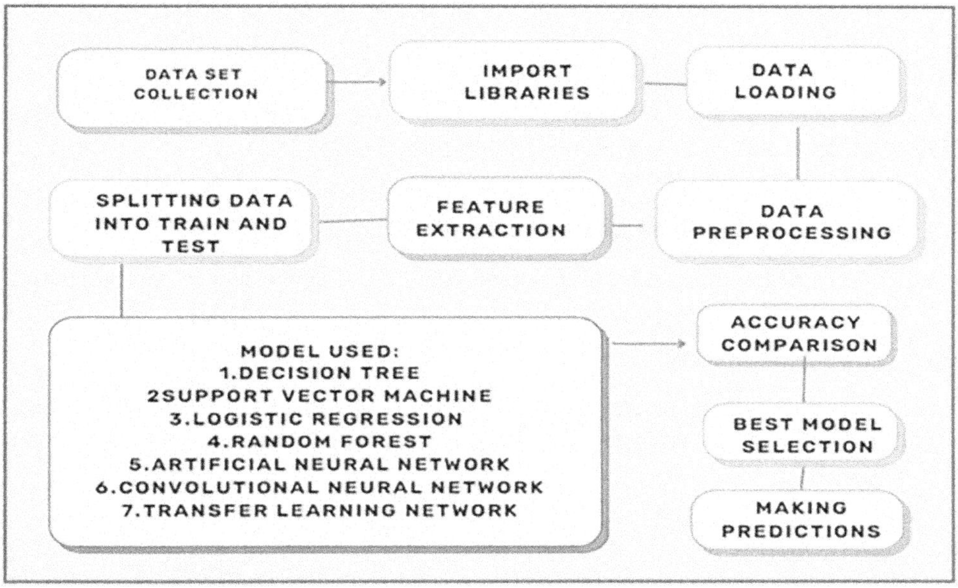

FIGURE 19.1 Proposed system architecture.

FIGURE 19.2 Plant disease dataset images [8].

19.3.1 Dataset Collection and Description

The dataset is downloaded from Kaggle [8]. The given dataset (see Figure 19.2) contains 70,000 images of plant leaves, which have 38 different class labels.

19.3.2 Importing Libraries

To create and train the model, Keras is required. Matplotlib and Seaborn are needed for data visualization.

19.3.3 Loading the Data

Here, we will define the path to our data. We will create our training and validation datasets. OpenCV's imread function is used to read the image in RGB and resize it to 256.

19.3.4 Data Preprocessing

Conversion of each image label to binary levels is done. Then 38 unique classes are printed using a label binarizer. After that, scaling is done on points from 0 to 255.

19.3.5 Splitting Data into Training and Test Data

Here, 80% of the images are used for training and 20% for testing the splitting of the datasets.

19.3.6 Data Augmentation

An image is generated with an image generator that performs shifts, sheers, random rotations, flips, shifts, and crops over several image datasets so more images are created to have more data and get better outcomes.

19.3.7 Model Selection

Various models are used for plant disease classification. Deep learning and transfer learning algorithms are primarily used.

Deep learning is a machine learning technique that constructs artificial neural networks to mimic the structure and function of the human brain.

Transfer learning approach allows using of various pre-trained models directly on several datasets. Also integration of new models as for pre-processing, feature extraction is allowed by transfer learning approach of its flexible nature. A pre-trained model is trained with a large dataset such as an ImageNet dataset:

- **Input layer—**It receives the data as input values from the dataset.
- **Normalization layer**—Scaling is done to input data up to suitable intervals so that bias may be removed.
- **Convolutional layer**—This layer is used to filter the information and extract several features.
- **Pooling layer**—More generalized data is created for good outcomes.
- **Activation layer—**This use of activation layer highlights the linearity and nonlinearity into the network. Also, it requires to learn complex patterns and relations among various data.
- **Dropout layer—**This layer is used to nullify random data to get more general data and reduce overfitting problems.
- **Output layer—**This layer is an outcome-based layer that gives outputs.

The Adam optimizer is used as a better option than any specific algorithm, and it is useful for suitable problems.

19.3.7.1 Convolutional Neural Networks

A CNN is applied for image classification and image processing by processing pixel data. It is used for image processing as an image classification AI that uses the deep learning concept to perform classification and image and video recognition, along with recommender systems and natural language processing. Figure 19.3 shows the CNN architecture.

19.3.7.1.1 Defining the Model

The CNN consists of convolution with max-pooling layers. A dropout layer is merged to avoid the problems of overfitting.

The CNN model processing consists of several layers: input layer, output layer, convolutional layer, full layer, SoftMax layer, connected layer, and pooling layer.

First, the first CONV => RELU => POOL is created. It contains 32 filters with a 3 × 3 kernel and rectified linear unit (ReLU) activation function. Then batch normalization is applied, and max pooling and dropout are used. Next, the creation of (CONV => RELU) * 2 => POOL blocks is performed.

19.3.7.1.2 Compiling and Fitting the Model

We compile the model using Adam as our optimizer and sparse categorical cross-entropy as the loss function. We are using a lower learning rate of 0.000001 for a smoother curve. We train our model for 50 epochs since our learning rate is very small.

The CNN model was able to achieve an accuracy of 95%.

19.3.7.2 Very Deep Convolutional Networks for Large-Scale Image Recognition

The VGG16 is a very important model for classification purposes. This VGG model is popular rather than AlexNet as it supports processing of large-scale of datasets with deeper layers and fewer filter for processing with higher performance rates. The layers that are applied are 13 convolutional layers, five pooling layers, and three dense layers. As it is sequential in nature, many layers used as filters are applied on it, as shown in Figure 19.4. So, at each stage, small 3 × 3 filters are required to

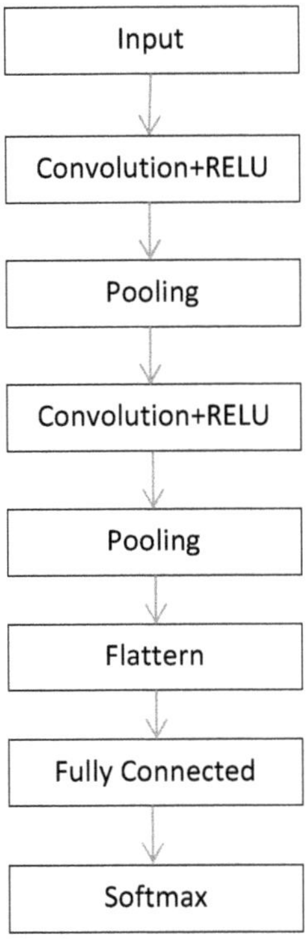

FIGURE 19.3 CNN architecture [9].

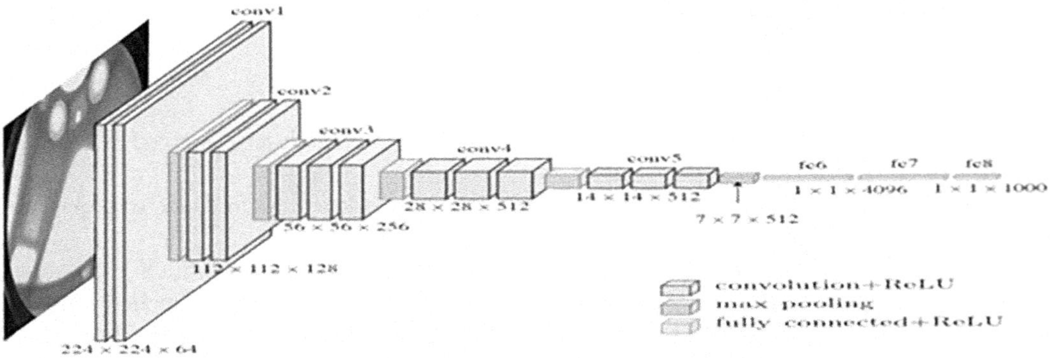

FIGURE 19.4 VGG16 architecture [10].

reduce the number of parameters in the hidden layers, those are then passed throughout the ReLU function properly.

19.3.7.2.1 Importing the Model

First, the VGG16 model is created as a baseline, and then transfer learning is applied.

19.3.7.2.2 Defining the Model

Base_model is defined with layers, and features are transferred to single vectors per the image. So dropout is implied to the final dense layer, followed by SoftMax activation function.

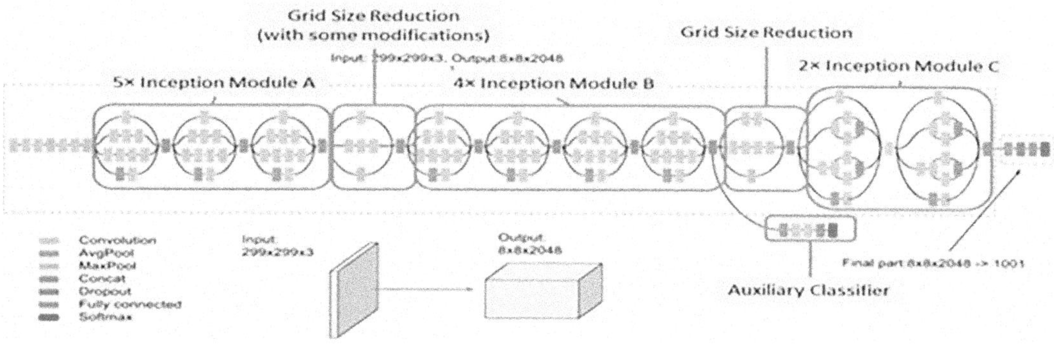

FIGURE 19.5 InceptionV3 architecture [11].

19.3.7.2.3 Compiling and Fitting the Model

We use the Adam optimizer with a low learning rate and then fit the model with 50 epochs. VGG16 achieved a 97.82% accuracy rate.

19.3.7.3 InceptionV3

VGG16 have already received 2nd place in the ILSVRC in recent years, in other case 1st place was given to GoogleNet (Inception) due to its use for reduction of dimensionality. This was named GoogleNet (Inception v1). The Inception module performs convolutions of several filters, performs max pooling, and joins the outcomes with a further Inception module. Using a 1 × 1 convolution operation drastically reduces several parameters. GoogleNet linearly stacks modules. It is 22 layers deep (27 including the pooling layers). Global average pooling is used in the final inception module, as shown in Figure 19.5.

19.3.7.3.1 Importing the Model

A baseline structure is created and then pre-trained on particular dataset.

19.3.7.3.2 Defining the Model

The model is defined as a base structure by layers and activation functions after converting the features to single vector per image. A dropout of 0.2, a dense layer with two neurons, and SoftMax activation are added.

19.3.7.3.3 Compiling and Fitting the Model

We compile with the Adam optimizer with a lower rate of learning and fit the model with 50 epochs. The Inceptionv3 model has an accuracy of 95.94%.

19.3.7.4 Mobile NetV-2

MobileNetV2 is a CNN-based transfer learning model that works well for mobile devices. The model of MobileNetV2 consists of 32 fully convolutional layers and 19 other residual bottleneck layers, per the non-linear types of data shown in Figure 19.6.

128*128*3 128*128 64*64 32*32 4*4
Pre-Processing Conv Max Pool Classifier

19.3.7.4.1 Importing the Model

The MobileNetV2 model is created as base model. This is used on the ImageNet dataset to import the datasets.

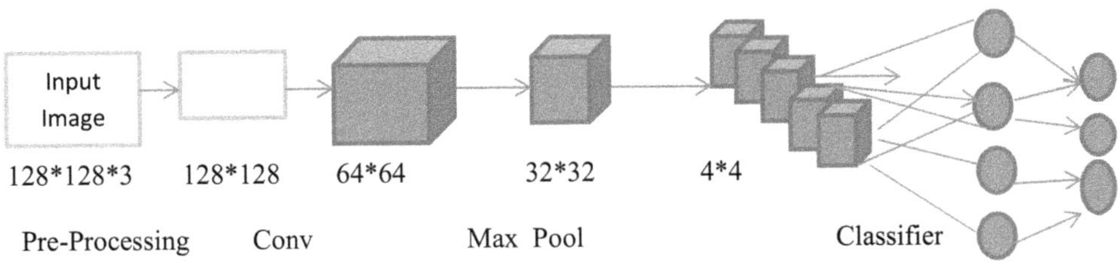

FIGURE 19.6 MobileNetV2 architecture [12].

19.3.7.4.2 Defining the Model

Next, we define our model using base_model followed by some layers and activation functions to convert the features into a single vector per image. We add a dropout of 0.2 and the final dense layer with two neurons and SoftMax activation.

19.3.7.4.3 Compiling and Fitting the Model

We compile using the Adam optimizer with a lower learning rate and fitting the model to proper datasets from a single iteration according to batch-wise.

19.3.7.5 ResNet50

ResNet50 is a transfer learning–based model that consists of 48 convolution layers, 1 MaxPool layer, and 1 average pool layer. It is based on the ResNet model, and ResNet50 is proposed for this research work.

19.3.7.5.1 Importing the Model

Initially, a ResNet50 model is created. Also, an ImageNet dataset is needed for the pre-trained structure.

19.3.7.5.2. Defining the Model

Next, those features are converted into a single vector for each image by modification of several layers. A dropout of 0.2, a dense layer, and SoftMax activation are added accordingly.

19.3.7.5.3 Compiling and Fitting the Model

Compilation is done with the Adam optimizer with a lower learning rate, and the model is fitted with 50 epochs.

19.4 RESULTS AND DISCUSSION: PLANT DISEASE CLASSIFICATION SYSTEM

We used deep learning and transfer learning algorithms for the classification of plant diseases using an image data set. It is found that the MobileNetV2 gives the best accuracy (98%) compared to the other models. The accuracy is compared in the following.

19.4.1 Results of Different Algorithms

The results of various algorithms are shown here.

19.4.1.1 CNN

```
Epoch 45/50
47/47 [==== ] - 40s 855ms/step - loss: 0.1613 - accuracy: 0.9466 - val_loss: 1.4525 - val_accu-
   racy: 0.6987
```

Epoch 46/50
47/47 (====] 405 852ms/step - loss: 0.1671 - accuracy: 0.9412 - val_loss: 0.6955 - val_accuracy: 0.8237
Epoch 47/50
47/47 (====] - 415 871ms/step - loss: 0.2027 - accuracy: 0.9331 - val_loss: 1.9181 - val_accuracy: 0.6066
Epoch 48/50
47/47 [====) - 415 864ms/step - loss: 0.1656 - accuracy: 0.9489 - val_loss: 1.0178 - val_accuracy: 0.7487
Epoch 49/50
47/47 (====) - 40s 855ms/step - loss: 0.1293 - accuracy: 0.9546 - val_loss: 1.0331 - val_accuracy: 0.7645
Epoch 50/50
47/47(====] - 40s 842ms/step - loss: 0.1387 - accuracy: 0.9546 - val_loss: 1.8336 - val_accuracy: 0.6842

The accuracy in different epochs is as follows.

Here we have plotted the training and validation accuracy graph and training and validation loss graph for the CNN, as shown in Figure 19.7.

We can see training accuracy increases with the number of epochs. It starts from 30% and reaches 95% at epoch 50.

Training loss decreases with the number of epochs. At the start of epoch training, loss is high, and as the number of epochs increases, training loss decreases. At epoch 50, training loss is 13%.

For the CNN model, a validation accuracy of 95% was reached in the training process.

19.4.1.2 VGG16

Epoch45/50
47/47 (======= - 455 955ms/step - loss: 0.1310 - acc: 0.9664 - val_loss: 0.2789 - val_acc: 0.9966
Epoch46/50
47/47(====== - 445 921ms/step - loss: 0.1176 - acc: 0.9748 - val_loss: 0.2732 - val_acc: 0.9013
Epoch47/50
47/47 (======== - 43s 925ms/step - loss: 0.1022 - acc: 0.9778 - val_loss: 0.2803 - val_acc: 0.9118
Epoch48/50
47/47 (==== - 445 939ms/step - loss: 0.1169 - acc: 0.9684 - val_loss: 0.3243 - val_acc: 0.8974
Epoch49/50
47/47 (=====-=] - 43s 905ms/step - loss: 0.1146 - acc: 0.9708 - val_loss: 0.3574 - val_acc: 0.8974
Epoch 50/50
47/47 (======] - 445 934ms/step - loss: 0.1014 - acc: 0.9782 - val_loss: 0.3038 - val_acc: 0.9053

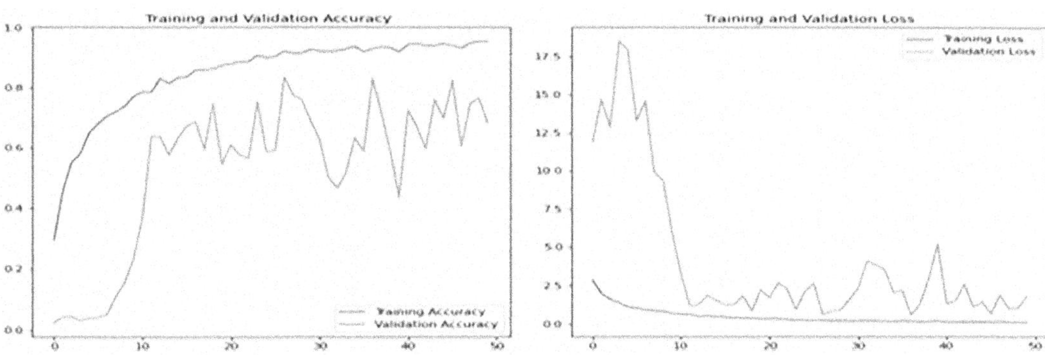

FIGURE 19.7 Training and validation performance graph (CNN).

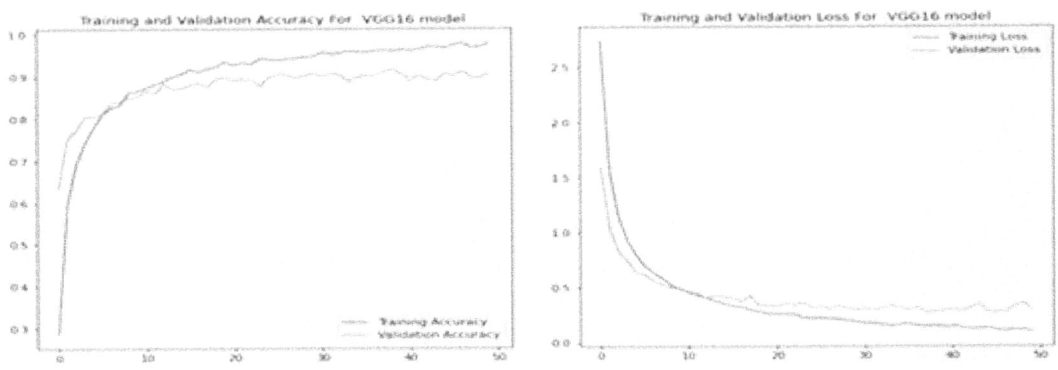

FIGURE 19.8 Training and validation performance graph of VGG16.

The results for VGG16 are given in the following.

The training and validation accuracy graph and training and validation loss graph for VGG16 are shown in Figure 19.8.

We can see the training accuracy increases with the number of epochs. It starts from 30% and reaches 97% at epoch 50.

Training loss decreases with the number of epochs. At the start of epoch training, loss is high, and as the number of epochs increases, training loss decreases. At epoch 50, training loss is 10%.

In the case of the VGG16 model, it achieves a 97.82% accuracy rate of classifying data.

19.4.1.3 InceptionV3

```
Epoch 45/50
47/47 (======= ] - 425 886ms/step - loss: 0.1349 - acc: 0.9573 - val_loss: 0.4464 - val_acc: 0.8566
Epoch 46/50
47/47 (= ===== ] - 425 898ms/step - loss: 0.1371 - acc: 0.9560 - val_loss: 0.4072 - val_acc: 0.8658
Epoch 47/50
47/47 [=======] - 415 872ms/step - loss: 0.1436 - acc: 0.9506 - val_loss: 0.4440 - val_acc: 0.8737
Epoch 48/50
47/47 =======] - 415 869ms/step - loss: 0.1435 - acc: 0.9533 - val_loss: 0.4614 - val_acc: 0.8500
Epoch 49/50
47/47 (=======] - 425 904ms/step - loss: 0.1483 - acc: 0.9516 - val_loss: 0.4112 - val_acc: 0.8789
Epoch 50/50
47/47 [=======] - 425 902ms/step - loss: 0.1272 - acc: 0.9594 - val_loss: 0.4068 - val_acc: 0.8842
```

The results for InceptionV3 are given as follows.

The training and validation accuracy graph and training and validation loss graph for InceptionV3 are given in Figure 19.9.

We can see training accuracy increases with number of epochs. It starts from 35%, and at epoch 50, it reaches 96% training accuracy.

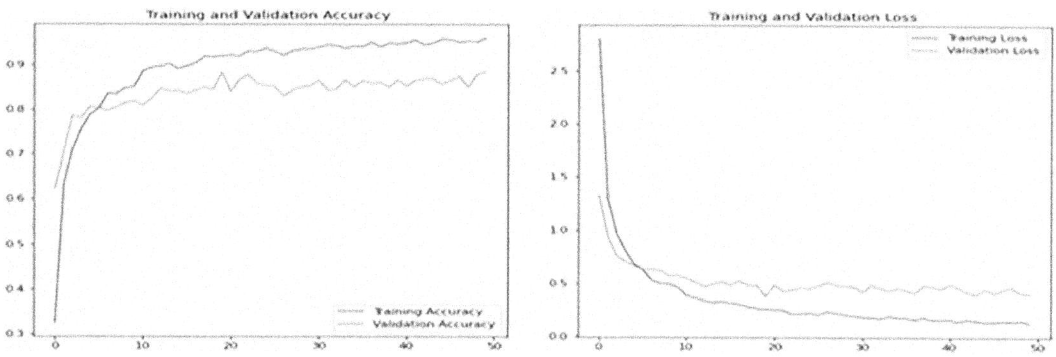

FIGURE 19.9 Training and validation performance graph (InceptionV3).

Training loss decreases with number of epochs. At the start of epoch training, loss is high, and as the number of epochs increases, training loss decreases. At epoch 50, training loss is 12%.

The InceptionV3 model is able to get 95.94% accuracy for classification.

19.4.1.4 MobileNetV2

```
Epoch 45/50
47/47 ==== ===] - 395 829ms/step - loss: 0.0735 - acc: 0.9741 - val_loss: 0.4181 - val_acc: 0.8842
Epoch 46/50
47/47 ( ========= ) - 40s 861ms/step - loss: 0.0653 - acc: 0.9785 - val_loss: 0.5135 - val_acc: 0.8763
Epoch 47/50
47/47[=========] - 405 838ms/step - loss: 0.0646 - acc: 0.9765 - val_loss: 0.5126 - val_acc: 0.8750
Epoch 48/50
47/47 [==== ====] - 395 836ms/step - loss: 0.0636 - acc: 0.9785 - val_loss: 0.6022 - val_acc: 0.8605
Epoch 49/50
47/47 [=========] - 40s 842ms/step - loss: 0.0577 - acc: 0.9778 - val_loss: 0.7140 - val_acc: 0.8474
Epoch 50/50
47/47 [=========] - 405 853ms/step - loss: 0.0507 - acc: 0.9819 - val_loss: 0.4931 - val_acc: 0.8750
```

The accuracy in different epochs is shown ssin the following.

The training and validation accuracy graph and training and validation loss graph for MobileNetV2 are shown in Figure 19.10.

We can see training accuracy increases with number of epochs. It starts from 40%, and at epoch 50, it reaches almost 98% training accuracy.

Training loss decreases with number of epochs. At the start of epoch training, loss is high, and as the number of epochs increases, training loss decreases. At epoch 50, training loss is 5%.

The simple MobileNetV2 model has an accuracy of 98.01%.

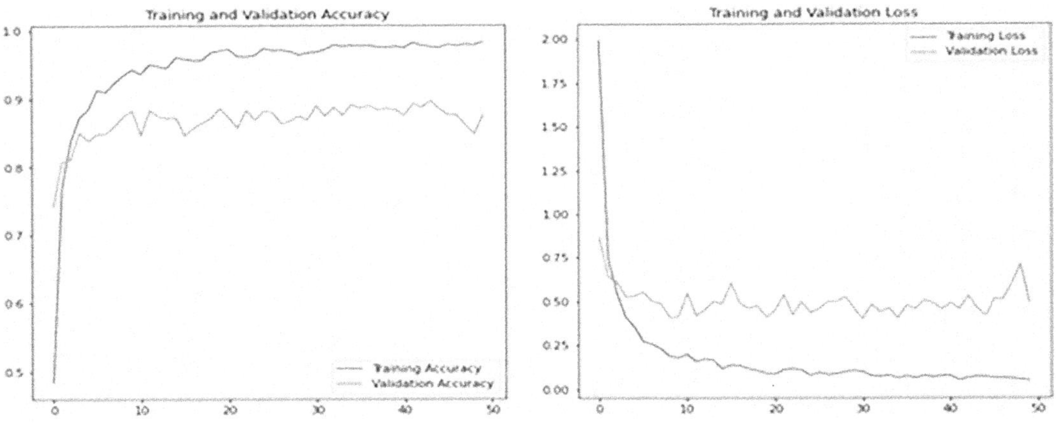

FIGURE 19.10 Training and validation performance graph (MobileNetV2).

19.4.1.5 ResNet50

```
Epoch 45/50
26/26 [=========] - 125 461ms/step - loss: 2.4756 - acc: 0.1895 - val_loss: 2.3877 - val_acc: 0.2438

Epoch 46/50
26/26 (=========] - 125 458ms/step - loss: 2.4472 - acc: 0.1858 - val_loss: 2.4069 - val_acc: 0.2062
Epoch 47/50
26/26 (======= ==] - 125 461ms/step - loss: 2.4388 - acc: 0.2090 - val_loss: 2.4567 - val_acc: 0.2062
Epoch 48/50
26/26 [=========] - 125 468ms/step - loss: 2.4146 - acc: 0.2249 - val_loss: 2.4922 - val_acc: 0.2000
Epoch 49/50
26/26 [======= =] - 125 462ms/step - loss: 2.4392 - acc: 0.2017 - val_loss: 2.4298 - val_acc: 0.2062
Epoch 50/50
26/26 [========= ] - 125 462ms/step - loss: 2.4294 - acc: 0.2042 - val_loss: 2.3677 - val_acc: 0.2062
```

The accuracy in different epochs is shown in the following.

The training and validation accuracy graph and training and validation loss graph for ResNet50 are shown in Figure 19.11.

We can see training accuracy increases with the number of epochs. It starts from 40%, and at epoch 50, it reaches 20% training accuracy.

The basic ResNet50 model achieves a 20.42% accuracy rate.

19.4.2 Accuracy Comparison

The following shows an accuracy comparison of different deep learning algorithms (see Figure 19.12).

We used deep learning and transfer learning approaches for classification, starting with a CNN approach, which gives 95% accuracy.

Then we moved to transfer learning, with various models like VGG 16, MobileNetV2, InceptionV3, and ResNet50.

VGG16 gives 97% accuracy, which is more than CNN. InceptionV3 gives almost 96% accuracy, which is slightly better than CNN but not as good as VGG16 or MobileNetV2. MobileNetV2 gives the highest accuracy at 98%, but the result of ResNet50 is very poor (only 20%).

We can see that MobileNetV2 gives the highest accuracy: 98.19%.

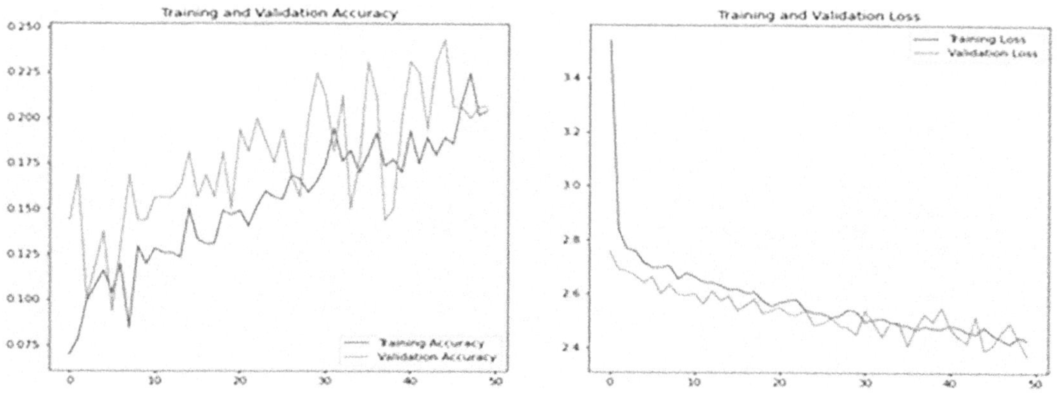

FIGURE 19.11 Training and validation performance graph (ResNet50).

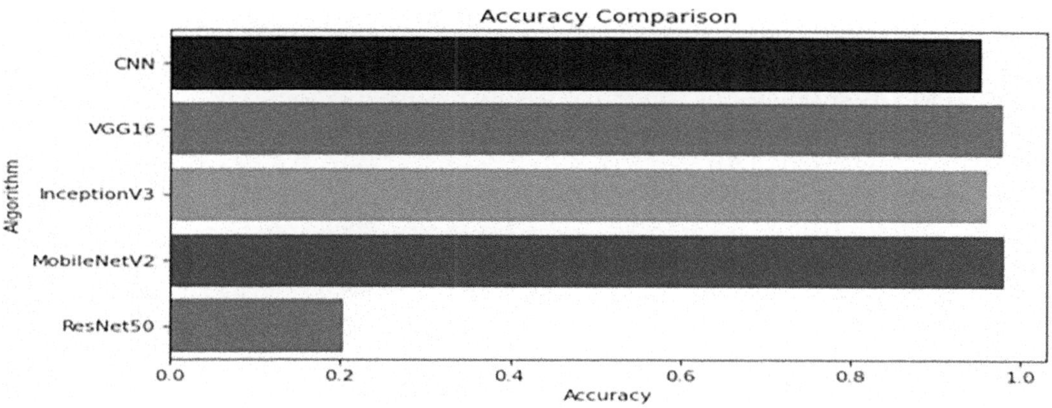

FIGURE 19.12 Accuracy comparison graph.

TABLE 19.1
Accuracy Comparison of Various Algorithms in Different Epochs

Epochs	CNN	VGG16	InceptionV3	MobileNetV2	ResNet50
45	0.9466	0.9664	0.9573	0.9741	0.1895
46	0.9412	0.9748	0.9560	0.9785	0.1858
47	0.9331	0.9778	0.9506	0.9765	0.2090
48	0.9489	0.9684	0.9533	0.9785	0.2249
49	0.9546	0.9708	0.9516	0.9778	0.2017
50	0.9546	0.9782	0.9594	0.9819	0.2042

19.4.3 Making Predictions

Now we will predict plant disease with the help of the deep learning model. We create a disease prediction function for classification that takes an image from a test folder, converts it into a numpy array, and predicts disease for the particular image.

```
]: predict_disease('C:/Users/Acer/Downloads/dataset1/dataset/train/Tomato___healthy/000bf685-b305-408b-

Tomato___healthy
```

FIGURE 19.13 Prediction of healthy tomato.

```
: predict_disease('C:/Users/Acer/Downloads/dataset1/dataset/test/PotatoEarlyBlight1.JPG')

WARNING:tensorflow:From <ipython-input-22-9ed41462558f>:6: Sequential.predict_classes (from tensorflow.python.keras.engine.sequ
ential) is deprecated and will be removed after 2021-01-01.
Instructions for updating:
Please use instead:* `np.argmax(model.predict(x), axis=-1)`,   if your model does multi-class classification   (e.g. if it uses
a `softmax` last-layer activation).* `(model.predict(x) > 0.5).astype("int32")`,   if your model does binary classification
(e.g. if it uses a `sigmoid` last-layer activation).
Potato___Early_blight
```

FIGURE 19.14 Prediction of potato early blight.

19.4.3.1 Prediction of Healthy Tomato

We take an image, Tomato_Healthy, from the test folder, and our model was able to predict the correct class for this image (Figure 19.13).

19.4.3.2 Prediction of Potato Early Blight

We take an image of potato early blight from the test folder, and our model was able to predict the correct class (Potato_Early _Blight) for this image (Figure 19.14).

19.4.3.3 Prediction of Healthy Pepper

We take an image of a healthy pepper from the test folder, and our model was able to predict the correct class (Pepper Healthy) for this image (see Figure 19.15).

19.4.3.4 Prediction of Tomato Early Blight

We take an image of tomato early blight from the test folder, and our model was able to predict the correct class (Tomato Early Blight) for this image (see Figure 19.16).

```
[53] predict_disease('/content/drive/dataset/train/Pepper__healthy.JPG')
```

FIGURE 19.15 Prediction of healthy pepper.

```
[40] predict_disease('/content/drive/dataset/test/TomatoEarlyBlight4.JPG')
```

FIGURE 19.16 Prediction of tomato early blight.

```
[47] predict_disease('/content/drive/test/TomatoYellowCurlVirus6.JPG')
```

FIGURE 19.17 Prediction of tomato yellow leaf curl virus.

19.4.3.5 Prediction of Tomato Yellow Leaf Curl Virus

We take an image of tomato yellow leaf curl virus from the test folder, and our model was able to predict the correct class (Tomato Yellow Leaf Curl Virus) for this image (see Figure 19.17).

19.4.3.6 Prediction of Cherry Powdery Mildew

We take an image of cherry powdery mildew from the test folder, and our model was able to predict the correct class for this image (see Figure 19.18).

FIGURE 19.18 Prediction of cherry powdery mildew.

TABLE 19.2
Comparative Study between Existing Work and Our Work

Research Study (Authors)	Methods	Results
Li et al. [13] (2020)	Shallow CNN with SVM and RF	94% accuracy
Geetharamani et al. [14] (2019)	Nine-layer deep CNN	96.46% accuracy
Oyewola et al. [15] (2021)	Deep residual neural network (DRNN)	96.75% accuracy
Ramcharan et al. [16] (2017)	Inception V3 based on GoogleNet	93% accuracy
Mohanty et al. [17] (2016)	AlexNet and GoogleNet	99.27% in AlexNet 99.34% in GoogleNet
Ferentinos et al. [18] (2018)	AlexNetOWTBn and VGG	99.49% in AlexNetOWTBn 99.53% in VGG
Our approach (deep and transfer learning)	CNN, VGG16, MobileNetV2	95.46% in CNN 97.82% in VGG16 98.19% in MobileNetV2

In our work, the MobileNetV2 model is able to give 98.19% accuracy, whereas CNN gives 95.46% accuracy, and VGG16 gives 97.82% accuracy.

19.4.3.7 Comparative Study between Existing Work and Our Work

A comparative study between existing work and our work follows. In existing work, the CNN model is mostly used for classification, but in some cases, transfer learning approaches are used for better accuracy. InceptionV3, GoogleNet, VGG16, and AlexNet are mostly used for transfer learning. These models are pretrained with huge databases like ImageNet which contain more than a million images and more than a thousand classes.

19.5 CONCLUSION AND FUTURE WORK

As significant loss can occur if there is infection in crops, it can cause great damage to agriculture. This affects farmers, and their income may be reduced. Several transfer learning and deep learning–based models help to accurately classify diseases and reduce the harm to crops, and chances of loss may be reduced. As now there is a higher dependency on the agricultural sector, the economy is dependent on production of crop yields. Faster detection can reduce damage, and a deep learning–based framework has become essential to recognize leaf diseases to reduce losses and improve the economy. The work depends highly on deep learning models like CNN, VGG16,

ResNet, MobileNet, and InceptionV3. Except for ResNet, these models provide more than 95% accuracy of classification of diseases.

We will run the system on a cloud server using Ngrok so that the application will be accessible to everybody. Anybody can access the application anywhere.

REFERENCES

1. S. Babu and Prof. R. Shirsath, 'A Software Model for Precision Agriculture for Small and Marginal Farmers,' *IEEE Global Humanitarian Technology Conference: South Asia Satellite (GHTC-SAS)*, 2013, doi: 10.1109/GHTC-SAS.2013.6629944.
2. S. Pudumalar, Rakeshkumar, M. P. Singh, Prabhat Kumar, and J. P. Singh, 'Crop Selection Methode to Maximize Crop Yield Rate Using Machine Learning Technique,' *International Conference on Smart Technologies and Management for Computing, Communication, Controls, Energy and Materials (ICSTM-2015)*, 2015, doi: 10.1109/ICSTM.2015.7225403.
3. S. D. Khirade and A. B. Patil, 'Plant Disease Detection Using Image Processing,' *2015 International Conference on Computing Communication Control and Automation*, 2015, pp. 768–771, doi: 10.1109/ICCUBEA.2015.153.
4. S. C. Madiwalar and M. V. Wyawahare, 'Plant Disease Identification: A Comparative Study,' *2017 International Conference on Data Management, Analytics and Innovation (ICDMAI)*, 2017, pp. 13–18, doi: 10.1109/ICDMAI.2017.8073478.
5. P. Moghadam, D. Ward, E. Goan, S. Jayawardena, P. Sikka, and E. Hernandez, 'Plant Disease Detection Using Hyperspectral Imaging,' *2017* International Conference on Digital Image Computing: Techniques and Applications (DICTA), 2017, pp. 1–8, doi: 10.1109/DICTA.2017.8227476.
6. G. Shrestha, Deepsikha, M. Das, and N. Dey, 'Plant Disease Detection Using CNN,' *2020 IEEE Applied Signal Processing Conference (ASPCON)*, 2020, pp. 109–113, doi: 10.1109/ASPCON49795.2020.9276722.
7. Savita N. Ghaiwat and Parul Arora, 'Detection and Classification of Plant Leaf Diseases Using Image processing Techniques,' *International Journal of Recent Advances in Engineering & Technology (IJRAET)*, Vol. 2, No. 3, 2014, ISSN (Online): 2347 – 28122014.
8. Plant Disease Dataset: www.kaggle.com/datasets/new-plant-diseases-dataset
9. Neural Network Applications: www.smartsheet.com/neural-network-applications
10. VGG16 Overview: https://medium.com/mlearning-ai/an-overview-of-vgg16-and-nin-models-96e4bf398484
11. Inceptionv3 Architecture: www.apriorit.com/dev-blog/606-inception-v3-for-video-classification
12. Obilenet: www.researchgate.net/figure/The-proposed-MobileNetV2-network-architecture_figure 1_350152088
13. Y. Li, J. Nie, and X. Chao, 'Do We Really Need Deep CNN for Plant Diseases Identification?' *Comput. Electron. Agric.*, 2020, 178, 105803.
14. G. Geetharamani and A. Pandian, 'Identification of Plant Leaf Diseases Using a Nine-layer Deep Convolutional Neural Network,' *Comput. Electr. Eng.* 2019, 76, 323–338.
15. D. O. Oyewola, E. G. Dada, S. Misra, and R. Damaševiˇcius, 'Detecting Cassava Mosaic Disease Using a Deep Residual Convolutional Neural Network with Distinct Block Processing,' *Peer J Comput Sci.*, 2021, 7, e352. [CrossRef] [PubMed]
16. A. Ramcharan, K. Baranowski, P. McCloskey, B. Ahmed, J. Legg, and D. P. Hughes, 'Deep Learning for Image-Based Cassava Disease Detection,' *Front. Plant Sci.*, 2017, 8, 1852.
17. S. P. Mohanty, D. P. Hughes, and M. Salathé, 'Using Deep Learning for Image-based Plant Disease Detection,' *Front. Plant Sci.*, 2016, 7, 1419.
18. K. P. Ferentinos, 'Deep Learning Models for Plant Disease Detection and Diagnosis,' *Comput. Electron. Agric.*, 2018, 145, 311–318.

20 Enabling Blockchain in Digital Twins for Decentralized Supply Chain Traceability and Crop Growth Optimization

Naveen Raj Y, Sudha G, and J. Jayanthi

20.1 INTRODUCTION

The global agricultural landscape stands at a pivotal juncture. The escalating pressures of a proliferating population, shifting dietary preferences, and the ever-looming threat of climate change are converging to create a complex and interconnected web of challenges that demand innovative and sustainable solutions. While traditional farming practices have long served as the backbone of food production, they are increasingly strained under the weight of these modern-day issues.

In this era of unprecedented challenges, the concept of "smart agriculture" is emerging as a beacon of hope, offering a path towards a more efficient, resilient, and sustainable food system. This paradigm shift is characterized by the integration of cutting-edge technologies into every facet of agricultural operations, from planting and harvesting to supply chain management and consumer engagement. Central to this revolution is the synergistic convergence of two transformative technologies: blockchain and digital twins. These technologies, once confined to the realms of finance and manufacturing, are now poised to revolutionize agriculture, promising a future where food production is not only abundant but also traceable, transparent, and environmentally responsible.

20.1.1 The Growing Pains of the Modern Food System

To fully comprehend the transformative potential of blockchain and digital twins, it is imperative to understand the multifaceted challenges plaguing modern agriculture:

- **Food Security and Demand:** The global population is projected to reach nearly 10 billion by 2050. This exponential growth, coupled with rising incomes and changing dietary preferences, is placing unprecedented demand on food production systems. Meeting this escalating demand while ensuring equitable access to nutritious food for all is a formidable task.
- **Resource Scarcity:** Agriculture is a resource-intensive industry, consuming vast quantities of water, land, and energy. The intensifying effects of climate change, such as droughts, floods, and extreme temperatures, are further exacerbating the scarcity of these vital resources, threatening crop yields and disrupting established farming practices.
- **Environmental Degradation:** Conventional agricultural methods have long been associated with negative environmental impacts, including deforestation, soil erosion, water pollution, and greenhouse gas emissions. The urgency to transition towards sustainable practices that minimize ecological footprints and promote regenerative agriculture is undeniable.

DOI: 10.1201/9781003507390-20

- **Supply Chain Complexities:** The agricultural supply chain is often characterized by fragmentation, opacity, and a lack of traceability. This makes it difficult to ensure the authenticity, quality, and safety of food products, leaving consumers vulnerable to fraud and foodborne illnesses.
- **Economic Sustainability:** Farmers, particularly small-scale producers in developing countries, face numerous economic hurdles, including volatile market prices, limited access to finance, and vulnerability to climate risks. These challenges often perpetuate poverty and inequality, hindering rural development.

20.1.2 Blockchain: The Unbreakable Chain of Trust

Blockchain technology, renowned for its role in cryptocurrencies like Bitcoin, offers a unique set of features that are remarkably well-suited to address the challenges faced by the agricultural sector:

- **Decentralization:** The blockchain functions on a decentralized network of computers, removing the necessity for a central authority or intermediary. This amplifies transparency and diminishes the potential for data tampering or deceit.
- **Immutability:** Once information is documented on the blockchain, it becomes incredibly challenging to modify or erase. This establishes an unalterable record of transactions, upholding the honesty and legitimacy of data all along the supply chain.
- **Transparency:** Every transaction on the blockchain is observable to all participants within the network, nurturing transparency and answerability. This empowers consumers to validate the source, quality, and safety of their food, cultivating trust and assurance in the food industry.
- **Smart Contracts:** Smart contracts based on the blockchain are self-operating agreements where the contract's conditions are directly inscribed into code. These contracts can automate various processes in the agricultural supply chain, such as payments to farmers, quality control checks, and supply chain logistics, reducing costs, increasing efficiency, and ensuring fair and transparent transactions.

20.1.3 Digital Twins: The Virtual Greenhouse for Crop Optimization

Digital twins, by creating virtual replicas of crops, fields, or entire farms, offer a powerful tool for precision agriculture and data-driven decision-making. By integrating real-time data from a multitude of sources—including Internet of Things (IoT) sensors, drones, satellites, and weather stations—digital twins can model and simulate crop growth under various conditions, providing farmers with valuable insights and predictive capabilities.

- **Precision Agriculture:** Digital twins empower farmers to practice precision agriculture by tailoring irrigation, fertilization, and pest control strategies to the specific needs of each crop and field. This optimizes resource utilization, minimizes environmental impact, and maximizes yields.
- **Yield Prediction:** By simulating crop growth under different scenarios, digital twins can predict yields with remarkable accuracy, allowing farmers to plan their harvesting, storage, and marketing activities more effectively.
- **Disease and Pest Management:** Digital twins can analyze data from sensors and imagery to detect early signs of crop diseases or stress, enabling timely interventions that can save crops and reduce losses.
- **Scenario Planning:** Farmers can use digital twins to simulate the impact of various decisions, such as changing planting dates, irrigation schedules, or crop varieties, before implementing them in the real world. This allows for informed decision-making and risk mitigation.

20.1.4 Converging Technologies for a Sustainable Future

The integration of blockchain and digital twins in agriculture represents a synergistic union with the potential to transform the entire agricultural value chain. The secure and transparent data exchange facilitated by blockchain ensures the integrity and authenticity of the information used to build and update digital twin models.

The convergence of blockchain technology and digital twin applications is poised to revolutionize the agricultural sector, promising unprecedented levels of efficiency, transparency, and data-driven decision-making. Blockchain, with its decentralized and immutable ledger, can enhance traceability, secure transactions, and foster trust within complex agricultural supply chains. Meanwhile, digital twins, virtual replicas of physical assets or processes, can enable real-time monitoring, predictive analytics, and optimized resource management in farming operations.

However, as these transformative technologies permeate the agricultural landscape, it is imperative to critically examine their ethical and social implications. While blockchain and digital twins offer immense potential for positive impact, they also raise concerns related to data privacy, ownership, algorithmic bias, power imbalances, and the equitable distribution of benefits. Furthermore, the integration of these technologies into traditional farming practices may disrupt existing social structures, alter labor dynamics, and exacerbate existing inequalities within the agricultural sector.

20.2 LITERATURE SURVEY

20.2.1 Improving Innovation, Reducing Crime, and Empowering Consumers

Kamble et al. (2022) conducted a comprehensive study on blockchain's impact on innovation, crime reduction, and consumer empowerment in agriculture. Their research highlights several key findings:

- **Traceability and Transparency:** Blockchain enables the creation of a transparent and traceable digital record of every transaction within the supply chain, from farm to fork. This enhanced traceability can improve food safety by identifying contamination sources quickly and accurately.
- **Smart Contracts:** Blockchain technology enables the automation of processes in the supply chain through smart contracts, eliminating the need for intermediaries and boosting operational efficiency. For instance, these contracts have the ability to automatically initiate payments to farmers once their produce meets specified quality benchmarks.
- **Data Integrity:** The inherent immutability of blockchain technology guarantees the accuracy and tamper-resistant nature of data, thereby aiding in the mitigation of fraud and enhancing consumer trust within the food industry.
- **Empowering Consumers:** Blockchain has the potential to give consumers access to comprehensive details regarding the source, manufacturing processes, and quality of their food. Having this knowledge enables consumers to make well-informed decisions when selecting and consuming food products.

20.2.2 Building Trust and Transparency in Food Systems

Peters and Panayi (2016) focused their research on the importance of trust and transparency in the food system. They argue that blockchain can play a crucial role in building consumer trust by providing a verifiable and transparent record of food production and distribution. Some key points from their research include:

- **Food Safety:** Blockchain can enhance food safety by tracking the movement of food products throughout the supply chain. This can help identify potential contamination sources and prevent outbreaks of foodborne illnesses.

- **Sustainability:** Blockchain can promote sustainable agriculture practices by incentivizing farmers to adopt environmentally friendly methods. For example, farmers could be rewarded for using less water or pesticides.
- **Fair Trade:** Blockchain can ensure fair compensation for farmers by providing a transparent record of transactions. This can prevent exploitation and ensure that farmers receive a fair price for their products.
- **Consumer Empowerment:** Blockchain can empower consumers by giving them access to detailed information about their food. This can help consumers make informed choices about the food they purchase and consume.

20.2.3 Smart Farming and Precision Agriculture

Sreedevi and Kumar (2023) conducted a comprehensive review of digital twin applications in smart farming. Their research highlights the importance of digital twins in various stages of agriculture, including:

- **Crop Monitoring:** Digital twins can monitor crop growth and development in real time, providing farmers with valuable insights into crop health and potential issues.
- **Irrigation Management:** Simulating soil moisture levels and weather conditions, digital twins help with managing irrigation effectively to optimize schedules and reduce water waste while increasing crop yields.
- **Production Forecasting:** Digital twins analyze historical data and real-time conditions to forecast production accurately, enabling farmers to plan for harvesting and marketing activities efficiently.
- **Disease Detection:** By monitoring plant health and identifying early signs of disease, digital twins can help farmers prevent outbreaks and minimize crop losses.

20.2.4 Addressing Agricultural Challenges with Data Models and Interoperable Platforms

Wolfert et al. (2022) focused their research on the challenges facing agriculture and the potential of digital twins to address these challenges. They emphasize the importance of developing accurate data models and interoperable platforms to maximize the benefits of digital twin technology. Some key takeaways from their research include

- **Data-Driven Decision-Making:** Digital twins can provide farmers with valuable data-driven insights into their operations, enabling them to make informed decisions about crop management, resource allocation, and marketing strategies.
- **Resource Optimization:** By simulating real-world conditions, digital twins can help farmers optimize the use of water, fertilizers, and other resources, reducing waste and minimizing environmental impact.
- **Risk Mitigation:** Digital twins can help farmers anticipate and mitigate risks such as pests, diseases, and extreme weather events.
- **Collaboration:** Interoperable digital twin platforms can facilitate collaboration between farmers, researchers, and other stakeholders, fostering innovation and knowledge sharing.

20.2.5 Integration of Blockchain and Digital Twins

Mistry et al. (2023) explored the integration of blockchain and digital twins in the context of agricultural decision-making. Their research focused on the transfer of data between digital twins using a blockchain framework. This approach has several key advantages:

- **Secure and Reliable Data Exchange:** Blockchain ensures that the data exchanged between digital twins is secure, tamper-proof, and verifiable. This is crucial for maintaining the integrity of the digital twin models and the decisions based on them.
- **Decentralized Data Management:** By using blockchain, the study avoids the need for a central authority to control the data. This enhances transparency and trust among stakeholders.
- **Enhanced Decision-Making:** The combination of real-time data from digital twins and the secure data exchange facilitated by blockchain provides farmers with a more comprehensive and reliable basis for making critical decisions.
- **Improved Productivity:** The study demonstrated that this integrated approach could lead to significant improvements in agricultural productivity, as farmers can make more informed decisions about crop management and resource allocation.

20.2.6 Blockchain Enhancements for Agricultural Supply Chains

20.2.6.1 Traceability and Provenance

- **Wang et al. (2023): A Real-Time Tracking System with QR Codes and NFC Tags:** This research delves into a sophisticated blockchain-based system that utilizes QR codes and NFC tags. The system is designed to offer real-time tracking of agricultural products, recording every transaction and movement from the farm to the consumer. This not only ensures transparency for consumers but also acts as a powerful deterrent against counterfeit products, which are a significant concern in the agricultural sector. The use of readily available technologies like QR codes and NFC tags makes this system particularly accessible for implementation.
- **Wu et al. (2022): Monitoring Perishable Goods with IoT Sensors and Blockchain:** This study explores the synergy between blockchain technology and IoT sensors in monitoring the transportation and storage conditions of perishable goods. By recording temperature, humidity, and other relevant data on the blockchain, the system ensures that these goods are maintained under optimal conditions throughout the supply chain. This is crucial for ensuring product quality and safety, reducing spoilage, and minimizing food waste.

20.2.6.2 Smart Contracts and Fair Trade

- **Zhang et al. (2022): Automated Payments with Smart Contracts:** Smart contracts, a key feature of blockchain technology, are leveraged in this research to automate payments to farmers. The smart contracts are programmed to trigger payments based on pre-defined quality and quantity metrics of the produce delivered. This system removes the need for intermediaries, streamlines transactions, and ensures that farmers receive fair compensation for their products, promoting trust and transparency in the agricultural supply chain.
- **Xu (2024): Incentivizing Sustainable Practices with Carbon Credits:** This study focuses on a novel blockchain-based system designed for carbon credits in agriculture. Farmers who adopt sustainable farming practices, such as reducing emissions or sequestering carbon, are rewarded with carbon credits. These credits can be traded on the blockchain, creating a financial incentive for farmers to adopt environmentally friendly practices. This approach aligns economic interests with environmental goals, making sustainable agriculture more attractive and feasible.

20.2.7 Digital Twin Advancements for Crop Optimization

20.2.7.1 Precision Agriculture and Yield Prediction

- **Benrajesh et al. (2023): Precise Control of Greenhouse Environments:** This research presents a comprehensive digital twin of a greenhouse environment. The twin integrates data from various sensors (temperature, humidity, light), weather forecasts, and crop growth models. By simulating different scenarios and analyzing real-time data, farmers can precisely control the greenhouse environment, optimizing conditions for plant growth and maximizing yield. The accuracy of the digital twin in predicting yields is a significant step towards data-driven decision-making in greenhouse farming.

- **Wang et al. (2023): Farm Management with Drone Imagery and Machine Learning:** This study takes the concept of digital twins to the open field. Drones capture high-resolution imagery of the farm, and machine learning algorithms process this data to create a digital representation of the entire farm. The digital twin predicts crop yields with remarkable accuracy and identifies specific areas within the farm that require attention due to factors like nutrient deficiencies or pest infestations. This allows for targeted interventions, improving resource efficiency and reducing overall costs.

20.2.7.2 Disease Detection and Pest Management

- **Liu et al. (2021): Early Warning System for Crop Diseases:** This research utilizes a digital twin to simulate the spread of crop diseases. By incorporating real-time data from sensors monitoring plant health and weather stations tracking environmental conditions, the digital twin can predict the likelihood of disease outbreaks. Early warnings allow farmers to implement preventive measures, minimizing crop losses and reducing the need for chemical interventions.
- **Kumar et al. (2024): Precision Pest and Disease Identification with Computer Vision:** This study combines the power of digital twins with computer vision technology. Images of crops are captured and analyzed by the digital twin, which is trained to identify specific pests and diseases. This enables early detection and targeted application of pesticides, reducing the overall amount of chemicals used and minimizing their environmental impact.

20.2.8 Synergistic Integration of Blockchain and Digital Twins

20.2.8.1 Data Integrity and Trust

- **Chen et al. (2023). Immutable Crop Records with Blockchain:** This research integrates blockchain with digital twins to create an immutable and transparent record of crop data. From planting to harvest, every stage of the crop's life cycle is recorded on the blockchain, ensuring the integrity of the data. This is particularly important for traceability and quality assurance [16], as it provides a verifiable history of the crop's journey.
- **Singh et al. (2024). Secure Data Sharing among Farms:** This study proposes a framework where blockchain technology is used to facilitate secure data sharing between digital twins of different farms. By enabling farmers to share anonymized data about their crops, soil conditions, and farming practices, the framework fosters collaborative decision-making. Farmers can learn from each other's experiences, access a broader pool of data for their digital twins, and make more informed choices for their own operations.

20.2.8.2 Resource Optimization and Sustainability

- **Zhao et al. (2022): Blockchain-Based Water Management:** This research introduces a blockchain-based platform for managing water resources in agriculture. The system employs digital replicas to generate virtual representations of agricultural fields, considering aspects such as soil moisture levels, weather conditions, and crop water needs. By examining this information, the digital replica can optimize watering schedules, ensuring that crops receive the appropriate amount of water at the precise time. This method encourages water preservation, decreases expenses, and enhances crop productivity.

- **Nguyen et al. (2024): Tracking Fertilizer Usage with Blockchain:** Fertilizer overuse is a major environmental concern in agriculture. This study addresses this issue by developing a blockchain-based system to track fertilizer usage. Every application of fertilizer is recorded on the blockchain, ensuring transparency and accountability. This data can be analyzed to optimize fertilizer application, minimizing waste and reducing the environmental impact of agriculture.

20.2.9 Blockchain's Expanding Role in Agricultural Supply Chains

20.2.9.1 Traceability and Provenance

- **Liang et al. (2024): Livestock Tracking with RFID and Blockchain:** This research pioneers the use of radio-frequency identification (RFID) technology in conjunction with blockchain for comprehensive livestock tracking. Each animal is tagged with an RFID chip at birth, and its movement, health records, and treatment history are recorded on the blockchain. This creates an immutable and transparent record that can be accessed by stakeholders throughout the supply chain, including farmers, veterinarians, slaughterhouses, and consumers. This system not only ensures the authenticity of meat products but also provides valuable insights into the animal's well-being, fostering ethical and sustainable livestock practices.
- **Tian et al. (2024): Blockchain-Powered Platform for Organic Produce:** This study addresses the challenge of verifying the authenticity of organic produce. The proposed blockchain platform allows farmers to register their organic products, recording details such as cultivation practices, certifications, and inspections. Consumers can scan a QR code on the product to access this information, confirming the product's organic status and building trust in organic farming.
- **Saberi et al. (2022): Combating Food Fraud with Blockchain-Based Traceability:** Food fraud, which involves deliberately misrepresenting food products for financial gain, is a worldwide concern. This research investigates the potential of blockchain technology in combating food fraud by establishing a secure and transparent supply chain. Through the blockchain, details such as the origin, ingredients, and processing history of food items are documented, making it challenging for deceptive individuals to alter information. This not only safeguards consumers but also protects genuine producers from fraudulent activities.

20.2.9.2 Smart Contracts and Fair Trade

- **Cai et al. (2023): Decentralized Marketplace for Farmers:** This research introduces a decentralized marketplace built on blockchain technology. Farmers can list their produce directly on the marketplace, connecting with consumers without the need for intermediaries. Smart contracts facilitate secure and transparent transactions, ensuring fair prices for farmers and timely payments. This approach not only empowers farmers but also gives consumers access to fresher, locally sourced produce.
- **Yang et al. (2024): Blockchain-Based Crop Insurance:** Weather-related risks are a constant threat to farmers. This study presents a blockchain-based insurance platform designed to provide farmers with protection against crop losses due to adverse weather events. Smart contracts are triggered automatically based on verifiable weather data obtained from reliable sources, ensuring swift and transparent payouts to farmers in need. This innovation can help farmers manage risks more effectively and ensure their financial security.

20.2.10 Digital Twin Advancements for Crop Optimization

20.2.10.1 Precision Agriculture and Yield Prediction

- **Sharma and Kumar (2023): Comprehensive Crop Growth Simulation:** This research develops a sophisticated digital twin that integrates various data sources to create a holistic model of crop growth. Data from soil sensors (measuring moisture, nutrients, and pH), weather stations (providing temperature, humidity, and rainfall data), and satellite imagery (monitoring vegetation health) are fed into the digital twin. The twin analyzes this data in real time and generates precise recommendations for irrigation, fertilization, and pest control, enabling farmers to optimize resource usage and maximize crop yields.

- **Kim et al. (2022): Crop Variety Selection with Digital Twins:** Choosing the right crop variety for a specific location and climate is crucial for agricultural success. This study demonstrates how digital twins can be used to simulate the growth of different crop varieties under varying environmental conditions. By analyzing the results, farmers can identify the varieties best suited to their specific needs, improving overall agricultural productivity and resource efficiency.

20.2.10.2 Disease Detection and Pest Management

- **Hu et al. (2021): Early Stress Detection with Hyperspectral Imaging:** This research showcases the potential of hyperspectral imaging combined with digital twins for early detection of crop stress. Hyperspectral cameras capture detailed information about plant health, including subtle changes that may indicate disease or nutrient deficiencies. The digital twin processes this data, providing early warnings to farmers, who can then take preventive measures before significant damage occurs.
- **Gupta et al. (2024): Pest Behavior Simulation and Control:** This study focuses on the development of a digital twin that simulates the behavior and spread of pests in agricultural fields. By understanding pest dynamics, farmers can predict outbreaks and implement targeted interventions at the optimal time. The digital twin also aids in determining the appropriate pesticide dosage, minimizing the environmental impact of pest control measures.

20.2.11 Synergistic Integration of Blockchain and Digital Twins

- **Data Integrity and Trust**
    - **Zhou et al. (2023):** This research proposes a hybrid blockchain-digital twin system for supply chain management in the agriculture industry. This system provides transparency and traceability of agricultural products throughout the supply chain.
    - **Wang et al. (2021):** This research focused on building a decentralized agricultural supply chain management system based on a consortium blockchain, which includes digital twins to monitor and predict the condition of agricultural products during transportation and storage, ensuring their quality and safety.
- **Resource Optimization and Sustainability**
    - **Lin et al. (2022):** This study explores the use of digital twins and blockchain for intelligent irrigation management. The digital twin simulates the water cycle in the field, and blockchain ensures the security and reliability of the irrigation data, promoting water conservation in agriculture.
    - **Zhang et al. (2024):** This research proposes a blockchain-based platform for tracking and managing agricultural waste, utilizing digital twins to model and optimize waste recycling and treatment processes to promote circular economy in agriculture.

20.2.12 Challenges and Future Directions

While the integration of blockchain and digital twins in agriculture holds great promise, several challenges remain to be addressed:

- **Scalability:** Blockchain technology can be resource-intensive, and its scalability needs to be improved to handle the vast amounts of data generated by agricultural digital twins.

- **Interoperability:** Different blockchain platforms and digital twin technologies may not be compatible, hindering the seamless exchange of data.
- **Adoption:** Farmers need to be educated and trained on how to use these technologies effectively.
- Research is increasingly focusing on addressing the scalability limitations of blockchain in handling large agricultural datasets. Layer 2 solutions and off-chain data storage are being explored.
- Economic Impact: A growing body of research is quantifying the economic benefits of integrating blockchain and digital twins in agriculture. Studies have shown that these technologies can lead to significant cost savings in terms of reduced input costs, optimized resource utilization, and minimized crop losses. They can also open up new market opportunities for farmers by providing transparency and quality assurance to consumers, potentially leading to premium prices for their products.
- Social Impact: The adoption of blockchain and digital twins can have a profound social impact, particularly for small-scale farmers in developing countries. By providing access to information, financial services, and markets, these technologies can empower farmers, improve their livelihoods, and contribute to poverty reduction. Furthermore, the increased transparency and traceability can help consumers make more informed choices about the food they purchase, promoting a healthier and more sustainable food system.
- Interoperability between different blockchain platforms and digital twin technologies remains a challenge. Standardization efforts are underway to facilitate seamless data exchange.
- Developing user-friendly interfaces and providing training to farmers is crucial for widespread adoption.
- More research is needed to explore the ethical and regulatory implications of blockchain and digital twins in agriculture, particularly regarding data privacy and ownership.

20.2.13 Blockchain in Agriculture: Promises and Perils

Blockchain's decentralized and immutable nature offers numerous benefits in agriculture, including improved traceability, transparency, and trust in supply chains. Studies like Peters et al. (2016) and Kamilaris et al. (2019) highlight blockchain's potential to combat food fraud, ensure fair trade, and empower smallholder farmers.

However, concerns about data privacy, ownership, and accessibility arise. For instance, who owns the data generated on a blockchain-enabled platform? What happens if a farmer's data is compromised or misused? These questions are addressed in works by De Vos et al. (2020) and Caro et al. (2018), who emphasize the need for robust governance frameworks and data protection mechanisms.

Another ethical consideration is the potential for exclusion. Small-scale farmers with limited access to technology may be left behind in a blockchain-driven agricultural ecosystem. Research by Eastwood et al. (2019) highlights this digital divide and calls for inclusive solutions.

20.2.14 Digital Twins in Agriculture: Opportunities and Challenges

Digital twins, virtual replicas of physical assets or systems, have immense potential for optimizing agricultural practices. They can simulate scenarios, predict outcomes, and enable data-driven decision-making. Studies by Tao et al. (2019) and Barricelli et al. (2019) demonstrate the use of digital twins for precision agriculture, pest management, and resource optimization.

However, digital twins raise concerns about algorithmic bias and the potential for unintended consequences. If the data used to train these models is biased or incomplete, the resulting decisions may be discriminatory or harmful. Furthermore, over-reliance on digital twins could lead to a loss of traditional ecological knowledge and farming practices.

The issue of data ownership and control is also relevant in the context of digital twins. Who owns the data generated by these virtual models, and how is it used? These questions are explored by Slob et al. (2022), who advocate for transparent and equitable data governance mechanisms.

20.2.15 Ethical Frameworks and Governance

To address these ethical and social considerations, several studies propose the development of ethical frameworks and governance mechanisms for blockchain and digital twin applications in agriculture. For instance, Wheeler (2020) outlines a set of ethical principles for blockchain in food supply chains, including transparency, accountability, and fairness. Similarly, Kamilaris et al. (2020) propose a governance framework for digital twins in agriculture, emphasizing data ownership, privacy, and interoperability.

20.2.16 Social Impact and Equity

The social impact of blockchain and digital twins in agriculture is a critical area for further research. These technologies have the potential to both empower and marginalize different stakeholders. While they can offer smallholder farmers access to global markets and financial services, they can also exacerbate existing power imbalances and inequalities.

Research by Eastwood et al. (2021) highlights the need to consider the social context of technology adoption and to ensure that the benefits are equitably distributed. This includes engaging with diverse stakeholders, addressing cultural sensitivities, and providing training and support for technology adoption.

20.2.17 Research Gaps and Opportunities

This literature review highlights a number of research gaps and opportunities for further exploration. More empirical research is needed to verify the effectiveness of blockchain and digital twin integration in real-world agricultural settings. Furthermore, research should focus on developing user-friendly and scalable solutions to meet the needs of different stakeholders in the agricultural sector. Addressing data privacy, security, and regulatory compliance issues are also important for the successful implementation of these technologies.

20.3 LIMITATIONS

Even though there are many advantages of blockchain and digital twin in agriculture there are limitations which stops their usage. Limitations such as data fragmentation, lack of standardization and less interoperability between different systems are found in the existing system. The cost of implementation of blockchain and digital twin is very high for farmers and developing countries. The limitations of blockchain and digital twin is categorized and described as follows:

20.3.1 Technical Challenges

- **Scalability:** Blockchain's decentralized nature, while ensuring security and immutability, can lead to scalability issues when dealing with the massive volumes of data generated by agricultural sensors and devices. As the number of transactions increases, the blockchain network can become congested, resulting in slower processing times and higher costs. Researchers are exploring solutions like Layer 2 scaling, sharding, and off-chain data storage to address this limitation.

- **Interoperability:** The agricultural sector utilizes a diverse range of blockchain platforms and digital twin technologies. The lack of standardization and interoperability between these different systems poses a significant challenge for seamless data exchange and integration. Efforts are underway to develop common standards and protocols that would enable different platforms to communicate and share information effectively.
- **Data Privacy and Security:** Even though blockchain provides improved security through its decentralized nature, it is still susceptible to vulnerabilities. Smart contracts, which are self-executing algorithms on the blockchain, may contain errors or weaknesses that could be taken advantage of by malicious parties. It is crucial to prioritize safeguarding the privacy of sensitive agricultural data stored on the blockchain, as any breaches could have serious repercussions for farmers and other individuals involved.
- **Complexity:** The implementation and upkeep of blockchain and digital twin systems can be intricate, demanding specialized technical knowledge. Small-scale farmers and individuals in developing nations may face challenges in adopting these technologies due to limited resources and expertise, unless they receive adequate assistance and training.
- **Energy Consumption:** Certain blockchain networks, especially those utilizing Proof-of-Work consensus mechanisms, consume significant amounts of energy for mining and transaction validation. This raises concerns regarding the environmental impact of blockchain technology, particularly in sustainable agriculture practices.

20.3.2 Adoption Barriers

- **Cost:** The initial investment required for implementing blockchain and digital twin infrastructure, including hardware, software, sensors, and network connectivity, can be substantial. Small-scale farmers and those in developing countries may find it difficult to afford these technologies, hindering widespread adoption. Innovative financing models, government subsidies, and public-private partnerships could play a crucial role in making these technologies more accessible.
- **Lack of Awareness and Understanding:** Many farmers are not yet aware of the potential benefits of blockchain and digital twins or lack a clear understanding of how these technologies can be applied to their specific agricultural practices. Comprehensive educational programs, workshops, and demonstration projects are needed to raise awareness and build capacity among farmers.
- **Regulatory and Legal Frameworks:** The regulatory landscape for blockchain and digital twins in agriculture is still evolving. Clear guidelines and regulations regarding data ownership, privacy, and liability are necessary to ensure the responsible and ethical use of these technologies. Additionally, international collaboration and coordination are needed to develop harmonized standards and regulations that facilitate cross-border trade and data sharing.
- **Resistance to Change:** The agricultural sector is often characterized by traditional practices and resistance to change. Introducing new technologies like blockchain and digital twins requires a cultural shift and a willingness to embrace innovation. Building trust among farmers and demonstrating the tangible benefits of these technologies through pilot projects and success stories is essential for overcoming this barrier.

20.3.3 Digital Twin Model Limitations

- **Data Quality and Availability:** The accuracy and effectiveness of digital twin models depend heavily on the quality and availability of data. In many agricultural settings, collecting reliable and comprehensive data on soil conditions, weather patterns, crop growth, and other factors can be challenging due to limited infrastructure, inconsistent data collection practices, and cost constraints.

- **Model Complexity:** Agricultural systems are inherently complex, with numerous variables and interactions influencing crop growth and productivity. Developing digital twin models that accurately capture this complexity requires sophisticated modeling techniques, extensive data analysis, and continuous calibration and validation.
- **Generalizability:** Digital twin models developed for one specific crop or geographic region may not be easily transferable to other contexts due to variations in soil types, climate conditions, and farming practices. Developing adaptable and customizable models that can be applied across diverse agricultural settings remains a challenge.
- **Uncertainty and Risk:** While digital twins can provide valuable insights and predictions, they are not infallible. There is always a degree of uncertainty associated with any model, and unforeseen events like extreme weather or pest outbreaks can significantly impact crop outcomes. Farmers need to be aware of the limitations of digital twin predictions and use them in conjunction with their own knowledge and experience.

20.3.4 Nascent Research Landscape

The field of blockchain and digital twin applications in agriculture is relatively new, and research on their ethical and social implications is still in its early stages. There is a limited body of empirical evidence to fully understand the long-term consequences of these technologies in real-world agricultural settings. Much of the existing literature is theoretical or based on small-scale pilot projects, making it difficult to generalize findings.

20.3.5 Exacerbation of Power Imbalances

The adoption of blockchain and digital twins in agriculture could further concentrate power in the hands of large agribusinesses and technology companies. Smallholder farmers may lack the resources and knowledge to fully participate in these technology-driven systems, leading to potential exclusion and marginalization.

20.3.6 Potential for Unintended Consequences

The widespread deployment of blockchain and digital twins in agriculture could have unforeseen social and ecological consequences. For example, over-reliance on digital models could erode traditional farming practices and knowledge systems. Furthermore, the optimization of agricultural processes driven by these technologies may lead to unforeseen environmental impacts, such as increased resource consumption or pesticide use

20.3.7 Cultural Sensitivity and Contextual Relevance

The application of blockchain and digital twins in agriculture must be sensitive to local cultural contexts and values. What works in one region may not be appropriate or effective in another. Ignoring cultural differences can lead to the rejection of these technologies or, worse, the disruption of traditional farming practices and social structures.

20.3.8 Difficulty in Measuring Social Impact

Assessing the social impact of blockchain and digital twin applications in agriculture is a complex task. The benefits and drawbacks may not be immediately apparent or easily quantifiable. Furthermore, the impact may vary across different groups and contexts. Developing robust evaluation frameworks and indicators is crucial for understanding the true social implications of these technologies.

20.4 PROPOSED SOLUTIONS AND MODELS

20.4.1 Addressing Technical Challenges

- **Scalability:**
  - **Layer 2 Solutions:** Implementing Layer 2 scaling solutions, such as the Lightning Network or Plasma, can significantly improve transaction throughput and reduce costs. These solutions enable off-chain transactions that are periodically settled on the main blockchain, relieving congestion and enhancing scalability.
  - **Sharding:** Partitioning the blockchain into smaller, more manageable shards can distribute the computational load and increase transaction processing capacity. Each shard can process transactions independently, improving overall efficiency.
  - **Directed Acyclic Graphs (DAGs):** DAG-based blockchain architectures, like Hashgraph or IoTA, offer an alternative to the traditional blockchain structure. DAGs can achieve high scalability by allowing parallel transaction processing and eliminating the need for block confirmation.
- **Interoperability:**
  - **Cross-Chain Bridges:** Developing cross-chain bridges that enable interoperability between different blockchain platforms is essential. These bridges allow for the transfer of assets and data between blockchains, facilitating seamless communication and collaboration.
  - **Standardized Protocols:** Establishing standardized protocols for data formats, APIs, and smart contract interactions can promote interoperability between different digital twin platforms and blockchain networks.
  - **Consortium Blockchains:** Creating consortium blockchains, where multiple stakeholders in the agricultural sector collaborate and agree on common standards and protocols, can foster interoperability and facilitate data sharing.
- **Data Privacy and Security:**
  - **Zero-Knowledge Proofs (ZKPs):** Zero-knowledge proofs enable verification of data without exposing the actual information, safeguarding the confidentiality of sensitive agricultural data while facilitating transparent transactions on the blockchain.
  - **Homomorphic Encryption:** This method of encryption enables calculations to be carried out on encrypted data without the need to decrypt it. It can boost the security of data stored on the blockchain and facilitate secure data sharing and analysis.
  - **Privacy-Preserving Smart Contracts:** Designing smart contracts with built-in privacy features, such as selective disclosure of information and access controls, can protect sensitive data while maintaining transparency.
- **Complexity:**
  - **User-Friendly Interfaces:** Developing intuitive and user-friendly interfaces for blockchain and digital twin platforms can make these technologies more accessible to farmers with varying levels of technical expertise.
  - **Training and Education:** Providing comprehensive training programs and workshops to farmers and agricultural stakeholders can empower them to effectively utilize these technologies and maximize their benefits.
  - **Simplified Deployment Models:** Creating pre-configured, modular blockchain and digital twin solutions that can be easily deployed and customized for specific agricultural use cases can reduce the complexity of implementation.
- **Energy Consumption:**
  - **Proof-of-Stake (PoS):** Transitioning to Proof-of-Stake consensus mechanisms, which are more energy efficient compared to Proof-of-Work, can significantly lessen the environmental impact of blockchain networks.

- **Renewable Energy Integration:** Integrating blockchain networks with renewable energy sources like solar or wind power can further decrease their carbon footprint.
- **Optimized Consensus Algorithms:** Developing new consensus algorithms that prioritize energy efficiency while maintaining security can contribute to sustainable blockchain solutions.

20.4.2 Overcoming Adoption Barriers

- **Cost:**
  - **Government Subsidies and Incentives:** Governments can play a crucial role in promoting the adoption of blockchain and digital twins by providing subsidies, tax incentives, or grants to farmers and agricultural businesses.
  - **Microfinance and Crowdfunding:** Innovative financing models like microloans and crowdfunding platforms can enable small-scale farmers to access the necessary funds for implementing these technologies.
  - **Shared Infrastructure:** Establishing shared blockchain and digital twin infrastructure, such as regional data centers or cloud-based platforms, can reduce individual costs for farmers and make these technologies more affordable.

  Lack of Awareness and Understanding:
  - **Outreach and Education:** Government agencies, agricultural extension services, and non-profit organizations can organize workshops, training sessions, and awareness campaigns to educate farmers about the benefits and practical applications of blockchain and digital twins.
  - **Demonstration Projects:** Implementing pilot projects that showcase the successful use of these technologies in real-world agricultural settings can serve as powerful examples and motivate other farmers to adopt them.
  - **Knowledge Sharing Platforms:** Creating online forums, communities, and knowledge-sharing platforms can facilitate peer-to-peer learning and provide farmers with access to resources and support.
- **Regulatory and Legal Frameworks:**
  - **International Collaboration:** Governments and international organizations should collaborate to develop harmonized regulations and standards for blockchain and digital twins in agriculture. This can facilitate cross-border trade and data sharing and ensure a level playing field for all stakeholders.
  - **Clear Guidelines and Standards:** Developing clear guidelines and standards for data ownership, privacy, and security can build trust among stakeholders and ensure the ethical and responsible use of these technologies.
  - **Sandbox Environments:** Creating regulatory sandboxes, where new technologies can be tested in a controlled environment before full-scale deployment, can help policymakers and regulators gain a better understanding of the risks and benefits, leading to more informed regulations.
- **Resistance to Change:**
  - **Change Management Strategies:** Implementing effective change management strategies, such as involving farmers in the design and implementation process, addressing concerns, and highlighting the benefits of new technologies, can help overcome resistance to change.
  - **Success Stories and Case Studies:** Sharing success stories and case studies of farmers who have successfully adopted blockchain and digital twins can inspire and motivate others to follow suit.
  - **Phased Implementation:** Introducing these technologies gradually, starting with small-scale pilot projects and scaling up as farmers gain confidence and experience, can be a more effective approach than a sudden, large-scale rollout.

20.4.3 Addressing Digital Twin Model Limitations

- **Data Quality and Availability:**
  - **Investing in Data Collection Infrastructure:** Governments and agricultural organizations should invest in developing robust data collection infrastructure, including sensors, weather stations, and satellite imagery, to ensure high-quality and reliable data for digital twin models.
  - **Data Sharing and Collaboration:** Promoting data sharing and collaboration between farmers, researchers, and data providers can expand the pool of available data and improve the accuracy and comprehensiveness of digital twin models.
  - **Data Standardization:** Developing standardized data formats and protocols can facilitate data integration from various sources, making it easier to build and calibrate digital twin models.
- **Model Complexity:**
  - **Advanced Modeling Techniques:** Investing in research and development of advanced modeling techniques, such as agent-based modeling, machine learning, and artificial intelligence, can enhance the ability of digital twins to capture the complexity of agricultural systems.
  - **Hybrid Models:** Combining different modeling approaches, such as mechanistic models and data-driven models, can leverage the strengths of each approach and improve the overall accuracy and predictive power of digital twins.
  - **Model Validation and Calibration:** Continuous validation and calibration of digital twin models using real-world data are essential to ensure their accuracy and relevance in different agricultural contexts.
- **Generalizability:**
  - **Modular Design:** Developing modular digital twin frameworks that can be easily customized and adapted for different crops, regions, and farming practices can improve their generalizability.
  - **Collaborative Development:** Encouraging collaboration between researchers, farmers, and technology providers in different regions can lead to the development of more adaptable and context-specific digital twin models.
- **Uncertainty and Risk:**
  - **Scenario Analysis:** Incorporating scenario analysis into digital twin models can help farmers explore the potential impact of different risk factors, such as extreme weather events, pest outbreaks, or market fluctuations, and develop mitigation strategies.
  - **Risk Assessment Tools:** Integrating digital twins with risk assessment tools can provide farmers with a more comprehensive understanding of the risks associated with their operations and enable them to make more informed decisions.

20.5 CONCLUSION

The integration of blockchain and digital twins revolutionizes smart agriculture, offering solutions to global food system challenges. Blockchain's decentralized ledger enables end-to-end traceability, combating fraud and empowering consumers with transparent information. Smart contracts automate processes, ensuring fair compensation for farmers and promoting sustainable practices. Blockchain's data integrity protects the agricultural ecosystem.

Simultaneously, digital twins, virtual replicas of agricultural systems, empower farmers with data-driven insights for precision agriculture. These twins optimize resource allocation, maximize yields, and enable early disease detection. Scenario planning with digital twins enhances adaptability to environmental changes.

Integrating blockchain and digital twins creates a synergy, ensuring trusted data exchange and fostering innovation. Blockchain's immutability promotes transparency, while decentralization encourages equitable decision-making. This synergy addresses food security, resource scarcity, and climate change by optimizing resource utilization and promoting sustainability. It empowers small-scale farmers, driving economic growth.

Realizing the full potential requires addressing technical challenges like scalability and interoperability, fostering collaboration among stakeholders, and developing user-friendly platforms. Future research should focus on ethical and regulatory implications, ensuring responsible and equitable use.

In conclusion, this integration is a new frontier in smart agriculture. The path is challenging, but the rewards are substantial. By embracing these technologies, a more sustainable, resilient, and equitable agricultural system can be achieved, benefiting all stakeholders and contributing to a healthier planet.

REFERENCES

Barricelli, B. R., Casiraghi, E., & Fogli, D. (2019). A survey on digital twin: Definitions, characteristics, applications, and design implications. *IEEE Access, 7*, 167653–167671. doi: 10.1109/ACCESS.2019.2953499.

Benrajesh, P., & Rajan, A. J. (2019, November). Optimizing the exhaust emission from logistics and packing industries, by means of green logistics. In *Journal of Physics: Conference Series* (Vol. 1355, No. 1, p. 012019). IOP Publishing.

Cai, J., Xu, X., & Yu, T. (2023). Performance evaluation of resource utilization with environmental externality: Evidence from Chinese agriculture. *Journal of Cleaner Production, 397*, 136561.

Caro, M. P., Ali, M. S., Vecchio, M., & Giaffreda, R. (2018). Blockchain technology and its potential applications in the food sector. *Food Control.*

Chen, H. Y., Sharma, K., Sharma, C., & Sharma, S. (2023). Integrating explainable artificial intelligence and blockchain to smart agriculture: Research prospects for decision making and improved security. *Smart Agricultural Technology, 6*, 100350.

De Vos, M., Kromhout, S., & De Vriend, H. (2020). Blockchain and food safety: A systematic review. *Food Control.*

Eastwood, C., Brookes, M., & Jessiman, P. (2019). Blockchain for sustainable supply chains: An exploratory study. *Supply Chain Management: An International Journal.*

Eastwood, C., Brookes, M., & Jessiman, P. (2021). The social impact of blockchain: An exploratory study of the food industry. *Technology in Society.*

Gupta, R., Malik, P., Rani, R., Solanki, R., Ameta, R. K., Malik, V., & Mukherjee, T. K. (2024). Recent progress on nanoemulsions mediated pesticides delivery: Insights for agricultural sustainability. *Plant Nano Biology*, 100073.

Hu, Y., Ye, B., & Im, H. (2021). Hope and post-stress growth during COVID-19 pandemic: The mediating role of perceived stress and the moderating role of empathy. *Personality and Individual Differences, 178*, 110831.

Kamble, S. S., Gunasekaran, A., & Gawankar, S. A. (2022). Blockchain technology in the food supply chain: A literature review. *International Journal of Production Economics.*

Kamilaris, A., Fonts, A., & Prenafeta-Boldú, F. X. (2019). The rise of blockchain technology in agriculture and food supply chains. *Trends in Food Science & Technology, 91*, 640–652.

Kamilaris, A., Prenafeta-Boldú, F. X., & Fonts, A. (2020). A review on blockchain technologies for smart agriculture: Benefits, challenges, and future research directions. *Computers and Electronics in Agriculture.*

Kim, W., Kim, S., Jeong, J., Kim, H., Lee, H., & Youn, B. D. (2022). Digital twin approach for on-load tap changers using data-driven dynamic model updating and optimization-based operating condition estimation. *Mechanical Systems and Signal Processing, 181*, 109471.

Kumar, P. R. V. S., Shrivastava, A., Prasanthi, R., Bindu, K. H., K. J., & Yadav, K. (2024). Optimizing pest detection and management in precision agriculture through deep learning approaches. *Library Progress International, 44*(3), 5681–5691.

Liang, J., Wu, C., Zhu, J., Tao, B., & Yin, Z. (2024). A lightweight segmented RFID tag antenna with a combined ferrite-core for underground applications. *IEEE Transactions on Antennas and Propagation.*

Lin, S. Y., Zhang, L., Li, J., Ji, L. L., & Sun, Y. (2022). A survey of application research based on blockchain smart contract. *Wireless Networks*, *28*(2), 635–690.

Liu, X., Min, W., Mei, S., Wang, L., & Jiang, S. (2021). Plant disease recognition: A large-scale benchmark dataset and a visual region and loss reweighting approach. *IEEE Transactions on Image Processing*, *30*, 2003–2015.

Mistry, I., Tanwar, S., Tyagi, S., Kumar, N., & Obaidat, M. S. (2023). Blockchain-based solutions for agriculture: Applications, challenges, and future research directions. *IEEE Access*.

Nguyen, N. D. K., Chaudhuri, A., & Ullah, A. (2024). Achieving business sustainability with blockchain technology. In *Handbook of Digital Innovation, Transformation, and Sustainable Development in a Post-Pandemic Era* (p. 78). CRC Press. https://doi.org/10.1201/9781003438748-5

Peters, G. W., & Panayi, E. (2016). Understanding the role of blockchain technology in enhancing trust and transparency in food supply chains. *British Food Journal*.

Saberi, S., Herold, D. M., Kouhizadeh, M., & Wilde, S. (2022). Categorizing transaction costs outcomes under uncertainty: A blockchain perspective for government organizations. *Journal of Global Operations and Strategic Sourcing*, *15*(3), 431–448.

Sharma, P. K., & Kumar, S. (2023). Soil temperature and plant growth. In *Soil Physical Environment and Plant Growth: Evaluation and Management* (pp. 175–204). Cham: Springer International Publishing.

Singh, S. K., Kumar, M., Khanna, A., & Virdee, B. (2024). Blockchain and FL-based secure architecture for enhanced external Intrusion detection in smart farming. *IEEE Internet of Things Journal*.

Slob, N., & Hurst, W. (2022). Digital twins and industry 4.0 technologies for agricultural greenhouses. *Smart Cities*, *5*(3), 1179–1192.

Sreedevi, A. G., & Kumar, M. P. (2023). Applications of digital twin technology in smart farming: A comprehensive review. *Computers and Electronics in Agriculture*.

Tao, F., Cheng, J., Qi, Q., Zhang, M., Zhang, H., & Sui, F. (2019). Digital twin-driven smart manufacturing. *International Journal of Production Research*.

Tian, X., Zhu, J., Zhao, X., & Wu, J. (2024). Improving operational efficiency through blockchain: evidence from a field experiment in cross-border trade. *Production Planning & Control*, *35*(9), 1009–1024.

Wang, H., Li, T., Nishida, E., Kato, Y., Fukano, Y., & Guo, W. (2023). Drone-based harvest data prediction can reduce on-farm food loss and improve farmer income. *Plant Phenomics*, *5*, 0086.

Wang, Y., Chen, C. H., & Zghari-Sales, A. (2021). Designing a blockchain enabled supply chain. *International Journal of Production Research*, *59*(5), 1450–1475.

Wheeler, S. (2020). The ethics of blockchain in food supply chains. *Agriculture and Human Values*.

Wolfert, S., Ge, L., Verdouw, C., & Bogaardt, M. J. (2022). Big data in smart farming—a review. *Agricultural Systems*.

Wu, Y., Jin, X., Yang, H., Tu, L., Ye, Y., & Li, S. (2022). Blockchain-based internet of things: Machine learning tea sensing trusted traceability system. *Journal of Sensors*, *2022*, 8618230. https://doi.org/10.1155/2022/8618230

Xu, S. (2024). Forestry offsets under China's certificated emission reduction (CCER) for carbon neutrality: Regulatory gaps and the ways forward. *International Journal of Climate Change Strategies and Management*, *16*(1), 140–156. https://doi.org/10.1108/IJCCSM-04-2022-0047

Yang, T., Ma, C., & Mi, X. (2024). The transformative potential of blockchain technology in developing green supply chain: An evolutionary perspective on complex networks. *Computers & Industrial Engineering*, *197*, 110548.

Zhang, L., Zhang, Z., Wang, W., Jin, Z., Su Y., & Chen, H. (2022). Research on a covert communication model realized by using smart contracts in blockchain environment. *IEEE Systems Journal*, *16*(2), 2822–2833. https://doi.org/10.1109/JSYST.2021.3057333

Zhang, X., Ren, Y., Xu, J., & Chi, C. (2024, August). Research and application framework for trusted circulation of food industry data based on blockchain and federated learning. In *2024 IEEE International Conference on Blockchain (Blockchain)* (pp. 530–535). IEEE.

Zhao, P., Li, S., Hu, P. J. H., Gu, C., Lu, S., Ding, S., . . . & Xiang, Y. (2022). Blockchain-based water-energy transactive management with spatial-temporal uncertainties. *IEEE Transactions on Smart Grid*, *14*(4), 2903–2920.

Zhou, J., Yang, M., Zhan, Y., & Xu, L. (2023). Digital twin application for reinforcement learning based optimal scheduling and reliability management enhancement of systems. *Solar Energy*, *252*, 29–38.

Index

0–9

3D representation, 238

A

access control mechanisms, 122
adoption barriers, 45
all pairs shortest path, 199
application programming interfaces, 127, 154
artificial intelligence, 17, 52, 69, 94, 113, 133, 179, 237, 251, 300, 313
automated machinery, 74, 251
automated processes, 115
automation, 4, 114
autonomous underwater vehicles, 276
autoregressive integrated moving average, 285

B

battery drainage, 82
big data analytics, 251
blockchain, 1, 154, 251, 308
blockchain technology, 75, 215

C

Climate Control, 115
climate fieldview, 143
climate-smart agriculture, 226
cloud computing, 92
cloud-edge indicator of farming anomalies, 204
community-supported agriculture, 223
computer-aided design, 32
constrained application protocol, 12
controlled environment agriculture, 5
convolutional neural networks, 161, 185
crop insurance schemes, 123
cryptocurrencies, 6
cryptographic, 118
customer relationship management, 154

D

data acquisition system, 240
data analytics, 75, 241
data anonymization, 60
data encryption, 77
data flow and integration, 120
data integrity and security, 310
data minimization, 78
data model, 238
data processing, 116
data security, 23
data silos, 119
decentralization, 7
decentralized autonomous organizations, 314
Decision support, 133
decision support systems, 117, 241, 266
deep diagnosis, 161
Denial-Of-Service (Dos) Attacks, 77, 78, 81, 208
Design and architecture, 59
device connectivity, 114
digital agriculture, 132
digital rupee, 154, 155, 157
digital twin, 24, 29, 237, 308, 332
digital twin creation, 133
directed acyclic graphs, 344
disease prevention, 54
distributed denial of service, 19
Drones, 73, 251

E

eavesdropping, 81
ecological sustainability, 102
edge computing, 313
educating consumers, 223
effects of climate change, 112
encryption, 60
end-to-end transparency, 119
enterprise resource planning, 37, 154
environmental data, 121
environmental degradation, 332
environmental monitoring, 114
environmental, social, and governance, 13
ethical considerations, 45
exploratory data analysis, 282

F

farm management software, 292
farmer-buyer agreements, 124
feedforward neural network, 289
field monitoring, 243
firmware attacks, 81
food and agriculture organization, 215
food supply chain, 22
Fourier transform infrared, 224

G

general data protection regulation, 60, 106, 128
generalizability, 343
genetically modified organisms, 301
geographic information system, 38, 71, 74, 98, 180, 193, 295
geographical information systems, 218
global positioning system, 4, 99, 136, 295
greater resilience, 4
gross domestic product, 192

H

hardware trojan, 77
health insurance portability and accountability, 60

health monitoring, 114
homomorphic encryption, 344
hypertext transfer protocol, 114

I

identity management and access control, 60
Indian council of agricultural research, 197
inefficient processes, 119
integration with digital twins, 124
internet of things, 4, 17, 31, 58, 69, 94, 113, 132, 218, 237, 251, 292, 313, 333
interoperability, 45, 114
interplanetary file system, 12

K

K-Nearest Neighbor, 160
knowledge gap, 305

L

label propagation algorithm, 200
labor efficiency, 27
lack of transparency, 119
laser-induced breakdown spectroscopy, 224
life cycle assessment, 233
livestock farming, 54
logistics and payments, 23
logistics and supply chains, 37
long short-term memory models, 285
long-term investment, 271

M

machine learning, 133, 182, 237, 251, 276, 295
malware and ransomware, 79
Man-In-The-Middle (Mitm) Attacks, 79, 208
Marine protected areas, 277
marker-assisted selection, 268
Mean Absolute Error (MAE), 184, 285
medicinal plant cultivation, 251
model synchronization, 238
moocall, 137

N

natural language processing, 55
near-infrared, 224

P

physical destruction, 77
physical tampering, 80
precision agriculture, 4, 64, 178, 333
precision irrigation, 251, 292
precision livestock farming, 56
predictive analytics, 27, 54, 63, 117
principal component analysis, 281
privacy concerns, 122
profitability, 2
proof of stake, 60, 344
proof of work, 5, 60

Q

quality assurance, 8
quality control, 109

R

radio frequency identification, 53, 204
real-time data, 115
real-time monitoring, 115
rectified linear unit, 280
Relative Absolute Error (RAE), 184
resilience and decentralization, 310
resource constraints, 112
resource optimization, 30, 335
resource scarcity, 332
return on investment, 302
risk mitigation, 30
robotics, 4, 300
role in automation, 123
role of smart contracts, 120
Root Mean Square Error (RMSE), 184, 285
Root Relative Square Error (RRSE), 184

S

scalability, 45
scalability and interoperability, 120
scenario analysis, 117
seasonal autoregressive integrated moving average, 285
security, 45
security risks, 122
sensor data acquisition, 116
sensor technology, 292
service-level agreements, 97
side-channel attacks, 77
simulation, 238
simulation and analysis, 117
single source shortest path, 199
smart agriculture, 1
smart farming, 251
soil monitoring, 114
solid-state drive, 207
species distribution models, 276
spoofing, 81
streamlining processes, 119
strongly connected component, 200
supply chain attacks, 77
supply chain automation, 123
supply chain data, 121
supply chain efficiency, 310
supply chain inefficiencies, 112
supply chain management, 38
supply chain optimization, 27
supply chain traceability, 332
supply chain transparency, 251
sustainability, 2
sustainable development goals, 14, 92
sustainable food supply chain, 215
sustainable sourcing, 257
synthetic minority over-sampling technique, 280

T

tampering, 76
Trimble agriculture, 135
trusted platform modules, 78

U

unmanned aerial vehicles, 74

V

variable rate technology, 71, 180, 295
virtual modeling, 117
visualization, 238

W

wireless sensor networks, 136, 204
world wildlife fund, 8

Y

yield data, 121
yield mapping, 114
yield prediction, 333

Z

zero-knowledge proofs, 344

For Product Safety Concerns and Information please contact our EU
representative GPSR@taylorandfrancis.com
Taylor & Francis Verlag GmbH, Kaufingerstraße 24, 80331 München, Germany

www.ingramcontent.com/pod-product-compliance
Ingram Content Group UK Ltd.
Pitfield, Milton Keynes, MK11 3LW, UK
UKHW050057200926
475702UK00019B/2887